W9-ACS-107

Bioinformatics for Geneticists

Second Edition

Bioinformatics for Geneticists

Second Edition

A bioinformatics primer for the analysis of genetic data

Michael R. Barnes
Bioinformatics, GlaxoSmithKline Pharmaceuticals, Harlow, Essex, UK

John Wiley & Sons, Ltd

Telephone (+44) 1243 779777

Email (for orders and customer service enquiries): cs-books@wiley.co.uk
Visit our Home Page on www.wiley.com

Other Wiley Editorial Offices

John Wiley & Sons Inc., 111 River Street, Hoboken, NJ 07030, USA

Jossey-Bass, 989 Market Street, San Francisco, CA 94103-1741, USA

Wiley-VCH Verlag GmbH, Boschstr. 12, D-69469 Weinheim, Germany

John Wiley & Sons Australia Ltd, 42 McDougall Street, Milton, Queensland 4064, Australia

John Wiley & Sons (Asia) Pte Ltd, 2 Clementi Loop #02-01, Jin Xing Distripark, Singapore 129809

John Wiley & Sons Canada Ltd, 6045 Freemont Blvd, Mississauga, ONT, L5R 4J3, Canada

Wiley also publishes its books in a variety of electronic formats. Some content that appears in print may not be available in electronic books.

Anniversary Logo Design: Richard J. Pacifico

Library of Congress Cataloging-in-Publication Data

Bioinformatics for geneticists : a bioinformatics primer for the analysis of genetic data/[edited by] Michael Barnes. – 2nd ed.
p. ; cm.
Includes bibliographical references and index.
ISBN-13: 978-0-470-02619-9 (cloth : alk. paper)
ISBN-13: 978-0-470-02620-5 (pbk. : alk. paper)
1. Genetics–Data processing. 2. Bioinformatics. I. Barnes, Michael R.
[DNLM: 1. Computational Biology–methods. 2. Genomics–methods. 3. Databases, Genetic. 4. Software. QU 26.5 B6147 2007]
QH441.2.B56 2007
576.50285–dc22 2006036197

A catalogue record for this book is available from the British Library

ISBN HB 9780470026199
ISBN PB 9780470026205

Typeset in 10.5/13pt Minion by Aptara Inc., New Delhi, India
Printed and bound in Great Britain by Antony Rowe Ltd, Chippenham, Wiltshire
This book is printed on acid-free paper responsibly manufactured from sustainable forestry in which at least two trees are planted for each one used for paper production.

Contents

Foreword

Despite a relatively short existence, bioinformatics has always seemed an unusually multidisciplinary field. Fifteen years ago, when sequence data were still scarce and only a small fraction of the power of today's pizza-box supercomputers was available, bioinformatics was already entrenched in a diverse array of topics. Database development, sequence alignment, protein structure prediction, coding and promoter site identification, RNA folding, and evolutionary tree construction were all within the remit of the early bioinformaticist.[1,2] To address these problems, the field drew from the foundations of statistics, mathematics, physics, computer science and, of course, molecular biology. Today, predictably, bioinformatics still reflects the broad base on which it started, comprising an eclectic collection of scientific specialists.

As a result of its inherent diversity, it is difficult to define the scope of bioinformatics as a discipline. It may be even fruitless to try to draw hard boundaries around the field. It is ironic, therefore, that even now, if one were to compile an intentionally broad list of research areas within the bioinformatics purview, it would often exclude one biological discipline with which it shares a fundamental basis: Genetics. On one hand, this seems difficult to believe, since the fields share a strong common grounding in statistical methodology, dependence on efficient computational algorithms, rapidly growing biological data, and shared principles of molecular biology. On the other hand, this is completely understandable, since a large part of bioinformatics has spent the last few years helping to sequence a number of genomes, including that of man. In many cases, these sequencing projects have focused on constructing a single representative sequence—the consensus—a concept that is completely foreign to the core genetics principles of variability and individual differences. Despite a growing awareness of each other, and with a few clear exceptions, genetics and bioinformatics have managed to maintain separate identities.

Geneticists needs bioinformatics. This is particularly true of those trying to identify and understand genes that influence complex phenotypes. In the realm of human genetics, this need has become particularly clear, so that most large laboratories now have one or two bioinformatics 'specialists' to whom other lab members turn for computing matters. These specialists are required to support a dauntingly wide assortment of applications: typical queries for such people might range from how to find instructions for accessing the internet, to how to disentangle a complex database schema, to how to optimize numerically intensive algorithms on parallel computing farms. These people, though somewhat scarce, are essential to the success of the laboratory.

With the ever-increasing volume of sequence data, expression information and well characterized structures, as well as the imminent genotype and haplotype data on large and diverse human populations, genetics laboratories now must move beyond singular dependence on the bioinformatics handyman. Some level of understanding and ability to use bioinformatics applications is becoming necessary by everyone in the lab. Fortunately, bioinformaticians have been particularly successful in developing user-friendly software that renders complex statistical methods accessible to the bench scientists who generated and should know most about the data being analysed. To further these analyses, ingenious software applications have been constructed to display the outcomes and integrate them with a host of useful annotation features such as chromosome characteristics, sequence signatures, disease correlates and species comparisons[3]. With these tools freely available and undergoing continued development, mapping projects that make effective use of genetic and genomic information will naturally enjoy greater success than those less equipped to do so. Simply put, genetics groups that cannot capitalize on bioinformatics applications will be increasingly scooped by those who can.

The emerging requirement of broader understanding of bioinformatics within genetics is the focus of this text, as easily appreciated by a quick glance at the title. Equally obvious is that geneticists are the editors' target audience. Still, one might ask 'toward what specific group of geneticists is this text aimed?' The software and computational backbone of bioinformatics is shared most noticeably with the areas of statistical and population genetics, so the statistical specialists would seem a plausible audience. By design, however, this text is not aimed at these specialists so much as at those with broader backgrounds in molecular and medical genetics, including both human and model organism research. The content should be accessible by skilled bench scientists, clinical researchers and even laboratory heads. Computationally, one needs only basic computing skills to work through most of the material. Biologically, appreciation of the problems described requires general familiarity with genetics research and recognition of the inherent value in careful use of *in silico* genetic and genomic information.

By necessity, the bioinformatics topics covered in this text reflect the diversity of the field. In order to obtain some order in this broad area, the editors have focused on computer applications and effective use of available databases. This concentration on applications means that descriptions of the statistical theory, numerical algorithms and database organization are left to other texts. The editors have intentionally bypassed much of this material to emphasize applications in widespread use—the focus is on efficient use, rather than development, of bioinformatics methods and tools.

The data behind many of the bioinformatics tools described here are rapidly changing and expanding. In response, the software tools and databases themselves tend to be (infuriatingly) dynamic. A consequence of this fluid state is that learning to use existing programs by no means guarantees a knack for using those in the future. Thus, one cannot expect long-term consistency in the tools and data-types described here (or in most any other contemporary bioinformatics text). By learning to use current tools more effectively, however, geneticists can not only capitalize on technology available, but perhaps engage more bioinformaticians in the excitement of genetics research. Bringing bioinformatics to geneticists is a crucial first step towards integrating the kindred fields and characterizing the frustratingly elusive genes that influence complex phenotypes.

Lon R. Cardon
Professor of Bioinformatics
Wellcome Trust Centre for Human Genetics
University of Oxford

1. Doolittle, R. F. *Of URFs and ORFs: A primer on how to analyze derived amino acid sequences* (University Science Books, Mill Valley, California, 1987).
2. von Heijne, G. *Sequence analysis in molecular biology: Treasure trove or trivial pursuit* (Academic Press, London, 1987).
3. Wolfsberg, T. G., Wetterstrand, K. A., Guyer, M. S., Collins, F. S. & Baxevanis, A. D. A user's guide to the human genome. *Nature Genetics* **32 (suppl)** (2002).

Preface

> I say 'locus–locus' instead of 'gene–gene' because if you work in human genetics long enough, you realize that you may never have a gene. But you learn not to let that put you off.
>
> *Peter A. Holmans*

Making sense of the results of a genetic experiment is a challenge on any level. Writing a book about the use of bioinformatics to achieve this goal might seem like a somewhat vainglorious exercise. Individual perceptions of what is constituted by bioinformatics vary widely. However, in the context of this book, bioinformatics seeks to illuminate biological function, while disease genetics, our primary focus, is essentially about understanding biological dysfunction. With this in mind, please think of bioinformatics as a tool for improving the understanding of genetics.

Since the first edition of this book, the reasons for thinking this way have become more compelling. Human disease genetics is rapidly becoming a high-throughput activity, and that means that making sense of a genetic experiment now means making sense of millions of data points. Again this underlines the need for genetics-focused bioinformatics. Quite simply, we need the *informatics* to manipulate and analyse data on this scale, and we need the *bio* to make sense of it all in the holistic biological system that is a human being.

This book could not have been realised without the insightful contributions of all the chapter authors. I really feel they have helped to make this book worthy of both the *bio* and the *informatics* monikers. I would also like to send my warmest thanks to Ian C. Gray, who co-edited the first edition with me, for providing helpful input and support on this edition. All the exciting science you see here would not exist without the incredibly dedicated team at Wiley, who have always kept things on track, including Joan Marsh, Andrea Baier, Fiona Woods, Kate Pamphilon and Emilie McDonough. I have a day job besides editing books, and so I would also really like to thank Philippe Sanseau and David Searls at GSK for giving me the time, encouragement and support to get this done. Finally, I would like to thank my wife, Aruna, for her constant love, support, encouragement and superior punctuation. Without her, I would not have had the will or punctuation skills to complete this magnum opus.

Michael R. Barnes
August 2006, Harlow, UK

Contributors

Catherine A. Ball Department of Biochemistry, Stanford University Medical School, Stanford, CA, USA

Aruna Bansal Discovery and Pipeline Genetics, GlaxoSmithKline Pharmaceuticals, Third Avenue, Harlow, Essex, UK

Michael R. Barnes Bioinformatics, GlaxoSmithKline Pharmaceuticals, Third Avenue, Harlow, Essex, UK

Bryan J. Barratt Research and Development Genetics, AstraZeneca, Alderley Park, Macclesfield, Cheshire, UK

Matthew J. Betts Structural and Computational Biology Programme, EMBL, Meyerhofstrasse 1, 69117 Heidelberg, Germany

Diana Blaydon Centre for Cutaneous Research, Institute of Cell and Molecular Science, Queen Mary's School of Medicine and Dentistry, Whitechapel, London, UK

Karl W. Broman Department of Biostatistics, Johns Hopkins University, Baltimore, MD, USA

Ellen M. Brown Discovery Informatics, AstraZeneca, Alderley Park, Macclesfield, Cheshire, UK

James R. Brown Bioinformatics, GlaxoSmithKline Pharmaceuticals, Upper Providence, PA, USA

Elissa J. Chesler Oak Ridge National Laboratory, Biosciences Division, Oak Ridge, TN, USA

Richard R. Copley Wellcome Trust Centre for Human Genetics, University of Oxford, Oxford, UK

Barry Dancis Bioinformatics, GlaxoSmithKline Pharmaceuticals Upper Providence, PA, USA

Steve Deharo Bioinformatics, GlaxoSmithKline Pharmaceuticals, Third Avenue, Harlow, Essex, UK

Paul S. Derwent Bioinformatics, GlaxoSmithKline Pharmaceuticals, Third Avenue, Harlow, Essex, UK

Ian C. Gray Paradigm Therapeutics (S) Pte Ltd, 10 Biopolis Way, Singapore 138670

Joel Greshock Translational Medicine, Clinical Pharmacology Division, GlaxoSmithKline Pharmaceuticals, Upper Merion, PA, USA

Simon C. Heath Centre National de Genotypage, Evry Cedex, France

David P. Kelsell Centre for Cutaneous Research, Institute of Cell and Molecular Science, Queen Mary's School of Medicine and Dentistry, Whitechapel, London, UK

Ralph McGinnis Wellcome Trust Sanger Institute, Hinxton, Cambridge, UK

Charles A. Mein Genome Centre, Queen Mary's School of Medicine and Dentistry, Charterhouse Square, London, UK

Mary Plumpton Bioinformatics, GlaxoSmithKline Pharmaceuticals, Stevenage, Hertfordshire, UK

Robert B. Russell Structural and Computational Biology Programme, EMBL, Meyerhofstrasse 1, 69117 Heidelberg, Germany

Philippe Sanseau Bioinformatics, GlaxoSmithKline Pharmaceuticals, Stevenage, Hertfordshire, UK

Colin A. M. Semple Bioinformatics, MRC Human Genetics Unit, Edinburgh EH4 2XU, UK

Gavin Sherlock Department of Genetics, Stanford University Medical School, Stanford, CA, USA

Christopher Southan Global Compound Sciences, AstraZeneca R&D, Mölndal, Sweden

Martin S. Taylor Wellcome Trust Centre for Human Genetics, University of Oxford, Oxford, UK

Magnus Ulvsbäck Molecular Pharmacology, AstraZeneca R&D, Mölndal, Sweden

Charlotte Vignal Discovery and Pipeline Genetics, GlaxoSmithKline Pharmaceuticals, Third Avenue, Harlow, Essex, UK

Chaolin Zhang Department of Biomedical Engineering, State University of New York at Stony Brook, NY, USA

Michael Q. Zhang Cold Spring Harbor Laboratory, Cold Spring Harbor, NY, USA

Xiaoyue Zhao Cold Spring Harbor Laboratory, Cold Spring Harbor, NY, USA

Glossary of Bioinformatics

BLAST (Basic Local Alignment Search Tool) A tool for identifying sequences in a database that match a given query sequence. Statistical analysis is applied to judge the significance of each match. Matching sequences may be homologous to, or related to, the query sequence. There are several versions of BLAST:

- **BLASTP** compares an amino-acid query sequence with a protein-sequence database.
- **BLASTN** compares a nucleotide query sequence with a nucleotide-sequence database.
- **BLASTX** compares a nucleotide query sequence translated in all reading frames with a protein-sequence database.
- **TBLASTN** compares a protein query sequence with a nucleotide sequence database dynamically translated in all reading frames.
- **TBLASTX** compares the six-frame translations of a nucleotide query sequence with the six-frame translations of a nucleotide-sequence database.

BLAT (BLAST-Like Alignment Tool) BLAT might superficially appear to be like BLAST, also being a tool for detecting subsequences that match a given query sequence; however, BLAT and BLAST have a number of differences. BLAT was developed at the UCSC; it searches the human genome by keeping an index of the entire genome in memory. The index consists of all non-overlapping 11-mers except for repeat sequences. A BLAT search of the human genome will quickly find sequences of 95 per cent and greater similarity of length of at least 40 bases. It may miss more divergent or shorter sequence alignments (see the UCSC FAQ for more details on this tool: http://genome.ucsc.edu/FAQ.html).

CDS Coding sequence.

Contig map A map depicting the relative order of overlapping (contiguous) clones representing a complete genomic or chromosomal segment.

DAS (distributed annotation system) A protocol for browsing and sharing genome sequence annotations across the Internet, allowing users to search and compare annotations from several sources. Ensembl provides a DAS reference server giving access to a wide range of specialist annotations of the human genome (for more detail, see http://www.ensembl.org/das/).

Data mining The ability to query very large databases in order to satisfy a hypothesis ('top-down' data mining), or to interrogate a database in order to generate new hypotheses based on rigorous statistical correlations ('bottom-up' data mining).

Domain (protein) A region of special biological interest within a single protein sequence. However, a domain may also be defined as a region within the three-dimensional structure of a protein that may encompass regions of several distinct protein sequences that accomplish a specific function. A domain class is a group of domains that share a common set of well-defined properties or characteristics.

Electronic PCR (ePCR) An electronic process analogous to laboratory-based PCR. Two primers are used to map a sequence feature (such as a single nucleotide polymorphism). To validate the position, both primers must map in the same vicinity spanning a defined distance, effectively producing an electronic PCR product.

Expressed sequence tag (EST) A short sequence read from an expressed gene derived from a cDNA library. Databases storing large numbers of ESTs can be used to gauge the relative abundance of different transcripts in cDNA libraries and the tissues from which they are derived. An EST can also act as a physical tag for the identification, cloning and full-length sequencing of the corresponding cDNA or gene.

FASTA format FASTA (Fast-All), originally devised for Lipman and Pearson's sequence alignment algorithm, is one of the simplest and most widely accepted formats for sequences, taking the form of a simple header preceded by a greater than (>) sign and sequence on the following line; e.g. >sequence_id gataggctgagcgatgcgatgctagctagctagc.

Golden path The term applied to the first and subsequent assemblies of the human genome.

Hidden Markov model (HMM) A joint statistical model for an ordered sequence of variables. The result of stochastically perturbing the variables in a Markov chain (the original variables are thus 'hidden'), whereby the Markov chain has discrete variables that select the 'state' of the HMM at each step. The perturbed values can be continuous and are the 'outputs' of the HMM. An HHH is equivalently a coupled mixture model where the joint distribution over states is a Markov chain. HHHs are valuable in bioinformatics because they allow a search or alignment algorithm to be trained by unaligned or unweighted input sequences, and because they allow position-dependent scoring parameters such as gap penalties, thus more accurately modelling the effects of evolutionary events on sequence families.

Homology (strict) Two or more biological species, systems or molecules that share a common evolutionary ancestor (general), or two or more gene or protein sequences that share a significant degree of similarity, typically measured by the amount of identity (in the case of DNA), or conservative replacements (in the case of protein), that they register along their lengths. Sequence 'homology' searches are typically performed with a query DNA or protein sequence to identify known genes or gene products that share significant similarity and hence might clarify the ancestry, heritage and possible function of the query gene.

in silico (biology) (literally, computer mediated) The use of computers to simulate, process, or analyse a biological experiment.

NCBI National Center for Biotechnology Information, Washington, DC, USA.

Open reading frame (ORF) Any stretch of DNA that potentially encodes a protein. ORFs begin with a start codon and end with a termination codon. No termination codons may be present internally. The identification of an ORF is the first indication that a segment of DNA may be part of a functional gene.

Orthologue/paralogue Paralogues are genes related by duplication within a genome. Orthologues retain the same function in the course of evolution, whereas paralogues evolve new functions, even if these are related to the original one.

Perl (Practical Extraction and Report Language) Perl is relatively straightforward up to a certain level, and this has facilitated its development as the primary language of biological computing.

Relational database A database that follows E. F. Coddís' 11 rules, a series of mathematical and logical steps for the organization and systemization of data into a software system that allows easy retrieval, updating and expansion. A relational database management system (RDBMS) stores data in a database consisting of one or more tables of rows and columns. The rows correspond to a record (tuple); the columns correspond to attributes (fields) in the record. RDBMSs use structured query language (SQL) for data definition, data management, and data access and retrieval. Relational and object-relational databases are used extensively in bioinformatics to store sequence and other biological data.

Secondary structure (protein) The organization of the peptide backbone of a protein that occurs as a result of hydrogen bonds, such as alpha helix or beta pleated sheet.

Sequence tagged site (STS) A unique sequence from a known chromosomal location that can be amplified by PCR. STSs act as physical markers for genomic mapping and cloning.

Single nucleotide polymorphism (SNP) A DNA sequence variation resulting from substitution of one nucleotide for another.

Structured query language (SQL) A type of programming language used to construct database queries and perform updates and other maintenance of relational databases. SQL is not a fully fledged language that can create stand-alone applications, but it is powerful enough to create interactive routines in other database programs.

Substitution matrix A model of protein evolution at the sequence level resulting in the development of a set of widely used substitution matrices. These are frequently called Dayhoff, MDM (mutation data matrix), BLOSUM or PAM (percent accepted mutation) matrices. They are derived from global alignments of closely related sequences. Matrices for greater evolutionary distances are extrapolated from those for lesser ones.

Tertiary structure (protein) Folding of a protein chain via interactions of its side-chain molecules, including formation of disulphide bonds between cysteine residues.

UCSC (University of California, Santa Cruz) An excellent genome browser.

UTR (untranslated region) The non-coding region of an mRNA transcript flanking either side of the open reading frame.

Section I

An Introduction to Bioinformatics for the Geneticist

1

Bioinformatics Challenges for the Geneticist

Michael R. Barnes[1]

[1] *Bioinformatics, GlaxoSmithKline Pharmaceuticals, Harlow, Essex, UK*

1.1 Introduction

The first edition of this book was published in February 2003, and now it is reasonable to say that expectations in the field of human genetics are higher than ever. Research-funding bodies, such as the US National Institutes of Health (NIH) and the Wellcome Trust, are intensifying their focus on initiatives to study the genetic basis of complex diseases. Why is this happening? It would appear that genetics research is experiencing something equivalent to an alignment of the constellations. Quite simply, 6 years after the first draft, and 3 years after the completion of the genome, we have the HapMap to complement the genome, and we have technologies to genotype rapidly hundreds of thousands of single nucleotide polymorphisms (SNPs). Everything seems to be in the right place to make a real leap in our understanding of the genetic determinants of complex diseases. Clearly, there could not be a better time to publish the second edition of this book!

To call this new edition of *Bioinformatics for Geneticists*, the second edition is probably a misnomer, as this implies a great deal of continuity with the first. Generally, as is reflected by the field of genetics itself, this is not the case. The challenges for human genetics have changed almost beyond recognition between 2003 and July 2006, the date that this second edition went to press. In 2003, precisely 50 years after the landmark discovery of the structure of DNA, the entire human genome sequence was completed in a final, polished form. This fully indexed but semi-intelligible

Bioinformatics for Geneticists, Second Edition. Edited by Michael R. Barnes
 ISBN 978-0-470-02619-9 (HB) ISBN 978-0-470-02620-5 (PB)

'book of life' immediately began to serve as a valuable framework for integration of genetic and biological data. However, knowledge of the genome sequence did not immediately clarify the nature and structure of human genetic variation. While in terms of genome function, our understanding in 2003 was mainly limited to the 25 000 or so genes that we could determine encoded within the sequence, today (July 2006), with the help of HapMap, a human haplotype map, we have a much better understanding of the structure and complexity of genetic variation. Knowledge of variation and improvements in genotyping technology have led to a dramatic scaling up of genotyping experiments, generating in turn unprecedented volumes of genotyping data. While, in terms of function, our knowledge of the human genome is now enhanced by knowledge of at least 13 other vertebrate genomes, we are also clarifying previously unrecognized but numerous genomic elements, such as non-coding micro-RNA (miRNA). We are beginning to realize that these elements may be just as important as the coding RNA component of the genome. Finally, our understanding is starting to expand beyond the genome to the epigenome – heritable changes other than those in the DNA sequence. All these factors add up to a complete transformation of the genetic landscape. To address this, the second edition of this book has also undergone a complete transformation, adding many new authors and chapters and just a few critical, but completely revised chapters from the first edition. Altogether, we hope these new contributions will address the lion's share of the newer and long-standing challenges that face the human geneticist.

1.2 The role of bioinformatics in genetics research

The function of bioinformatics is now essential to the effective interrogation of genetic and genomic data as well as most other biological data. This makes expertise in bioinformatics a prerequisite for effectiveness in genetics. Expertise in bioinformatics is no mystery; the right bioinformatics tools, coupled with an enquiring mind and willingness to experiment (key requirements for any scientist, bioinformatician or not), can yield confidence and competence in handling bioinformatics data in a very short space of time. The objective of this book is not to provide an exhaustive guide to bioinformatics; other texts fulfil this role. Instead, it is intended as a specialist guide to help the human geneticist navigate the Internet to some of the best tools and databases for the job; that is, linking and associating genes with diseases and genetic traits. In this chapter, we give a flavour of the many processes in human genetics where bioinformatics can have a major impact, and refer to subsequent chapters for greater detail.

1.2.1 Gaining understanding of genetic traits

The process of understanding a genetic trait typically proceeds through three stages: first, recognition of the disease state or syndrome, including assessment of its hereditary character; second, discovery and mapping of the related polymorphism(s) or mutation(s); and third, elucidation of the biochemical/biophysical mechanism leading to the disease phenotype. Each of these stages proceeds with a variable degree of laboratory investigation and data analysis, often by bioinformatics methods. Both activities are complementary, bioinformatics without laboratory work is a sterile activity just as laboratory work without bioinformatics can be futile and inefficient. In fact, these two sciences are really one, genetics and genomics generate data, and bioinformatics allows efficient storage, access and analysis of the data – together, they constitute the most efficient manifestation of genetic research in action.

1.3 Genetics in the post-genome era

In the broadest sense, bioinformatics in a genetic research context covers the following aspects:

- knowledge management and expansion
- data management, integration and mining
- mastering genes, genomes and genetic variation data
- genetic study design and analysis
- determination of function (moving from candidate genes to disease alleles)
- analysis at the genetic and genomic data interface.

These categories are quite generic and could apply to most fields of biology, but are clearly applicable to genetics. Both genetics and bioinformatics are essentially concerned with asking the right questions, generating and testing hypotheses, and organizing and interpreting large numbers of data to detect biological phenomena.

1.3.1 Knowledge management and expansion

Genetics, as the innate code of an organism, largely defines biology. Consequently, few areas of biological research call for a broader background in biology than genetic research. This background is tested to the extreme in the selection of candidate genes to test for involvement in a disease process, or in identifying candidates from the results of a genome scan. Candidate genes need to be chosen and prioritized by many criteria. Often biological links may be very subtle. Candidate gene interactions might be considered similar to human interactions, bringing to mind the famous 'six degrees of separation' concept from an experiment by social psychologist Stanley

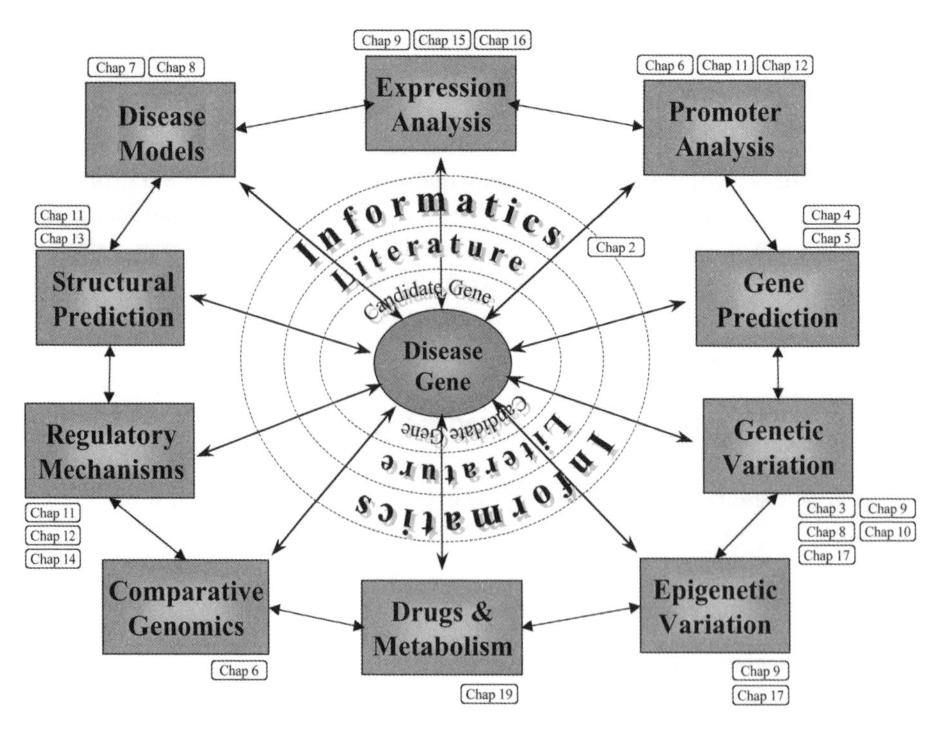

Figure 1.1 Approaches moving from linked or associated genes to validated disease genes. Chapters detailing each aspect are indicated

Milgram in 1967, which suggested that two random US citizens are connected, on average, by a chain of six acquaintances. For example, a candidate gene may regulate a gene that regulates a gene that, in turn, may act upon the target disease pathway. Faced with the complexity of relationships between genes, geneticists must be able to expand pathways and identify complex cross-talk between pathways. As this process can extend almost interminably to the point that virtually every gene is a candidate for every disease, knowledge management is important to help weigh up evidence to prioritize genes for either initial analysis or follow-up.

Geneticists can rarely afford to be authorities on every disease that they study, nor can they expect to know the details of all gene and pathway interactions. Therefore, bioinformatics and effective use of disease biology resources on the Web are needed for quick evaluation of the role of each candidate and its related pathways with respect to the target phenotype. Figure 1.1 illustrates some of the areas to be effectively utilized by geneticists to formulate the questions that need to be asked to move from candidate genes to disease genes. These areas of biology are touched on directly or indirectly throughout this book, so chapters that may help to formulate and perhaps answer these questions are indicated in the figure.

Literature, as an embodiment of (almost) all prior knowledge, is the most powerful resource to support this process, but it is also the most complex and confounding data source to search. To expedite this process, some databases have been constructed that attempt to encapsulate the available gene/disease-focused literature, such as Online Mendelian Inheritance in Man (OMIM) (http://www.ncbi.nlm.nih.gov/entrez/query.fcgi?db=OMIM). These centralized data resources can often be very helpful for gaining a quick overview of an unfamiliar pathway or gene, but inevitably one needs to re-enter the literature to build up a fuller picture and to answer the questions that are most relevant to the target phenotype or gene. The Internet is also an excellent resource to help in this process; this probably makes the ubiquitous search engine Google (http://www.google.com) one of the most powerful bioinformatics tools. Well-chosen Google keywords can usually return highly relevant information or links to tools and databases that contain information being sought, while Google Scholar (http://scholar.google.com/) can offer even more focused results. We offer pointers throughout this book to effective literature-searching strategies, and to some of the best tools and databases related to genes, proteins, pathways and disease biology on the Internet, but regular Google searches are also necessary to keep abreast of the latest tool and database developments.

1.3.2 Data management, integration and mining

Efficient application of knowledge relies on well-organized data. Dependent on statistical analysis, genetics is also highly dependent upon good data, increasingly in very large volumes. Accessing available data, particularly in bulk, is often the biggest informatic frustration for geneticists. In Chapter 2 of this book, dealing with data entry and manipulation, and taking the first steps in software coding, we have tried to give some pointers for overcoming some of these frustrations. Generally, we focus on accessing data from public databases and some of the more lightweight methods of analysing data in locally installed databases with Perl and similar coding languages. Methods of industrial-scale genetic data curation and analysis, in the form of either 'off the shelf' or custom-built laboratory information-management systems (LIMS), belong to a specialist area beyond the scope of this book.

1.3.3 Mastering genes, genomes and genetic variation data

A key problem that frequently hinders effective genetic data mining is the localization of data in many independent databases rather than a few centralized repositories. A clear exception to this is SNP data, which have now coalesced around a single central database – dbSNP at NCBI (Sherry *et al.*, 2001). This may have helped to stimulate the genetics research community to complete the HapMap (International HapMap Consortium, 2003), which has enabled the comprehensive characterization of linkage disequilibrium (LD) and haplotype relationships between SNPs in four population samples. As mentioned earlier, HapMap is revolutionizing genetic analysis, but this resource is not without caveats, so we provide a comprehensive review addressing some of these issues in Chapter 3.

Perhaps with the exception of dbSNP, most database development in bioinformatics has not been implicitly designed for geneticists; instead, genomic databases and genome viewers have generally been developed to aid the annotation of genes and the human genome. Of course, such data are vital for genetics, but this development may explain why the available tools often appear to lack important functionality for the geneticist. One has to make use of what functionality is available, although sometimes this means using tools in ways that were not originally intended (for example, many geneticists use BLAST to identify sequence primer homology in the human genome, but few realize that the default parameters of this tool are entirely unsuited for this task). We will attempt to address these issues throughout this book and offer practical methods to get the most value from existing tools wherever possible. In Chapter 4, we examine the use of human genome browsers for genetic research. Tools such as Ensembl and the UCSC human genome browser annotate important genetic information on the human genome, including SNPs, microsatellites and, of course, genes and regulatory regions. User-defined queries place genes and genetic variants in their full genomic context, giving very detailed information on nearby genes, promoters or regions conserved between species, including a number of vertebrate species that now have complete genome sequences. Sometimes this bewildering wealth of information might even be seen as a hindrance to the clear understanding of gene function. Therefore, in Chapter 5, we discuss defining the boundaries and full complexity of a gene from all available data so that genetic analysis can effectively evaluate it.

It is hard to overstate the value of genomic information for genetics. For example, cross-species genome comparison is invaluable for the analysis of function, as interspecies sequence conservation is generally thought to be restricted to functionally important gene or regulatory regions. This makes comparative genome analysis one of the most powerful tools for identifying potential regulatory regions or undetected genes. Chapter 6 deals with this whole area in detail, while several other chapters in this book cover related tools and databases to support these approaches (see Chapters 12 and 16).

1.3.4 Genetic study design and analysis

Despite the recent improvements in the throughput of genetic and genomic techniques, the genes that contribute to the most common human diseases are still elusive. By contrast, the identification of genes mutated in rare single-gene disorders (so-called Mendelian or monogenic disorders) is now relatively straightforward if suitable kindreds are available. The identification of the genes responsible for a plethora of monogenic disorders is one of the genetics success stories of the late 1980s and the 1990s; genes identified include, to name but a few, *CFTR* (cystic fibrosis; Riordan *et al.*, 1989), Huntingtin (Huntington's disease; Huntington's Disease Collaborative Research Group, 1993), Frataxin (Friedreich's ataxia; Campuzano *et al.*, 1996) and *BRCA1* in breast and ovarian cancer (Miki *et al.*, 1994). Some evidence suggests that an understanding of Mendelian phenotypes may also help to identify genes involved in complex disease; for example, PPARγ variants have been implicated in both monogenic and complex forms of type II diabetes (Altshuler *et al.*, 2000b; Savage *et al.*, 2003). Therefore, Chapter 7 addresses some of the unique issues raised during the process of identifying monogenic disease gene mutations.

Unfortunately, identification of genes with a role in complex (i.e. multigenic) disease has been far less successful. Notable examples are the involvement of *APOE* in late-onset Alzheimer's disease (Saunders *et al.*, 1993) and the role of *NOD2* in Crohn's disease (Hugot *et al.*, 2001). However, genes for most of the common complex diseases remain elusive. Our ability to detect disease genes is often dependent on the analysis method applied. Methods for the identification of disease genes can be divided neatly into two broad categories, linkage and association. Although many common principles apply to both of these study types, each approach has distinct informatics demands, which are reviewed in detail in Chapter 8.

Unlike single-gene Mendelian diseases, complex genetic diseases are caused by the combined effect of multiple polymorphisms in a number of genes, often coupled with environmental factors. The successes of linkage analysis in the rapid identification of Mendelian disease genes have spawned large-scale efforts to track down genes involved in the more common complex disease phenotypes. Unfortunately, these efforts have been largely unsuccessful to date, mainly because each gene with phenotypic relevance is thought to make a relatively small contribution to disease susceptibility. These small effects are likely to be below the threshold of detection by linkage analysis in the absence of unfeasibly large sample sizes (Risch, 2000; Wang *et al.*, 2005; see Chapter 18).

Association studies have three main advantages over linkage studies for the analysis of complex disease: (i) case-control cohorts are generally easier to collect than extended pedigrees; (ii) association studies have greater power to detect small genetic effects than linkage studies, a clear example being the insulin gene, which shows extremely strong association with type 2 diabetes, but very weak linkage (Spielman *et al.*, 1993); (iii) LD typically stretches over tens of kilobases rather than several megabases (Reich *et al.*, 2001), allowing focus on much smaller and more manageable loci. Among other reasons (discussed in Chapter 8), this is because an association-based approach exploits recombination in the context of the entire population, rather than within the local confines of a family structure.

Of course, this last point is the other side of the double-edged sword of marker density and resolution mentioned in the context of linkage analysis above. The trade-off is reduced range over which each marker can detect an effect, resulting in a need for increased marker density, scaling up to a genome-wide requirement of hundreds of thousands of SNPs. In terms of technical requirements, the new ultra-high-density, oligonucleotide-based SNP genotyping panels address these increased needs (Matsuzaki *et al.*, 2004). But, unfortunately, there is another trade-off, which is a dramatically increased level of testing, leading to a very high number of chance associations (see Chapter 18 for detailed discussion of these issues).

The distinct issues that linkage and association studies of complex diseases each raise can be addressed to varying degrees by bioinformatics analysis. In the case of linkage studies, the very large regions identified call for rigorous prioritization of genes and markers for further analysis based on biological rationale and disease understanding. In the case of association analysis, the problems stemming from lack of power and issues of multiple testing can also be ameliorated by filtering results with different forms of rationale. Whichever method is employed, comprehensive informatics input at each stage can contribute to the quality, efficiency and outcome of a study. Chapters 8 and 10 review the elements of experimental design and statistical analysis that can help to address intrinsically some of these issues, while Chapters 9 and 18 address in detail the bioinformatics approaches that can be used to define a locus or series of genome-wide associations, allowing a logical and systematic approach to marker and gene selection, prioritization, and subsequent genetic and biological analysis. This can simultaneously reduce the cost and complexity of a project and improve the chances of successfully discovering a phenotype-genotype correlation.

1.3.5 Moving from candidate genes to disease alleles

Ultimately, the biologist requires evidence of a change in function to support a hypothetical genetic association; bioinformatics has a role to play here, too. For example, DNA variants that alter subsequent amino-acid sequences can be checked for potential functional consequences by a range of software tools (Chapters 11 and 13). Similarly, a thorough bioinformatics characterization of putative regulatory elements can give an indication of the possible impact of polymorphisms on splicing and expression levels (Chapter 12). Finally, our understanding of the functional elements of the genome is still expanding; the most startling example of this is our knowledge of miRNA. This large class of small, non-coding RNAs was almost unheard of when the first edition of this book appeared (July 2003). Now, however, miRNAs are recognized as one of the major regulatory gene families in eukaryotic cells (Kim and Nam, 2006). We hope Chapter 14 atones for the shocking omission of coverage of this important domain of biology in the first edition!

1.3.6 Analysis at the genetic and genomic data interface

The final section of this book addresses some of the emerging issues that geneticists face as the mature fields of genomics and genetics become increasingly closely interfaced, mainly through the complementary application of microarray technology to some of the most complex genetic problems. Chapter 15 presents a general overview of the microarray as a genomics platform and some of the issues that may arise in dealing with data from this technology. Chapter 16 addresses one of the most exciting applications of microarray technology – analysis of gene expression as a quantitative trait. Studies in mice are identifying genetic variants that influence a wide range of gene-expression phenotypes; these are, in turn, identifying complex transcriptional regulatory modules that may be megabases away from the gene(s) being regulated (Li *et al.*, 2006). Similar studies are also yielding results in man. For example, Stranger *et al.* (2005) performed a genome-wide quantitative trait analysis of 630 genes in the Caucasian HapMap cell lines. Using HapMap genotypes, they identified many regions with statistically significant associations between specific SNPs and expression variation in the HapMap lymphoblastoid cell lines after correcting for multiple tests. Their results suggest that regulatory polymorphism is widespread in the human genome, not necessarily in the immediate 5′ region of genes. Such studies will significantly enhance our ability to annotate the non-coding part of the genome and interpret functional variation.

Another key application of oligonucleotide microarray technology is in the field of oncology. Chapter 17 reviews some of the distinct bioinformatics challenges that face geneticists studying cancer or, more appropriately, cancer genomes – each with its own unique array of point mutations, copy-number alterations and gross chromosomal changes.

Genome-wide association studies are perhaps the area of genetics where microarray technologies are making the biggest impact. The new generation of SNP-genotyping arrays, which allow simultaneous testing of hundreds of thousands of SNPs, is revolutionizing genome-scan analysis and the search for genes that influence common genetic traits. A number of major studies are now under way, and many more are in planning, to perform association scans with LD to detect risk-associated variants in large population-based sample collections (Thomas *et al.*, 2005; http://www.ncbi.nlm.nih.gov/WGA/). However, these studies are also creating problems for geneticists on an unprecedented scale. The foremost among these issues is probably type I error (false-positive association) due to multiple testing. More than ever, effective bioinformatics is required to help to filter and prioritize the outputs of these genome-wide scans. Chapter 18 examines these issues in detail and suggests some potential bioinformatics solutions for this entire area of research.

In the final chapter of this book, Chapter 19, we address one of the key end points of genetics research – the development of new drugs and therapeutics. This domain of genetics research was inexplicably missing from the first edition of this book, despite being so close to the heart of its editors and most of the contributors! The development of new therapeutics is often cited as one of the primary objectives of a genetic study, but, unfortunately, so far there have been very few published examples of genetic associations being translated into drugs, although this may already be changing (Roses *et al.*, 2005). Aside from drug discovery, genetics is also being used to clarify the basis of the observed interindividual variability in drug response, the nascent field of pharmacogenetics. We expect that the study of pharmacogenetics will expand further, as it is seen increasingly as a requirement for drug development by regulatory authorities, such as the US Food and Drug Administration (FDA) (Woodcock, 2005), and as public-domain databases are established to collect pharmacogenetic data (Gurwitz *et al.*, 2006). Bioinformatics has a great deal to offer genetics focused either on drug discovery or drug response. Both usually involve finite 'universes' of genes; in the case of drug discovery, it is the druggable components of the genome (Hopkins and Groom, 2002), and in the case of pharmacogenetics, it is genes known to be involved in drug absorption, dissemination, metabolism or excretion (collectively known as ADME genes).

1.4 Conclusions

On behalf of all the contributors, we sincerely hope this book will help geneticists to design and carry out effective genetic analyses. Effective bioinformatics can have a real impact on the success of laboratory research, but it is not intended as a replacement for the laboratory process. Misconceptions regarding the power of bioinformatics as a stand-alone science are perhaps among the biggest mistakes that bioinformatics specialists can make and may even explain a degree of prejudice against bioinformatics, which is perceived by some as an '*in silico*' science with little basis in reality. Taken to an extreme and without both a balanced understanding of the application of software tools and a good appreciation of basic biological principles, this is exactly what bioinformatics can be; but where bioinformatics proceeds as part of 'wet' and 'dry' cycles of investigation, both processes are stronger as a result.

1.4.1 New opportunities for the geneticist

Another criticism of bioinformatics also reveals a possible strength. Bioinformatics scientists often need to be generalists, covering a vast knowledge domain. This rarely allows time for the development of in-depth expertise in more than a very limited range of areas; however, it does offer great opportunities to spot potential synergies between different research domains. Genetics has the potential to affect just about every area of biology, so it might be worth highlighting a few of these for particular attention.

Epigenetics – 'It's the Epigenome, *stupid!'*

Good advice to any geneticist. During the 1992 US presidential campaign, James Carville, an adviser to Bill Clinton, decided that the push for the presidency needed focus. Drawing on electoral research, he came up with a simple focus for the campaign. At every opportunity, James Carville wrote four words – 'IT'S THE *ECONOMY*, STUPID!' – on a whiteboard for Bill Clinton to see every time he went out to speak.

Clearly, this worked for Bill Clinton, so it might just work for genetics. Epigenetics represents a secondary inheritance system that has so far been the subject of very limited investigation. Epigenetics is concerned with the study of heritable changes other than those in the DNA sequence and encompasses two major modifications of DNA or chromatin: DNA methylation and post-translational modification of histones (Callinan and Feinberg, 2006). These modifications are critical regulatory cues, making DNA more or less accessible to DNA-binding proteins. Preliminary evidence suggests that epigenetics is something that geneticists must think about in their genetic analysis. Flanagan *et al.* (2006) demonstrated evidence of significant epigenetic variability in human sperm cells, suggesting that epigenetic patterns can be efficiently transmitted across generations, possibly influencing phenotypic outcomes in health and disease. DNA methylation profiles are complex and dynamic, and can vary with developmental stage, tissue type, age, the alleles' parent-of-origin, and phenotype or disease state. This fits very well with many of the observed characteristics of diseases such as a defined or variable age of onset, variable penetrance, and variable tissue distribution.

In the absence of an entire chapter on this rapidly expanding area, Chapter 9 suggests some approaches that might help to incorporate this information into genetic analysis. A full chapter on epigenetics is a definite requirement for the third edition of this book, by which time the Human Epigenome Project (Rakyan *et al.*, 2004) will be complete, and consideration of epigenetics may have become an integral part of the way that geneticists work.

The HapMap – it's more than LD

Apologies for the flippancy, but the data generated by the HapMap really are more than LD. For example, it can clarify the demographic history and evidence of selection in human populations (Voight *et al.*, 2006) and of previously undetected regulatory relationships and gene networks (Petkov *et al.*, 2005). All of these properties make the HapMap no less important as a resource than the human genome sequence itself. Further investigations of these alternative applications are well under way, and they can be effectively monitored by a simple PubMed search using 'HapMap' as a keyword.

Let us not forget the 'unknown unknown' elements of the genome

Obviously, we will not find this category of genomic elements annotated anywhere in Ensembl or the UCSC genome browser, but they are undoubtedly important (http://en.wikipedia.org/wiki/Unknown_unknown). The potential value of these elements is informed by former members of this class, miRNAs being one of the most notable. Ten years ago, we had no idea that miRNAs existed, but today we know of more than 300 in man, and their role in global gene regulation appears to be critical (see below). But how can we identify these elements? Comparison of genomes between species is one way to highlight evolutionarily constrained (putatively functional) but otherwise unknown elements in the genome (see Chapter 6). But as miRNA are also illustrating, not all functional elements are conserved; for example, miRNA target sites show limited conservation between mouse and human genomes. This illustrates that, for example, the huge differences between mice and man may be due to genomic elements of which we are still completely ignorant.

miRNA!

We missed them the first time, but we are not going to let this happen again. MiRNAs appear to be a critical element of gene regulation that genetics needs to account for. Once the reader has finished Chapter 14, boring old 3′ UTR SNP associations will never seem the same again.

How much do we really know about gene regulation?

Just as our knowledge of the role of miRNA is revealing unknown mechanisms of gene regulation, the identification of *cis*-acting expression quantitative trait loci is starting to challenge the dogma of our knowledge of the promotion of gene expression (Chapter 16). Knowledge that regulatory control or promoter elements may be located more than a megabase away from a gene obviously makes genetic analysis potentially very difficult. This makes detailed bioinformatics characterization of genetic association data all the more important. Tools such as GeneNetwork.org are starting to address some of these issues, but this is still an area that all geneticists should watch closely.

Carriage return

These are just a few of the issues that geneticists may have to address in the next few years. In this introduction, we have briefly examined ways in which genetics can be assisted by bioinformatics; we now invite the reader to more detailed coverage of each of these areas in the remaining chapters of this book.

Acknowledgements

I extend warmest thanks to Ian C. Gray for his generous help and support in the drafting of the second edition of this book, and for his substantial contribution to the first-edition version of this chapter.

References

Altshuler, D., Pollara, V. J., Cowles, C. R. *et al.* (2000a). An SNP map of the human genome generated by reduced representation shotgun sequencing. *Nature* **407**, 513–516.

Altshuler, D., Hirschhorn, J. N., Klannemark, M. *et al.* (2000b). The common PPARgamma Pro12Ala polymorphism is associated with decreased risk of type 2 diabetes. *Nat Genet* **26**, 76–80.

Callinan, P. A. and Feinberg, A. P. (2006). The emerging science of epigenomics. *Hum Mol Genet* **15** (Spec No 1), R95-101.

Campuzano, V., Montermini, L., Molto, M. D. *et al.* (1996). Friedreich's ataxia: autosomal recessive disease caused by an intronic GAA triplet repeat expansion. *Science* **271**, 1423–1427.

Flanagan, J. M., Popendikyte, V., Pozdniakovaite, N. *et al.* (2006). Intra- and interindividual epigenetic variation in human germ cells. *Am J Hum Genet* **79**(1), 67–84.

Fredman, D., Siegfried, M., Yuan, Y. P. *et al.* (2002). HGVbase: a human sequence variation database emphasizing data quality and a broad spectrum of data sources. *Nucleic Acids Res* **30**, 387–391.

Gurwitz, D., Lunshof, J. E. and Altman, R. B. (2006) A call for the creation of personalized medicine databases. *Nat Rev Drug Discov* **5**(1), 23–26.

Hopkins, A. L. and Groom, C. R. (2002). The druggable genome. *Nat Rev Drug Discov* **1**(9), 727–730.

Hugot, J. P., Chamaillard, M., Zouali, H. *et al.* (2001). Association of NOD2 leucine-rich repeat variants with susceptibility to Crohn's disease. *Nature* **411**, 599–603.

Huntington's Disease Collaborative Research Group (1993). A novel gene containing a trinucleotide repeat that is expanded and unstable on Huntington's disease chromosomes. *Cell* **72**, 971–983.

International HapMap Consortium (2003). The International HapMap Project. *Nature* **426**, 789–796.

Kim, V. N. and Nam, J. W. (2006). Genomics of microRNA. *Trends Genet* **22**(3), 165–173.

Li, H., Chen, H., Bao, L. *et al.* (2006). Integrative genetic analysis of transcription modules: towards filling the gap between genetic loci and inherited traits. *Hum Mol Genet* **15**(3), 481–492.

Matsuzaki, H., Dong, S., Loi, H. *et al.* (2004). Genotyping over 100,000 SNPs on a pair of oligonucleotide arrays. *Nature Methods* **1**, 109–111.

Miki, Y., Swensen, J., Shattuck-Eidens, D. *et al.* (1994). A strong candidate for the breast and ovarian cancer susceptibility gene BRCA1. *Science* **266**, 66–71.

Petkov, P. M., Graber, J. H., Churchill, G. A. *et al.* (2005). Evidence of a large-scale functional organization of mammalian chromosomes. *PLoS Genet* **1**, e33.

Rakyan, V. K., Hildmann, T., Novik, K. L. *et al.* (2004). DNA methylation profiling of the human major histocompatibility complex: a pilot study for the human epigenome project. *PLoS Biol* **2**(12), e405.

Reich, D. E., Cargill, M., Bolk, S. *et al.* (2001). Linkage disequilibrium in the human genome. *Nature* **411**, 199–204.

Risch, N. J. (2000). Searching for genetic determinants in the new millennium. *Nature* **405**, 847–856.

Riordan, J. R., Rommens, J. M., Kerem, B. *et al.* (1989). Identification of the cystic fibrosis gene: cloning and characterization of complementary DNA. *Science* **245**, 1066–1073.

Roses, A. D., Burns, D. K., Chissoe, S. *et al.* (2005). Disease-specific target selection: a critical first step down the right road. *Drug Discov Today* **10**(3), 177–189.

Saunders, A. M., Strittmatter, W. J., Schmechel, D. *et al.* (1993). Association of apolipoprotein E allele E4 with late-onset familial and sporadic Alzheimer's disease. *Neurology* **43**, 1467–1472.

Savage, D. B., Tan, G. D., Acerini, C. L. *et al.* (2003). Human metabolic syndrome resulting from dominant-negative mutations in the nuclear receptor peroxisome proliferator-activated receptor-gamma. *Diabetes* **52**(4), 910–917.

Sherry, S. T., Ward, M. H., Kholodov, M. *et al.* (2001). dbSNP: the NCBI database of genetic variation. *Nucleic Acids Res* **29**, 308–311.

Spielman, R. S., McGinnis, R. E. and Ewens, W. J. (1993). Transmission test for linkage disequilibrium: the insulin gene region and insulin-dependent diabetes mellitus (IDDM). *Am J Hum Genet* **52**(3), 506–516.

Stenson, P. D., Ball, E. V., Mort, M. *et al.* (2003). Human Gene Mutation Database (HGMD): 2003 update. *Hum Mutat* **21**(6), 577–581.

Stranger, B. E., Forrest, M. S., Clark, A. G. *et al.* (2005). Genome-wide associations of gene expression variation in humans. *PLoS Genet* **1**(6), e78.

Thomas, D. C., Haile, R.W. and Duggan, D. (2005). Recent developments in genome-wide association scans: a workshop summary and review. *Am J Hum Genet* **77**, 337–345.

Voight, B. F., Kudaravalli, S., Wen, X. *et al.* (2006). A map of recent positive selection in the human genome. *PLoS Biology* **4**, e7.

Wang, W. Y., Barratt, B. J., Clayton, D. G. *et al.* (2005).Genome-wide association studies: theoretical and practical concerns. *Nat Rev Genet* **6**(2), 109–118.

Woodcock, J. (2005). Pharmacogenetics: on the road to 'personalized medicine'. *FDA Consum* **39**(6), 44.

2

Managing and Manipulating Genetic Data

Karl W. Broman[1] and Simon C. Heath[2]

[1] *Department of Biostatistics, Johns Hopkins University, Baltimore, MD, USA*

[2] *Centre National de Genotypage, Evry Cedex, France*

2.1 Introduction

Geneticists must learn to program: for efficiency, to avoid introducing errors into data, and to make simple what would otherwise be unfeasible. If a geneticist were to learn just one programming language, Perl would be an excellent choice; it is especially valuable for the manipulation of text files, which are the input and output of most statistical genetic software.

Our ability to learn from data relies upon the accuracy and integrity of such data. Thus, it is critical that data be stored and managed with great care. The continual growth in the size and complexity of genetic data has led to an increasing need for a formal approach to data management.

Many data are in the form of a rectangle: many individuals measured at many variables. Genetic data, however, are generally of more complex form, including pedigree information and genetic maps. Moreover, no standard data format has emerged, nor does there exist a comprehensive statistical genetic software package. The analysis of genetic data generally requires the use of multiple computer programs, each having a unique data input format.

A fundamental task in statistical genetic analyses is thus the manipulation of data files in order to conform to the variety of input formats required by the variety of software tools that must be used. Such data manipulation is cumbersome,

Bioinformatics for Geneticists, Second Edition. Edited by Michael R. Barnes
 ISBN 978-0-470-02619-9 (HB) ISBN 978-0-470-02620-5 (PB)

time-consuming, and error-prone, if not impossible, without the ability to program in a language like Perl. Programming also provides the ability to automate analyses and to perform computer simulations.

In this chapter, we describe the essential issues in the management and manipulation of genetic data, focusing on the case of human linkage data, although the basic principles apply to all types of data. Towards the end of the chapter, we provide some sample snippets of Perl code, to give the reader a flavour of the language and to emphasize certain features of Perl that are especially valuable for this type of work. We include examples of code with some trepidation, as we fear that readers will run in fright from learning to program. And so we hope that if the code frightens readers, they will ignore it, initially, and focus on the essential ideas. But we also hope that readers will be persuaded by our argument that geneticists must learn to program (or hire a programmer).

2.2 Basic principles

We begin with a brief set of guiding principles for the manipulation of genetic data. Our goals are, first, to maintain the integrity of the data; second, to be as efficient as possible; and third, to ensure that results are reproducible.

2.2.1 Never modify data 'by hand'

If certain genotypes are to be removed as likely to be in error, create a file of such, and write a program that creates a new version of the data with those genotypes removed. If the data must be reformatted for a particular software package, do not edit the files directly; write a program to do so. Why? One then avoids the introduction of errors, results can be easily reproduced, and the process can be automated so that, if the primary data should change, essentially no further effort must be expended to get back to the same point. Moreover, the computer program provides a record of what was done.

We would like to emphasize the value of command-line programs over point-and-click programs for this reason. Pointing and clicking can be useful for the occasional user of software, or for preliminary, interactive analyses, but if automation is needed, pointing and clicking is far too cumbersome, and if the analysis is to be repeated (and it usually is), how much easier is the repeated run of a program than repeated pointing and clicking!

2.2.2 Be organized; keep notes

When one leaves the laboratory and sits down in front of a computer, the importance of a laboratory notebook should not be forgotten. The procedures in data analysis are not unlike those of a laboratory experiment: there are often many steps to be taken and many choices to be made at each step. Careful account must be taken of the particular steps and the particular choices, so that the results obtained may be understood, trusted, and reproduced. Such organization requires the investment of some effort, but this is made in order to minimize future effort.

Computer programs can serve as a useful record of one's analyses. However, it is often the case that multiple short programs are written, and that each includes some flexibility (and, indeed, we will emphasize the importance of both of these features subsequently). And so further notes on the particulars of one's analyses will be desired. If copious printouts are to be avoided, a short electronic notebook might be recommended.

It is unfortunate that statisticians have not adopted the laboratory notebook tradition, especially given the growth in the size and complexity of their computer simulations. (Statisticians' simulation results are notoriously irreproducible.) We hope that they soon do.

2.2.3 Reuse code

Few tasks are performed just once in a career, and so in writing a computer program, one should consider the possibility that it may be of some use in the future. Programs should be written in a modular and reasonably general form, and explanations ('comments') should be included in order to clarify any aspects of the program that are not obvious.

One must balance current versus future effort. If a program is written that is quite specific to the current task, it cannot be reused without modification. If the program is made somewhat more general (so that, for example, file names and parameter values are specified on the command line rather than within the program), there is a greater chance that it will be reused without modification in the future. But to write the program in more complete generality may require considerably more current effort without any guarantee that the added features will ever be put to use.

Modularity of software can increase the chance that one's programming effort will be put to future use. All of one's tasks might be solved by a single long, strung-out program, but it is unlikely that the same long sequence will be required unchanged in the future. If the long program is split into many small, independent modules, it is much more likely that some individual module will be of future use, unchanged.

Documentation of software is critical, even for code that is intended only for the programmer's own use. Think of yourself 3 months or 3 years hence; will you remember what you did and be able to modify or fix your code? That the program is written with some clarity is as important as proper documentation. If extensive explanations are required, perhaps it is best that the code be rewritten so that its use is more transparent. It is important that the documentation describe not only the operation of the program, but also the assumptions that the program makes about the input data. It is all too easy to write a program, that relies on a particular feature of a data set (for example, that the records are sorted, or that the columns in the data file are in a particular order). If the software is subsequently reused on new data that do not have this feature, the results will be incorrect. Ideally, programs should perform extensive checking of any input data, particularly if the programs are intended for reuse, but further comparisons of input and output data are recommended to ensure that the data have not become garbled due to some subtle change in data format.

2.2.4 There's probably an easier way, but...

The first priority in programming should be to write code that works. There are generally many approaches to any program; do not concern yourself initially (if at all) with finding the optimal solution. Another trade-off arises here: time to construct the program versus time to run the program. For tasks in data manipulation, efficiency of computation is seldom of much importance. First solve the problem. If it is later seen to be important to reduce computation time, seek a more optimal solution, but retain your initial solution as a benchmark.

2.3 Data entry and storage

Data seldom begin their life within a computer; ideally, they are transmitted directly from the measuring instrument to the computer. If data are to be entered into the computer by hand, it is best done independently by at least two people, in order to reduce the possibility of errors. Any discrepancies between the two data sets may be checked against the original data.

Data sets of small or moderate size can reasonably be stored in an office spreadsheet program, such as Microsoft Excel. It is best to insert a value in every cell, using a standardized code (such as NA) in any cells for which the data are missing, rather than leave some cells empty. Empty cells are ambiguous: was the value missing, or was an error in data entry made? It is best not to use special fonts (such as boldface) or colours to encode important information, as such codes will be difficult to extract from the software. Consistency in the coding throughout the data will, of course, simplify its later use.

We routinely receive data as Excel files, but convert them to comma- or tab-delimited text files prior to their use, as such text files are easily manipulated via computer programs and are generally needed for input into statistical genetic software. For much of our work, it is sufficient to maintain the data in such text files.

The increasing size and complexity of genetic data argue for the abandonment of Excel or other spreadsheets as a solution for data storage, especially as Excel is limited in the number of columns (256) and rows (about 65 000) that are allowed. We continue to use plain text files for storing extremely large data sets (e.g. genotype data on 500K SNPs), but for complex data (particularly for the maintenance of multiple projects whose data may be pooled, or for a project with a large number of individuals measured at many phenotypes longitudinally), a formal database may be preferred. The choice of database software depends on the size and complexity of the data (as well as the budget). For smaller projects, open-source solutions, such as MySQL or PostgreSQL, can work very well. For very large collections of data, however, it might be better to use one of the commercial offerings, such as Oracle or Sybase. In any case, if data storage and handling requirements are such that a database is required, it will generally be necessary either to hire a dedicated employee who is proficient in the design, implementation, and maintenance of databases, or to buy a complete solution where the database application has already been developed. The advantage of the latter solution is that these packages generally come with support from the supplier. The disadvantage is the cost, which in many cases can be substantial (although the cost of hiring a database programmer for the first solution must not be forgotten).

We hope it is unnecessary to emphasize that all data should be backed up regularly (and automatically), with backups kept off site so that, should a catastrophe occur, minimal data are lost.

2.4 Data manipulation

The analysis of genetic linkage data involves a sequence of tasks: verify and correct relationships between individuals, identify and resolve genotyping errors, identify and resolve errors in the phenotypes and any covariates, and perform the actual analysis. Sometimes one may then conduct computer simulations to assess the performance of the statistical methods or to obtain P values that properly account for test multiplicity.

As the different tasks involve the use of different programs, and as each such program may have its own data input format, the central problem concerns the manipulation of the data files to conform to the necessary input formats. The program Mega2 (Mukhopadhyay *et al.*, 2005) can be useful in this regard: once the data are put into Mega2, the program can be used to create files conforming to most, if not all, statistical genetic software of interest. We, however, have not made use of Mega2, but instead have written our own Perl programs to convert data between formats.

It is essential, for the manipulation of genetic data files, to define a single standard format for one's work. For almost every linkage project we are involved in, the primary data arrive in a unique format. One might be tempted to write new Perl programs to convert data from each such format into that needed for each analysis program of interest. If we are involved in 20 projects and there are 12 analysis programs we wish to use, we would then need to write 240 different Perl programs. A better approach is to define our own standard format, and write Perl programs to convert data from that format to each of the 12 analysis programs, and then for each project, we write just one Perl program to convert the data to our standard form. With 20 projects and 12 analysis programs, we then have 32 Perl programs. And for each additional project, we write just one new Perl program, rather than 12.

A second important use of Perl in genetics is the automation of analyses. A particularly important example of this concerns single-marker linkage analysis (so-called two-point analysis), in which each of about 400 markers is investigated, one at a time, for linkage to a putative disease gene. We are aware of cases in which an investigator created, 'by hand', 400 input files (one for each marker), and then ran a linkage program 400 times, again 'by hand', writing down the one or two numbers that characterize the results for each marker. The problem with this approach should be obvious. More important than the enormous waste of effort is that the manual manipulation of data files, and the transcription of the results, can be extremely error-prone. With proficiency in Perl, it is a simple matter to write a program that reads all of the genetic data, steps through the markers one at a time, creates the required input files, runs the linkage program and extracts the essential pieces of information, and finally produces a table of the results for all markers.

Finally, Perl is extremely valuable for performing computer simulations with other genetic software, either to explore the performance of an analysis method or to obtain P values that make proper adjustment for the multiplicity of tests performed. This task is much like that of automating analyses: one simulates data (either with Perl or someone else's program), sends it through an analysis program, extracts the interesting bits from the output, and repeats the entire process many times. The greatest advantage of Perl for simulations is in the extraction and tabulation of the one or two interesting numbers at each replicate from the copious output produced by most analysis programs. This approach can be applied to essentially any statistical genetic software.

2.5 Examples of code

In this section, we provide some examples of Perl code, in order to give the reader a flavour of the language and to emphasize certain features of Perl that are especially useful for our work. We are unsure of the value of this section for a reader with no prior Perl programming experience; such readers may wish to skip this section.

Perl programs are generally run from a terminal window in Mac OS X or Unix, or from a command shell in Windows. The Perl interpreter will be pre-installed in Mac OS X and most Unix distributions. A Windows version of Perl may be obtained from http://www.activestate.com/ActivePerl.

2.5.1 The traditional first example

A traditional first example, and closest to the simplest possible Perl program, is displayed in Figure 2.1. This program simply prints 'Hello, world!' to the screen. The

```
#!/usr/bin/perl -w

print("Hello, world!\n");
```

Figure 2.1 A simple but complete Perl program

first line is necessary for Unix and MacOS, and indicates where the Perl interpreter is located. The `-w` indicates that the Perl interpreter should provide warnings regarding various constructions in Perl that, while being strictly legal, are more likely than not to be errors.

The second line prints the desired phrase. Note that `\n` is the 'newline' character. The semicolon indicates the end of the Perl statement.

One must create a text file containing the above code. To run the program in Unix or MacOS, the file must be made 'executable', by typing, from a terminal window, `chmod +x` *`filename`*, where *`filename`* is the name of the file. The program is then run by typing the name of the file. In Windows, `chmod` is not needed. Instead, the program file must be given a name of the form *`filename`*`.pl`. The program is then run from a command shell by typing the name of that file or by typing `perl filename.pl`.

2.5.2 Combining marker data

A common issue in genetic data manipulation is the combination of genotype data from multiple input files. In an extreme case, one may be confronted with a single file for each genetic marker. In Figure 2.2, we present a Perl program for reading all files in a directory in order to combine genotype data. We are imagining here that there is a single directory containing one file for each marker, with each file having a name like `D10S1123.txt`, where D10S1123 is the marker. The files are in LINKAGE PRE format, that is to say, each line contains the family identifier, individual identifier, dad, mom, sex, and disease status and then the two alleles for that subject at that marker. The aim of the first program is to read in all of the data, to store them in such a way that we can easily work with them. This may not appear so useful in itself, but we will show in subsequent examples how the program can be extended to perform recoding of marker alleles, estimation of allele frequencies and generation of input files for the LINKAGE programs.

The first line is the usual first line for a Perl program. The second and third lines instruct Perl to be stricter in terms of what it accepts and to issue warnings for unsafe code. This is highly recommended, as without these it is very easy to make errors that can be very difficult to detect.

The main inconvenience of this is that it is now necessary to *declare* each variable before use using the `my` command. For example, in line 5, `my $dir` declares that `$dir`

```
1   #!/usr/bin/perl
    use strict;
    use warnings;

5   my $dir = "data";
    opendir DIR, $dir or die "Cannot open directory $dir:$!\n";
    my (%ped, %gtypes, @markers);
    while(my $file=readdir(DIR)) {
        next unless $file =~ /(.+)\.txt$/;
10      my $mark = $1;
        push @markers, $mark;
        my $idx = $#markers;
        my $infile = "$dir/$file";
        open IN, $infile or die "Cannot open $infile:$!\n";
15      my $line = 0;
        while(<IN>) {
            $line++;
            my @v=split;
            if(@v<8) {
20              print "Short line at $line!\n";
                next;
            }
            my ($fam,$ind,$father,$mother,$sex,$status,$g1,$g2)=@v;
            my $id="$fam\_$ind";
25          $ped{$ind} = [$fam,$ind,$father,$mother,$sex,$status];
            $gtypes{$ind}[$idx] = "$g1 $g2";
        }
        close IN;
    }
```

Figure 2.2 A Perl program to read data from all data files with a `.txt` extension

is a scalar variable, indicated by the dollar sign, which is here assigned the character string `data`. The content will be just the bit between the double quotation marks. The advantage of having Perl enforce pre-declaration of variables is that it is very easy to mistype a variable name, and, by default, Perl will not complain but silently create a new variable with the mistyped name. This can lead to some extremely subtle and difficult to track down bugs in programs. For all but very short programs, therefore, it is generally advised to follow the practice here of adding the `use strict;` and `use warnings;` statements to the start of your programs.

In line 6, we open a directory using a 'directory handle' `DIR`. This allows us, from line 8, to 'loop' through each file in the directory; within this `while` loop, we read one file name at a time from the directory until there are no files remaining to be read. Note that if the `opendir` command fails, the `die` statement will be executed, which stops the program and prints the message `Could not open directory $dir: $!`. The variable `$dir` is expanded in the message to give the value we assigned in line 5. The odd-looking variable `$!` is a system variable, which gives the last error message from a system command, in this case `opendir`.

In line 9, we use pattern matching to check that the file name ends in `.txt`; otherwise, that file is skipped. (This is important, because the directories '`.`' and '`..`' will be included, but should be skipped.) The code for the pattern matching is a bit complicated at first glance. The first thing to note is that a period (.) matches any character and a plus sign (+) means one or more of the previous match. To match a literal period, it is necessary to escape the period with a backslash. The dollar sign at the end of the pattern matches the end of the string. If we ignore the brackets for the moment, the code in line 9 will therefore match one or more characters terminated by `.txt`. The brackets around the first part `.+` direct Perl to store the part of the input string which matched this part of the pattern, and store it in the variable `$1`, which is assigned to the variable `$mark` in line 10.

In line 11, the marker name is appended to the end of an array of all marker names, `@markers`. The `@` symbol indicates an array: an ordered list of values, indexed by 0, 1, 2, ... The index of the last item in an array is given by `$#`*name_of_array*, so line 12 sets `$idx` to the index of the last marker added, i.e., the current marker.

In line 13, `$infile` is assigned the full file name: the directory name followed by a / followed by the simple part of the file name. Note how we can use variables inside a quoted string, and they will be expanded to give the resulting string. We then open this file in line 14, producing a 'file handle', `IN`.

In line 15, we initialize the variable `$line` to zero; this will be used to track the line number of the input file, so that errors can be reported.

From line 16, we loop through each line in the input file. In a similar way to the while loop starting at line 8, this loop will exit when there are no more lines to be read.

In lines 17–18, we increment the line number and split the line into fields separated by white space (any combination of non-printing characters such as spaces or tabs), storing the results in the array `@v`. Lines 19–22 then check that there are at least eight columns of data; if there are fewer, we print an error message and skip to the next line.

In line 23, we assign the contents of the array `@v` to the individual variables, `$fam`, `$ind`, etc.

In lines 24–26, we store the information on the individuals' parents and sex, using 'hashes'. (This is rather difficult for beginning Perl programmers, but hashes are extremely valuable for this sort of work, as we will see in the next example.) A hash is like an array, but the hash is keyed by an arbitrary character string rather than indexed by numbers 0, 1, 2, ... Here we create a unique identifier for an individual by concatenating the family and individual identifiers together with an underscore character between them in line 24. Note here that we escape the underscore after `$fam` because otherwise Perl would take it as part of the variable name. We then store the pedigree information and genotype information in lines 25–26 keyed by his unique identifier. Note that for the genotype, we also index with the variable `$idx` (from line 8), which indicates which marker we are working on. We use braces {} for the variable `$id` and square brackets `[]` for the marker index at line 26 to indicate to Perl that `$id` should be treated as a hash key and `$idx` should be treated as a conventional numeric index. It is not important to understand the details of how the data are stored in lines 25–26; the key point is that with the individuals' identifiers, we can access their pedigree information and genotype data.

We could avoid using hashes if we could assume that the same individuals appear in each input file in the same order. We could then just index the data by the line number. However, it is not always safe to make this assumption; in general, it is safer to use hashes.

At line 28, the input file has all been read in, so we close the file, and continue with the next file, if present.

2.5.3 Recoding alleles

The program in Figure 2.2 would be more useful if it could do some basic data manipulation. One such manipulation that is often required is allele recoding. Many programs for genetic analysis expect marker alleles to be coded from 1 up to the number of alleles present. The raw data, however, rarely come in this form. Microsatellite data come as allele sizes such as 180 or 225, and SNP data typically come as a series of nucleic acid codes (A, C, G or T). It is simple to use hashes in Perl to recode alleles, and this is a good illustration of the power of hashes. The strategy is to use the original allele code as the key to the hash. We can use this to check whether a numeric code has already been assigned to this allele and, if not, assign it the next available code.

In Figure 2.3, we provide a modification of the program in Figure 2.2 which will enable the program to recode the marker alleles into consecutive numeric codes starting from 1. The key additions are from lines 26–37. We start at line 26 by checking that the first allele is non-zero. (Zero typically indicates a missing value.) We then check whether this allele has already been encountered for this marker by checking the array `@recode`, which is indexed by the marker index `$idx` and the allele `$g1`. If not, then at line 28 we assign the next available code for this marker (stored in the array `@n_alleles`, and then at line 29 we change the original allele code to the numeric code. The same procedure is then followed for the second allele `$g2`. Note that doing this procedure without hashes would be a much more complex operation involving sorting and searching through the list of marker alleles.

2.5.4 Estimating allele frequencies

Another useful function of the program is to estimate allele frequencies, as most genetic analysis programs require these, and good estimates matched with the data set are not always available. In this case we can obtain allele estimates by simply counting the alleles in observed individuals. While marker allele frequencies are best estimated on the basis of unrelated individuals, such as the founding individuals in a set of pedigrees, genotypes of such founders are sometimes not available, and simple allele

```
1   #!/usr/bin/perl
    use strict;
    use warnings;

5   my $dir = "data";
    opendir DIR, $dir or die "Cannot open directory $dir:$!\n";
    my (%ped, %gtypes, @markers, @n_alleles, @recode);
    while(my $file=readdir(DIR)) {
        next unless $file =~ /(.+)\.txt$/;
10      my $mark = $1;
        push @markers, $mark;
        my $idx = $#markers;
        my $infile = "$dir/$file";
        open IN, $infile or die "Cannot open $infile:$!\n";
15      my $line = 0;
        while(<IN>) {
            $line++;
            my @v=split;
            if(@v<8) {
20              print "Short line at $line!\n";
                next;
            }
            my ($fam,$ind,$father,$mother,$sex,$status,$g1,$g2)=@v;
            my $id="$fam\_$ind";
25          $ped{$ind} = [$fam,$ind,$father,$mother,$sex,$status];
            if($g1 != 0) {
                if(!$recode[$idx]{$g1}) {
                  $recode[$idx]{$g1} = ++$n_alleles[$idx];
                }
30              $g1 = $recode[$idx]{$g1};
            }
            if($g2 != 0) {
                if(!$recode[$idx]{$g2}) {
                  $recode[$idx]{$g2} = ++$n_alleles[$idx];
35              }
                $g2 = $recode[$idx]{$g2};
            }
            $gtypes{$ind}[$idx] = "$g1 $g2";
        }
40      close IN;
    }
```

Figure 2.3 A Perl program to read data from all data files with a `.txt` extension and recode marker alleles

counting provides unbiased estimates, without the great computational effort that can be required to account for the relationships between individuals (Broman, 2001).

Since we have already recoded the alleles to consecutive numbers in the previous example, it is simple to add a section to the program in Figure 2.3 to accumulate allele count information and to estimate allele frequencies. Figure 2.4 contains a snippet of Perl code which should go at the end of the previous program. It will estimate allele frequencies, and store them in the double indexed `@freq` so that `$freq[$i][$j]` will have the estimated frequency of allele `$j` of marker `$i`.

```
1  my (@freq, @count);
   for my $ind(keys %ped) {
       my $gt = $gtypes{$ind};
       for my $i(0..$#markers) {
5          my $g=$$gt[$i] || "0 0";
           my @all=split " ",$g;
           for my $j(0..2) {
               if($all[$j]) {
                   $freq[$i][$all[$j]]++;
10                 $count[$i]++;
               }
           }
       }
   }
15 for my $i(0..$#markers) {
       my @fq=@{$freq[$i]};
       for my $j(1..$#fq) {
           $fq[$j] /= $count[$i];
       }
20 }
```

Figure 2.4 A snippet of Perl for calculating marker allele frequencies

The first line of the snippet simply declares the arrays `@freq` and `@count`, where the former was described in the previous paragraph, and the latter will keep a count of the number of alleles observed for a given marker.

In line 2, we loop through all individuals for whom we have pedigree information, that is, every individual we read in previously, and then in line 4 we loop through the genotypes for each marker for this individual. If an individual did not appear in all of the input files, some of the genotypes will be undefined, and attempting to work with them will give a warning. We avoid this in line 5 by using the string '`0 0`' for any undefined genotype.

In line 6, we split the genotype on spaces to get the two alleles, and in lines 7–12 we loop through the two alleles, accumulating the counts for all non-zero alleles, and a total count for the marker. After this, it is just necessary to loop through each marker, and for each allele at each marker, and divide the allele counts by the total number of counts for that marker. This is done in lines 15–20.

We can see that the logic of the frequency estimation is very simple, but it is so simple because we have already recoded the alleles numerically, using hashes in the previous example. If we had not done this, the operation would have been much more complicated.

2.5.5 Automating single-marker analyses

Now that we have the alleles recoded and have obtained allele frequency estimates, there are many things we could do. For example, we could print out the number

```
1   my @lod;
    for my $i(0..$#markers) {
        my $datfile = "datafile.dat";
        open OUT, ">$datfile" or die "Cannot open $datfile for writing: $!\n";
5       print OUT "2 0 0 5\n0 0.0 0.0 0\n 1 2\n";
        print OUT "1 2 # Trait locus (2 alleles)\n";
        print OUT "0.999 0.001 # Disease allele frequency\n";
        print OUT "1 # Liability class\n";
        print OUT "0.0 0.0 1.0 # Recessive model\n";
10      print OUT "3 $n_alleles[$i] # $markers[$i]\n";
        my @fq=@{$freq[$i]};
        print OUT join (" ",@fq[1..$#fq]),"\n";
        print OUT "0 0\n0.0\n";
        print OUT "1 0.05 0.45 # Recombination varied, increment, last value\n";
15      close OUT;
        my $pedfile = "pedfile.pre";
        open OUT, ">$pedfile" or die "Cannot open $pedfile for writing: $!\n";
        for my $ind(keys %ped) {
            my $p = $ped{$ind};
20          print OUT join ("\t",@$p);
            my $gt = $gtypes{$ind}[$i] || "0 0";
            print OUT "\t$gt\n";
        }
        close OUT;
25      my $results_file = "tempout.txt";
        system("makeped $pedfile pedfile.dat n > $results_file");
        system("unknown >> $results_file");
        system("mlink >> $results_file");
        open IN, $results_file or die "Cannot open $results_file for input: $!\n";
30      my $theta;
        while(<IN>) {
            if(/^THETAS\s+(\S+)/) {
                $theta=$1;
            } elsif(/LOD SCORE =\s+(\S+)/) {
35              $lod[$i]{$theta} = $1;
                print "$markers[$i]\t$theta\t$1\n";
            }
        }
        close IN;
40  }
```

Figure 2.5 A snippet of Perl for running MLINK for each of many markers

of observations per marker, and obtain estimates of the success rate per marker. We could equally well count the number of observations per individual, and check whether a particular DNA sample appears to have worked less well than others. These are all important steps in the quality control of the genotyping process. We are not going to go into more details about these analyses, but instead we will finish with a demonstration of how we could use the previous examples to automate single-marker (i.e. two-point) linkage analysis with MLINK from the LINKAGE (Lathrop *et al.*, 1984) or FASTLINK (Cottingham *et al.*, 1993) packages.

The snippet of Perl in Figure 2.5 should go at the end of the previous examples in order to function properly. We first declare the array `@lod`, which will store the calculated LOD scores for each marker at each theta value. We loop over all possible markers (line 2), and then write the necessary information to a locus data file (lines

3–15), and a pedigree file (lines 16–24). In line 4, the greater-than sign in '`>$datfile`' is used to open the file for writing (as opposed to reading, as in Figure 2.2). In lines 13 and 20, `join` is used to write out each element of an array, in turn separated by a space character at line 13, and a tab character at line 20.

In lines 26–28, `system` is used to request the operating system to execute the specified commands; this is where the real work is done. Note that a greater-than sign is used to have the program output sent to a file, and two greater-than signs together indicate that the output should be appended to the file, rather than replace the file.

In the remainder of this Perl snippet, we read through the output of MLINK, pulling out the LOD score at each recombination fraction, and store this information in a hash. Hence we can run MLINK for each marker, one at a time, and distil and assemble the few essential numbers from its profuse output, which can then be written to a file, or form a part of subsequent calculations, as, for example, in the calculation of heterogeneity LOD scores.

We congratulate readers who have persevered through the sample Perl code and our brief explanations. We hope that several of the techniques and idioms that we have demonstrated in these examples can be adapted by readers for use in more general situations. While the code looks quite complicated, the language is not as difficult to learn as it may appear, and the great power that comes from knowledge of Perl well justifies the effort that must be made to acquire it.

2.6 Resources

There are numerous books on Perl; we recommend those published by O'Reilly: *Learning Perl* (Schwartz *et al.*, 2005) for the novice, *Programming Perl* (Wall *et al.*, 2000) as a reference, and *Perl Cookbook* (Christiansen and Torkington, 2003) for recipes encompassing many common tasks. These books, plus a couple of others, may be purchased together on a CD at a very good price: the *Perl CD Bookshelf.*

There are numerous online tutorials on Perl; links to some are available at http://www.biostat.jhsph.edu/~kbroman/perlintro. This web page also contains a sample Perl program for genetic data manipulation, with line-by-line explanations. The Cold Spring Harbor Laboratory (CSHL) held a bioinformatics course in autumn 2004 that included a great deal on Perl programming; all of the lecture notes are available online at http://stein.cshl.org/genome_informatics.

Enormous amounts of useful Perl code may be obtained from the Comprehensive Perl Archive Network (CPAN) at http://cpan.perl.org. The CSHL lecture notes (mentioned above) provide good explanations of how to find and install code from CPAN. The reader may also be interested in Bioperl (http://www.bioperl.org): Perl tools for bioinformatics and genomics research, mostly for sequence data. Readers interested in the use of Perl for sequence data may wish to look at Tisdall (2001, 2003). Moorhouse and Barry (2004) will also be of interest.

Mega2 (Mukhopadhyay *et al.*, 2005), a program to facilitate the handling of genetic linkage data, is available at http://watson.hgen.pitt.edu/register.

2.7 Summary

The ever-increasing size and complexity of genetic data has led to an increasing need for geneticists to learn computer programming. As the most fundamental task for the genetic data analysis involves the manipulation of data files, proficiency in a computer language, such as Perl, with which such manipulation of text files is most natural, is recommended. For large, complex data sets, the use of a formal database, such as MySQL, in place of spreadsheet software, such as Microsoft Excel, may be important for the maintenance of data integrity and fidelity. Never modify data by hand, be organized and keep notes, and plan for the future but get the job done. Learn Perl!

Acknowledgements

Andrew Broman, Ken Manly, and Fernando Pineda generously provided comments to improve the manuscript. This work was supported in part by NIH/NHGRI grant GM074244 (to K.W.B.).

References

Broman, K. W. (2001). Estimation of allele frequencies with data on sibships. *Genet Epidemiol* **20**, 307–315.

Christiansen, T. and Torkington, N. (2003). *Perl Cookbook*, 2nd edn. Sebastopol, CA: O'Reilly Media.

Cottingham, R. W., Idury, R. M. and Schaffer, A. A. (1993). Faster sequential genetic linkage computations. *Am J Hum Genet* **53**, 252–263.

Dwyer, R. A. (2003). *Genomic Perl*. Cambridge: Cambridge University Press.

Lathrop, G. M., Lalouel, J. M., Julier, C. *et al.* (1984). Strategies for multilocus linkage analysis in humans. *Proc Natl Acad Sci USA* **81**, 3443–3446.

Moorhouse, M. and Barry, P. (2004). *Bioinformatics Biocomputing and Perl*. Chichester: Wiley.

Mukhopadhyay, N., Almasy, L., Schroeder, M. *et al.* (2005). Mega2: data-handling for facilitating genetic linkage and association analyses. *Bioinformatics* **21**, 2556–2557.

Reese, G., Yarger, R. J. and King, T. (2002). *Managing and Using MySQL*, 2nd edn. Sebastopol, CA: O'Reilly Media.

Schwartz, R. L., Phoenix, T. and Foy, B. D. (2005). *Learning Perl*, 4th edn. Sebastopol, CA: O'Reilly Media.

Tisdall, J. (2001). *Beginning Perl for Bioinformatics*. Sebastopol, CA: O'Reilly Media.

Tisdall, J. (2003). *Mastering Perl for Bioinformatics*. Sebastopol, CA: O'Reilly Media.

Wall, L., Christiansen, T. and Orwant, J. (2000). *Programming Perl*, 3rd edn. Sebastopol, CA: O'Reilly Media.

Section II

Mastering Genes, Genomes and Genetic Variation Data

3 The HapMap – A Haplotype Map of the Human Genome

Ellen M. Brown[1] and Bryan J. Barratt[2]

[1] *Discovery Informatics and* [2] *Research and Development Genetics, AstraZeneca, Alderley Park, Macclesfield, Cheshire, UK*

3.1 Introduction

Since its inception in 2002, the International HapMap Project (International HapMap Consortium, 2003; 2005) has generated a vast number of data describing patterns of DNA sequence variation (linkage disequilibrium (LD)) in man. These data can be used to assist researchers in the mapping of loci affecting disease, drug response and other human traits. In addition, the data serve as a resource for research in other more general aspects of population genetics, such as investigations of population structure (Weir *et al.*, 2005) or aiding the identification of regions that may have been subject to evolutionary pressure in different populations (Nielsen *et al.*, 2005), and in molecular genetics, as in the identification of sequence elements associated with regional variations in recombination rate (Smith *et al.*, 2005). In this chapter, we will review the approaches to downloading and viewing of these data and provide an overview of factors affecting the choice of SNPs for genotyping in association studies.

3.1.1 Historical background

The unveiling of the first draft of the human genome in June 2000 (Yamey, 2000) enabled a rapid acceleration in research aimed at identifying the genetic variation

Bioinformatics for Geneticists, Second Edition. Edited by Michael R. Barnes
 ISBN 978-0-470-02619-9 (HB) ISBN 978-0-470-02620-5 (PB)

underlying human traits. The availability of a comprehensive reference sequence and the concomitant annotation of the human transcriptome permitted researchers engaged in mapping disease genes and other traits to identify more easily polymorphisms located within candidate genes or chromosomal regions identified through linkage studies. The availability of the human genome sequence and the continuing expansion of publicly available repositories of polymorphism data such as dbSNP (Sherry *et al.*, 2001), coupled with advances in chemistry and technology in relation to both sequencing and genotyping (and their associated cost reductions), have allowed genetic studies to be performed more efficiently and enabled researchers to tackle more complex problems. Nevertheless, until recently, the majority of human genetic variation was either unknown or poorly characterized, and an in-depth study of a candidate gene or region typically required a considerable investment of both time and financial resource in order to locate, characterize and select relevant polymorphisms for genotyping.

During 2001, some key observations regarding the structure of the human genome were published. Of these, two are particularly illustrative of the thinking in the period immediately preceding the initiation of the International HapMap Project. Firstly, within closely linked regions extending over tens to hundreds of kilobases (kb), the diversity of haplotypes (the alleles present on a single chromosome at a number of neighbouring polymorphic sites) was observed to be limited. It was hypothesized that the limited diversity within such 'haplotype blocks' was a result of a punctuation of the genome by recombination hotspots (Daly *et al.*, 2001). Secondly, it was observed that within regions of limited haplotype diversity a reduced number of polymorphisms (haplotype tag single-nucleotide polymorphisms (htSNPs)) were capable of defining the genetic variation present (Daly *et al.*, 2001; Johnson *et al.*, 2001; Patil *et al.*, 2001). Together, these observations led to hopes that, by the application of marker selection based upon haplotype patterns across the human genome, studies of association between genetic variants and human traits might be made much more cost-efficient than previously proposed. For limited regions, it was feasible to characterize genetic variation by re-sequencing and then implement 'tagging' methodology. However, in order to apply tagging in a cost-effective manner on a larger scale, a comprehensive map of human genetic polymorphisms and the interrelationships between their alleles (a haplotype map) would need to be generated. To this end, the International HapMap Project was officially initiated in October 2002, with the aim of generating a freely available haplotype map of the human genome (the HapMap), to provide a resource for researchers attempting to identify genes involved in human phenotypic variations such as complex diseases and responses to drugs and environmental factors (International HapMap Consortium, 2003).

3.1.2 Subjects, SNP selection and genotyping

The volunteer subjects selected for HapMap genotyping comprised samples from four populations, summarized in Table 3.1. It is important to note that the naming

Table 3.1 Recommended population descriptors and abbreviations

Population descriptor	Abbreviation	Subjects
Han Chinese in Beijing, China	(CHB)	45 unrelated individuals
Japanese in Tokyo, Japan	(JPT)	45 unrelated individuals
Yoruba in Ibadan, Nigeria	(YRI)	30 parent–offspring trios
CEPH* (Utah residents with ancestry from northern and western Europe)	(CEU)	30 parent–offspring trios

*Centre d'Etude du Polymorphisme Humain

convention chosen for the populations from which the samples were ascertained was not idly conceived, and its underlying rationale is based on cultural, ethical and scientific considerations. The International HapMap Consortium recommend that, to avoid over generalization, the full population descriptor (Table 3.1) be supplied in any article before any use of shorthand such as 'Yoruba', 'Japanese' or the three-letter abbreviations, and all authors who refer to HapMap data in their publications should adhere carefully to the latest guidelines and naming conventions (http://www.hapmap.org/citinghapmap.html). Where the JPT and CHB samples are analysed together, it is recommended that the term 'analysis panel' be used (International HapMap Consortium, 2005).

The criteria used to assign membership to the populations are briefly described on the project website as follows: 'For the Yoruba, donors were required to have four of four Yoruba grandparents. For the Han Chinese, donors were required to have at least three of four Han Chinese grandparents. For the Japanese, donors were simply told that the aim was to collect samples from persons whose ancestors were from Japan. The criteria used to assign membership in the CEPH population have not been specified, except that all donors were residents of Utah.' Additional background information on the populations is available *via* the project website. For researchers wishing to perform their own laboratory-based work on these panels, DNA samples and cell cultures for all the subjects genotyped in the project are available *via* the Human Genetic Cell Repository at the Coriell Institute for Medical Research (http://ccr.coriell.org/nigms/products/hapmap.html).

In phase I of the project, the aim was to achieve genotyping of SNPs at an approximate spacing of 5 kb across the human genome. At the outset of the project, the publicly available data describing the identity, validation status and frequency of SNPs was insufficient for the construction of such a map, and an extensive SNP discovery effort was undertaken. SNPs selected for genotyping in phase I were deliberately biased toward those with minor allele frequencies greater than 5 per cent, and SNPs in coding regions were prioritized within each 5 kb bin (International HapMap Consortium, 2005). In addition to the overall target of an average spacing of 5 kb, the selection of SNPs in phase I was augmented in 10 of the 500-kb regions studied as part of the ENCODE (ENCyclopedia of DNA Elements) project (ENCODE Project Consortium, 2004). These regions were re-sequenced in 48 unrelated subjects

(16 YRI, 16 CEU, eight CHB and eight JPT), and genotyping was attempted for all SNPs whether novel or publicly available in dbSNP.

The genotyping was conducted as a multicentre, international effort using a range of technology platforms. The target of one SNP per 5 kb was reached in March 2005 (one SNP per 279 bp was achieved in the ENCODE regions), and a final phase I data freeze made available in June 2005 (public release no. 16c.1). These data have been comprehensively described (International HapMap Consortium, 2005) and comprise 1 007 329 SNPs that were both polymorphic in each of the three analysis panels (YRI, CEU and CHB + JPT) and passed quality-control (QC) filters.

A second phase of the project, aimed at increasing the density of genotyped SNPs, has subsequently been completed in a remarkably short period of time. The combined phases I and II data (public release no. 19, October 2005) comprised QC-filtered genotypes for between 3 806 920 and 3 903 524 SNPs in each of the four populations. A small proportion of SNPs (31 000) have subsequently been excluded in the remapping to NCBI Build 35 coordinates (public release no. 20, January 2006). The latest data releases represent an average density of approximately one SNP per kb and provides greater coverage of rare SNPs (less than 5 per cent minor allele frequency) that were biased against during the phase I SNP ascertainment.

3.2 Accessing the data

When the International HapMap Project was initiated, it was decided that data would be released into the public domain as quickly as possible after their generation. However, it was initially protected by a licence to prevent users from filing patents that would restrict use of the data by others and, as an unavoidable consequence, this prevented the incorporation of the data into other public databases and tools. In December 2004, a decision was made to lift the licence restrictions, making the data freely available to all users for any purpose (http://www.genome.gov/12514423). Following this decision, HapMap data has become available *via* a number of different sources. More information about the HapMap data release policy can be found at http://www.hapmap.org/datareleasepolicy.html.

3.2.1 Downloading HapMap data

Genotype data, allele and genotype frequencies, LD data, phase information, SNP assay details, protocol and sample documentation are available for download from the primary sources: the HapMap website (http://www.hapmap.org) and its Japanese mirror site (http://hapmap.jst.go.jp). Novel SNPs identified by the HapMap Project have been submitted to the public domain variation databases dbSNP and JSNP, and the data incorporated into Ensembl and the UCSC Generic Genome Browser (Table 3.2).

Table 3.2 Examples of publicly available databases incorporating HapMap data

Database	Examples of HapMap information available	Reference	URL
HapMap	Primary data source	(Thorisson *et al.*, 2005)	http://www.hapmap.org mirrored at http://hapmap.jst.go.jp
dbSNP	Individual genotypes Allele and genotype frequencies LD plots	(Sherry *et al.*, 2001)	http://www.ncbi.nlm.nih.gov/SNP
JSNP	Genotype frequencies	(Hirakawa *et al.*, 2002)	http://snp.ims.u-tokyo.ac.jp
Ensembl	Individual genotypes Allele and genotype frequencies LD plots Tag SNP identification	(Hubbard *et al.*, 2005)	http://www.ensembl.org/index.html
UCSC Genome Browser	Recombination rates and hotspots hotspots Sequencing coverage and allele frequencies in ENCODE regions	(Kent *et al.*, 2002)	http://genome.ucsc.edu

A user's guide to the International HapMap Project website has recently been published (Thorisson *et al.*, 2005) and should be referred to for more details of the data and tools incorporated. Here we provide a brief summary of the three main routes for accessing individual genotype data from the HapMap website: bulk download, the Generic Genome Browser and HapMart. All routes can be accessed from http://www.hapmap.org/. All data downloaded from the HapMap site use refSNP identifiers (reference SNP identifiers assigned to non-redundant clusters of variations within dbSNP).

Bulk download

Bulk download can be used either to obtain the complete HapMap data set or data sets for specific chromosomes and populations. Three versions are available: non-redundant, redundant-filtered and redundant-unfiltered. The non-redundant data set eliminates duplicate genotypes (generated from sample duplicates within a plate or where a SNP has been genotyped in the same population by more than one centre) and records that fail QC filters. In the redundant-filtered data set, duplicate genotypes are retained, but records that fail QC filters are removed. The redundant-unfiltered data set is the unprocessed genotype data, in which records that fail QC filters are retained but flagged. All data are in a text format, arranged with one row per SNP and one column per individual/genotype and item of supporting information (chromosome, position, genome build, assay ID, strand, etc.).

The Generic Genome Browser (GBrowse)

The Generic Genome Browser (Stein *et al.*, 2002) is incorporated into the website and can be used to select a region of interest for download by searching for a chromosomal position, gene or SNP. Individual genotype data, allele and genotype frequency data, LD data and tag SNP data can be downloaded in text format. Downloaded genotype data is in the same format as the bulk download data and can be opened directly in a locally installed copy of HaploView (Barrett *et al.*, 2005).

HapMart

HapMart has been developed by BioMart (Gilbert, 2003) and enables the retrieval of genotype data, frequency data or assay details for HapMap SNPs. Filters allow data to be retrieved by population, minor allele frequency, monomorphic status, gene location, refSNP identifier, chromosomal region, gene name and ENCODE region. Fields to be exported can be specified and, in addition to standard text formats, Excel-formatted output is also supported.

3.2.2 Viewing HapMap LD data

As previously described, the Generic Genome Browser incorporated into the project website can be used to select regions of interest for data download. The browser can also be used to visualize LD plots of HapMap data with the integrated HaploView software (Barrett *et al.*, 2005), which can be configured to display D', r^2 or LOD score values between markers. Data for multiple populations can be viewed simultaneously (Figure 3.1). In addition to displaying LD plots, the browser can also be configured to display haplotypes generated by the PHASE program (Stephens and Donnelly, 2003) or to run the Tagger software (de Bakker *et al.*, 2005) for the selection of tag SNPs (see Section 3.3.2). HapMap LD data can be viewed alongside a variety of other features, including Entrez genes and RefSeq mRNAs. Additionally, it is possible to upload one's own annotations and share these with colleagues and collaborators.

The integration of the web-based Generic Genome Browser with software such as HaploView, Tagger and PHASE offers a range of functionality that may satisfy many small- or medium-scale project needs, without the necessity to implement local databases and tools. If large-scale projects are contemplated, users may find it advantageous to download the data and manipulate them locally (the Generic

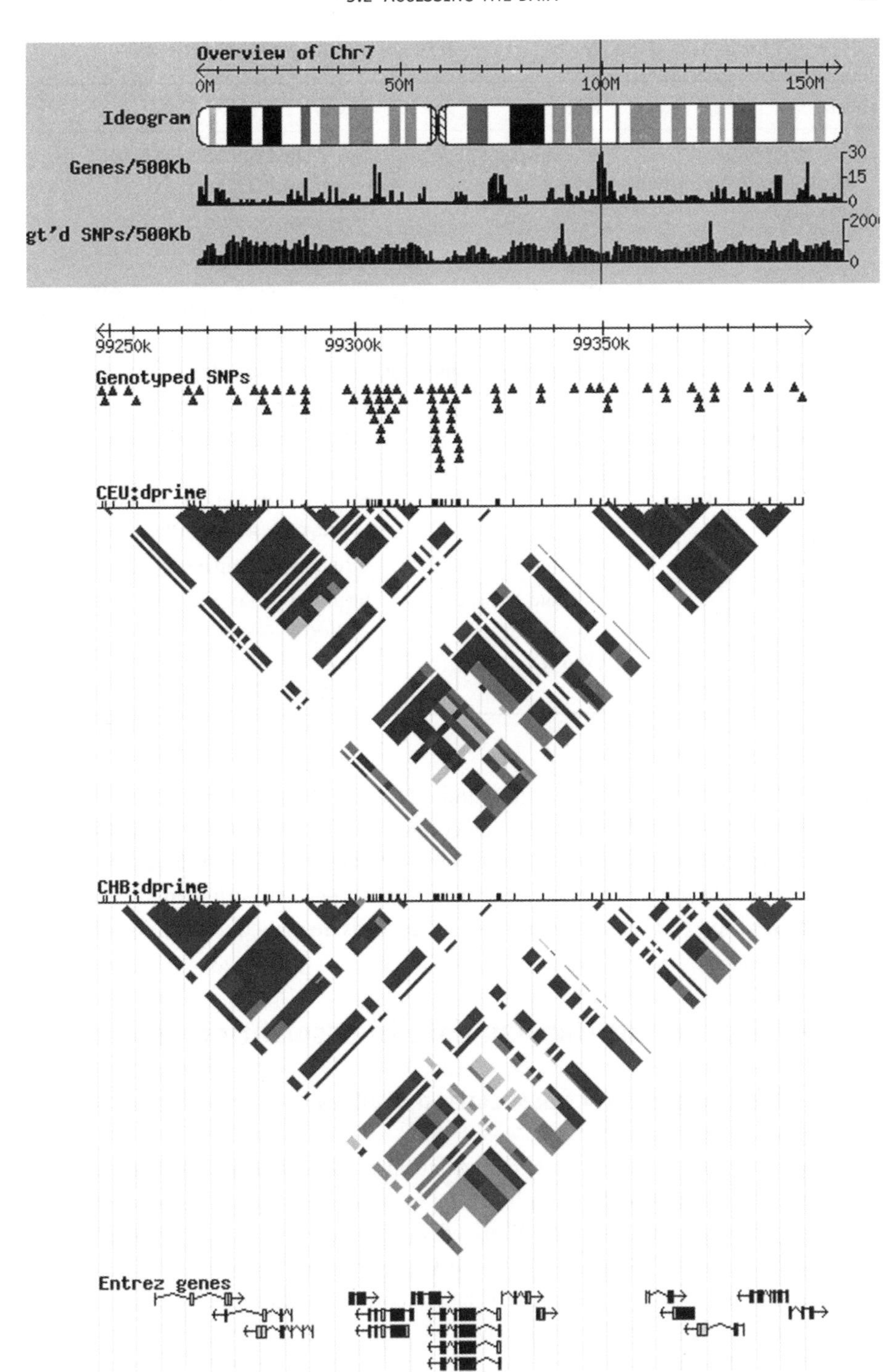

Figure 3.1 The Generic Genome Browser with incorporated HaploView LD plots and Entrez genes

Table 3.3 Examples of software for viewing linkage disequilibrium

Software	Reference	URL
GOLD (Graphical overview of linkage disequilibrium)	(Abecasis and Cookson, 2000)	http://www.sph.umich.edu/csg/abecasis/GOLD
GOLDsurfer	(Pettersson *et al.*, 2004)	http://www.umbio.com (part of Evince graphical software package)
HAPLOT	(Gu *et al.*, 2005)	http://info.med.yale.edu/genetics/kkidd/ programs.html (incorporates HaploView)
HaploView	(Barrett *et al.*, 2005)	http://www.broad.mit.edu/mpg/haploview/index.php
JLIN (Java LINkage disequilibrium plotter)		http://www.genepi.com.au/projects/jlin
LDheatmap (R package)		http://stat-db.stat.sfu.ca:8080/statgen/research/LDheatmap
Marker	(Forton *et al.*, 2005)	http://www.gmap.net/marker
PowerMarker	(Liu and Muse, 2005)	http://statgen.ncsu.edu/powermarker/index.html
SNPAnalyzer	(Yoo *et al.*, 2005)	http://www.istech.info/istech/board/login_form.jsp

Genome Browser, HaploView, Tagger and PHASE are among the freely available software applicable for this purpose). Of course, even for small-scale projects, if the user has a specific preference or need for an alternate LD viewer (Table 3.3) or tag SNP selection algorithm (Section 3.3), or wishes to perform other manipulations, data will need to be obtained *via* one of the methods described in Section 3.2.1.

3.3 Application of HapMap data in association studies

3.3.1 Direct and indirect association studies

The key factors required in genetic association studies that are likely to identify genuine loci contributing to complex traits include (i) sample collections that are adequately powered to detect the likely effect size expected to be exhibited by an individual genetic component; (ii) sample ascertainment that is relatively free from biases (such as population substructure in case-control studies which can lead to false associations); and (iii) availability of a well-ascertained and suitably powered sample collection in which initial findings are replicable. Ideally, findings from genetic association studies should also be supported by functional data, demonstrating the role of the genetic variant in question (Editorial, 1999; Cardon and Bell, 2001). However, even in the presence of adequately powered sample collections (and assuming a genuine locus resides in the gene/region studied), the likelihood of detecting an association can be heavily influenced by the choice of markers.

The choice of markers in association studies may be driven by the hypothesis that the variant studied is, itself, likely to influence a trait through influence on protein structure, expression levels or patterns, etc. (the direct approach), or by the hypothesis that the variant studied may be in LD with a functional variant (the indirect approach). Both approaches have certain advantages and disadvantages. In the direct approach the bias toward variants with a likely functional consequence typically results in a vastly reduced genotyping burden. However, the likely function of a given variant is not always easily predicted. At this time, this is especially true for variants affecting regulation of expression; as such, there is no guarantee of including all the potentially relevant polymorphisms in a study, even in a well-characterized region. In the indirect approach, no prior hypothesis is required in relation to the function of variants. The objective is, instead, to capture as much of the common genetic variation as possible. Although an informed selection of SNPs is used to keep genotyping costs within affordable limits, the more comprehensive nature of the indirect approach typically results in a greater genotyping burden than the direct approach.

In reality, most association studies performed represent a combination of both approaches. For example, in studies in which the indirect approach dominates, marker selection is often deliberately biased toward, or supplemented with, variants with clear potential for functional effects. Conversely, from the results of an association study motivated by the direct approach, it would not be possible without further research to establish whether an observed association was due to the variant genotyped, or was the result of LD with some other variant that had not previously been recognized as relevant.

There can be no doubt that the HapMap data are a valuable resource for the design of association studies motivated by the direct approach. For researchers working on the premise of the common disease/common variant hypothesis, coverage of common variation is comprehensive. For less frequent variants, although coverage is incomplete, genotyping was attempted for all SNPs annotated within dbSNP as non-synonymous. Overall, the project has delivered data on a staggering number of novel SNPs, many of which may be hypothesized to have a putative function and, therefore, be eligible for selection in such a study. However, one of the primary drivers for the HapMap Project was to inform, and make more efficient, the selection of SNPs on the basis of LD in indirect association studies. Therefore, it is the use of the HapMap data in the context of indirect association studies that will form the subject of the remainder of this section.

3.3.2 The use of LD to inform SNP selection

Within the framework of an indirect approach, a key factor affecting the power of an association study is the level of correlation between the markers selected for genotyping (tag SNPs) and un-genotyped markers. Correlation is often characterized by the coefficient of determination, r^2 (Hill and Robertson, 1968), which describes the proportion of information at one variant that may be captured by another single variant. If an individual variant is predicted by a group of variants, such as haplotypes or genotype data at multiple SNPs in which haplotypes are not resolved (unphased genotypes), the notation generally used is R^2 (Chapman *et al.*, 2003; Clayton *et al.*, 2004), although, for convenience, r^2 is used in some publications to describe both univariate and multivariate coefficients of determination. Both have the attractive property of having a direct relationship with statistical power. For example, if (for a given effect size, allele frequency and mode of inheritance) N individuals were required to achieve 80 per cent power to detect an effect at a directly genotyped causal SNP, then N/r^2 individuals would be required to achieve the same power *via* the indirect method, given the r^2 between the genotyped marker and the ungenotyped causal variant (Pritchard and Przeworski, 2001).

Univariate (pairwise) correlations between SNPs are exploited by programs such as ldSelect (Carlson *et al.*, 2004) and CLUSTAG (Ao *et al.*, 2005) to identify tag SNPs. ldSelect evaluates pairwise r^2 and forms 'bins' of SNPs, each of which has an r^2 greater than a user-specified threshold with one or more SNPs within that bin. Within each bin, ldSelect distinguishes between tag SNPs (those which are, in isolation, capable of capturing all the other SNPs in the bin at the specified level of r^2) and 'other' SNPs (those which may be captured by any of the tag SNPs in the bin, but are themselves not capable of capturing all SNPs in the bin). Hence, from the ldSelect output, one tag SNP is sufficient to capture the known variation within each bin.

Initially, the selection of tag SNPs was based on multivariate relationships and oriented toward the identification of SNPs that were, in combination, able to distinguish between the limited numbers of haplotypes observed within a region of strong LD (Johnson *et al.*, 2001); hence, the term 'haplotype tag SNP' (htSNP) was applied. In common with many other publications, we will use here the more general term 'tag SNP' to describe any SNPs chosen to capture information at other SNPs, regardless of whether the methodology is haplotype based or not. Originally, the criterion optimized during selection was a measure termed 'percentage of diversity explained' (PDE), which provided an estimate of the fraction of total haplotype diversity captured if only the tag SNPs were genotyped (Johnson *et al.*, 2001). If later association analyses are to be primarily focused on haplotypes as the 'unit of inheritance' which entails risk of a particular trait, this criterion is in many ways adequate. However, the PDE measure did not fully describe the ability of a set of tag SNPs to predict allele frequencies at individual ungenotyped loci. To facilitate the association testing of ungenotyped variants on a marker-by-marker basis, many of the subsequent developments of multivariate tagging methodology have focused on selecting tag SNPs either by the optimization of R^2-based metrics (or other measures of 'informativeness' which describe the accuracy of prediction), or by considering power more directly (Chapman *et al.*, 2003; Weale *et al.*, 2003; Clayton *et al.*, 2004; Halldórsson *et al.*, 2004a; de Bakker *et al.*, 2005; Rinaldo *et al.*, 2005).

Within this group of methods, differences exist in the way the alleles at ungenotyped loci are predicted, which (in addition to utilizing pair-wise relationships) can be via estimation of tag SNP haplotype frequencies or, more directly, obtained from multiple tag SNP allele frequencies without the need to resolve phase. Although the criteria used by these methods are oriented toward optimizing the prediction of alleles at ungenotyped loci, the selection methods used do not necessarily preclude the evaluation of association of a region in global tests (here we define global tests as haplotype analyses or multivariate tests of unphased genotypes that simultaneously evaluate the association of a given region, as opposed to on a marker-by-marker or single haplotype basis). It is important to note, however, that if global tests are to be performed, an element of dependency on local LD (block) structure will be introduced (see below under 'Performance of HapMap-derived tags in other populations').

In addition to R^2-based multivariate tag SNP selection methods aimed at predicting individual variants, R^2-based metrics have also been applied in the prediction of haplotypes, whereby the correlation between the haplotypes estimated from data including all SNPs and those estimated from a restricted set of tag SNPs is evaluated (Stram, 2004). Although primarily oriented toward the evaluation of haplotypes as the inherited risk factor, as in the original description of haplotype tagging, the use of R^2 values as criteria during selection makes evaluation of the likely power of association testing easier to assess than that of measures such as PDE.

Spectral decomposition/principal components analysis has been proposed by some authors as a means of tag SNP selection (Meng *et al.*, 2003; Horne and Camp, 2004; Lin and Altman, 2004). Although correlation matrices underlie such procedures, the choice of tag SNPs is based on the closeness of their relationship with a subset of eigenvectors (mathematical abstractions formed from weighted contributions of SNPs) that best describe the data. Other methods include those in which the haplotype diversity captured is evaluated on the basis of entropy (Nothnagel *et al.*, 2002; Ackerman *et al.*, 2003; Hampe *et al.*, 2003; Sebastiani *et al.*, 2003). As noted elsewhere (Halldórsson *et al.*, 2004b), the power to detect association that is likely to be achieved by entropy-based tag SNPs may be difficult to assess without additional evaluations unless the termination criteria used during the search is one where all information is retained.

The range of proposed methods for tag SNP selection is by no means limited to those described above. The development and fine-tuning of algorithms by multiple research groups has resulted in the availability of numerous software implementations, a selection of which are represented in Table 3.4. Details of additional software may be found in published reviews of tagging approaches, such as that of Halldórsson *et al.* (2004b). Despite the apparently bewildering choice of tag selection approaches, the majority can be characterized by one (or some combination) of three broad categories: methods that are dependent on pair-wise, multivariate-phased (haplotypic) or multivariate-unphased correlations. Multivariate approaches (phased or unphased) may be further classified into those primarily oriented toward global association tests (e.g., the criteria used assess the ability of tags to evaluate

Table 3.4 Examples of available software for Tag SNP selection

Software	Reference	URL
BEST	(Sebastiani *et al.*, 2003)	http://genomethods.org/best/
CLUSTAG	(Ao *et al.*, 2005)	http://hkumath.hku.hk/web/link/CLUSTAG/CLUSTAG.html
eigen2htSNP	(Lin and Altman, 2004)	http://htsnp.stanford.edu/PCA/
ENTROPY	(Ackerman *et al.*, 2003)	http://www.well.ox.ac.uk/~rmott/SNPS/
MLRtagging		http://alla.cs.gsu.edu/~software/tagging/tagging.html
HapBlock	(Zhang *et al.*, 2005)	http://www.cmb.usc.edu/msms/HapBlock/ Also incorporated in HAPLOT (Table 3.3)
Haploblockfinder	(Zhang and Jin, 2003)	http://cgi.uc.edu/cgi-bin/kzhang/haplo BlockFinder.cgi
Hclust.R	(Rinaldo *et al.*, 2005)	http://wpicr.wpic.pitt.edu/WPICCompGen/hclust.htm
HtSNP2	(Chapman *et al.*, 2003; Clayton *et al.*, 2004)	http://www-gene.cimr.cam.ac.uk/clayton/software/stata/
htSNPer	(Ding *et al.*, 2005)	http://www.chgb.org.cn/htSNPer/htSNPer.html
ldSelect	(Carlson *et al.*, 2004)	http://droog.gs.washington.edu/ldSelect.html
SNPSpD	(Nyholt, 2004)	http://genepi.qimr.edu.au/general/daleN/SNPSpD/
SNPtagger	(Ke and Cardon, 2003)	http://www.well.ox.ac.uk/~xiayi/haplotype/index.html
STAMPA	(Halperin *et al.*, 2005)	software on request from the authors
Tagger	(de Bakker *et al.*, 2005)	http://www.broad.mit.edu/mpg/tagger/ Also implemented in HaploView (Table 3.3)
Tag'n'Tell		http://snp.cgb.ki.se/tagntell/
TagIT	(Weale *et al.*, 2003)	http://www.genome.duke.edu/resources/computation/software
TagSNPs	(Stram *et al.*, 2003)	http://www-rcf.usc.edu/~stram/tagSNPs.html

simultaneously the diversity of a defined region in a single test) and those oriented toward the inference of alleles at individual ungenotyped loci. Several of the available software implementations enable more than one class of approach to be pursued. For example, the prediction of ungenotyped SNPs (via either phased or unphased multivariate data) or the maximization of haplotype diversity captured is possible with htSNP2, whereas Tagger enables the capture of ungenotyped SNPs by either pair-wise correlations or pair-wise correlations supplemented with haplotype-based predictions for greater efficiency, and also includes various options for approaches oriented toward capturing haplotypes for either specific or exhaustive testing.

It is beyond the scope of this chapter to review each of the available selection strategies and software in detail. Informative overviews of tag selection algorithms and some of their technical properties have been published (Halldórsson *et al.*, 2004b;

Ke *et al.*, 2005), and many of the papers in which approaches are originally described provide comparisons with other methods. Despite differences in the identity of the tag SNPs selected, the amount of genetic variation captured by different methods may be similar provided sufficiently high thresholds for quality criteria are specified (Ke *et al.*, 2005). However, aside from another obvious consideration, namely, the number of tag SNPs selected by a given approach, there are a number of theoretical factors that may influence the choice of algorithm and the manner in which it is most reliably applied in different scenarios.

Relevance of the statistical methods used to test for association

Association analyses of SNPs on a marker-by-marker basis have the advantage of minimal degrees of freedom, whereas global haplotype analyses have a larger number of degrees of freedom, but may be advantageous at loci where more than one variant in a region is contributing to the trait under study; for example, in the presence of strong haplotype-specific (*cis*) effects. Previously, haplotype analyses were also partly motivated by the fact that the majority of genetic variation was unidentified. Where disease alleles have arisen on, and been maintained as part of, ancestral haplotypes, these analyses were considered more likely to detect association at unobserved variants, or observed variants with which the correlation with the genotyped markers was unknown. With the rapidly increasing knowledge of common human genetic variation and its correlation structure, and with an increase in the density of markers typically genotyped, the rationale for global haplotype analyses may become somewhat eroded with respect to the presence of unidentified or poorly characterized common variation, although such tests may still be advantageous in the detection of unobserved rare variants with strong genetic effects (de Bakker *et al.*, 2005). However, the study of unobserved rare variants by haplotype analyses can be problematic; many of the haplotypes identified may not capture rare variants, and the number of degrees of freedom in global tests (or the number of tests if each haplotype is evaluated individually) may be greatly increased. Although the better prediction of rare alleles may sometimes offset the increase in degrees of freedom or number of tests (and the power may be increased), these factors will often negatively affect the power to detect common variants (de Bakker *et al.*, 2005). Other potential problems relating to the study of rare variants are discussed below. Global analyses based upon unphased genotype data (Chapman *et al.*, 2003; Halperin *et al.*, 2005) also have greater degrees of freedom than single locus analyses, although typically less than the corresponding haplotype-based tests. The loss in predictive ability which would otherwise be gained by the resolution of haplotype phase should often be compensated for by the reduced degrees of freedom, generally resulting in a more powerful test of association (Chapman *et al.*, 2003). Such analyses have the added advantage that dominance effects (deviation from the additive effects usually assumed in haplotype analyses) may be easily allowed for (Chapman *et al.*, 2003; Clayton *et al.*, 2004). A further implication of exclusively pursuing either global haplotype or unphased multivariate analyses is that the number of tests performed is fewer than when considering SNPs on a marker-by-marker basis, influencing the extent to which multiple testing corrections, if applied, will affect results.

Currently, there is some uncertainty in the field as to which approach to association testing is likely to be the most powerful. The performance of each is likely to vary depending on the characteristics of the locus to which they are applied, such as allele frequency, strength of effect and presence or absence of interactions. The approach taken is, therefore, largely determined by the assumptions made by individual research groups. Nevertheless, the type of association analyses intended is an important prior consideration when choosing a method by which to choose tag SNPs, as mismatching of selection and analysis strategies is likely to result in unintended loss of information and, therefore, power.

Clearly, tag selection approaches oriented toward characterizing haplotype diversity are primarily applicable to subsequent estimation and global testing of haplotypes. Numerous tag selection approaches are applicable to subsequent marker-by-marker testing, including pair-wise approaches and multivariate (either haplotype or unphased genotype) approaches optimized for inferring alleles at ungenotyped SNPs. Of course, multiple testing considerations aside, there is nothing preventing the data from tag SNPs chosen with a view to marker-by-marker analyses being used in additional global haplotype analyses. Similarly, tag SNPs chosen to describe haplotype diversity are often also tested individually for association. Nevertheless, in order to help extract maximum value from the genotyping performed, it is beneficial to decide in advance what the primary analysis strategy will be, and to choose an approach in which tag selection is made by appropriate criteria.

General limitations of tagging methodology

The general limitations of tagging methodology should be considered, and, in light of this, researchers may, under certain circumstances, wish to modify the criteria applied during tag SNP selection. One of the potential pitfalls of tagging methodology is the possibility of differences in LD patterns between the samples in which the tags are selected (the training set) and those in which the association study is conducted (the study population). This potential problem may be characterized by two main issues; inaccuracies in the estimates of LD within the training set itself and genuine differences in LD patterns between the population from which the training set is sampled and the study population. The latter is a key issue in the generalization of HapMap LD to other populations, which several studies have tried to address (see below under 'Performance of HapMap-derived tags in other populations').

The finite size of training sets means that LD measures have a degree of uncertainty due to sampling error, and this increases with decreasing allele frequency. Additionally, owing to indeterminate phase, haplotype frequencies are almost always inferred rather than observed directly from genotype data. From extended pedigree data, or when applying an expectation-maximization (EM)-based algorithm to unrelated subjects within regions of little recombination, estimated haplotype frequencies can be highly accurate (Stram, 2004). However, this will not always hold true in attempting to estimate the frequency of rare haplotypes, such as may be encountered in the presence of weak LD (which may occur simply as a consequence of considering numerous SNPs in an overextended region) or in the presence of missing data (Forton *et al.*, 2005). Furthermore, a bias in estimates of LD can occur under these circumstances. Some, or many, rare haplotypes present in the population are likely to remain unobserved in small training sets and, as a consequence, LD is more likely to be overestimated than underestimated (Stram, 2004). This can result in situations where the tag SNP selection appears to capture ungenotyped SNPs with a high degree of accuracy in the training set, but weaker LD in the study population means that the achieved power is less than intended. These errors and biases (and the likely absence of an adequate surrogate if exclusively pair-wise methods are used) pose problems for the tagging of observed rare variants (Weale *et al.*, 2003; Carlson *et al.*, 2004; Schulze *et al.*, 2004; Ahmadi *et al.*, 2005).

Using the TagIT method (Weale *et al.*, 2003) with a training set of 64 individuals (approximately equivalent to the number of independent subjects in either the HapMap CEU or YRI trios), one study reports that approximately 90 per cent of variants with allele frequencies of at least 20 per cent, and approximately 80 per cent of variants with allele frequencies between 5 per cent and 20 per cent, are well represented by the tag SNPs chosen, but below the 5 per cent threshold the performance of tags begins to decline more rapidly (Ahmadi *et al.*, 2005). In the same study, a training set of 32 individuals (smaller than that available for either the HapMap CHB or JPT samples) was also evaluated, and it performed almost as well. Thus, for common variants, although studies based on HapMap data are likely to be reasonably robust to training set sample size effects, researchers should be wary of the reliability of LD estimates relating to rare SNPs.

Performance of HapMap-derived tags in other populations

Aside from inaccuracies during estimation, genuine differences in LD between the training set and study population may exist. The transferability of tags selected from HapMap to study populations has already been the subject of a number of investigations. For example, a study of eight European populations indicated that although CEU-derived tag SNPs performed well in a number of genes across all populations, this was not a universal phenomenon; in certain regions of the genome, differences in LD between some populations may sometimes result in loss of information (Mueller *et al.*, 2005). A study in which the performance of YRI-derived tags was assessed in African-Americans also suggests that tag SNPs are not likely to generalize well between these two populations, this observation being consistent with the high level of genetic diversity expected to be present across Africa as a whole (Sawyer *et al.*, 2005). Other studies have indicated that CEU-derived tag SNPs do appear to perform well in other Caucasian populations, as when applied to a Finnish sample (Willer *et al.*, 2006), an Estonian sample (Montpetit *et al.*, 2006) and Australians of mostly northwest European ancestry (Stankovich *et al.*, 2006). Other recent reports are also encouraging regarding transferability to population isolates (Bonnen *et al.*, 2006), between the JPT and CHB analysis panel and Korean samples (Lim *et al.*, 2006) and, more generally, within continental groups (Gonzalez-Neira *et al.*, 2006).

The number of different tagging strategies available, combined with the number of populations between which it is possible to make comparisons and the fact that differences in LD between any two given populations (potentially even those from similar geographic regions (Liu *et al.*, 2004) are likely to vary from one genomic region to another, makes a comprehensive evaluation of the transferability of tag SNPs selected by different methods a formidable task. Intuitively, we might expect that the less efficient algorithms, which select more SNPs and therefore have more innate redundancy, may identify more transferable sets of tags, whereas a highly aggressive algorithm may select a very efficient set of tags that fit the training set perfectly well but which may be sensitive to relatively minor differences in LD. However, the situation is probably not so straightforward, as the performance of the tags is also likely to depend on the underlying methods on which the means of prediction are based. Owing to the scope and complexity of the issues, and the relatively recent availability of comprehensive reference data such as those released in HapMap phase II, we need a great deal of further study into the robustness of tag SNPs selected by different tagging algorithms when transferring between populations.

The discrepancies that may occur between the LD observed in a training set and that present in a study population will not always have a negative effect on the power of an association study at all variants. It is, of course, possible that LD may sometimes be underestimated in the training set due to sampling error, and, thus, more information than anticipated is captured when tag SNPs are applied in the study population. Similarly, it is possible that LD between certain sets of variants is genuinely stronger in the study population than the training set. In general, the mean proportion of information captured across a large number of variants is, however, expected to be lower than the thresholds specified during tag SNP selection, partly due to the upward bias of LD estimates in limited sample sizes and partly due to the proximity of the thresholds generally applied to the upper bounding of (complete) LD. Although the majority of studies indicate that if the training set and study population are well matched, the level of information loss is likely to be acceptable, researchers may, under certain circumstances, wish to increase the thresholds of criteria applied during tag SNP selection. Such scenarios may include those where doubts exist regarding the matching of the training set to the study population, or if statistical power is perceived to be limited due to the study population sample size but flexibility in genotyping capacity is available.

Dependency of methods on haplotype block structure or physical distance

The dependency of the tag SNP selection method on either haplotype block definition or physical distance should also be considered. In general, the tag SNP selection methods oriented toward global analyses of association require the presence of haplotype blocks to be taken into account. For instance, if a small genomic region selected for study is divided by an LD breakpoint and there are three common haplotypes in each of the resulting haplotype blocks, then there could be as many as nine haplotypes across the region as a whole. The implications for global haplotype-based methods are that association analyses are likely to lose power for the detection of common variants, owing to the transition from two tests, each with two degrees of freedom, to one test which may have as many as eight degrees of freedom. However, as discussed, there may be circumstances in which there are increases in power to detect rare variants (de Bakker *et al.*, 2005) or *cis*-interactions (see above under 'Relevance of the statistical methods used to test for association'). Indeed, it has been proposed that long haplotypes may be used to search for rare variants and shorter haplotypes be used for common variants (Lin *et al.*, 2004). Although the likely distribution of frequencies and risks conferred by variants relevant to human traits remains uncertain (Wang *et al.*, 2005), the popularity of the common disease/common variant hypothesis and the difficulties involved in tagging rare variants has meant that a large proportion of association studies are performed with common variants in mind. As such, the definition of haplotype blocks and the subsequent choice and testing of tag SNPs on a block-by-block basis is widely applied if global tests of association are to be performed.

Not all multivariate methods have a dependency on LD block structure. Algorithms oriented toward the testing of ungenotyped SNPs on an individual basis (either by carefully selected tests of specific haplotypes or unphased genotype combinations), as opposed to global testing, can circumvent such dependency (Weale *et al.*, 2003; Halldórsson *et al.*, 2004a; de Bakker *et al.*, 2005; Halperin *et al.*, 2005), thus avoiding the complex issue of how blocks should be delineated. Most boundaries of LD blocks are not clearly defined and interblock LD, sometimes referred to as long-range LD, is known to exist (Lawrence *et al.*, 2005). As such, block definition is an imprecise science and there are known dependencies on SNP density, sample sizes and ascertainment biases in the frequencies of the SNPs (Schwartz *et al.*, 2003; Wall and Pritchard, 2003). Many methods of block definition have been proposed, and results can vary substantially even when applied to the same data (Schulze *et al.*, 2004). Typically, methods also allow the user to specify the thresholds of the parameters used in the evaluation of block structure, creating yet more variability in the range of possible results. The apparent lack of an optimal solution to block definition, and the fact that block-independent tag selections may sometimes be more efficient owing to their ability to exploit interblock correlation, make their use attractive, despite the potential loss of some information that might otherwise only be captured in global analyses.

Methods based solely on pair-wise correlations between SNPs are also considered to be independent of prior block definition.

When applied in a limited region, such as a single candidate gene, block-independent tag selection methods may often be applied with no regard to physical distance. However, there is a potential pitfall when considering very large regions. The finite number of possible arrangements of alleles in a training set of limited size results in a situation in which it is possible to make observations of apparently strong LD over very long physical distances which are, in fact, due entirely to chance. For pair-wise methods, the probability of this event is relatively small except when variants are rare. For multivariate methods, the chances of an algorithm identifying a combination of SNPs that apparently have the ability to predict alleles at SNPs at any position on a chromosome may be substantially increased, owing to the increased number of possible arrangements of all alleles at different combinations of tag SNPs. For this reason (aside from the previously outlined EM considerations and processing burden when considering large regions), it is desirable to impose limits on the physical distance over which block-independent (either pair-wise or multivariate) tag selection algorithms operate. The tag selection algorithm incorporated into the HapMap website (Tagger), which can exploit either pair-wise or multivariate correlations, enables such a limitation on physical distance to be specified. An alternative to specifying physical distance is to specify 'neighbourhoods', either by defining distance in terms of LD units and limiting selection to regions over which useful levels of correlation are likely to be present (Morton *et al.*, 2001; Maniatis *et al.*, 2002), or by taking the union of multiple putative LD blocks (Halldórsson *et al.*, 2004a).

Processing burden

For studies of a limited number of candidate genes, almost all available methods are capable of selecting tags within an acceptable time frame. For studies of large regions, such as linkage peaks, whole chromosomes or genome-wide studies, the processing time of some algorithms (reviewed in Halldórsson *et al.* (2004b)) can be a limiting factor in their application, even when strict limits are imposed on the size of each subset of data considered. In such situations, it may be more practicable to use less intensive (possibly less efficient) algorithms or to make use of preselected tags available via HapMart.

Summary of theoretical considerations

Given all of the factors discussed (including the variability in the frequency and magnitude of effects of variants underlying human traits, variability in LD in different regions and between different populations, and the fact that in some methods optimizing performance for rare variants may reduce power to detect common variants and vice versa) and the interrelationships between them, there is currently no widely recognized optimal solution to tag SNP selection that is universally applicable. As such, judgements must be made on the part of individual research groups, which will also be dependent on the exact size and characteristics of the study population available and the assumptions made regarding the nature of the trait of interest. It is probable that the Tagger program will be a popular choice, owing to its integration with the HapMap data and a large degree of flexibility in the available modes of tag selection, such as the prediction of either individual variants or haplotypes, the choice of region delineation on the basis of physical distance or block boundaries, and the ability to deal with data from related or unrelated individuals and to force or exclude the selection of specified variants as tags.

Regardless of the software and options used, it is worth bearing in mind that the level of information captured by different sets of tag SNPs is likely to converge as the thresholds of evaluation criteria used during selection are increased (Ke *et al.*, 2005). Thus, choosing high values for thresholds may be more important than the evaluation criterion itself (Halldórsson *et al.*, 2004b). Where the information captured is comparable, the differences in the genotyping burden of each approach may appear to be the most relevant deciding factor. However, by (i) ensuring that the selection strategy exploits correlations in a manner consistent with the downstream association analyses, (ii) considering whether the manner in which correlations are exploited are likely to be robust to potential differences in LD between the relevant populations and (iii) taking due account of either block structure or physical distance as required by the chosen algorithm, it should be possible to minimize some of the avoidable losses of power that may otherwise be encountered.

3.3.3 The use of HapMap data to aid the design of fine-mapping experiments

If an initial observation of a trait association is considered sufficiently convincing to warrant further study, the HapMap data may assist in determining the physical extent of the relevant region in which the aetiological variant is likely to reside. Owing to multiple solutions to block partitioning (Schwartz *et al.*, 2003; Schulze *et al.*, 2004) and the possibility of long-range interblock correlations (Lawrence *et al.*, 2005), the accurate delineation of such a region is not likely to be straightforward, nor is it likely to rely solely on apparent block structure. Nevertheless, by examination of pair-wise r^2 values between the observed associated variant and surrounding markers in the HapMap data, or by consideration of measures such as LD units (Maniatis *et al.*, 2002), it should be possible to impose some informed limits on the next stage of fine mapping, which may, for example, involve genotyping of known markers within the region and reassessing both association and LD, before undertaking in-depth resequencing to exclude the presence of or identify novel variants for genotyping or functional study. Some of the practical steps in this process are reviewed in Chapter 9.

3.4 Future perspectives

There is little doubt that the HapMap project data has already had a major and beneficial impact in facilitating genetic association studies, as well as helping to address general questions regarding the LD structure of the human genome and enabling inferences regarding demographic history. Current research, aimed at clarifying issues such as the extent of LD differences between populations, and the reliability of different tag selection strategies when transferring between such populations, should facilitate greater understanding of the likely power that may be achieved in tag-based association studies. A third phase of HapMap is planned in which genotyping of additional populations will be undertaken, and the data generated should be highly informative in this respect. In-depth investigation of these issues should, in future, allow us to achieve the desired balance between statistical power and genotyping efficiency in a more informed manner than before and, ideally, improve both simultaneously.

Ultimately, the success of association studies based on HapMap LD data will not be solely determined by the characteristics of the HapMap data, such as the identity and number of samples in the training sets, SNP frequencies and density. The HapMap is a facilitating component, and only if it is used in conjunction with adequate sample sizes, well-defined case phenotypes, appropriate controls, carefully implemented tag selection, association analyses and interpretation is it likely to lead to acceleration in the identification of complex trait loci. If these other components of a well-designed association study are not in place, or if the assumptions regarding the nature of the genetic contribution to underlying biological processes are flawed, then any limitations that may be perceived to be present in the HapMap data become largely irrelevant (Clark *et al.*, 2005). With the massively increased potential for large-scale or genome-wide association studies also comes the potential for spectacular and hugely expensive failures. It is important, therefore, that such studies are designed and executed carefully with an appropriate focus on all relevant aspects of study design if the impressive resource of information made available via HapMap is to be utilized effectively (Clark *et al.*, 2005; Hirschhorn and Daly, 2005; Wang *et al.*, 2005).

References

Abecasis, G. R. and Cookson, W. O. (2000). GOLD – graphical overview of linkage disequilibrium. *Bioinformatics* **16**, 182–183.

Ackerman, H., Usen, S., Mott, R. *et al.* (2003). Haplotypic analysis of the TNF locus by association efficiency and entropy. *Genome Biol* **4**, R24.

Ahmadi, K. R., Weale, M. E., Xue, Z. Y. *et al.* (2005). A single-nucleotide polymorphism tagging set for human drug metabolism and transport. *Nat Genet* **37**, 84–89.

Ao, S. I., Yip, K., Ng, M. *et al.* (2005). CLUSTAG: hierarchical clustering and graph methods for selecting tag SNPs. *Bioinformatics* **21**, 1735–1736.

Barrett, J. C., Fry, B., Maller, J. *et al.* (2005). Haploview: analysis and visualization of LD and haplotype maps. *Bioinformatics* **21**, 263–265.

Bonnen, P. E., Pe'er, I., Plenge, R. M. *et al.* (2006). Evaluating potential for whole-genome studies in Kosrae, an isolated population in Micronesia. *Nat Genet* **38**, 214–217.

Cardon, L. R. and Bel, I. J. I. (2001). Association study designs for complex diseases. *Nat Rev Genet* **2**, 91–99.

Carlson, C. S., Eberle, M. A., Rieder, M. J. *et al.* (2004). Selecting a maximally informative set of single-nucleotide polymorphisms for association analyses using linkage disequilibrium. *Am J Hum Genet* **74**, 106–120.

Chapman, J. M., Cooper, J. D., Todd, J. A. *et al.* (2003). Detecting disease associations due to linkage disequilibrium using haplotype tags: a class of tests and the determinants of statistical power. *Hum Hered* **56**, 18–31.

Clark, A. G., Boerwinkle, E., Hixson, J. *et al.* (2005). Determinants of the success of whole-genome association testing. *Genome Res* **15**, 1463–1467.

Clayton, D., Chapman, J. and Cooper, J. (2004). Use of unphased multilocus genotype data in indirect association studies. *Genet Epidemiol* **27**, 415–428.

Daly, M. J., Rioux, J. D., Schaffner, S. F. *et al.* (2001). High-resolution haplotype structure in the human genome. *Nat Genet* **29**, 229–232.

de Bakker, P. I., Yelensky, R., Pe'er, I. *et al.* (2005). Efficiency and power in genetic association studies. *Nat Genet* **37**, 1217–1223.

Ding, K., Zhang, J., Zhou, K. *et al.* (2005). htSNPer1.0: software for haplotype block partition and htSNPs selection. *BMC Bioinformatics* **6**, 38.

Editorial (1999). Freely associating. *Nat Genet* **22**, 1–2.

ENCODE Project Consortium (2004). The ENCODE (*ENC*yclopedia *O*f *D*NA *E*lements) project. *Science* **306**, 636–640.

Forton, J., Kwiatkowski, D., Rockett, K. *et al.* (2005). Accuracy of haplotype reconstruction from haplotype-tagging single-nucleotide polymorphisms. *Am J Hum Genet* **76**, 438–448.

Gilbert, D. (2003). Shopping in the genome market with EnsMart. *Brief Bioinform* **4**, 292–296.

Gonzalez-Neira, A., Ke, X., Lao, O. *et al.* (2006). The portability of tagSNPs across populations: a worldwide survey. *Genome Res* **16**, 323–330.

Gu, S., Pakstis, A. J. and Kidd, K. K. (2005). HAPLOT: a graphical comparison of haplotype blocks, tagSNP sets and SNP variation for multiple populations. *Bioinformatics* **21**, 3938–3939.

Halldórsson, B. V., Bafna, V., Lippert, R. *et al.* (2004a). Optimal haplotype block-free selection of tagging SNPs for genome-wide association studies. *Genome Res* **14**, 1633–1640.

Halldórsson, B. V., Istrail, S., and De La Vega, F. M. (2004b). Optimal selection of SNP markers for disease association studies. *Hum Hered* **58**, 190–202.

Halperin, E., Kimmel, G., and Shamir, R. (2005). Tag SNP selection in genotype data for maximizing SNP prediction accuracy. *Bioinformatics* **21** Suppl 1, i195–i203.

Hampe, J., Schreiber, S., and Krawczak, M. (2003). Entropy-based SNP selection for genetic association studies. *Hum Genet* **114**, 36–43.

Hill. W. G., and Robertson, A. (1968). The effects of inbreeding at loci with heterozygote advantage. *Genetics* **60**, 615–628.

Hirakawa, M., Tanaka, T., Hashimoto, Y. *et al.* (2002). JSNP: a database of common gene variations in the Japanese population. *Nucleic Acids Res* **30**, 158–162.

Hirschhorn, J. N. and Daly, M. J. (2005). Genome-wide association studies for common diseases and complex traits. *Nat Rev Genet* **6**, 95–108.

Horne, B. D. and Camp, N. J. (2004). Principal component analysis for selection of optimal SNP-sets that capture intragenic genetic variation. *Genet Epidemiol* **26**, 11–21.

Hubbard, T., Andrews, D., Caccamo, M. *et al.* (2005). Ensembl 2005. *Nucleic Acids Res* **33**, D447–453.

International HapMap Consortium (2003). The International HapMap Project. *Nature* **426**, 789–796.

International HapMap Consortium (2005). A haplotype map of the human genome. *Nature* **437**, 1299–1320.

Johnson, G. C., Esposito, L., Barratt, B. J. *et al.* (2001). Haplotype tagging for the identification of common disease genes. *Nat Genet* **29**, 233–237.

Ke, X. and Cardon, L. R. (2003). Efficient selective screening of haplotype tag SNPs. *Bioinformatics* **19**, 287–288.

Ke, X., Miretti, M. M., Broxholme, J. *et al.* (2005). A comparison of tagging methods and their tagging space. *Hum Mol Genet* **14**, 2757–2767.

Kent, W. J., Sugnet, C. W., Furey, T. S. *et al.* (2002). The human genome browser at UCSC. *Genome Res* **12**, 996–1006.

Lawrence, R., Evans, D. M., Morris, A. P. *et al.* (2005). Genetically indistinguishable SNPs and their influence on inferring the location of disease-associated variants. *Genome Res* **15**, 1503–1510.

Lim, J., Kim, Y. J., Yoon, Y. *et al.* (2006). Comparative study of the linkage disequilibrium of an ENCODE region, chromosome 7p15, in Korean, Japanese, and Han Chinese samples. *Genomics* **87**, 392–398.

Lin, S., Chakravarti, A. and Cutler, D. J. (2004). Exhaustive allelic transmission disequilibrium tests as a new approach to genome-wide association studies. *Nat Genet* **36,** 1181–1188.

Lin, Z. and Altman, R. B. (2004). Finding haplotype tagging SNPs by use of principal components analysis. *Am J Hum Genet* **75**, 850–861.

Liu, K. and Muse, S. V. (2005). PowerMarker: an integrated analysis environment for genetic marker analysis. *Bioinformatics* **21**, 2128–2129.

Liu, N., Sawyer, S. L., Mukherjee, N. *et al.* (2004). Haplotype block structures show significant variation among populations. *Genet Epidemiol* **27**, 385–400.

Maniatis, N., Collins, A., Xu, C. F. *et al.* (2002). The first linkage disequilibrium (LD) maps: delineation of hot and cold blocks by diplotype analysis. *Proc Natl Acad Sci U S A* **99**, 2228–2233.

Meng, Z., Zaykin, D. V., Xu, C. F. *et al.* (2003). Selection of genetic markers for association analyses, using linkage disequilibrium and haplotypes. *Am J Hum Genet* **73**, 115–130.

Montpetit, A., Nelis, M., Laflamme, P. *et al.* (2006). An evaluation of the performance of Tag SNPs derived from HapMap in a Caucasian population. *PLoS Genet* **2**, e27.

Morton, N. E., Zhang, W., Taillon-Miller, P. *et al.* (2001). The optimal measure of allelic association. *Proc Natl Acad Sci U S A* **98**, 5217–5221.

Mueller, J. C., Lohmussaar, E., Magi, R. *et al.* (2005). Linkage disequilibrium patterns and tagSNP transferability among European populations. *Am J Hum Genet* **76**, 387–398.

Nielsen, R., Williamson, S., Kim, Y. *et al.* (2005). Genomic scans for selective sweeps using SNP data. *Genome Res* **15**, 1566–1575.

Nothnagel, M., Furst, R. and Rohde, K. (2002). Entropy as a measure for linkage disequilibrium over multilocus haplotype blocks. *Hum Hered* **54**, 186–198.

Nyholt, D. R. (2004). A simple correction for multiple testing for single-nucleotide polymorphisms in linkage disequilibrium with each other. *Am J Hum Genet* **74**, 765–769.

Patil, N., Berno, A., Hinds, D. A. *et al.* (2001). Blocks of limited haplotype diversity revealed by high-resolution scanning of human chromosome 21. *Science* **294**, 1719–1723.

Pettersson, F., Jonsson, O. and Cardon, L. R. (2004). GOLDsurfer: three dimensional display of linkage disequilibrium. *Bioinformatics* **20**, 3241–3243.

Pritchard, J. K. and Przeworski, M. (2001). Linkage disequilibrium in humans: models and data. *Am J Hum Genet* **69**, 1–14.

Rinaldo, A., Bacanu, S. A., Devlin, B. *et al.* (2005). Characterization of multilocus linkage disequilibrium. *Genet Epidemiol* **28**, 193–206.

Sawyer, S. L., Mukherjee, N., Pakstis, A. J. *et al.* (2005). Linkage disequilibrium patterns vary substantially among populations. *Eur J Hum Genet* **13**, 677–686.

Schulze, T. G., Zhang, K., Chen, Y. S. *et al.* (2004). Defining haplotype blocks and tag single-nucleotide polymorphisms in the human genome. *Hum Mol Genet* **13**, 335–342.

Schwartz, R., Halldórsson, B. V., Bafna, V. *et al.* (2003). Robustness of inference of haplotype block structure. *J Comput Biol* **10**, 13–19.

Sebastiani, P., Lazarus, R., Weiss, S. T. *et al.* (2003). Minimal haplotype tagging. *Proc Natl Acad Sci U S A* **100**, 9900–9905.

Sherry, S. T., Ward, M. H., Kholodov, M. *et al.* (2001). dbSNP: the NCBI database of genetic variation. *Nucleic Acids Res* **29**, 308–311.

Smith, A. V., Thomas, D. J., Munro, H. M. *et al.* (2005). Sequence features in regions of weak and strong linkage disequilibrium. *Genome Res* **15**, 1519–1534.

Stankovich, J., Cox, C. J., Tan, R. B. *et al.* (2006). On the utility of data from the International HapMap Project for Australian association studies. *Hum Genet* **119**, 220–222.

Stein, L. D., Mungall, C., Shu, S. *et al.* (2002). The generic genome browser: a building block for a model organism system database. *Genome Res* **12**, 1599–1610.

Stephens, M. and Donnelly, P. (2003). A comparison of Bayesian methods for haplotype reconstruction from population genotype data. *Am J Hum Genet* **73**, 1162–1169.

Stram, D. O. (2004). Tag SNP selection for association studies. *Genet Epidemiol* **27**, 365–374.

Stram, D. O., Haiman, C. A., Hirschhorn, J. N. *et al.* (2003). Choosing haplotype-tagging SNPS based on unphased genotype data using a preliminary sample of unrelated subjects with an example from the Multiethnic Cohort Study. *Hum Hered* **55**, 27–36.

Thorisson, G. A., Smith, A. V., Krishnan, L. *et al.* (2005). The International HapMap Project Website. *Genome Res* **15**, 1592–1593.

Wall, J. D. and Pritchard, J. K. (2003). Assessing the performance of the haplotype block model of linkage disequilibrium. *Am J Hum Genet* **73**, 502–515.

Wang, W. Y., Barratt, B. J., Clayton, D. G. *et al.* (2005). Genome-wide association studies: theoretical and practical concerns. *Nat Rev Genet* **6**, 109–118.

Weale, M. E., Depondt, C., Macdonald, S. J. *et al.* (2003). Selection and evaluation of tagging SNPs in the neuronal-sodium-channel gene SCN1A: implications for linkage-disequilibrium gene mapping. *Am J Hum Genet* **73**, 551–565.

Weir, B. S., Cardon, L. R., Anderson, A. D. *et al.* (2005). Measures of human population structure show heterogeneity among genomic regions. *Genome Res* **15**, 1468–1476.

Willer, C. J., Scott, L. J., Bonnycastle, L. L. *et al.* (2006). Tag SNP selection for Finnish individuals based on the CEPH Utah HapMap database. *Genet Epidemiol* **30**, 180–90.

Yamey, G. (2000). Scientists unveil first draft of human genome. *BMJ* **321**, 7.

Yoo, J., Seo, B. and Kim, Y. (2005). SNPAnalyzer: a web-based integrated workbench for single-nucleotide polymorphism analysis. *Nucleic Acids Res* **33**, W483–488.

Zhang, K. and Jin, L. (2003). HaploBlockFinder: haplotype block analyses. *Bioinformatics* **19**, 1300–1301.

Zhang, K., Qin, Z., Chen, T. *et al.* (2005). HapBlock: haplotype block partitioning and tag SNP selection software using a set of dynamic programming algorithms. *Bioinformatics* **21**, 131–134.

4

Assembling a View of the Human Genome

Colin A. M. Semple

Bioinformatics, MRC Human Genetics Unit, Edinburgh EH4 2XU, UK

4.1 Introduction

The miraculous birth of the draft human genome sequence took place against the odds. It was only made possible by parallel revolutions in the technologies used to produce, store and analyse the sequence data, and by the development of new, large-scale consortia to organize and obtain funding for the work (Watson, 1990). The initial flood of human sequence has subsided as the sequencing centres have sequenced genomes from other mammalian orders and beyond. The steady progress of the cloned fragments of more than 1000 genomes toward a finished state can be observed in the Genomes OnLine Database (Liolios *et al.*, 2006; http://www.genomesonline.org/), but although we can examine these sequences in public databases, we have yet to interpret them comprehensively. There is a need to relate the raw sequence data to what we already know about genetics and biology in general – this is the process of genome annotation. Preliminary annotation of a genome is usually a semi-automated process, with human curators interpreting the results of various computer programs. In practical terms, preliminary annotation currently consists of determining the position of known markers, known genes and repetitive sequence in combination with efforts to delineate the structure of novel genes. Eventually, we would like to know much more, including the multifarious interactions of the genome's contents with one another and the environment, their expression in the biology of the cell and their physiological roles. These additional

Bioinformatics for Geneticists, Second Edition. Edited by Michael R. Barnes
 ISBN 978-0-470-02619-9 (HB) ISBN 978-0-470-02620-5 (PB)

layers of annotation will come from the patient laboratory work of the next several decades, but a prerequisite for this work is a complete (or nearly complete) genome sequence, and an accurate preliminary annotation that is available to the total scientific community. This chapter will aim to describe the sources of freely available annotation, their strengths, their shortcomings and some likely future developments.

4.2 Genomic sequence assembly

Any discussion of computational sequence annotation should begin with a consideration of the sequence data itself. Genomic sequence data have traditionally come from many sources: studies of transcribed sequences, studies of individual genes, and genetic/physical markers from mapping studies. Over the past decade, we have entered the era of large-scale efforts to sequence entire genomes, and the most abundant sources of sequence have become the sequencing vectors from these efforts. In practical terms, this has meant that we acquire many fragments, from a few hundred bases to a few hundred kilobases in length, of a genome that must then be assembled computationally to produce a continuous sequence. In the case of the human genome, two unfinished 'draft' sequences were produced by different methods, one by the International Human Genome Sequencing Consortium (IHGSC) and one by Celera Genomics (CG).

The IHGSC began with a BAC (bacterial artificial chromosome) clone-based physical map of the genome (IHGSC, 2001). This map was constructed by digesting each clone with restriction enzymes and deriving a characteristic pattern or fingerprint. All of the fingerprints are then processed by a program called FPC (Soderlund *et al.*, 2000) that produces BAC clone contigs on the basis of the shared fragments in their fingerprints (International Human Genome Mapping Consortium (IHGMC), 2001). A selection of clones from this map, covering the vast majority of the genome, was then 'shotgun sequenced' (Sanger *et al.*, 1982). The fragments of each clone were then assembled into initial sequence contigs based upon overlaps between shotgun sequencing reads. The collection of initial sequence contigs from a single clone makes up the sequence data for a clone in GenBank. As more shotgun sequencing of the clone is done, the initial sequence contigs are reassembled with the new sequences, and the database sequence entry for the clone is updated accordingly. Gradually, the initial sequence contigs increase in length and decrease in number, until the sequence of the clone is finished and is represented by a single contig 100–200 kb in length. The program used to assemble the initial sequence contigs is called Phrap (Green, unpublished; http://bozeman.genome.washington.edu/index.html) and takes sequencing quality estimates for each base into account. CG used the whole-genome shotgun method where the entire genome is randomly fragmented and each of the cloned fragments is sequenced (Venter *et al.*, 2001). Sequences from these cloned fragments are produced as mate-pairs: 150–800 bp sequencing reads from either end of the clone with known relative orientation and approximate spacing. A mixture of clones of different sizes was used: 2, 10, 50 and 100 kb. CG assembled their sequence data with that produced by the IHGSC and published an analysis of this early CG draft genome assembly (Venter *et al.*, 2001). In spite of the differences between the two efforts to sequence the human genome, both groups had to address the fundamental problem of assembling incomplete data. In both cases, the strategy was broadly to merge overlapping sequences into contigs and then to order contigs relative to one another using various types of mapping data.

The published IHGSC assembly was produced by the program GigAssembler devised at the University of California at Santa Cruz (UCSC) (Kent and Haussler, 2001). GigAssembler began with initial sequence contigs from GenBank at a given point (a 'freeze' data set). All sequences were repeat masked by the RepeatMasker program (Smit and Green, unpublished; http://www.repeatmasker.org/) to highlight known repetitive sequence. Within each IHGMC physical map contig (IHGMC, 2001), the initial sequence contigs from BAC clones belonging to it were assembled into consensus 'raft' sequences using sequence overlaps between fragments. The first joins were made between the best matching fragments. These rafts were ordered and orientated relative to one another with bridging sequences from other sources (mRNA, EST, plasmid and BAC end pairs) and FPC contig data. For instance, the 5′ end of a single mRNA may be found within one raft while the 3′ end matches another raft. Repeated tracts of the letter 'N' were inserted between rafts to give a sequence for each IHGMC map contig. The published version of the UCSC assembly and all subsequent versions were made freely available online (http://genome.ucsc.edu/) and helped to set the standard for public access to subsequent genome sequence data.

The CG draft genome assembly was carried out by a program described as a 'compartmentalized shotgun assembler' (CSA) (Huson *et al.*, 2001), using both CG sequence data and IHGSC initial sequence contigs from GenBank (as of 1 September 2000 for the published CG assembly) fragmented into smaller sequences a few hundred base pairs long. The CSA began by comparing all CG mate-pair fragments with all the initial sequence contig fragments and avoiding matches based upon repetitive sequence. Repetitive sequence was identified by comparisons to a library of known repeats (analogously to RepeatMasker) but also by additional procedures to detect sequence likely to represent unknown repeat sequences. The mate-pair fragment pairs matching more than one initial sequence contigs were then used as bridging sequences to order and orientate the initial sequence contig fragments within and between BAC clones. Essentially, the paired CG fragments are used as high-resolution mapping data to reassemble both IHGSC BAC sequences and the broader genomic regions they originate from. The result was a set of 'scaffolds' consisting of ordered, oriented sequence contigs separated by gaps of estimated sizes. CG fragments not matching IHGSC initial sequence contigs were also assembled with a different algorithm (Myers *et al.*, 2000) to give additional scaffolds containing sequence not represented in IHGSC data. Scaffolds were then positioned relative to one another based upon sequence overlaps and bridging mate-pair fragments. The derived order of scaffolds was then manually curated to identify mistakes by examining sequence alignments by eye and confirming or rejecting orders based on external physical mapping data such as those from the IHGMC. Although originally there was only restricted access to this assembly, it was eventually deposited in the public sequence databases.

A third assembly method, using repeat masked data from the IHGSC, was produced by the National Centre for Biotechnology Information (NCBI), using a computational protocol (NCBI, unpublished; http://www.ncbi.nlm.nih.gov/genome/guide/build.html) based upon the BLAST algorithm (Altschul *et al.*, 1997). The NCBI approach also began by finding an order for adjacent BACs, but in this case it was derived from BAC sequence overlaps (detected with a variant of BLAST), fluorescence *in situ* hybridization (FISH) chromosome assignment, and STS content. The sequence fragments from these overlapping BACs were then merged into consensus 'meld' sequences. As with the UCSC method, these melds were then ordered and orientated, on the basis of ESTs, mRNAs and paired plasmid reads, before being combined into a single NCBI genomic sequence contig with melds separated by runs of the letter 'N'. NCBI contigs were ordered and oriented relative to one another according to matches to mapped STS markers and paired BAC end sequences.

Since the assembly protocols used by UCSC, CG and NCBI differed in terms of the number and variety of input data and the algorithms used, it would have been surprising if they gave identical assemblies as output. Of particular interest are the relative rates of misassembly (sequences assembled in the wrong order and/or orientation) and the relative coverage achieved by the three protocols. Unfortunately, the UCSC group were alone in having published assessments of the rate of misassembly in the contigs they produced. Using artificial data sets, they found that, on average, ~10 per cent of assembled fragments were assigned the wrong orientation and ~15 per cent of fragments were placed in the wrong order by their protocol (Kent and Haussler, 2001). Two independent assessments of UCSC assemblies have come to similar conclusions. Katsanis *et al.* (2001) examined various UCSC consecutive draft genome assembly releases and reported that 10–15 per cent of EST sequences identified within them appeared to be on wrongly assembled genomic sequences. In agreement with this, Semple *et al.* (2002) observed 19 per cent and 11 per cent of erroneously ordered marker sequences in two consecutive UCSC assemblies for a ~5.8 Mb region of chromosome 4. The latter study also found wide variation in coverage (23–59 per cent of the available IHGSC sequence data included) and rates of misassembly (2.08–4.74 misassemblies per Mb) between consecutive UCSC and NCBI assemblies and the published CG assembly for the same region. These analyses indicated that the lowest rate of misassembly was produced by the CG protocol, followed by the UCSC and lastly the NCBI protocols. However, the CG protocol also produced the lowest coverage, including only around half of the sequence data recruited into the UCSC and NCBI assemblies. Olivier *et al.* (2001) compared orders of TNG radiation hybrid map STSs produced by UCSC and CG protocols. They found widespread differences, such that 36 per cent of TNG STS pairs were present in orders that differed between UCSC and CG assemblies. The TNG order was consistent with the CG assembly order slightly more often than with the UCSC assembly order. The UCSC website provided a variety of comparisons of its assemblies to genetic, physical and cytogenetic mapping data, and these comparisons represented a useful resource for users to assess the likely degree of misassembly in a region of interest. However, subsequent genomes from other species have generally appeared without detailed assessments of the quality of draft assemblies.

Unsurprisingly, it has been shown that differences between assemblies do indeed result in differences in annotation. Semple *et al.* (2002) found variable amounts of tandemly duplicated and interspersed repeat sequence between UCSC, NCBI and CG assemblies and more striking differences in annotation were also identified by Hogenesch *et al* (2001) between CG and UCSC assemblies. Hogenesch *et al.* (2001) found large differences between the genes found in CG and UCSC assemblies, such that more than a third of the genes identified in one assembly were not found in the other. Thus, genomic sequence annotation can be only as good as the underlying genomic sequence assembly, and, as we have seen, accurate assembly of draft sequence fragments is far from error free. Genome assembly continues to be an important issue for bioinformatics, since in spite of the availability of generally reliable assembly algorithms (e.g., Batzoglou *et al.*, 2002) nature has continued to surprise us. Certain species have turned out to be unexpectedly polymorphic and can confound the most sophisticated attempts to assemble them, as the sequencing of the sea squirt *Ciona savignyi* has shown (Vinson *et al.*, 2005).

After the publication of the publicly available human genome draft in 2001, the IHGSC undertook the arduous task of 'finishing': producing a genome sequence covering 99 per cent of the euchromatic regions sequenced to an accuracy of 99.99 per cent. On 14 April 2003, the IHGSC announced that this target had been reached; leaving less than 400 persistent gaps where highly repetitive sequences evaded current sequencing technology. A steady trickle of papers in the journal *Nature* has marked the emergence of each finished human chromosome sequence, along with the annotation describing its notable features. It now seems that a significant fraction of the genome (perhaps 5 per cent) consists of large (>10 kb) duplicated segments that share 90-98 per cent sequence identity. Regions containing such duplicated segments are notoriously difficult to assemble accurately and are found not only in pericentromeric and subtelomeric regions but also across the rest of the genome, including the gene-rich regions that sequence annotators are primarily interested in (Eichler, 2001). A comparison of the completed sequence of chromosome 20 with the preceding public CG and UCSC draft assemblies of the same chromosome identified 'major discrepancies' (Hattori and Taylor, 2001). These authors concluded that the draft assemblies were probably confounded by large duplicated regions. Such problems do not entirely disappear with 'finished' sequence, as the recent publication of human chromosome 8 has shown. This chromosome contains a large region with an unexpectedly high mutation rate, and rich in segmental duplications, flanking a persistent assembly gap (Nusbaum *et al.*, 2006). In the time between the publication of the draft human genome in 2001 and the present finished chromosomes, we have entered the era of shallow genome sequencing. Although the human and mouse genome projects sequenced each base at least seven times

(7 × coverage), Craig Venter's poodle warranted only 1.5× coverage (Kirkness *et al.*, 2003), and now more than 20 other mammals are being sequenced at 2× coverage (http://www.genomesonline.org/). These data are intended only for comparative genomics, to shed light on novel functional regions conserved across mammals, and would not be a good basis on their own for the detailed laboratory work required to investigate gene function.

4.3 Annotation from a distance: the generalities

If some troublesome regions of the genome are set to continue as problems for cloning, sequencing and assembling, this is a minor concern in comparison to the comprehensive annotation of genomic sequence. At almost every level, computational annotation of genomic sequence is error prone and incomplete. Of course, the aim of computational annotation, in common with much of bioinformatics, is to provide a preliminary set of predictions that must then be tested by 'wet' laboratory work. The aim is a rapid first-pass or 'baseline' annotation, as the most popular genomic annotation resource Ensembl (Hubbard *et al.*, 2002) puts it. From the computational point of view, this enterprise is hugely successful: merely by considering the statistical qualities of the raw sequence data, we can detect the presence of most protein-coding human genes. We can then identify the presence of known, structural domains within the conceptually translated products of these predicted genes and make informed guesses about functional roles and subcellular localization. Looking at a raw BAC sequence entry from GenBank, we may easily appreciate the scale of these achievements, but the view from the wet laboratory bench can be different. The broad success of computational gene prediction is little consolation to the bench geneticist who has to sift through numerous artefactual exon predictions only to find later that his gene of interest was not detected by any of the algorithms used. What is broadly impressive to the bioinformaticist can be just plain wrong to those dealing with specifics. In an excellent introduction to genomic sequence annotation, Lincoln Stein has defined three hierarchical levels of annotation: (i) the most fundamental nucleotide level; (ii) protein level; (iii) process level (Stein, 2001).

4.3.1 Nucleotide level

Nucleotide level is the point at which the raw genomic sequence is analysed and forms the basis for subsequent levels of interpretation. The first step is to identify as many known genomic landmarks as possible; these are generally markers from previous mapping studies, repeats and known genes already in public databases. This can be done quickly and accurately by a variety of programs. Markers from previous genetic, physical and cytogenetic maps are placed upon the genomic sequence by algorithms designed to find short, almost exact sequence matches, such as the ePCR program (Schuler, 1997; http://www.ncbi.nlm.nih.gov/sutils/e-pcr/), BLASTN (Altschul *et al.,* 1990), SSAHA (Ning *et al.*, 2001; http://www.sanger.ac.uk/Software/analysis/SSAHA/) and BLAT (Kent, 2002; http://genome.ucsc.edu/cgi-bin/hgBlat?command=start). Identifying these markers is essential to allow the genomic sequence to be seen in relation to the previous, pre-genome sequence literature, such as that on human disease genetics. The newest type of markers, single-nucleotide polymorphisms (SNPs), are also identified in the sequence to facilitate the next generation of disease gene-mapping studies. Similar algorithms, extended to incorporate information on gene structure, are used to identify the positions of known mRNAs within the genomic sequence; examples of these are as follows: Spidey (Wheelan *et al.*, 2001; http://www.ncbi.nlm.nih.gov/IEB/Research/Ostell/Spidey/), SIM4 (Florea *et al.*, 1998; http://bio.cse.psu.edu/) and est2genome, which is available from the EMBOSS package (Rice *et al.*, 2000; http://emboss.sourceforge.net/). Just as the efforts to assemble genomic sequence take measures to identify and exclude repetitive sequence, an important part of annotation is to identify interspersed and simple repeats. The most widely used program for this task is RepeatMasker (http://www.repeatmasker.org/).

The central problem of nucleotide level annotation is the prediction of gene structure. Ideally, we would like to delineate correctly every exon of every gene, but in large, repeat-rich eukaryotic genomes, liberally scattered with long genes with many exons, this task has turned out to be more difficult than expected. *Ab initio* gene prediction algorithms (that rely only on the statistical qualities of genomic sequence data) identify most protein-coding genes reliably in prokaryotic genomes, but the task is more complex in eukaryotic genomes (Burge and Karlin, 1998). Fundamentally, the problem is gene density; whereas in prokaryotic genomes and yeast more than two-thirds of the genome is protein-coding sequence, only a low percentage of the human genome fits this description. Additional problems are added by overlapping genes, alternatively spliced exons and the paucity of differences between intergenic sequence and introns. The gene prediction literature is full of metaphors involving needles and haystacks, and with good cause. The 13-Mb *S. cerevisiae* yeast genome provides a sobering example; completed in 1996 and initially thought to contain 6274 genes, the sequence has provided a steady trickle of additional genes that had been overlooked. Since publication of the yeast genome, a further 202 genes have been discovered; most appear to have been missed because they are relatively short or overlap a previously annotated gene on the opposite strand (Kumar *et al.*, 2002). At the same time, later analyses of these yeast sequences by a variety of statistical analyses and comparative genomics approaches have suggested that several hundred of the originally annotated genes may be spurious (Malpertuy *et al.*, 2000; Zhang and Wang, 2000).

This brings us to the use of sequence similarity in gene prediction. In practice, genome annotators use a combination of information to make predictions of gene structures: *ab initio* exon predictions (predictions of coding sequence made by a program on the basis of statistical measures of features such as codon usage, initiation signals, polyA signals and splice sites), repetitive sequence content, and similarity to expressed sequences and proteins. These different strands of evidence are usually combined and evaluated by human annotators who use graphical interfaces, such as those provided by ACEDB (Eeckman and Durbin, 1995; http://www.acedb.org/) or Otter (Searle *et al.*, 2004), to view all the evidence simultaneously. A recent trend in gene prediction is the design of programs that automatically incorporate evidence based on sequence similarity into their predictions. Among the best and most widely used *ab initio* algorithm is GENSCAN (Burge and Karlin, 1997; http://genes.mit.edu/GENSCAN.html). Guigo *et al.* (2000) tested its success in artificially produced sequence data designed to mimic human BAC sequences. At the same time, they tested algorithms that use sequence similarity to make their predictions, such as GENEWISE (Birney *et al.*, 2004; http://www.ebi.ac.uk/Wise2/). The results showed a clear advantage to including evidence from sequence similarity where the similarity was strong. In such cases, GENEWISE could correctly identify 98 per cent of coding bases present, while generating a comparatively low level of artefactual exons (2 per cent) and missing 6 per cent of real exons. Where levels of similarity were more modest, however, the performance of algorithms such as GENEWISE declined to below that of GENSCAN. GENSCAN was found to identify 89 per cent of coding bases at the cost of a rather high level of artefactual exons (41 per cent) and 14 per cent of real exons missed. Guigo *et al.*(2000) suggest that the success of all the programs tested is expected to be lower in real genomic sequence. Another comparison of gene-prediction programs using *D. melanogaster* genomic sequence identified similar levels of performance for the programs tested and also indicated an advantage to algorithms including similarity-based evidence in predictions (Reese *et al.*, 2000). Shortcuts to the structures of many genes have come from large collections of full-length mouse (Carninci *et al.*, 2006) and human cDNA sequences (Kikuno *et al.*, 2002), which have grown rapidly over the last few years. However, these collections are time-consuming and costly to produce; thus, for most organisms, we must still wrestle with the problems of computational gene prediction.

As we amass genomic sequence data from many organisms, the reach of computational annotation based upon sequence similarity is increasing. Methods aimed at the prediction of non-coding features in the genome, such as regulatory regions and non-coding RNAs (ncRNAs), are evolving rapidly. Whereas protein-coding exons have a distinctive statistical fingerprint, ncRNAs do not, or at least they do not appear to from our present, limited knowledge of them (Eddy, 2001). For better understood classes of ncRNAs, such as tRNAs, prediction methods involving secondary structure prediction have been successful (Lowe and Eddy, 1997), but for novel ncRNAs the only effective methods are based on comparative genomics (Rivas *et al.*, 2001). The great recent success story in ncRNA prediction has been for microRNAs that inhibit translation of target genes by binding to their mRNAs (Bentwich, 2005), but the majority of the RNA universe undoubtedly remains hidden. The same is true for novel regulatory sequences, where only a fraction of transcription factor-binding sites have been identified to date (Wingender *et al.*, 2001). Even incomplete, fragmentary sequence data from other organisms have been used with some success to predict putative regulatory regions (Chen *et al.*, 2001).

4.3.2 Protein level

Once we have a gene prediction that we believe, the next step is to assign a possible function to the encoded protein; this is the central task of protein level annotation. Most computationally assigned functions are derived from sequence similarity. A pair of proteins that align along 60 per cent or more of their lengths with significant similarity (e.g., $E < 0.01$ in a BLASTP search of a large public database) are very likely to be homologous – that is, derived from a common ancestor. Such a pair of sister proteins may be paralogues, derived from a duplication event, or orthologues, which exist as a result of a speciation event. For every homologous pair identified in this way, additional searches may verify that each member of the pair identifies the other member as the best match within the organism of interest. This makes it likely that the pairs identified are likely to be orthologues (Huynen and Bork, 1998), as is desirable, since orthologues are likely to share the same function (Jordan *et al.*, 2001) whereas functional diversification between paralogues is thought to be common (Li, 1997). In most cases, this strategy of reciprocal sequence similarity searches to identify orthologues is successful (Chervitz *et al.*, 1998) and is the rationale that underlies the construction of the Clusters of Orthologous Groups of proteins (COGs) database (Tatusov *et al.*, 2000; http://www.ncbi.nlm.nih.gov/COG/). However, caution is necessary when dealing with the results of such analyses. For example, a novel human gene may be directly descended from a common ancestor of a yeast gene (in which case the two genes are orthologues and are likely to share the same function), or it may be descended from a duplicated sister yeast gene (and the two genes are really paralogues) with a different function. Without a complete picture of the related family of proteins we are dealing with, it can be difficult to decide. Definitive evidence of orthology versus paralogy can come from comprehensive phylogenetic analysis, but even then, with larger families and/or incomplete data, it can be difficult. As a result, it is not uncommon to find mistaken computational predictions of function that are not supported by further experiment (Iyer *et al.*, 2001).

In the absence of any detailed knowledge about the evolutionary pedigree of the protein under study, similarity may sometimes still imply functional similarity. For example, two proteins only 30 per cent identical may share much of their biochemistry but have different substrates (Todd *et al.*, 2001). In spite of their divergence, they may share a common functional domain. There are a variety of protein domain databases, and they are widely used in genome annotation. For example, version 19 of the Pfam database contains 8183 domains that match 75 per cent of proteins in public sequence databases, with domains represented by alignments between regions of proteins containing them (Finn *et al.*, 2006; http://www.sanger.ac.uk/Software/Pfam/). Statistical models of these alignments are constructed and searched against new protein sequences using the elegant HMMER software package (Eddy, 1998; http://hmmer.wustl.edu/). The Interpro database (Mulder *et al.*, 2005; http://www.ebi.ac.uk/interpro/), which amalgamates several databases (including Pfam) covering protein domains, families and functional sites, is now a standard annotation source for each new draft genome sequence that appears. Interpro entries provide links to additional information including functional descriptions, references to the literature and structural data. Since the IHGSC draft genome publication, the EBI (European Bioinformatics Institute; http://www.ebi.ac.uk/proteome/) has maintained and updated annotation for the set of known and predicted proteins with Interpro, and their most recent analyses match around 77 per cent of the proteins in public databases. Thus, even our most strenuous efforts to gain clues to protein function, often based upon rather distant homology, tell us nothing about a quarter of known proteins.

4.3.3 Process level

Ultimately, the goal of genetics is to understand the relationship between genotype and phenotype. There is a large gap between annotation at the nucleotide or protein level and an understanding of how a given protein influences phenotype. Even in the best case, with a known gene coding for a protein containing well-studied domains, there are always questions that remain to be asked. How does the protein interact or complex with other proteins? Where does it localize within the cell? Which cellular processes and organelles is it involved with? In which tissues and at which developmental stages does it act? The answers to these questions provide process level annotation. The most important applications of our knowledge about the human genome are in medicine, to discover the variations and aberrations that underlie disease. Process level annotation provides a rational way to select the best candidate genes for involvement in disease. For example, when it was first submitted to GenBank in 1997, a certain gene (accession no. U80741) was annotated as 'Homo sapiens CAGH44 mRNA' and 'polyglutamine rich'. Due to the painstaking work of Lai *et al* (2001) on a region associated with speech disorders, we now know this gene as FOXP2, the first gene found to be involved in human language-acquisition disorders. Before their work, FOXP2 appeared to be one of many transcription factors, expressed in many tissues and best studied in *D. melanogaster.* With better process level annotation, FOXP2 may have been identified earlier as a good candidate for involvement in disease.

The main source of process level annotation is the scientific literature, but, even with modern access through the Web, this literature is a twentieth-century resource unsuited to twenty-first-century needs. What we have is a dizzying array of terms for a single gene, function or process, and no accepted way of organising this information. Added to this are all the vagaries and idiosyncrasies of human language. What is needed is a structured resource with a limited number of terms for genes and descriptions of their functions, organized so that it is easily processed automatically by computer programs. An important initiative, the Gene Ontology (GO) project, has provided a framework to achieve this (Gene Ontology Consortium, 2001; http://www.geneontology.org/). GO consists of a hierarchical set of structured vocabularies to describe the molecular functions, biological processes, and cellular components associated with gene products. With the known and predicted genes in a genome annotated by GO, we can retrieve quickly, for example, all genes encoding transmembrane receptors, all genes involved in apoptosis, or all genes encoding products localized to the cytoskeleton. The hierarchical nature of GO means that subsets of these categories can also be retrieved, such as all G-protein-coupled receptors within the transmembrane receptor category. GO annotation was quickly adopted by databases for several model organism genomes, including the Saccharomyces Genome Database (Dwight *et al.*, 2002; http://genome-www.stanford.edu/Saccharomyces/), FlyBase (FlyBase Consortium, 2002; http://flybase.org/) and the Mouse Genome Database (Blake *et al.*, 2002; http://www.informatics.jax.org/). Often GO annotations are added to genes in these databases manually by trained biologist curators browsing the scientific literature. However, with the rapidly increasing number of completed genomes, this process has become increasingly automated. Efforts continue to develop better software for automatic extraction of information from the literature to be incorporated into the GO annotation of a gene (Blaschke *et al.*, 2005).

The scale of the problem of providing process-level annotation for every human gene is prompting the development of large-scale technologies to generate data on many genes at once. Large-scale parallel measurement of gene expression for entire genomes is now possible and should give good data on the developmental timing and tissue specificity of many human genes, from which it is possible to infer process level annotation (Noordewier and Warren, 2001). An important step on the way to designating the processes a protein is involved in is to define the proteins with which it interacts, and work is well under way to elaborate the web of interacting proteins and complexes that define the proteome in organisms from *S. cerevisiae* (Gavin *et al.*, 2002; Ho *et al.*, 2002) to man (Rual *et al.*, 2005). However, these high-throughput methods are known to generate false-positives and false-negatives; that is, they identify some artefactual interactions and miss some genuine interactions. Thus, high-throughput technologies may eventually provide useful process-level annotation for many, if not most, human genes, but there will always be an indispensable role for conventional, detailed laboratory studies of smaller scale. New databases and analyses will be necessary to make sense of the network of genetic interactions that underlie the phenotype. A good example is the Mouse Atlas and Gene Expression Database Project (Baldock *et al.*, 2001; http://genex.hgu.mrc.ac.uk/), which aims to describe the patterns of gene expression responsible for the emergence of anatomical structure during mouse development. It will enable gene expression data to be viewed in the context of three-dimensional embryo sections.

4.4 Annotation up close and personal: the specifics

Despite the difficulties and shortcomings in computational annotation discussed above, several well-resourced groups have undertaken the task of compiling, maintaining and updating freely accessible annotation for the entire human genome. There are now three well-designed websites (Table 4.1) offering users the chance to browse annotation of the draft human genome. All three sites offer a graphical interface to display the results of various analyses, such as gene predictions and similarity searches, for draft and finished genomic sequence. These interfaces are indispensable for rapid, intuitive comparisons between the features predicted by different programs. For instance, one can see at once where an exon prediction overlaps with interspersed repeats or a SNP. But the four sites are not equivalent; there are important distinctions between them in terms of the data analysed, the analyses carried out and the way the results are displayed.

4.4.1 Ensembl

Ensembl is a joint project between the EBI (http://www.ebi.ac.uk/) and the Wellcome Trust Sanger Institute (http://www.sanger.ac.uk/). The Ensembl database (Hubbard *et al.*, 2002; http://www.ensembl.org/), launched in 1999, was the first to provide a window on the draft genome, curating the results of a series of computational analyses. Until January 2002 (Release 3.26.1), Ensembl used the UCSC draft sequence assemblies as its starting point, but it is now based upon NCBI assemblies. The Ensembl analysis pipeline consists of a rule-based system designed to mimic decisions made by a human annotator. The idea is to identify 'confirmed' genes that are computationally predicted (by the GENSCAN gene prediction program) and also supported by a significant BLAST match to one or more expressed sequences or proteins. Ensembl also identifies the positions of known human genes from public sequence database entries, usually using GENEWISE to predict their exon structures. The total set of Ensembl genes should therefore be a much more accurate reflection of reality than *ab initio* predictions alone, but it is clear that some novel genes are missed (Hogenesch *et al.*, 2001). Of the many novel genes that are detected, some are expected to be incomplete for two main reasons. Firstly, as we have seen, while GENSCAN can detect the presence of most genes in a genomic sequence, it is substantially less successful in predicting their correct exonic structures (as with other *ab initio* gene predictions). Secondly, any prediction is entirely dependent upon the quality of the genomic sequence, and where the sequence is gapped or wrongly assembled, the missing exons may not be present for the software to find. However, in the finished

Table 4.1 The websites referred to in the text

Site description	URL
Genomic sequence assemblies	
NCBI Human Genome Assemblies	http://www.ncbi.nlm.nih.gov/Genomes/
UCSC Human Genome Assemblies	http://genome.ucsc.edu/
Annotation browsers	
Ensembl at EBI/Sanger Institute	http://www.ensembl.org/
Human Genome Browser at UCSC	http://genome.ucsc.edu/
Map Viewer at NCBI	http://www.ncbi.nlm.nih.gov/mapview/
Data sources	
ArrayExpress at EBI	http://www.ebi.ac.uk/arrayexpress/
COGs database at NCBI	http://www.ncbi.nlm.nih.gov/COG/
dbSNP at NCBI	http://www.ncbi.nlm.nih.gov/SNP/index.html
DOTS at University of Pennsylvania	http://www.allgenes.org/
Entrez Gene at NCBI	http://www.ncbi.nlm.nih.gov/entrez/query.fcgi?db=gene
FlyBase	http://flybase.org/
Genomes OnLine Database	http://www.genomesonline.org/
GEO at NCBI	http://www.ncbi.nlm.nih.gov/geo/
IHGMC FPC map at Washington University in St Louis	http://genomeold.wustl.edu/cgi-bin/ace/GSCMAPS.cgi?
InterPro at EBI	http://www.ebi.ac.uk/interpro/
Mouse Genome Database at Jackson Laboratory	http://www.informatics.jax.org/
Mouse Atlas Database at MRC Human Genetics Unit	http://genex.hgu.mrc.ac.uk/
OMIM at NCBI	http://www.ncbi.nlm.nih.gov/Omim/
Pfam at Sanger Institute	http://www.sanger.ac.uk/Software/Pfam/
Proteome Analysis at EBI	http://www.ebi.ac.uk/proteome/
RefSeq at NCBI	http://www.ncbi.nlm.nih.gov/LocusLink/refseq.html
Saccharomyces Genome Database at Stanford University	http://genome-www.stanford.edu/Saccharomyces/
UniGene at NCBI	http://www.ncbi.nlm.nih.gov/UniGene/
Software	
ACEDB at Sanger Institute	http://www.acedb.org/
Acembly at NCBI	http://www.ncbi.nlm.nih.gov/IEB/Research/Acembly/index.html
Apollo at Ensembl	http://www.fruitfly.org/annot/apollo/
BioMart	http://www.biomart.org/
BLAST at NCBI	http://www.ncbi.nlm.nih.gov/BLAST/
BLAT at UCSC	http://genome.ucsc.edu/cgi-bin/hgBlat?command=start)
DAS at Cold Spring Harbor Laboratory	http://biodas.org/
EMBOSS at EMBnet	http://emboss.sourceforge.net/
ePCR at NCBI	http://www.ncbi.nlm.nih.gov/genome/sts/epcr.cgi
GBrowse	http://www.gmod.org/
Gene Ontology Consortium	http://www.geneontology.org/
GENEWISE at EBI	http://www.ebi.ac.uk/Wise2/

Continues overleaf

Table 4.1 *(continued)*

Site description	URL
GENSCAN at MIT	http://genes.mit.edu/GENSCAN.html
HMMER at Washington University in St Louis	http://hmmer.wustl.edu/
Phrap at University of Washington	http://bozeman.genome.washington.edu/index.html
RepeatMasker	http://www.repeatmasker.org/
SIM4 at Pennsylvania State University	http://bio.cse.psu.edu/
Spidey at NCBI	http://www.ncbi.nlm.nih.gov/IEB/Research/Ostell/Spidey/
SSAHA at Sanger Institute	http://www.sanger.ac.uk/Software/analysis/SSAHA/

human and mouse genomes, where there are large full-length cDNA collections to guide the hunt for genes, Ensembl should be very reliable.

From the beginning, many genomic features other than predicted genes were included in Ensembl: different repeat classes, cytological bands, CpG island predictions, tRNA gene predictions, expressed sequence clusters from the UniGene database (Wheeler *et al.*, 2002; http://www.ncbi.nlm.nih.gov/UniGene/), SNPs from the dbSNP database (Sherry *et al.*, 2001; http://www.ncbi.nlm.nih.gov/SNP/index.html), disease genes found in the draft genome from the OMIM database (Online Mendelian Inheritance in Man database; Wheeler *et al.*, 2002; http://www.ncbi.nlm.nih.gov/Omim/) and regions of homology to other draft genomic sequences. More recent innovations have seen the annotation of a large range of non-coding RNAs (ncRNAs) from the Rfam database (Griffiths-Jones *et al.*, 2005) and predicted regulatory sites from the cisRED database (Robertson *et al.*, 2006). There is much to do in both of these emerging areas but even preliminary data have already given new insights into mammalian biology: it seems there is high lineage specific expansion of some ncRNA classes relative to protein-coding genes (Birney *et al.*, 2006). Another growing area of activity is in cataloguing the genetic variation present in human populations as Ensembl reflects the progress of the International Haplotype Map Project (Thorisson *et al.*, 2005).

More speculative data, such as GENSCAN-predicted exons that have not been incorporated into Ensembl-confirmed genes, may also be viewed. This means that the display can be used as a workbench for the user to develop personalized annotation. For example, one may discover novel exons by finding GENSCAN exon predictions which coincide with good matches to a fragment of the draft mouse genome, or novel promoters by finding matches to the draft mouse genome that occur upstream of the 5′ end of a gene. Once we have identified a gene of interest, we can link to a wealth of information at external sites such as the Interpro protein domains it encodes and its expression profile according to the SAGEmap repository (Lash *et al.*, 2000). Eventually, Ensembl aims to become a platform for studies in comparative genomics, and already it is possible, while browsing the human genome, to jump to a homologous region of another organism's genome via a match to a genomic sequence fragment. Substantial thought and effort have evidently gone into the Ensembl site design. The result is certainly a user-friendly experience, and not just by the standards of computational biology. The Web interface to the database achieves the laudable aim of providing seamless access to the human genome. The user can sink down through cytogenetic ideograms of whole chromosomes, to large sequence contigs many megabases long and then to the single base pair level. Along the way, a graphical display shows the relative positions of genes and other features.

Figure 4.1 shows the Ensembl display for the genomic region around the FOXP2 gene mentioned earlier. The region is shown at three levels of resolution. The upper panel shows the position of the region as a small red box on a cytogenetic ideogram of chromosome 7. The middle panel shows an exploded view of this box, including the structure of the draft genome assembly; the extent of synteny (conserved gene order) with other organisms; the relative positions of various markers; and a simple overview of the gene content. The bottom panel gives a detailed view of a subsection (indicated again by a red box) of the middle panel. This detailed view is the business end of the browser and is easily customized, via pull-down menus, to display any desired combination of the available features. In Figure 4.1, the combination chosen shows the positions of similarity to a variety of other vertebrate genomes (Rn is rat, Pt is chimpanzee, Mm is mouse, Gg is chicken and Cf is dog) in relation to predicted exons and similarities to protein and cDNA sequences, allowing a user to define non-coding conserved regions that may be of regulatory importance. Using this display, one could also select SNPs that have been shown to be genuine ('genotyped SNPs') and that also lie outside repetitive sequences; both are important considerations for PCR-based SNP assays.

Data retrieval is extremely well catered for in Ensembl, with text searches of all database entries, BLAST searches of all sequences archived, and the availability of bulk downloads of all Ensembl data and even software source code. Ensembl annotation can also be viewed interactively on one's local machine with the Apollo viewer (Lewis *et al.*, 2002; http://www.fruitfly.org/annot/apollo/).

4.4.2 The UCSC Human Genome Browser

The UCSC Human Genome Browser (UCSC) bears many similarities to Ensembl; it, too, provides annotation of the NCBI assemblies, and it displays a similar array of features, including confirmed genes from Ensembl. The range of features displayed in UCSC (and Ensembl) often changes between releases, but usually there are additional features of UCSC that are not found in Ensembl, and vice versa. For example, at the time of writing, UCSC includes predictions from a wider range of *ab initio* gene-prediction programs. This could help the user to identify false-positives (i.e., artefactual exons) from particular programs, and concentrate on exons predicted by

⊟ Chromosome 7

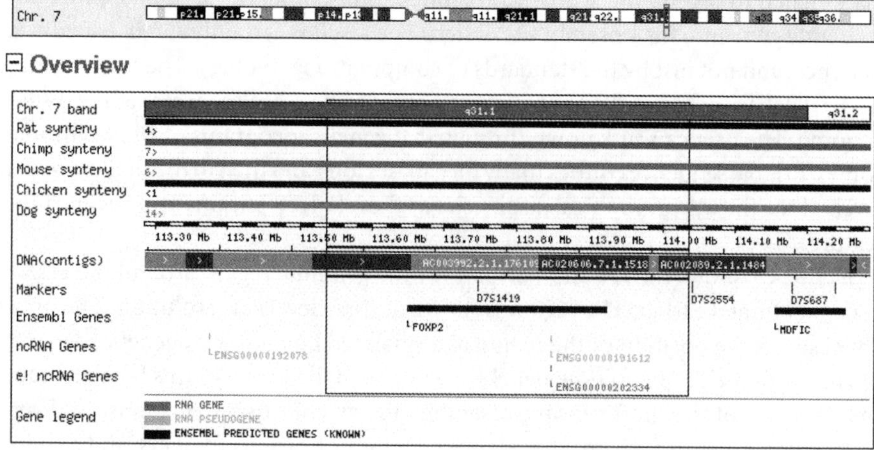

⊟ Detailed view

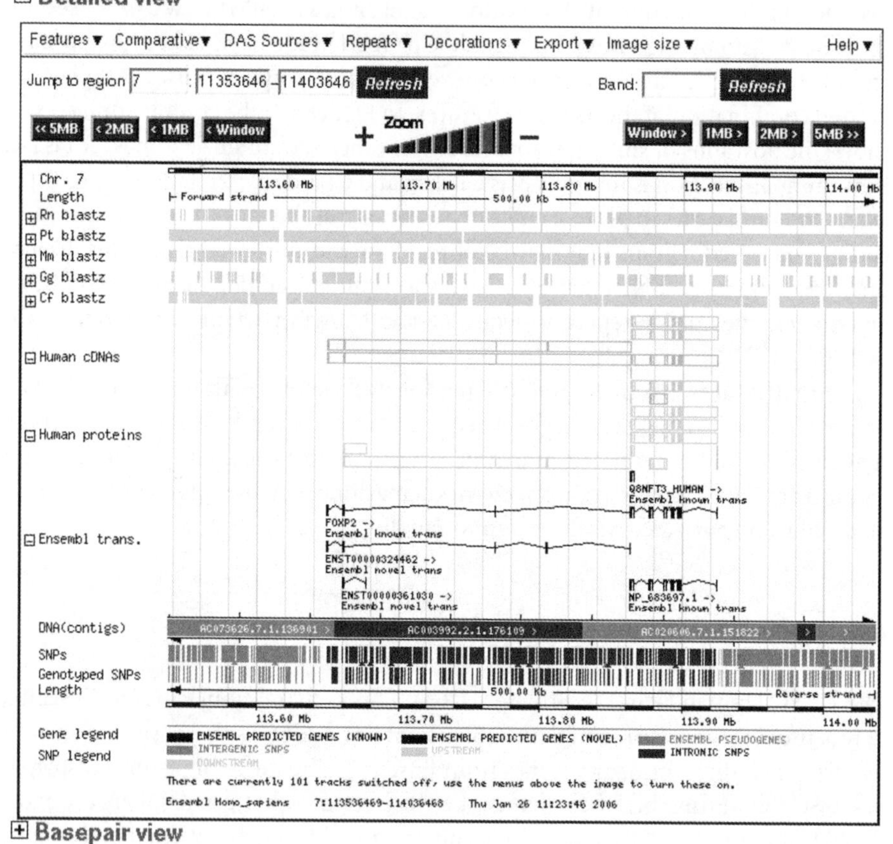

Figure 4.1 The genomic region around the FOXP2 gene according to Ensembl

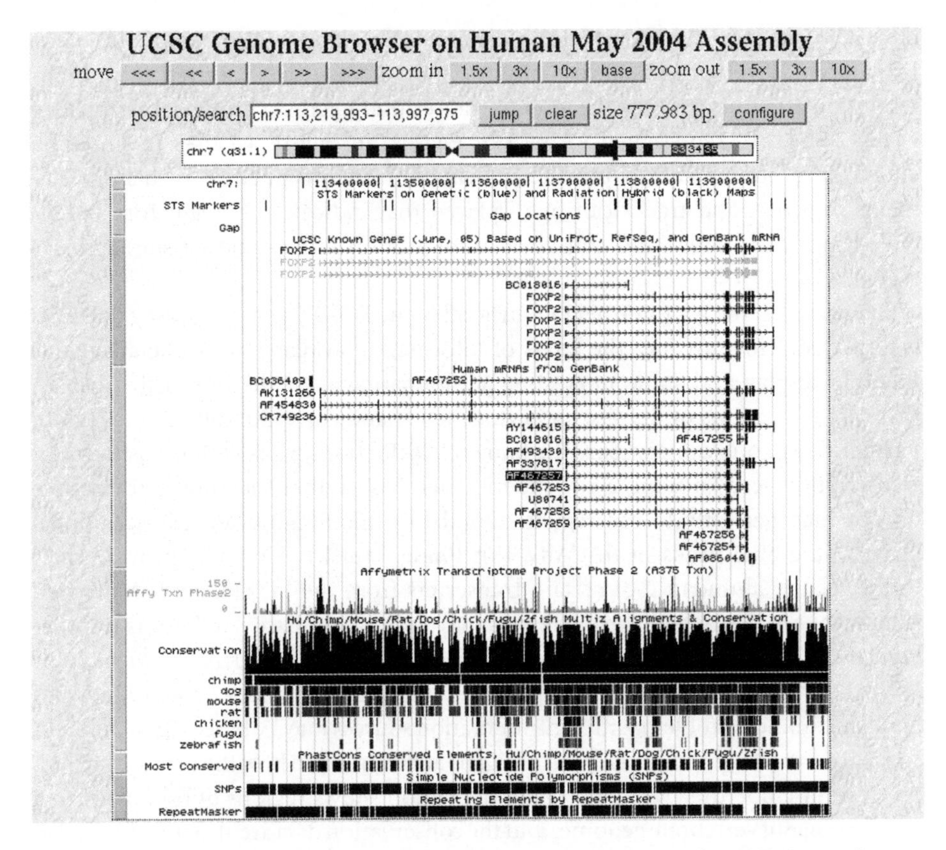

Figure 4.2 The genomic region around the FOXP2 gene according to the UCSC Human Genome Browser. Generated using the UCSC Human Genome Browser, http://genome.ucsc.edu

more than one program that are most likely to be real. UCSC also currently indicates regions with significant homology to the various vertebrate genomes as in Ensembl, but it displays the data quite differently, using summary tracks to indicate overall conservation across several genomes ('Conservation' and 'Most Conserved' in Figure 4.2). These UCSC-specific features can provide useful information when one is dealing with gene predictions that are not well supported by similarity to expressed sequence. Another useful feature of UCSC is the detailed description of the genomic sequence assemblies. Graphical representations of the fragments making up a region of draft genome can be displayed, showing the relative size and overlaps of each fragment and also whether any gaps between fragments are bridged by mRNAs or paired BAC end sequences. This means one can get an idea of the likely degree of misassembly in a draft region. There are now a large number of data available from large-scale gene expression studies, and public repositories have emerged for their curation, such as the NCBI Gene Expression Omnibus (http://www.ncbi.nlm.nih.gov/geo/) and ArrayExpress at the EBI (http://www.ebi.ac.uk/arrayexpress/). At the moment, the UCSC is the browser which incorporates the largest number of this data. Even recent developments such as transcriptome tiling data, derived from high-resolution attempts to assay the level of transcription across chromosomes (Cheng *et al.*, 2005), are represented ('Affy Txn Phase2' in Figure 4.2). As in Ensembl, efforts have been made to provide information on the putative regulatory elements of genes, and tracks can be displayed that indicate the 'regulatory potential' (King *et al.*, 2005) and conserved transcription factor-binding sites across a region.

In Figure 4.2, the genomic neighbourhood of the FOXP2 gene according to UCSC is displayed. This provides the kinds of information available from the analogous Ensembl display and some interesting additional data. At the top of the display, there are indications of the size and cytogenetic band corresponding to the region. Further down, one can compare the known FOXP2 transcripts with the patterns of transcription seen in tiling array experiments. Notice that the known transcripts do not map perfectly to the regions found to exhibit significant transcriptional activity (the blue peaks in the 'Affy Txn Phase2' track). This may provide clues to the relative abundance of certain transcripts from the FOXP2 gene. Moreover, significant activity outside known exons may indicate undiscovered exons or other regulatory RNA species. It is also notable that the number of cDNA sequences for FOXP2 differs between Ensembl (Figure 4.1) and UCSC (Figure 4.2). This illustrates another common problem: different annotation sources may be based upon different sequence data, depending on what is available at the time and how the data are filtered. As with Ensembl, the UCSC display of the region shows regions of homology to a similar range of vertebrate genomes, but the conservation data are also summarized in an intuitive graph ('Conservation'). More importantly, a statistically valid indication of the best conserved regions ('Most Conserved') is provided, using output from the PhastCons program (Siepel *et al.*, 2005). The relative scores of these regions (which can also be displayed) would be a reasonable criterion to rank non-coding regions for further study as regulatory elements.

Data retrieval at UCSC is facilitated by text and BLAT (Kent, 2002; a BLAST-like algorithm) searches and bulk downloads of annotation or sequence data. Other complementary tools at UCSC have extended the functionality of UCSC. For instance, the Proteome Browser graphically displays protein properties such as hydrophobicity, charge and structural features across any publicly available protein sequence (Hinrichs *et al.*, 2006). As with Ensembl, the UCSC website has been well designed and is sympathetic to the naive user, but the UCSC graphical interface is more Spartan. If Ensembl is Disney, then UCSC is *South Park*. The positive side of this is that UCSC will usually display a region on your local web browser more quickly than Ensembl can. Both the Ensembl and UCSC interfaces offer users the ability to jump between their respective views of a region, and so, when they are both annotating the same version of the same NCBI assembly, they can easily be used as complementary resources.

4.4.3 NCBI Map Viewer

As a wider range of organisms are subject to genome sequencing, the problems of dealing with draft sequence data have remained, but an additional task has arisen: curation of the finished sequences representing each complete chromosome. This task is undertaken at the NCBI in the form of Entrez Genome, the section of the NCBI sequence retrieval system concerned with genomes and individual genome assembly versions, and the sequences of individual whole genome shotgun reads are also available.

As the name suggests, the NCBI Map Viewer (NMV; http://www.ncbi.nlm.nih.gov/mapview/) evolved to allow graphical depictions of, and comparisons between, a wide range of genetic and physical maps in parallel with NCBI draft and finished sequence contigs. The locations of genes, markers, and SNPs are indicated on the assembled sequences. As with Ensembl, there is a NCBI analysis protocol which aims to predict gene structures based upon EST and mRNA alignments with the draft genome. This is carried out by a program called Acembly (unpublished; http://www.ncbi.nlm.nih.gov/IEB/Research/Acembly/index.html), which aims to derive gene structure from these alignments alone. The program also attempts to give alternative splice variants of genes where its alignments suggest them. These gene structures and transcripts end up as records in the NCBI RefSeq database, which aims to compile a non-redundant, curated data set representing current knowledge of known genes (Wheeler *et al.*, 2002; http://www.ncbi.nlm.nih.gov/entrez/query.fcgi?db=gene). Like the Ensembl protocol, many Acembly-predicted structures (the NCBI estimate 42 per cent) are incomplete. These structures can be displayed alongside *ab initio* gene models, Ensembl-predicted genes, and matching UniGene clusters to allow users to make their own conclusions about the likeliest gene structure.

Figure 4.3 shows the FOXP2 gene as it appears in the NMV, which shows features on a vertical rather than horizontal display. The familiar chromosome ideogram is shown in the leftmost frame, followed by BLAST matches to four UniGene-expressed sequence clusters (in the 'HsUniG' column). This gene is typical in having more than one UniGene cluster representing it, particularly at the 3' end, as ESTs are more commonly sequenced from the 3′ ends of mRNAs. The next columns depict various human cDNA sequence matches. SNPs from the NCBI dbSNP database are also displayed (in the 'Variation' column) with susceptibility loci for various disorders (the 'Pheno' column) from the NCBI OMIM database. In the right-most column, the FOXP2 gene structure is displayed according to the NCBI RefSeq database model. In contrast to the Ensembl and UCSC displays, it is not possible to depict comparative genomics data or putative regulatory regions.

The NMV offers tabulated downloads of data, and it is possible to BLAST-search genome assemblies (via the NCBI BLAST site: http://www.ncbi.nlm.nih.gov/BLAST/) and view the matching regions with the NMV. All annotated genes

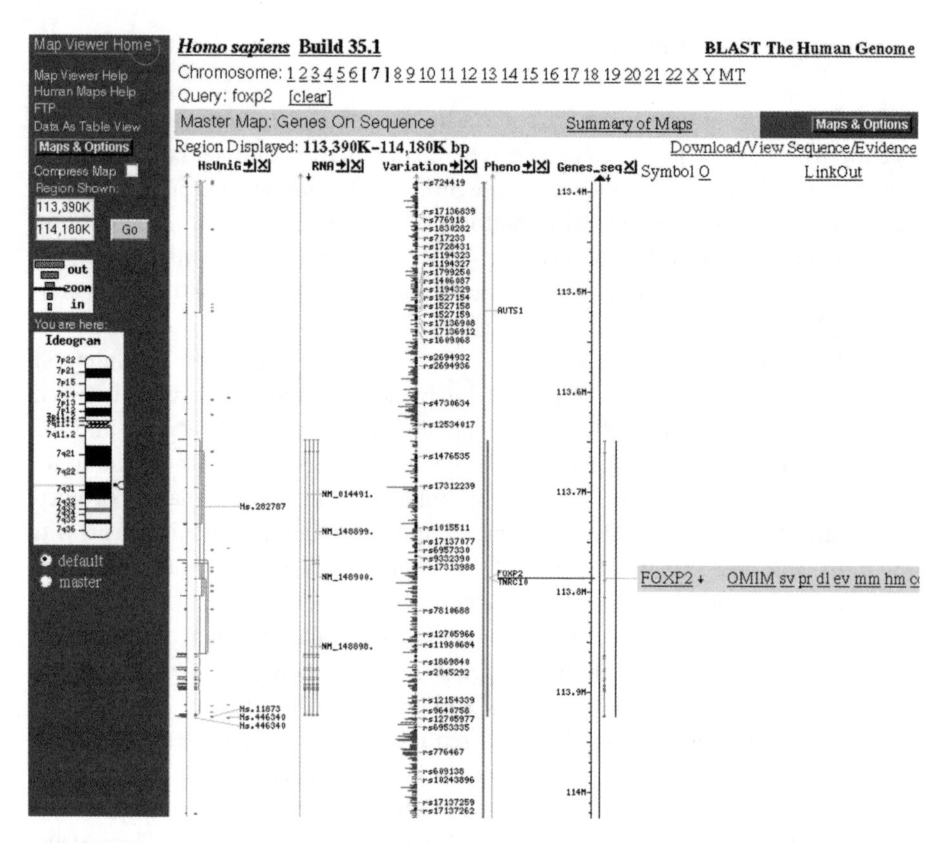

Figure 4.3 The genomic region around the FOXP2 gene according to the NCBI Map Viewer

are connected to NCBI Entrez Gene, which provides links to associated information such as related sequence accession numbers, expression data, known phenotypes and SNPs.

4.5 Annotation: the next generation

In spite of difficulties with the quality of genomic sequence assemblies and the errors and omissions of computational annotation, the browsers discussed above remain extremely useful tools for the cautious biologist. They undoubtedly indicate the presence of most coding sequence in a given fragment of genomic sequence and indicate their location in the genome based on the best genomic sequence available. In addition, they attempt to predict gene structures for novel genes and should be accurate if the gene in question is known or has a close homologue which is known. Most aspects of the analysis carried out are the subjects of active research, and improvements in performance due to the inclusion of new sequence data and annotation software will continue. The downside of these developments is that all annotation of genomic sequence is potentially in flux, and one should not assume that the representation of a region will remain the same between different software or data releases.

Although the finished human genome sequence is now the subject of curation rather than successive draft sequence assemblies, annotation of these sequences is still at a relatively early stage. Even at nucleotide level, there is much to be done, particularly in fully exploiting the data available from other genome-sequencing projects. The cutting edge of nucleotide level annotation is in defining regulatory regions: transcription start sites (TSSs), transcription factor-binding sites and promoter modules (Werner, 2001). Here again, comparative genomics is already a rich source of information, simply using local alignment programs' output, as in Ensembl and UCSC. At a higher level, gene expression is also regulated by the large-scale topology of chromosomes, and annotation may eventually indicate features such as chromatin structure, chromosome domains (genomic regions that bind histone modifying proteins) and matrix attachment sites (regions that facilitate the organization of DNA within a chromosome into loops). However, defining the genes whose transcription is regulated from such features may be an insoluble problem computationally, since they may regulate transcription from a given TSS, or from several different TSSs of the same gene or multiple genes in a region.

At the protein and process levels of annotation, there is also progress, as, for instance, in our ability to detect more remote homologies and gain clues about function. Homologous proteins, sharing a common three-dimensional structure and function, need not share detectable, sequence similarity. There is therefore increasing interest in annotation by similarity at the level of protein structure (Gough and Chothia, 2002). The genome sequence has already changed the way we study biology as we start to fill in the gaps between genetics, cellular function and development. Rather than studying a particular gene or protein, we are increasingly able to study all elements in a system of interest, a group of proteins that participate in a complex, for example. We might start with a single protein and identify others in the proteome that potentially interact with it, on the basis of the presence of domains known to interact. In the process, we may discover previously unknown connections with other complexes or biochemical pathways that can be included in the annotation of the relevant sequences. Studies on this scale are prompting the development of multidisciplinary groups that study the behaviour and perturbation of entire biological systems: the new field of systems biology (Ideker *et al.*, 2001). Recent studies in computational systems biology seek to extend the reach of our predictions beyond the human genome to the interactions with systems within other organisms such as pathogens (Uetz *et al.*, 2006). Over the next decade or two, these efforts should provide a genome sequence with rich annotation that can be browsed at the level of a gene's genomic neighbourhood but also at the level of the interactions, complexes and processes that it participates in and the phenotypes it influences.

This review has provided only a brief introduction to the fields of computational draft genome assembly and annotation, but it should be evident that what has already been achieved has involved innovations as great as those in the biotechnology that led to the production of the sequence data itself. At the same time, problems remain at every level and are the subjects of active research. As a result, many different groups around the world are working on interpreting the data avalanche that is modern genetics, and communication and comparison of results are often difficult. In response, prominent members of the bioinformatics community (such as those behind Ensembl and the UCSC) have steadily developed freely available generic tools to allow the organization, display and exchange of annotation. The Distributed Annotation System (DAS) (Dowell *et al.*, 2001; http://biodas.org/) aims to provide a framework for people to exchange data easily over the Web. Two other notable developments are BioMart and GBrowse. The BioMart project (http://www.biomart.org/), originally a spin-off from Ensembl, offers a generic data management system that allows complex searches of biological data such as sequence annotation. The GBrowse project (Stein *et al.*, 2002; http://www.gmod.org/) has produced a generic genome browser that can be customized to organize, display and query a new genome scale data set. These tools promise a future without the current confusion of incompatible interfaces and data formats, and an increase in the open exchange of data and ideas.

Acknowledgements

Colin A. M. Semple enjoys the financial support of the UK Medical Research Council. Martin S. Taylor provided comments on an early version of this chapter.

References

Altschul, S. F., Gish, W., Miller, W.*et al.* (1990). Basic local alignment search tool. *J Mol Biol* **215**, 403-410.

Altschul, S. F., Madden, T. L., Schaffer, A. A. *et al.* (1997). Gapped BLAST and PSI-BLAST: a new generation of protein database search programs. *Nucleic Acids Res* **25**, 3389–3402.

Baldock, R., Bard, J., Brune, R. *et al.* (2001). The Edinburgh Mouse Atlas: using the CD. *Brief Bioinform* **2**, 159–169.

Batzoglou, S., Jaffe, D. B., Stanley, K. *et al.* (2002). ARACHNE: a whole-genome shotgun assembler. *Genome Res* **12**, 177–189.

Bateman, A., Birney, E., Cerruti, L. *et al.* (2002). The Pfam protein families database. *Nucleic Acids Res* **30**, 276–280.

Beck, S. and Sterk, P. (1998). Genome-scale DNA sequencing: where are we? *Curr Opin Biotechnol* **9**, 116–120.

Bentwich, I. (2005). Prediction and validation of microRNAs and their targets. FEBS Lett **579**, 5904–5910.

Birney, E., Andrews, D., Caccamo, M. *et al.* (2006). Ensembl 2006. *Nucleic Acids Res* **34**, D556–561.

Birney, E., Clamp, M. and Durbin, R. (2004). GeneWise and Genomewise. *Genome Res* **14**, 988–995.

Blake, J. A., Richardson, J. E., Bult, C. J. *et al.* (2002). The Mouse Genome Database (MGD): the model organism database for the laboratory mouse. *Nucleic Acids Res* **30**, 113–115.

Blaschke, C., Leon, E. A., Krallinger, M. *et al.* (2005). Evaluation of BioCreAtIvE assessment of task 2. *BMC Bioinformatics* **6**, S16.

Burge, C. B. and Karlin, S. (1998). Finding the genes in genomic DNA. *Curr Opin Struct Biol* **8**, 346–354.

Carninci, P., Kasukawa, T., Katayama, S. *et al.* (2006). The transcriptional landscape of the mammalian genome. *Science* **309**, 1559–1563.

Chen, R., Bouck, J. B., Weinstock, G. M. *et al.* (2001). Comparing vertebrate whole-genome shotgun reads to the human genome. *Genome Res* **11**, 1807–1816.

Cheng, J., Kapranov, P., Drenkow, J. *et al.* (2005). Transcriptional maps of 10 human chromosomes at 5-nucleotide resolution. *Science* **308**, 1149–1154.

Chervitz, S. A., Aravind, L., Sherlock, G. *et al.* (1998). Comparison of the complete protein sets of worm and yeast: orthology and divergence. *Science* **282**, 2022–2028.

Dowell, R. D., Jokerst, R. M., Day, A. *et al.* (2001). The Distributed Annotation System. *BMC Bioinformatics* **2**, 7.

Dwight, S. S., Harris, M. A., Dolinski, K. *et al.* (2002). Saccharomyces Genome Database (SGD) provides secondary gene annotation using the Gene Ontology (GO). *Nucleic Acids Res* **30**, 69–72.

Eddy, S. R. (1998). Profile hidden Markov models. *Bioinformatics* **14**, 755–763.

Eddy, S. R. (2001). Non-coding RNA genes and the modern RNA world. *Nat Rev Genet* **2**, 919–929.

Eeckman, F. H. and Durbin, R. (1995). ACeDB and macace. *Methods Cell Biol* **48**, 583–605.

Eichler, E. E. (2001). Segmental duplications: what's missing, misassigned, and misassembled – and should we care? *Genome Res* **11**, 653–656.

Finn, R. D., Mistry, J., Schuster-Bockler, B. *et al.* (2006). Pfam: clans, Web tools and services. *Nucleic Acids Res* **34**, D247–251.

FlyBase Consortium (2002). The FlyBase database of the *Drosophila* genome projects and community literature. *Nucleic Acids Res* **30**, 106–108.

Gavin, A. C., Bosche, M., Krause, R. *et al.* (2002). Functional organization of the yeast proteome by systematic analysis of protein complexes. *Nature* **415**, 141–147.

Gene Ontology Consortium (2001). Creating the gene ontology resource: design and implementation. *Genome Res* **11**, 1425–1433.

Gough, J. and Chothia, C. (2002). SUPERFAMILY: HMMs representing all proteins of known structure. SCOP sequence searches, alignments and genome assignments. *Nucleic Acids Res* **30**, 268–272.

Griffiths-Jones, S., Moxon, S., Marshall, M. *et al.* (2005). Rfam: annotating non-coding RNAs in complete genomes. *Nucleic Acids Res* **33**, D121–124.

Guigo, R., Agarwal, P., Abril, J. F. *et al.* (2000). An assessment of gene prediction accuracy in large DNA sequences. *Genome Res* **10**, 1631–1642.

Hattori, M. and Taylor, T. D. (2001). Part three in the book of genes. *Nature* **414**, 854–855.

Hinrichs, A. S., Karolchik, D., Baertsch, R. *et al.* (2006). The UCSC Genome Browser Database: update 2006. *Nucleic Acids Res* **34**, D590–598.

Ho, Y., Gruhler, A., Heilbut, A. *et al.* (2002). Systematic identification of protein complexes in *Saccharomyces cerevisiae* by mass spectrometry. *Nature* **415**, 180–183.

Hogenesch, J. B., Ching, K. A., Batalov, S. *et al.* (2001). A comparison of the Celera and Ensembl predicted gene sets reveals little overlap in novel genes. *Cell* **106**, 413–415.

Hubbard, T., Barker, D., Birney, E. *et al.* (2002). The Ensembl genome database project. *Nucleic Acids Res* **30**, 38–41.

Huson, D. H., Reinert, K., Kravitz, S. A. *et al.* (2001). Design of a compartmentalized shotgun assembler for the human genome. *Bioinformatics* **17** Suppl 1, S132–139.

Huynen, M. A. and Bork, P. (1998). Measuring genome evolution. *Proc Natl Acad Sci U S A* **95**, 5849–5856.

Ideker, T., Galitski, T. and Hood, L. (2001). A new approach to decoding life: systems biology. *Annu Rev Genomics Hum Genet* **2**, 343–372.

International Human Genome Sequencing Consortium (2001). Initial sequencing and analysis of the human genome. *Nature* **409**, 860–921.

Iyer, L. M., Aravind, L., Bork, P. *et al.* (2001). Quod erat demonstrandum? The mystery of experimental validation of apparently erroneous computational analyses of protein sequences. *Genome Biol* **2**, research0051.

Jordan, I. K., Kondrashov, F. A., Rogozin, I. B. *et al.* (2001). Constant relative rate of protein evolution and detection of functional diversification among bacterial, archaeal and eukaryotic proteins. *Genome Biol* **2**, research0053.

Katsanis, N., Worley, K. C. and Lupski, J. R. (2001). An evaluation of the draft human genome sequence. *Nat Genet* **29**, 88–91.

Kent, W. J. (2002). BLAT – the BLAST-like alignment tool. *Genome Res* **12**, 656–664.

Kent, W. J. and Haussler, D. (2001). Assembly of the working draft of the human genome with GigAssembler. *Genome Res* **11**, 1541–1548.

Kikuno, R., Nagase, T., Waki, M. *et al.* (2002). HUGE: a database for human large proteins identified in the Kazusa cDNA sequencing project. *Nucleic Acids Res* **30**, 166–168.

King, D. C., Taylor, J., Elnitski, L. *et al.* (2005). Evaluation of regulatory potential and conservation scores for detecting *cis*-regulatory modules in aligned mammalian genome sequences. *Genome Res* **15**, 1051–1060.

Kumar, A., Harrison, P. M., Cheung, K. H. *et al.* (2002). An integrated approach for finding overlooked genes in yeast. *Nat Biotechnol* **20**, 58–63.

Lai, C. S., Fisher, S. E., Hurst, J. A. *et al.* (2001). A forkhead-domain gene is mutated in a severe speech and language disorder. *Nature* **413**, 519–523.

Lash, A. E., Tolstoshev, C. M., Wagner, L. *et al.* (2000). SAGEmap: a public gene expression resource. *Genome Res* **10**, 1051–1060.

Lewis, S. E., Searle, S. M. J., Harris, N. *et al.* (2002). Apollo: a sequence annotation editor. *Genome Biol* **3**, research0082.

Li, W. H. (1997). *Molecular Evolution.* Sunderland, MA: Sinauer Associates.

Liolios, K., Tavernarakis, N., Hugenholtz, P. *et al.* (2006). The Genomes On Line Database (GOLD) v.2: a monitor of genome projects worldwide. *Nucleic Acids Res* **34**, D332–334.

Lowe, T. M. and Eddy, S. R. (1997). tRNAscan-SE: a program for improved detection of transfer RNA genes in genomic sequence. *Nucleic Acids Res* **25**, 955–964.

Malpertuy, A., Tekaia, F., Casaregola, S. *et al.* (2000). Genomic exploration of the hemiascomycetous yeasts. 19. Ascomycetes-specific genes. *FEBS Lett* **487**, 113–121.

Mulder, N. J., Apweiler, R., Attwood, T. K. *et al.* (2005). InterPro, progress and status in 2005. *Nucleic Acids Res* **33**, D201–205.

Ning, Z., Cox, A. J. and Mullikin, J. C. (2001). SSAHA: a fast search method for large DNA databases. *Genome Res* **11**, 1725–1729.

Noordewier, M. O. and Warren, P. V. (2001). Gene expression microarrays and the integration of biological knowledge. *Trends Biotechnol* **19**, 412–415.

Nusbaum, C., Mikkelsen, T. S., Zody, M. C. *et al.* (2006). DNA sequence and analysis of human chromosome 8. *Nature* **439**, 331–335.

Olivier, M., Aggarwal, A., Allen, J. *et al.* (2001). A high-resolution radiation HYBRID map of the human genome draft sequence. *Science* **291**, 1298–1302.

Reese, M. G., Hartzell, G., Harris, N. L. *et al.* (2000). Genome annotation assessment in *Drosophila melanogaster. Genome Res* **10**, 483–501.

Rice, P., Longden, I. and Bleasby, A. (2000). EMBOSS: the European Molecular Biology Open Software Suite. *Trends Genet* **16**, 276–277.

Rivas, E., Klein, R. J., Jones, T. A. *et al.* (2001). Computational identification of noncoding RNAs in *E. coli* by comparative genomics. *Curr Biol* **11**, 1369–1373.

Rual, J. F., Venkatesan, K., Hao, T. *et al.* (2005). Towards a proteome-scale map of the human protein–protein interaction network. *Nature* **437**, 1173–1178.

Sanger, F., Coulson, A. R., Hong, G. F. *et al.* (1982). Nucleotide sequence of bacteriophage lambda DNA. *J Mol Biol* **162**, 729–773.

Schuler, G. D. (1997). Sequence mapping by electronic PCR. *Genome Res* **7**, 541–550.

Semple, C. A. M., Morris, S. W., Porteous, D. J. *et al.* (2002). Computational comparison of human genomic sequence assemblies for a region of chromosome 4. *Genome Research*, **12**, 424–429.

Searle, S. M., Gilbert, J., Iyer, V. *et al.* (2004). The otter annotation system. *Genome Res* **14**, 963–970.

Sherry, S. T., Ward, M. H., Kholodov, M. *et al.* (2001). dbSNP: the NCBI database of genetic variation. *Nucleic Acids Res* **29**, 308–311.

Siepel, A., Bejerano, G., Pedersen, J. S. *et al.* (2005). Evolutionarily conserved elements in vertebrate, insect, worm, and yeast genomes. *Genome Res* **15**, 1034–1050.

Soderlund, C., Humphray, S., Dunham, A. *et al.* (2000). Contigs built with fingerprints, markers, and FPC V4.7. *Genome Res* **10**, 1772–1787.

Stein, L. (2001). Genome annotation: from sequence to biology. *Nat Rev Genet* **2**, 493–503.

Stein, L. D., Mungall, C., Shu, S. *et al.* (2002). The generic genome browser: a building block for a model organism system database. *Genome Res* **12**, 1599–1610.

Tatusov, R. L., Galperin, M. Y., Natale, D. A. *et al.* (2000). The COG database: a tool for genome-scale analysis of protein functions and evolution. *Nucleic Acids Res* **28**, 33–36.

Thorisson, G. A., Smith, A. V., Krishnan, L.*et al.* (2005). The International HapMap Project Web site. *Genome Res* **15**, 1592–1593.

Todd, A. E., Orengo, C. A. and Thornton, J. M. (2001). Evolution of function in protein superfamilies, from a structural perspective. *J Mol Biol* **307**, 1113–1143.

Uetz, P., Dong, Y. A., Zeretzke, C. *et al.* 2006. Herpesviral protein networks and their interaction with the human proteome. *Science* **311**, 239–242.

Venter J. C. Adams, M. D., Myers, E. W. *et al.* (2001). The sequence of the human genome. *Science* **291**, 1304–1351.

Vinson, J. P., Jaffe, D. B., O'Neill, K. *et al.* (2005). Assembly of polymorphic genomes: algorithms and application to *Ciona savignyi. Genome Res* **15**, 1127–1135.

Werner, T. (2001). Cluster analysis and promoter modelling as bioinformatics tools for the identification of target genes from expression array data. *Pharmacogenomics* **2**, 25–36.

Wheelan, S. J., Church, D. M. and Ostell, J. M. (2001). Spidey: a tool for mRNA-to-genomic alignments. *Genome Res* **11**, 1952–1957.

Wheeler, D. L., Church, D. M., Lash, A. E. *et al.* (2002). Database resources of the National Center for Biotechnology Information: (2002) update. *Nucleic Acids Res* **30**, 13–16.

Wingender, E., Chen, X., Fricke, E. *et al.* (2001). The TRANSFAC system on gene expression regulation. *Nucleic Acids Res* **29**, 281–283.

Zhang, C. T. and Wang, J. (2000). Recognition of protein coding genes in the yeast genome at better than 95 per cent accuracy based on the Z curve. *Nucleic Acids Res* **28**, 2804–2814.

5

Finding, Delineating and Analysing Genes

Christopher Southan[1] and Michael R. Barnes[2]

[1] *Global Compound Sciences, AstraZeneca R&D Mölndal, Sweden*

[2] *Bioinformatics, GlaxoSmithKline Pharmaceuticals, Harlow, Essex, UK*

5.1 Introduction

This chapter will describe ways to interrogate the human genome with the results of genetic experiments in order to locate and delineate known genes. It will also describe the assessment of evidence for genes that do not yet have experimental support and some analytical choices that may reveal more about them. In addition to some general aspects of gene detection, some specific examples will be worked through in some detail. This illustrates technical subtleties that are not easy to capture at the overview level. A caveat needs to be added here that many roads lead to Rome. Some particular ways of hacking through the genome jungle are implicitly recommended by being used for the examples in this chapter. They will also be restricted to public databases and Web tools. These are the personal choices of the authors based on an assessment of their availability and utility. Other experts may propose alternative routes to the same information, using different public resources, locally downloaded datasets, Unix-based tools, commercial software or subscription databases.

Genetic investigations are concerned with discerning the complex relationships between genotype and phenotype. The statement that phenotype is determined by the biochemical consequences of gene expression is equally obvious. However, the reason for making this explicit is to recommend that those performing and interpreting genetic experiments may find it more useful to conceptualize the gene as a cascade

Bioinformatics for Geneticists, Second Edition. Edited by Michael R. Barnes
 ISBN 978-0-470-02619-9 (HB) ISBN 978-0-470-02620-5 (PB)

of evidence that connects DNA to a protein product rather than abstract ideas about what might constitute a gene locus. The idea of focusing on gene products also makes it easier to design experiments to verify predicted transcripts and proteins. It must also be remembered that many gene products are non-message RNA molecules, but they will not be covered in this chapter (see Chapter 14 for a detailed review of this area). Before describing the evidence used to classify gene products, we must define some of the terminology encountered in the literature and database descriptions. These are variously classified as known, unknown, hypothetical, model, predicted, virtual, or novel. There are no widely accepted definitions of these terms, but their usage in this chapter will be as follows. A known gene product is experimentally supported and would be expected to give close to a 100 per cent identity match to a unique genome location. The term 'unknown' is typically applied to gene products that are supported experimentally, but lack any detectable homology or experimentally determined function. The term 'predicted', also referred to as 'model' or 'hypothetical' by the NCBI, will be reserved for an mRNA or protein open reading frame (ORF) predicted from genomic DNA. Virtual mRNAs will refer to constructs assembled from overlapping expressed sequence tags (ESTs) that exceed the length of any single component. The term 'novel' has diminishing utility and will simply refer to a protein with no extended identity hits in the major protein databases.

5.2 Why learn to predict and analyse genes in the complete genome era?

Might we question at the outset of this chapter the need for the geneticist to learn the art of gene prediction and analysis? The answer to this question might sound a little equivocal. There are certainly plenty of public resources available which offer high-quality annotation and analysis of the gene complement of the human genome (Table 5.1). Where possible, it is worth using these resources, because the results are generally of high quality. However, there are caveats:

1. *Most gene models are automated and therefore many are incomplete.* By necessity, data on human genes are generated by automated analysis methods based on gene prediction and the combined evidence of existing mRNA, cDNA and EST data. Going a step further to curate a gene taking in all the evidence can reveal extra information, including weaker evidence that automated processes necessarily miss.

2. *Automating curation of splice variants is technically difficult.* Information on splice variants is particularly difficult to capture by automated efforts, especially if the splice variant is evidenced only by ESTs.

3. *Genes may be expressed only under very specific conditions.* Genes with tight regulatory mechanisms may be expressed transiently in very specific tissue locations, developmental stages or cellular conditions, causing their expression to be undetectable by standard methods.

Table 5.1 Useful resources for gene finding and analysis

Site description	URL
Genome-focused tools	
Ensembl	http://www.ensembl.org/
UCSC genome browser	http://genome.ucsc.edu/
Map Viewer at NCBI	http://www.ncbi.nlm.nih.gov/cgi-bin/Entrez/map_search
DAS – distributed annotation	http://biodas.org/
Gene/transcript-focused tools	
Entrez Gene	http://www.ncbi.nlm.nih.gov/entrez/query.fcgi?db=gene
Unigene EST clusters	http://www.ncbi.nlm.nih.gov/UniGene/
CCDS project	http://www.ncbi.nlm.nih.gov/projects/CCDS/
RefSeq at NCBI	http://www.ncbi.nlm.nih.gov/RefSeq/
TIGR Gene Index	(http://www.tigr.org/tigr-scripts/tgi/T_index. cgi?species=human
Protein-focused tools	
Proteome analysis at EBI	http://www.ebi.ac.uk/proteome/
Uniprot	http://www.uniprot.org
InterPro at EBI	http://www.ebi.ac.uk/interpro/
International Protein Index	http://www.ebi.ac.uk/IPI/IPIhelp.html
SWISS-2DPAGE database	http://ca.expasy.org/ch2d/
Gene-prediction tools	
GENEWISE at Sanger Institute	http://www.sanger.ac.uk/Software/Wise2/
GENSCAN at MIT	http://genes.mit.edu/GENSCAN.html
Fgenesh at Sanger Institute	http://genomic.sanger.ac.uk/gf/Help/fgenesh.html
Homology searching and analysis	
BLAST at NCBI	http://www.ncbi.nlm.nih.gov/BLAST/
BLAT at UCSC	http://genome.ucsc.edu/cgi-bin/hgBlat?command=start)
SSAHA at Sanger Institute	http://www.sanger.ac.uk/Software/analysis/SSAHA/
Miscellaneous gene analysis	
Expasy translation tool	http://ca.expasy.org/tools/dna.html
Derwent sequence patent databases	http://www.derwent.com/geneseq/index.html
MatchMiner (gene aliases)	http://discover.nci.nih.gov/matchminer
Google literature search portal	http://scholar.google.com/

4. *Genes might be expressed at vanishingly low levels.* For example, many G-protein-coupled receptors (GPCRs) are completely absent from EST data and cDNA libraries. Most GPCRs have been identified by the combination of gene prediction and homology-based searches. Genes with no known homologues and very low expression are likely to be absent from the current complement of human genes.

The importance of possessing a correct, complete gene model is entirely dependent on the use case of this information. It may be important for setting up a screen for variation in the gene to ensure that all exons are screened, including untranslated exons and alternatively spliced exons. This also follows through to selection of variants if genotyping of all potentially functional variants is considered important. Understanding the full complexity of a gene is also important for functional analysis of genetically associated variants; for example, there may be evidence to support the assertion that a SNP annotated in dbSNP as intronic might turn out to be exonic (and functional) if there is evidence to support an alternatively spliced exon. However, just to put this into perspective, in the vast majority of cases, this kind of gene analysis may not be important. For example, if a genetic experiment employs a haplotype tag-based approach to capture variation across a gene, the precise boundaries of the gene may not be important, unless unknown exons are outside the scope of the tag SNPs.

5.3 The evidence cascade for gene products

So what kinds of evidence need to be considered before we assess the likelihood of a stretch of genomic DNA giving rise to a gene product and what kinds of numbers can be assigned to these evidence levels? In the following section, we review these sources of information and give figures for these evidence levels, based on queries completed in June 2006.

The NCBI Entrez Gene database is probably the most comprehensive non-redundant source of known gene loci. There are currently 32 014 (excluding pseudogenes) human genes in Entrez Gene (www.ncbi.nlm.nih. gov/projects/Gene/gentrez_stats.cgi?SNGLTAX=9606) (Maglott *et al.*, 2005). These loci include protein-coding loci and also non-coding loci, such as micro-RNAs (see Chapter 14). If a gene locus is unknown or further evaluation of a known gene is needed to ensure that the gene and transcript model are as accurate as possible (e.g., to assess the impact of functional variation in the gene), the entire cascade of evidence for a gene and its products may need to be reviewed. We review each of these steps in the following sections.

5.3.1 Experimentally determined protein sequence

The most solid evidence of a gene is the experimental verification of the protein product by mass spectrometry and/or Edman sequencing. Although these techniques are commonly used to analyse proteins produced by heterologous expression *in vitro*, surprisingly few genes from *in vivo* or cell-line sources have been verified at this level. From the entire SP/TR collection of human proteins, only 420 are cross-referenced as having at least a fragment of their primary structure identified directly from a 2DPAGE experiment (http://ca.expasy.org/ch2d/) (Hoogland *et al.*, 2004). Numerous mass spectrometry-based identifications and peptide sequences from human proteins are reported in the literature, but few of these data have been formally submitted to the public databases, and therefore they have not been captured by SwissProt or other secondary databases (see Webster and Oxley, 2005, for a review of these methods). However, even this most direct of gene product verifications is rarely sufficient to confirm the entire ORF. For example, secreted proteins are characterized by the removal of signal peptides and frequent C-terminal processing. This precludes defining the N and C translation termini by protein chemical means.

5.3.2 Messenger RNA (mRNA) databases

The next level down in the evidence cascade is, of course, an extended mRNA. There are a bewildering range of sources of mRNA sequences, ranging from experimentally verified sequences to *in silico* predicted mRNA with no other supporting evidence. Most of these sources can be viewed in the three main genome browsers. Figure 5.1 displays some of the sources discussed below across the BACE1 gene in the UCSC genome browser. Starting at the top of the mRNA evidence cascade, the consensus CDS (CCDS) project (www.ncbi.nlm.nih.gov/projects/CCDS/CcdsBrowse.cgi) contains the most stable group of transcripts, which are now completely stabilized between the NCBI, UCSC and Ensembl genome viewers (Figure 5.1). In June 2006, there were 14 795 transcripts in the CCDS database. The NCBI RefSeq collection is probably the next most reliable link in the cascade (www.ncbi.nlm.nih.gov/RefSeq/). Refseq currently lists 49 565 human transcripts, including transcript variants (Pruitt *et al.*, 2005). Although this collection attempts to provide a non-redundant snapshot of gene transcription, it must be remembered that they are not all full-length transcripts, nor do they represent all known splice variants. If the databases do not contain an extended mRNA, the assembly of overlapping and/or clone-end clustered ESTs can be considered as a virtual mRNA (Schuler, 1997). The ESTs have the additional utility that many of them can be ordered as clones. Alternatively, the virtual consensus sequence, backed up by comparisons to the genomic DNA, can be used for PCR cloning. ESTs are one of the most prolific sources of evidence of mRNA, which makes them one of the commonest sources of supporting evidence for a transcript, especially if they include a plausible splice junction and are derived from multiple clones from different tissue cDNA libraries. There are a few sources of pre-assembled EST clusters; the NCBI Unigene database (www.ncbi.nlm.nih.gov/UniGene/) currently contains 86 806 human EST clusters. Another resource, the TIGR human gene index, contains over 200 000 tentative human consensus sequences (THCs). These are a useful source of pre-assembled virtual mRNA (www.tigr.org/tigr-scripts/tgi/T_index.cgi?species=human) (Quackenbush *et al.*, 2001). Both Unigene and the TIGR Gene Index can also be viewed as UCSC genome browser tracks. The use of unspliced ESTs as evidence for a transcribed gene is generally unreliable, as they can arise from genomic contamination of cDNA libraries. However, human EST-to-genome matches for exon detection can be further supported where orthologous ESTs from other vertebrates, such as mouse or rat, match uniquely in the same section of the genome. If an assembly of mouse ESTs is consistent with a human gene model, the existence of an orthologous human

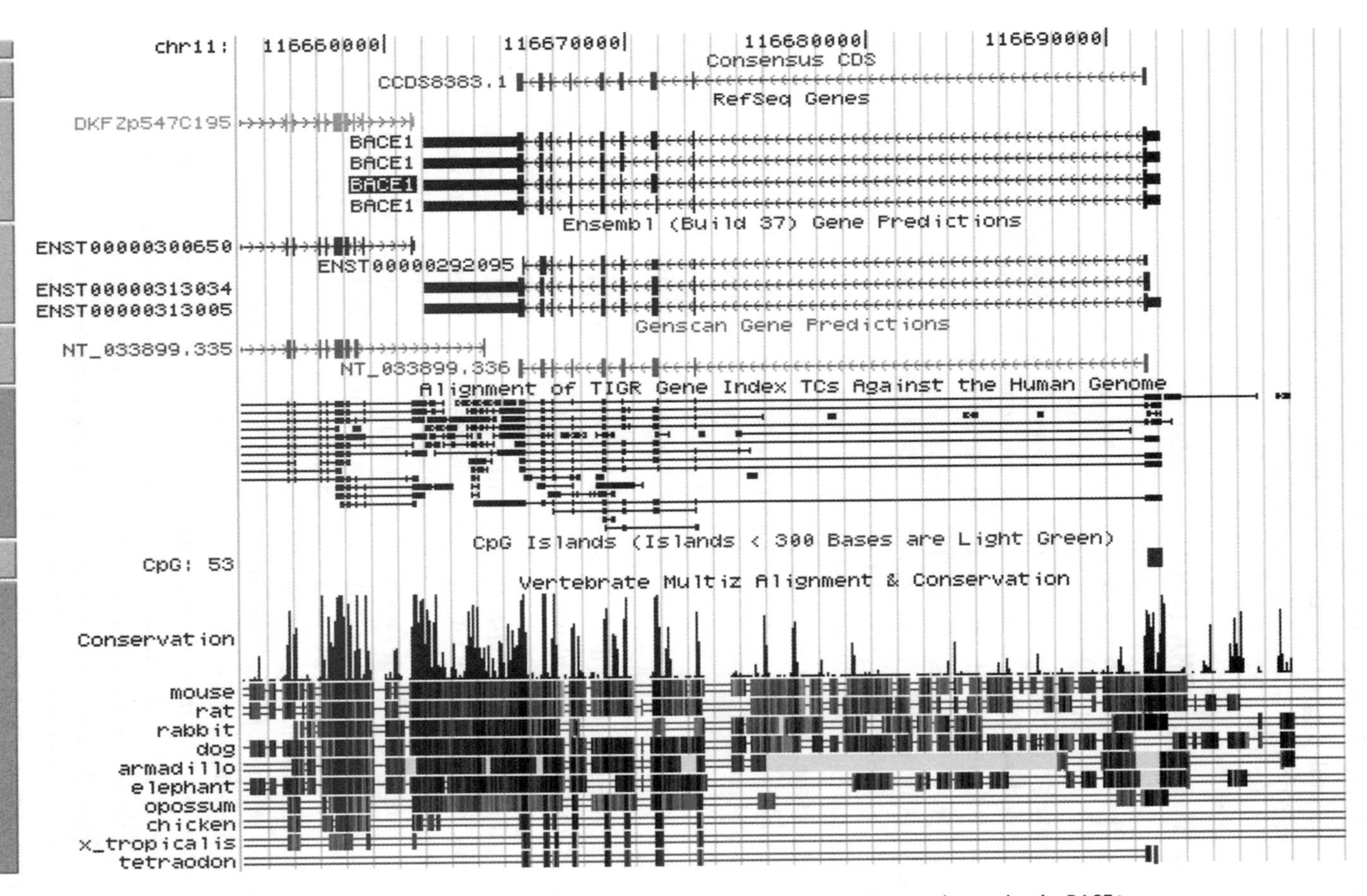

Figure 5.1 A UCSC view of the evidence cascade for genes, transcripts and proteins in BACE1

transcript is strongly suggested. Support can also be provided by evidence of conservation across vertebrate genomes; this can be assessed quite rapidly with the 'conservation' track in the UCSC genome browser.

5.3.3 Protein databases

The protein databases occupy the centre of the evidence cascade for gene products. Those mRNAs or full-length cDNAs that contain a large ORF tend to be viewed as potential gene transcripts even if they are not full length and/or there is ambiguity about the choice of potential initiating methionines. However, the fact that the protein databases have now expanded to include human ORFs derived solely from genomic predictions (described in the next section) means that the evidence supporting them as gene products becomes circular. The highest curation level is provided by SwissProt sequences, a manually curated dataset from the Human Proteomics Initiative (HPI) (http://ca.expasy.org/sprot/hpi/hpi_stat.html). The June 2006 SwissProt release comprised 14 094 unique gene products and 7707 isoforms arising from alternative promoter use or alternative splicing (O'Donovan *et al.*, 2001). The next highest curation level is provided by UniProt (SP/TR), an automated dataset combined with the manually curated SwissProt. The total for human proteins in June 2006 was 38382, including splice variants (http://www.ebi.ac.uk/integr8/OrganismStatsAction.do?orgProteomeId=25). The International Protein Index (IPI) maintains a database of cross-references between the data sources UniProt, RefSeq and Ensembl (Kersey *et al.*, 2004). This provides a minimally redundant yet maximally complete set of human proteins with one sequence per transcript (http://www.ebi.ac.uk/IPI/IPIhuman.html). The June 2006 release contains 60 090 protein sequences, but this includes a number of predicted ORFs from transcript models which are not supported by mRNAs.

5.3.4 *Ab initio* gene prediction

The next level of evidence can be classified as genomic prediction; that is, where a cDNA, a translated ORF and a plausible gene splice pattern can be predicted from a stretch of genomic DNA (Burge and Karlin, 1997). This is done after filtration of repeats, which can be considered as another link in the evidence chain. A very high local repeat density certainly suggests where exons are unlikely, but the converse is not true; that is, the absence of repeats does not prove the presence of genes. The shortcomings of *ab initio* gene prediction have been pointed out, but the geneticist should at least be aware of possible false-positives and false-negatives (Guigo *et al.*, 2000). The Ensembl statistics of the ratio of genes predicted by Genscan to genes with a high evidence-supported threshold is currently 3.2:1 (http://www.ensembl.org/Homo_sapiens/index.html). Although this clearly represents over-prediction, some may be 'genes-in-waiting' that more accumulated evidence may verify, as by the cloning of an extended mRNA. Looking for a consensus or at least common exons from a number of gene prediction programs with different underlying gene model assumptions can strengthen this type of evidence, but this can become a circular argument where the programs are both trained and benchmarked with known genes. The most effective way of filtering down genomic predictions without experimental evidence is homology support; that is, the predicted protein shows extended similarity with other proteins. This is described in detail in the Ensembl documentation, but, in essence, all possible protein similarity sections from translated DNA are identified and used to build homology-supported gene predictions by Genewise (Birney and Durbin, 2000). The advantage of gene detection by homology is that the entirety of protein sequence space can be used. The caveat is that predicted gene products with low similarity to extant proteins would be discarded in this filter, although the entire set of Genscan predictions are preserved for searching in Ensembl and can also be displayed at UCSC.

5.3.5 Comparative genomics

The next link in the evidence chain is a special case of the similarity principle, but in this case utilizing comparisons between the genomes of other vertebrates, many of which are now complete or close to completion such as dog, mouse, chicken, frog and fish. The Ensembl and UCSC sites now display at least 16 vertebrate genome assemblies; these can either be viewed directly or aligned against the human genome. Cross-species data can be assessed at several levels. Comparison of DNA similarity between (vertebrate) genomes is termed 'phylogenetic footprinting' (Susens and Borgmeyer, 2001; see Chapter 6 for a detailed review of this approach). This is a valuable technique for the detection of vertebrate genes and conserved regulatory regions, but the problem for gene product detection is that this is too sensitive; that is, mouse/human syntenic regions have many conserved similarity 'patches' outside the boundaries of known exons. Conserved regions are likely to be important for functions not yet understood, but it is difficult to discriminate superficially between potential coding and potential regulatory regions. Often these regions need to be identified in a relatively high throughput manner to allow primer design or SNP selection. The ECR browser (Ovcharenko *et al.*, 2004) has been specifically designed for visualizing and accessing evolutionarily conserved region (ECR) data from comparisons of multiple vertebrate genomes, and it suits the needs of geneticists very well (Figure 5.2). The ECR browser annotates ECRs across a query region, using a user-configurable set of alignment conditions. ECR and known exon-annotated DNA sequences corresponding to the entire genomic region can be displayed. In Figure 5.2, there appear to be several mouse ECRs in intron 1 of BACE1. These might correspond to alternative exons or regulatory regions. There is a strong argument to support investigations across these regions, such as SNP selection for genotyping or screening

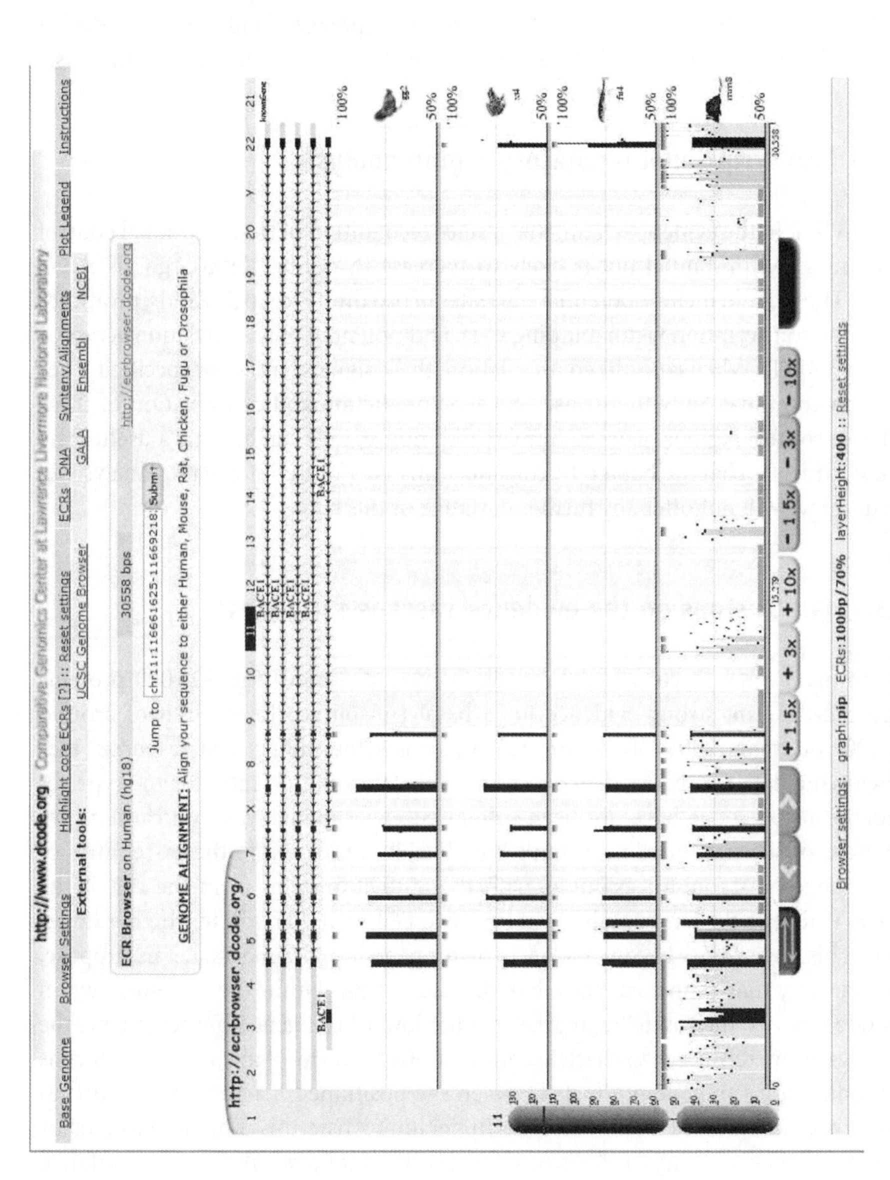

Figure 5.2 The EGR comparative genome browser

for polymorphisms. The ECR browser expedites this process, by providing access to the sequences of ECRs detected and a list of their positions in the displayed region. A detailed ECR description page contains ECR sequences from both species in any pairwise comparison, and a display of the underlying DNA sequence alignment. In addition, sequence characteristics are combined with pipelined links to primer design tools and the rVista program for transcription factor-binding site (TFBS) analysis.

5.3.6 Transcriptional regulatory region analysis

The last link in the evidence chain, the *in silico* recognition of transcriptional control regions, is circumstantial but is likely to increase in utility (Kel-Margoulis *et al.*, 2002). These could include potential start sites in proximity to CpG islands, promoter elements, transcription factor-binding sites, and potential polyadenylation acceptor sites in 3′ UTR. When considered in isolation, these signals have poor specificity, but taken in combination with a consensus gene prediction and conservation of these putative control regions between human and mouse, they can become a useful part of the evidence chain. Chapter 12 considers this area of bioinformatics analysis in detail, so we will not offer any further coverage of this here.

5.3.7 Conclusions on the evidence cascade for genes

In summary, there is currently direct experimental support for ~15 000 protein-coding genes and strong evidence for a basal (unspliced) lower limit of around 25 000 (Southan, 2004). In Figure 5.3 we review the public data resources that provide the data for this evidence cascade. The confirmation rates for the types of evidence listed above have not been calibrated experimentally, so we cannot give any kind of scoring function to rank gene likelihood. Going to the extremities of the evidence cascade, for example, with the 60 090 proteins from the IPI or the 86 806 UniGene clusters containing at least two ESTs, would result in a higher upper limit. This uncertainty becomes a key issue for genetic experiments. Let us suppose, for example, that a linkage study has defined a trait within the genomic region bounded by two microsatellite markers. If the lower limit gene number is true, the investigator merely needs to check the annotations from any of the three gene portals to produce a list of gene products between the positioned markers from which to choose candidates for further work. If the upper limit is true, this approach has a major limitation because many of the genes between the markers will not be annotated. However, the different levels of gene evidence described above can be visualized in the display tracks of the genome viewers. Consideration of the evidence will enable the geneticist to decide what experiments need to be designed to confirm potential novel gene products. An example of working through this evidence is given in the examples below.

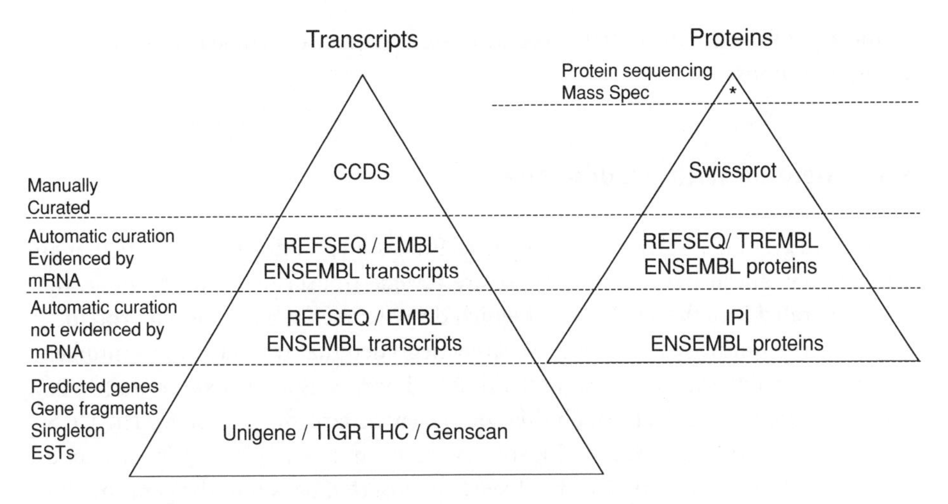

Figure 5.3 The evidence cascade for genes, transcripts and proteins

5.4 Dealing with the complexities of gene models

One of the conclusions that is clear from the analysis of the human genome sequence is that gene models are not a simple cascade of a defined gene locus → a single mRNA species → a single protein. This rarely describes the complex relationship between the genome and its products. Attempts to fit transcript data into this kind of view highlight a number of inconvenient 'grey' areas that just might mean the difference between success and failure in a genetic experiment.

5.4.1 Delineating the 5′ and 3′ extent of an mRNA transcript

The first of these grey areas is the delineation of the extreme 5′ and 3′ ends of the mRNA transcripts (Pesole *et al.*, 2002; Suzuki *et al.*, 2002). The fact that many mRNAs are labelled as partial is testimony to the difficulty of finding library inserts that are complete at the 5′ end. In many cases, the mRNAs are considered finished when a plausible ORF has been delineated. However, very few cDNAs are full-length in that they have been 'walked out' to determine the true 5′-most initiation of transcription in the 5′ UTR. The same problem applies to the UTR at the 3′ end. There may be substantial stretches of 3′ UTR extending downstream of the first polyadenylation position at which further cloning attempts have ceased. If we overlook this UTR sequence in a genetics experiment, we might be overlooking functionally important micro-RNA-binding sites with a key role in gene regulation (see Chapter 14). The problem is compounded by the poor performance of gene-prediction programmes for 5′ and 3′ ends. The first step toward resolving the uncertainty about the extremities of a transcript is to survey the coverage of all available cDNA sequences, whether nominally full-length or partial, ESTs and patent sequences. These can often extend the UTR sections.

5.4.2 Dealing with pseudogenes

The second grey area concerns pseudogenes. Processed pseudogenes in particular are common (Shemesh *et al.*, 2006). These are reverse-transcribed mRNA copies that have integrated into the genome, but which do not code for a functional protein. In some cases, genomic sequence is so severely degraded that transcription is unlikely; however, in many other cases, transcription still occurs. Entrez Gene contains 7202 human pseudogene loci, although this count may be far from complete. The UCSC presents several tracks with pseudogene information; one track, 'retroposed genes', shows 16 731 processed mRNAs that have been inserted back into the genome since the mouse/human split. These can be either functional genes that have acquired a promoter from a neighbouring gene, non-functional pseudogenes or transcribed pseudogenes.

5.4.3 Dealing with gene-product heterogeneity

The third grey area is gene-product heterogeneity. In some cases, there may be alternative upstream initiation methionines or alternatively spliced exons in the 5′ UTR. The causes of 3′ heterogeneity include variations in the pattern of intron splicing from a pre-mRNA, as well as alternative polyadenylation positions inside the 3′ UTR. Potential for such gene-product heterogeneity can often be rapidly evaluated with genome viewers, and evidence of alternatively spliced exons or alternative first exons may be identified in spliced EST data or after comparison with other vertebrate mRNAs or genomic regions. Unfortunately, getting beyond this potential evidence to a robust gene model can be one of the most complex and confounding bioinformatics analysis tasks, so again it is important to determine what level of detail is required for the task in hand. For example, a complete transcript model may not be required to evaluate the impact of a SNP – in this case, the exon and its immediately preceding exon are all that are needed to determine a possible coding change.

5.4.4 Dealing with overlapping and embedded genes

The fourth grey area concerns overlapping genes. As genomic annotation proceeds, we can find more examples of this from both gene products reading from opposite strands and same-strand genes in close proximity (see Makalowska *et al.*, 2005, for a review). Embedded genes are another consideration. These are small, often intronless expressed genes (possibly with similar origins to pseudogenes) that are located in the intronic regions of 'host' genes. Transcription of embedded genes may often be driven by the promoter of the host gene, so this can make the determination of function (e.g., expression) of an embedded gene and its host very difficult to separate. A good example of an embedded gene is CHML in intron 1 of the OPN3 gene (Halford *et al.*, 2001).

5.5 Locating known genes in the human genome

Genes can be located by one of the following; a section of raw sequence data, a primary accession number, a secondary accession number, a similarity search, a gene product name, or a set of genome coordinates. Each of these has advantages and disadvantages, and, although the main genome portals are generally consistent, they may not give the same answers in every case. Bearing in mind that only the first two of these gene location methods are based on stable (almost) unambiguous information, it is better to use at least two ways to define and store the results: for example, a section of raw sequence and a gene name, or a primary accession number and a set of genome coordinates. The BACE1 gene will be used as an example of a known gene to locate. The potential complexity of this task is illustrated if we view the Ensembl gene report for BACE1, which is often a good place to start to get a feel for the data relating to a gene (http://www.ensembl.org/Homo_sapiens/geneview?gene=ENSG00000186318).

5.5.1 Using raw sequence data to locate genes

The availability of the human genome sequence means that most features can now be unambiguously located in the genome with as little as 100 bp of sequence. This means that storing a sequence string, preferably with a longer sequence context of 200–1000 bp, is a useful, future-proofed, method of locking-on to a genomic location. Sequences are more or less immune to the vagaries of shifting secondary accession numbers, naming ambiguities or GP sequence finishing that can change the genomic coordinates. Performing nucleotide searches against the genome using tools such as BLAT (UCSC), SAHA (Ensembl), or BLAST (NCBI) means that sequence matches can be quickly located. The disadvantage for raw sequence is that it has to be stored in its entirety, it may contain errors, it needs the operation of a similarity search to be located, and similarity matches across repeat-containing sections or duplicated regions of the genome need close inspection to sort out. This can be a particular problem for sequence-tagged sites (STSs) and SNPs if the GP match is in the region of 98 per cent to 95 per cent identity. Within this range, it is difficult to discriminate technical sequencing errors from multiple genomic locations, assembly duplication errors or even copy number polymorphisms. The genome portals capture mRNA entries for most gene products; however, because of the thin annotation, they do not capture sequences from the patent divisions of GenBank. An NCBI BLAST search of the gbPAT database with any BACE1 mRNA returns hundreds of high-identity DNA matches (http://www.ncbi.nlm.nih.gov/BLAST/Blast.cgi). These are clearly mRNAs that could be usefully compared with all other mRNA sequences for polymorphisms, splice variants or UTR differences. However, users should be aware that not only are some of these entries identical versions of the same sequence derived from multiple claims in the patent documents but also they may be identical to a public accession number if the authors and inventors are from the same institution. Another possible reason for using raw sequence data for gene-product checking is that all secondary databases suffer from the snapshot effect whereby updates lag behind the content of the primary databases. For example, the SNP or EST assignments made for BACE in the secondary databases (see below) could be checked by BLAST searches against the updates of dbSNP or dbEST (the latest EST data need to be searched in 'month' as well as dbEST).

5.5.2 Using primary accession numbers

A primary accession number (or primary database record) is assigned to a DNA or protein sequence or other genomic entity when it is first entered in a database. This accession should be related to a specific experiment, and it should contain contact details for the investigator that carried out the experiment. Primary sequences should be treated with some care, especially if they are particularly old, as they may often contain sequencing errors or possibly polymorphisms. These are usually corrected (or annotated as polymorphic) based on a consensus alignment of all primary accessions in the secondary sequence record, such as RefSeq. In the case of BACE1, AF204943 is one of the primary accessions for this gene. Because these uniquely define stretches of sequence, they are stable except where genomic DNA, and occasionally mRNAs, undergo version changes. They can be used in any of the major genome query portals to go directly to a genomic location. The disadvantage is redundancy for mRNAs, short sequence context for some STSs, and both redundancy and large multigene sequence tracts for genomic DNA, and very recent accessions may not be indexed in genome builds. If the query fails to connect to a genome feature, the sequences can be searched as raw sequence. In the BACE1 example, interrogating the UCSC browser with BACE1 retrieves three primary Genbank IDs. Users need to be aware that although an mRNA accession number can provide a specific route into the genome, the variable number of links to the genome portals is related to their update frequency.

5.5.3 Using secondary accession numbers

Secondary database records do not directly relate to a specific sequence submission; instead, they usually represent a consensus view of the primary data to capture representative versions of each splice variant or alternative initiation. This may include extending the sequence based on an alignment of all primary sequences. Some examples of secondary databases are RefSeq, SwissProt and RefSNP. If we view BACE1 in Ensembl or the UCSC, there are several secondary accession numbers that designate BACE1 mRNAs and proteins. Although secondary accession records have the advantage of capturing a consensus across all the available information, they do have some problems in use for gene localization. Firstly, the sequence linked to a secondary accession number may not be stable, as new information arises, and these can be merged, split or retired. Recent improvements mean that retired IDs are now linked to new IDs and version change histories but this can be confusing. However, SwissProt and RefSeqNP protein IDs can be considered stable even if there can be minor changes in the linked sequence.

5.5.4 Using gene names and symbols

The whole area of gene symbols and aliasing can be fraught with confusion, and it is probably fair to say that this is one of the commonest sources of error in bioinformatics searches. If we take BACE1 as an example again, there are four synonyms or aliases (BACE, ASP2, HSPC104 and KIAA1149). Using ASP2 as a search query for the UCSC browser, we retrieve BACE1 but also ASP2 (aspartate aminotransferase 2), a completely unrelated human gene. This illustrates the problem when gene products are given different names by different authors. A good tool for checking gene aliases either individually or in batches is MatchMiner (http://discover.nci.nih.gov/matchminer/). In an attempt to avoid this confusion, the Human Gene Nomenclature Committee is trying to establish official HUGO gene symbols for all human genes. Where possible, these should always be used when referring to a gene, and many journals now require the use of these symbols for publications. It is possible to check HUGO gene symbols at the organization website (http://www.gene.ucl.ac.uk/nomenclature/). The complexity of the aliases for just one gene product makes it clear that any gene name lists, such as candidate genes to be screened for mutations, should be backed up by accession numbers, raw sequence or chromosome locations. It also illustrates the need to cross-check aliases and their spellings when attempting a comprehensive literature search on a particular gene product. The formal sequence-literature links that can be followed in Entrez Gene or SwissProt are not comprehensive because they are dependent on the journal-author-database system that usually only makes these links explicit for a new accession number. Much important literature remains outside this system. Review articles, for example, do not typically include primary accession numbers when describing genes, so the specificity of literature searches remains dependent on the name links. Information trawling with gene names can also be done with the standard Internet search portal. Putting the term 'beta-site app cleaving enzyme' into the Google Scholar literature mining engine gave 408 hits (http://scholar.google.com/). The listing included duplicates but very few false positives.

5.5.5 Using genome coordinates

Since the adoption of defined releases of human genome assemblies, this method of genomic location has become more reliable, but users are strongly advised to check the version of the genome assembly that their coordinates are derived from. At the time of writing (June 2006), the May 2004 (NCBI35) human genome assembly was still in most frequent use by the majority of applications. A March 2006 (NCBI36) release is just beginning to be incorporated into the UCSC and pre-Ensembl servers. This creates a potential problem, of which the user must always be aware. When genomic coordinates of different data types are compared, it is critical to ensure that they are both based on the same NCBI genome build. Considering this, it is good practice to record the genome build with any data set containing genomic coordinates. Data mapped against different assemblies can be compared by the UCSC Batch coordinate conversion tool(http://genome.ucsc.edu/cgi-bin/hgLiftOver).

5.6 Genome portal inspection

From the descriptions above, it should be possible to locate any known gene or genetic marker such as an STS or a SNP. Descriptions of the genome viewer features for Ensembl, UCSC and NCBI are covered in detail in Chapter 4. However, we give one specific example below (Figure 5.4) because it effectively illustrates some of the issues in gene analysis. The UCSC genome browser view of the 3′ portion of the BACE1 gene (Figure 5.4) shows that there are significant differences in the lengths of the 3′ ends of some of the primary mRNA records. Clearly, AF201468 (5878 bp) and AB032975 (5814 bp) are the longest reads, but, in fact, AB032975 is labelled as a partial coding sequence because of what may be a sequencing error at the 5′ end. A detailed analysis of the 3′ ends by EST and mRNA distribution profiles indicates that the different UTR lengths in this case arise not from incomplete cloning but from three alternative polyadenylation positions (Southan, 2001). Further heterogeneity is illustrated by three splice variants affecting exons 3 and 4 (the furthest exons on the right of Figure 5.4). The representative mRNAs are AB050436, AB050437 and AB050438. There is also an alternative protein reading frame from AF161367, a partial mRNA cloned from $CD34^+$ stem cells. Opening the spliced EST tracks in the viewer shows individual ESTs corresponding to these splice forms and others that correspond to potentially novel exons. This suggests the possibility of further BACE1 splice forms, but to provide further evidence to support this, the EST to genome alignment would need to be inspected for the presence of a canonical splice site (see Chapter 11). Beyond this, experimental verification of this variant would be recommended.

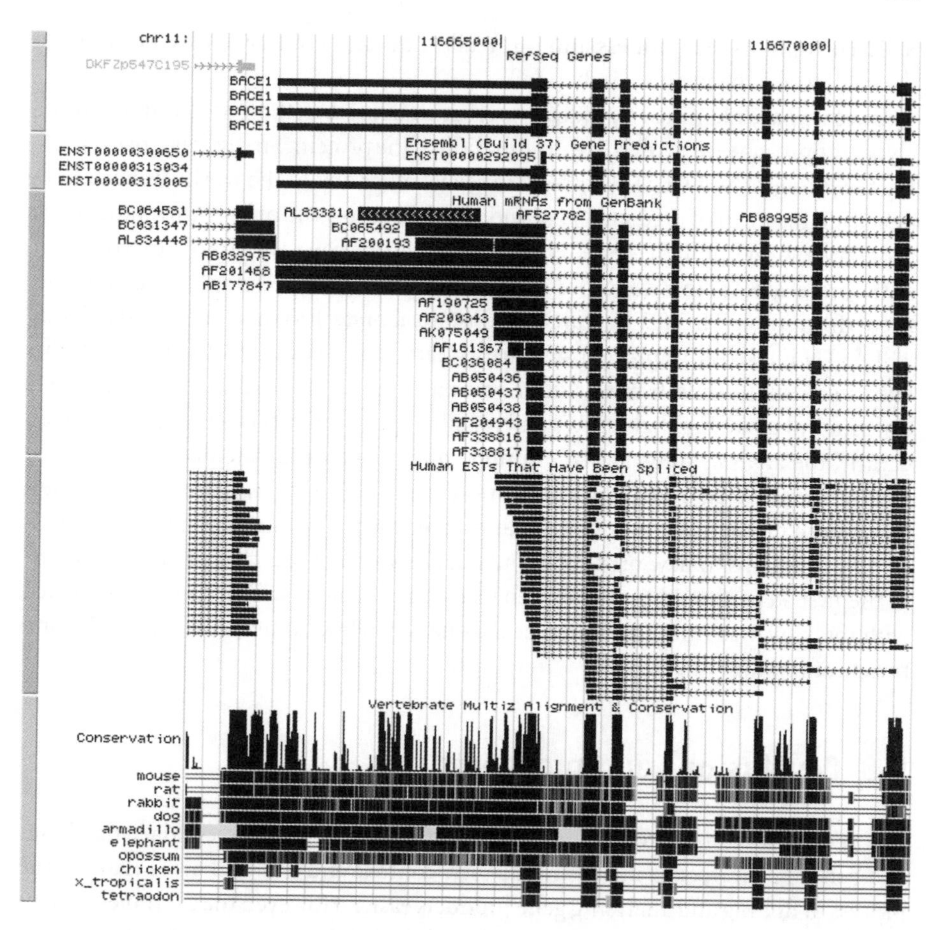

Figure 5.4 Reviewing the evidence cascade in the UTR region of BACE1

5.7 Analysing novel genes

In many cases, experimental results will locate a genomic region where there are no annotated genes. Inspection of all three genome browsers might indicate possible novel gene products with a variety of supporting evidence. This evidence might extend from *ab initio* gene predictions through evidence of genomic conservation to EST evidence. Building up a robust package of evidence around an unannotated gene in genomic sequence can be a bit of an art, and we discuss this further in Chapter 9. Again it is worth considering what level of detail is required for genetic analysis. For genotyping across a gene, a well-selected set of tag SNPs should suffice. Even if an association is localized to a SNP in a possible novel gene, it may not be necessary to build a full gene model. If there is clear evidence of exons, it may be possible to obtain a multiframe translation across several exons to evaluate the impact of the

SNP. The options are limitless and it is really a question of how far we need to take the analysis for the purpose in hand.

Once we have identified a putative novel protein *in silico*, there are a number of analytical approaches to investigate the function of the protein. Firstly, it is worth checking the Ensembl and UCSC browsers to see whether any of the work has been done already in the form of predicted gene models. Both browsers offer highly informative protein views. Without the benefit of this information, you really are out on your own, so the first step is to cross-check for reading frame consistency and species orthologues by performing a TBLASTN search against vertebrate EST and mRNA databases. TBLASTN is a sensitive protein similarity query used against DNA databases translated in all six reading frames (http://www.ncbi.nlm.nih.gov/BLAST/). The results can rapidly help to highlight ORFs and possible splice variants in a predicted protein. Clearly, the analysis of what, for example, might be a candidate disease-associated gene has to move on from the identification of an ORF to the assignment of function that is both mechanistically plausible and experimentally testable. The subject of assigning functions to new proteins is outside the scope of this chapter. However, one of the first steps should be a comprehensive motif analysis. This can be completed with tools such as InterPro Scan (see Southan, 2000; Kriventseva *et al.*, 2001, for a review of this area).

5.8 Conclusions and prospects

The geneticist is in the fortunate position of having access to secondary databases and genomic viewers of increasing quality, content and utility. This is making the process of finding and analysing gene products easier. However, the examples used in this chapter also show that there are many subtle details in genomic annotation, and the implications of these will take some time to unravel. This requires comprehensive inspection and may ultimately need experimental verification. The expansion of Web-linked interoperativity and interrogation tools means that new options will already be available by the time this book is in print. One consequence of these advances could be the perception of a diminished necessity to perform bioinformatic analysis. Although this is true in the sense that secondary database include an increasing amount of 'precooked' bioinformatic data, there is a paradox in that the more sophisticated the public annotation becomes, the more important it is to understand the underlying principles. For example, it is important to be able to distinguish between gene products defined by *in vitro* data or only by *in silico* prediction.

So what of the future? There are a few developments worth highlighting, all of which are covered in detail in later chapters. The first is that the combination of increasing transcript coverage and the availability of multiple complete vertebrate genomes will further diminish the uncertainty limits of gene numbers. Secondly, with the HapMap in hand and the ability to carry out genome-wide genetic association scans across a comprehensive set of gene products, we should be in a better position than ever to determine which genes are associated with common diseases. It is fair to say that we still have no idea what many thousands of genes in the human genome actually *do*. This presents us with a prospect that is both exciting and daunting – do these genes play a role in human diseases, and, if so, what is this role? Answers to these questions may take a while longer in coming than the initial associations, but the end result might well be the illumination of entire new pathways with direct relevance to human disease.

References

Birney, E. and Durbin, R. (2000). Using GeneWise in the *Drosophila* annotation experiment. *Genome Res* **10**, 547–548.

Burge, C. and Karlin, S. (1997). Prediction of complete gene structures in human genomic DNA. *J Mol Biol* **268**, 78–94.

Guigo, R., Agarwal, P., Abril, J. F. *et al.* (2000). An assessment of gene prediction accuracy in large DNA sequences. *Genome Res* **10**, 1631–1642.

Halford, S., Bellingham, J., Ocaka, L. *et al.* (2001). Assignment of panopsin (OPN3) to human chromosome band 1q43 by in situ hybridization and somatic cell hybrids. *Cytogenet Cell Genet* **95**(3-4), 234–235.

Hoogland, C., Mostaguir, K., Sanchez, J.-C. *et al.* (2004). SWISS-2DPAGE, ten years later. *Proteomics* **4**(8), 2352–2356.

Kel-Margoulis, O. V., Kel, A. E., Reuter, I. *et al.* (2002). TRANSCompel: a database on composite regulatory elements in eukaryotic genes. *Nucleic Acids Res* **30**, 332–334.

Kersey, P. J., Duarte, J., Williams, A. *et al.* (2004) The International Protein Index: an integrated database for proteomics experiments. *Proteomics* **4**(7), 1985–1988.

Kriventseva, E. V., Biswas, M. and Apweiler, R. (2001). Clustering and analysis of protein families. *Curr Opin Struct Biol* **11**, 334–339.

O'Donovan, C., Apweiler, R. and Bairoch, A. (2001). The human proteomics initiative (HPI). *Trends Biotechnol* **19**, 178–181.

Maglott, D., Ostell, J., Pruitt, K. D. *et al.* (2005). Entrez Gene: gene-centered information at NCBI. *Nucleic Acids Res* 33(Database issue), D54–58.

Makalowska, I., Lin, C. F. and Makalowski, W. (2005). Overlapping genes in vertebrate genomes. *Comput Biol Chem* **29**(1), 1–12.

Ovcharenko, I., Nobrega, M. A., Loots, G. G. *et al.* (2004). ECR browser: a tool for visualizing and accessing data from comparisons of multiple vertebrate genomes. *Nucleic Acids Res* **32**(Web Server issue), W280–286.

Pesole, G., Liuni, S., Grillo, G. *et al.* (2002). UTRdb and UTRsite: specialized databases of sequences and functional elements of 5′ and 3′ untranslated regions of eukaryotic mRNAs. Update 2002. *Nucleic Acids Res* **30**, 335–340.

Pruitt, K. D., Tatusova, T. and Maglott, D. R. (2005). NCBI Reference Sequence (RefSeq): a curated non-redundant sequence database of genomes, transcripts and proteins. *Nucleic Acids Res* **33**(1), D501–D504.

Quackenbush, J., Cho, J., Lee, D. *et al.* (2001). The TIGR Gene Indices: analysis of gene transcript sequences in highly sampled eukaryotic species. *Nucleic Acids Res* **29**, 159–164.

Schuler, G. D. (1997). Pieces of the puzzle: expressed sequence tags and the catalog of human genes. *J Mol Med* **75**, 694–698.

Shemesh, R., Novik, A., Edelheit, S. *et al.* (2006). Genomic fossils as a snapshot of the human transcriptome. *Proc Natl Acad Sci U S A*, **103**, 1364–1369.

Southan, C. (2000). Website review: interPro (the integrated resource of protein domains and functional sites). *Yeast* **17**, 327–334.

Southan, C. (2001). A genomic perspective on human proteases as drug targets. *Drug Discov Today* **6**, 681–688.

Southan, C. (2004). Has the yo-yo stopped? An assessment of human protein-coding gene number. *Proteomics* **4**(6), 1712–26.

Susens, U. and Borgmeyer, U. (2001). Genomic structure of the gene for mouse germ-cell nuclear factor (GCNF). II. Comparison with the genomic structure of the human GCNF gene. *Genome Biol* **2**, research no. 0017.

Suzuki, Y., Yamashita, R., Nakai, K. *et al.* (2002). DBTSS: Database of human transcriptional start sites and full-length cDNAs. *Nucleic Acids Res* **30**, 328–331.

Webster, J. and Oxley, D. (2005). Peptide mass fingerprinting: protein identification using MALDI-TOF mass spectrometry. *Methods Mol Biol* **310**, 227–240.

6

Comparative Genomics

Martin S. Taylor and Richard R. Copley

Wellcome Trust Centre for Human Genetics, University of Oxford, Oxford, UK

6.1 Introduction

The geneticist is typically concerned with the investigation of genetic variation between individuals of a population. In contrast, much of comparative genomics is based on the differences that have accumulated between species. More specifically, comparative genomics uses the signal of past selection as a highly sensitive assay for function in genome sequences. Unlike experimental approaches, it does not require a prior hypothesis of that function. The realization that comparative sequence analysis is crucial to understanding the functions encoded in the human and other genomes is driving a major comparative sequencing effort. The fruits of this labour are a rapidly expanding number of whole-genome sequences and new computational methods to analyse these data in efficient and meaningful ways.

For the geneticist, one of the great attractions of comparative genomics is the potential to focus in on functional polymorphisms from a list of tens or possibly hundreds of candidates based on genetic evidence alone. Comparative genomics can also provide clues to the function of a polymorphic site and can lead to the generation of experimentally testable hypotheses for the investigation of that function. By including DNA sequence from genetically tractable model organisms in comparative analyses, this approach can also provide a route into model organism studies through the identification of orthologous sites in the target genome.

In this chapter, we introduce the concepts and techniques of comparative genomics; in doing so, we also venture into the topics of sequence alignment and phylogenetics. In general terms, the approaches we describe can be applied to sequence data from any collection of organisms, but our emphasis here is primarily on

Bioinformatics for Geneticists, Second Edition. Edited by Michael R. Barnes
 ISBN 978-0-470-02619-9 (HB) ISBN 978-0-470-02620-5 (PB)

questions of relevance to human genetics. We begin, in Section 6.2 by presenting an overview of genome structure and content, providing a context for the subsequent discussions. We then introduce the concepts of natural selection, the neutral theory of evolution, homology and phylogenetic distance that underlie comparative genomic analyses. In Section 6.4, we consider the types of questions that can be addressed and the strategies that can be employed to address them. We also consider the availability and accuracy of genomic sequence data. With Section 6.5, we introduce the three main technical challenges of comparative genomic sequence analysis: genomic sequence alignment, the visualization of sequence relationships and detecting the signal of selection. We review the methods employed to meet these challenges and discuss the most popular and the most promising new tools. In Section 6.6, we illustrate the utility of comparative genomic studies with recent applications that have given new insights into human biology. Finally, in Section 6.7, we highlight some resources that are likely to have a profound impact on future comparative genomic studies and identify future research challenges.

6.2 The Genomic landscape

The human genome is approximately 3 200 000 000 (3.2 gigabases (Gb)) nucleotides long (Lander *et al.*, 2001; Venter *et al.*, 2001). At first sight, a monotonous repetition of A, T, C and G representing the four nucleotides of DNA, it is, in fact, a diverse and still in many ways mysterious landscape. Of the total 3.2 Gb, 2.85 Gb has been sequenced to high accuracy (IHGSC, 2004); the remainder has not, largely because of heterochromatic regions (centromeres and telomeres) that are highly repetitive and refractory to current sequencing technology.

6.2.1 Gene content

The definition of a gene depends upon the context of its use. To a classical geneticist, it is a unit of inheritance; to many biologists, it is a DNA sequence that encodes a protein; and to the popular media, it is something, which causes a disease! For the purposes of genomic annotation, it is often practical to think in terms of a transcription unit: a set of overlapping transcripts from the same template DNA strand. Chapter 11 (Figure 11.2) outlines the typical genomic structure of a eukaryotic transcription unit, including the presence of a core promoter region immediately upstream of the transcription unit, and more distantly located *cis*-regulatory elements mediating transcriptional control and punctuation of the transcribed region by introns which are spliced from the transcript during RNA maturation. Even though introns are removed and degraded, in higher eukaryotes such as mammals, the length of introns often far exceeds that of exons.

Proteins are often thought of as the principal functional product of a genome. Consequently, protein-coding sequences are the first place screened for disease-associated mutations and functionally significant polymorphisms. The human genome encodes approximately 22 000 protein-coding genes (http://www.ensembl.org), although the total diversity of proteins produced is likely to be several times this thanks to alternate transcript initiation and processing (Maniatis and Tasic, 2002; Carninci *et al.*, 2005). However, it appears that this protein-coding sequence accounts for less than 1.5 per cent of the human genome sequence (Lander *et al.*, 2001). The situation is similar to that in rodents (Waterston *et al.*, 2002; Gibbs *et al.*, 2004), other mammals (Lindblad-Toh *et al.*, 2005) and, to varying degrees, other vertebrates (Aparicio *et al.*, 2002; Hillier *et al.*, 2004). These estimates of coding sequence content appear to be robust, as they are supported by multiple lines of evidence, including the integration of comparative data (Roest Crollius *et al.*, 2000) with transcript evidence (Potter *et al.*, 2004), and they are also consistent with extrapolation from targeted regions investigated in considerable detail (Miller *et al.*, 2004). This finding does, of course, raise the question, what is the function of the remainder of the genome?

It is clear that protein-coding genes are not the complete story. There are also many transcription units with specific functions other than the encoding of a protein, ribosomal and transfer RNAs being classic examples. More recently, the abundance and importance of micro-RNAs that act to regulate the expression of other genes have come to the fore (Lim *et al.*, 2005; see Chapter 14). In addition to these known 'functional RNAs', there is considerable evidence for the existence of many RNA transcripts that have no known function (Carninci *et al.*, 2005).

6.2.2 Repetitive elements

A major component of the human and many other higher eukaryotic genomes is sequence derived from interspersed repetitive elements (IRE) such as endogenous retroviruses, retrotransposons and DNA transposons. At least 45 per cent of the human genome is identifiably derived from IREs (Lander *et al.*, 2001), although this almost certainly underestimates their true contribution, as older, more divergent repeat-derived sequences are unlikely to be identified. These elements are often considered 'junk' DNA, and rarely have organism-level biological functions been attributed to them, although a small number of exceptions are known (Kowalski *et al.*, 1999; Kapitonov and Jurka, 2005). It is interesting to note that some vertebrate lineages, most notably that of the pufferfish, are almost devoid of such IRE-derived sequence and have a genome approximately eightfold smaller than the human despite encoding a similar, or slightly greater number of protein-coding genes (Aparicio *et al.*, 2002).

An interesting consequence of a genome rich in repetitive elements, particularly those that replicate through the process of reverse transcription (duplication via an RNA intermediate), is the abundance of processed pseudogenes. Occasionally, rather than the enzyme responsible for reverse transcription (reverse transcriptase) driving the replication of an IRE, this enzyme will reverse-transcribe the mRNA of a gene. This leads to the integration of a processed duplicate of the gene into the genome. The result is a processed pseudogene, the copy of an mRNA integrated into the genome, which bears the hallmarks of transcript processing, such as the removal of introns and 3′ polyadenylation (Zhang *et al.*, 2004). Processed pseudogenes are often incomplete at the 5′ end, a consequence of reverse-transcriptase reading the 3′ end of the mRNA first. Some genes, such as those encoding the ribosomal proteins, are particularly susceptible to generating new processed pseudogenes (Zhang *et al.*, 2004), probably reflecting in part the level germ-line transcript of the gene.

6.2.3 A varied landscape

In probably every measure that has been made of the human genome sequence, it has been found to be far from homogeneous. We have already touched on the distinction between heterochromatic regions that perform roles in the packaging and segregation of chromosomes, from the remaining (euchromatic) regions. Throughout the rest of the euchromatic genome, there is considerable variation in gene density (the number of genes per unit sequence), IRE content, nucleotide and dinucleotide frequency, and the observed rates of genetic recombination, nucleotide substitution, insertions and deletions. Many of these attributes have been found to co-vary across the genome (Hardison *et al.*, 2003; Gibbs *et al.*, 2004; Singh *et al.*, 2005), but currently the basis of their interrelationships is not well understood. Of particular relevance to comparative genomic studies is the fluctuation of substitution, insertion and deletion rates across the genome (Wolfe *et al.*, 1989), which suggest there may be regional variation in the rate at which mutations occur. At least in rodents, the scale of this variation is of the order of 1 Mb, so that the substitution rates for two neutrally evolving regions of sequence are generally well correlated if they lie within this distance of each other, but the correlation decreases rapidly with increasing genomic distance (Gaffney and Keightley, 2005).

The rate of sequence mutation is dependent not only on the large-scale region of a genome, but also on the sequence and composition of neighbouring sites (Hardison *et al.*, 2003; Taylor *et al.*, 2004). For example, tandemly repeated sequences and mononucleotide tracts are prone to insertion and deletion mutation (Taylor *et al.*, 2004). The epigenetic methylation of cytosine nucleotides, when they are located directly upstream of a guanine (CG), is a common occurrence in mammalian genomes and to a lesser extent in other metazoans (Bird, 2002). This nucleotide modification has had a major influence in shaping mammalian genomes. Thanks to a quirk of biochemistry, a methylated C can mutate to T at a much higher frequency than all other nucleotide substitutions occur. As a result, CG dinucleotides are grossly under-represented across the majority of the human genome, relative to chance expectation given the frequency of C and G nucleotides (approximately 20 per cent of the expected frequency (Sved and Bird, 1990; Lander *et al.*, 2001)), and CG mutation rates tend to be substantially higher than those of other dinucleotides. However, within specific islands of sequence (commonly known as CpG or CG islands), CGs are not methylated, at least in the germ line (Bird *et al.*, 1985), and so are not under-represented. CG islands are often associated with the 5′ end and promoters of some genes (Bird *et al.*, 1985), and so represent sequences that are often of particular interest in genetic and functional studies.

6.2.4 Segmental duplication

Segmental duplications are a genomic feature that can often cause problems for sequence assembly such that they are frequently overlooked. These are large (typically 5 kb is taken as an arbitrary, minimum threshold in their definition) tracts of sequence that occur multiple times in a genome, often as tandem repeats. The duplicated regions can encompass whole genes or even multiple genes. Recent segmental duplications will share a high degree of nucleotide identity and are likely to be polymorphic in the population. The mechanism of segmental duplication provides a rapid means of divergence between species (Law *et al.*, 1992; Nguyen *et al.*, 2006).

6.3 Concepts

The replication of DNA is imperfect; new mutations are continually arising with each generation. In the absence of selection (neutrality), the eventual fate of a new mutation will be determined by genetic drift, chance fluctuations in frequency that result from sampling a finite population. For most mutations, this will result in their loss from the population, but some will drift to fixation. As this is a random process, any observed sequence changes can be considered an unbiased sample of all mutations that occurred. However, natural selection disrupts this unbiased sampling of mutations.

If we assume that a region of an organism's DNA has a biological function that contributes to the survival of that organism, it is probable that a random mutation in this region will disrupt that function. This is analogous to someone randomly connecting a pair of wires in a computer – it may make the computer work better, but most likely it will have a detrimental effect on function. Consequently, the majority of mutational changes within functional elements are likely to be detrimental and removed by the process of selection, whereas there is no such driving force to eliminate mutations within non-functional elements. As a result, functionally important sequences are expected, in general, to accumulate fewer mutational changes than neutrally evolving DNA. This is the same as saying that two functional regions of sequence diverged from a common ancestor are expected to be more similar than a pair of non-functional regions that diverged at the same time. Local regions of sequence similarity resulting from selective constraint are often referred to as a phylogenetic footprint (Tagle *et al.*, 1988). This selective constraint is often referred to as negative or purifying selection.

However, it is clear that species adapt and evolve (Darwin, 1859). Especially in response to changing environmental conditions, there is a selective pressure driving change rather than conservation. If the function of a DNA sequence is subject to such adaptive pressure, it may be expected to accumulate changes at a faster rate than expected for neutrally evolving sequences. This is often referred to as positive or diversifying selection. There are instances, such as sexual selection and host-pathogen arms races, of sustained selective pressure for diversification (Nielsen *et al.*, 2005), but in the majority of cases, a period of diversification will be both preceded and succeeded by longer periods of purifying selection. As such, diversifying selection can be difficult to identify unambiguously, and the majority of comparative studies outside protein-coding sequences currently focus on the identification of purifying selection. For an in-depth discussion of genetic drift, selection and the influence of population size, see Lynch (2006).

Both purifying and diversifying selection result in a departure from the neutral rate of sequence evolution; this departure is diagnostic and can be considered the signature of selection. Natural selection can act only on genetic variation that manifests as phenotypic differences between individual organisms of a population. It is a stringent filter: even a 0.001 per cent reduction in fitness will result in a polymorphism being efficiently removed from most mammalian populations (Ohta, 1976; Piganeau and Eyre-Walker, 2003). Therefore, the signature of selection defines a sequence as significantly contributing to the biology of the organism. As we have discussed (Section 6.2), vertebrate and many other higher eukaryotic genomes are dominated by sequences that appear to have no biological function. This means that although the human genome is approximately 3.2 Gb in size (Lander *et al.*, 2001; Venter *et al.*, 2001), most of the biological functions and, consequently, disease-associated polymorphisms and biological insight are concentrated into as little as 0.16 Gb of sequence (Gibbs *et al.*, 2004; Lunter *et al.*, 2006) (Section 6.6.1). Comparative genomics provides a means of identifying that rich vein of functional sequence, and, unlike laboratory-based approaches, it does so without requiring prior assumptions of what that function may be.

6.3.1 Homologues, orthologues and paralogues

The rate of sequence evolution is measured from an alignment between sequences that have diverged from a common ancestor; that is, they are homologous. If the point of divergence for two homologous sequences was a speciation event, they are referred to as orthologues. Otherwise, they are paralogues of one another. The distinction between orthology and paralogy is important for two reasons. Orthologues are more likely than paralogues to have conserved the same function since divergence, because the processes giving rise to paralogues, such as intragenome duplication and horizontal gene transfer, provide an opportunity for functional diversification through the relaxation of selective constraint (Gogarten and Olendzenski, 1999). Secondly, when we compare loci of multiple orthologues between the same range of organisms, a common phylogenetic relationship and divergence times can be assumed for all of the loci, enabling direct comparison between loci. No such assumptions can be made for comparisons of paralogous loci. For these reasons, the majority of studies are based on alignments of orthologue sequences.

'Phylogenetic scope', a term introduced by Cooper *et al.* (2003), defines the range of organisms being considered in an analysis, denoted by their most recent common ancestor. For example, a study involving sequences from zebrafish, chicken, frog, mouse and man is vertebrate in scope, whereas one looking at man, chimpanzee and macaque is primate in scope. The phylogenetic scope of a study must be matched to the biological questions being asked. In general, more closely related species are more likely to have similar biology than distantly related species. The *Sonic hedgehog* gene discussed later (Section 6.6.3) provides a good example of the potential pitfalls of an inappropriate phylogenetic scope.

The number of expected differences between sequences has important implications for the utility of a particular sequence in a comparative analysis, and how the analysis should be performed. It is useful then to have some standard measure of the expected degree of sequence divergence. For orthologous sequences, a widely used measure has been divergence time in millions of years, estimated through the integration of fossil records and molecular data. The greater the divergence time, the greater the number of changes that are likely to have accumulated. However, these date estimates vary wildly with the methods used and assumptions made; for example, the divergence between rodent and primate lineages has been estimated as occurring between 75 and 121 MYA (Waterston *et al.*, 2002; Gibbs *et al.*, 2004; Glazko *et al.*, 2005).

A more useful measure for comparative genomics analysis is that of branch length, sometimes simply referred to as distance. This measure denotes the number of mutational changes per unit of sequence, such as substitutions per nucleotide, deletions per amino acid or inversions per kilobase. The most useful and widely used measure when considering comparative genomics is that of substitutions per nucleotide, as it is readily calculated and is reasonably robust to alignment methodology. As a measure, it also relates directly to the amount of information present in aligned sequences and also how accurate an alignment between those sequences is likely to be (see below). In the phylogenetic tree shown in Figure 6.1, the total branch length between man and mouse is $D = 0.63$ substitutions per site in a neutrally evolving sequence, calculated by summing branch lengths between the human and mouse terminal nodes $(0.025 + 0.12 + 0.399 + 0.083)$. It should be noted that branch length is often not the same as the sum of sequence differences, as the methods used to calculate substitution rates typically take into account the likelihood of multiple changes at the same site.

In theory, the power of a study to distinguish non-neutral from neutral evolution is proportional to the total divergence (branch length) of the analysis; in the case of Figure 6.1, this would be the sum of each value shown on the tree (total: 0.989).

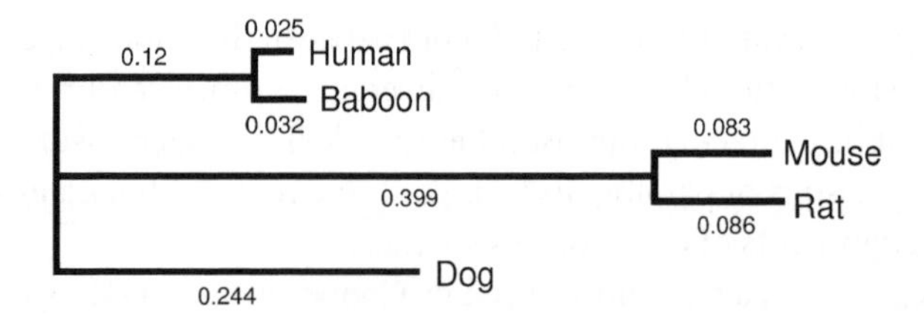

Figure 6.1 Phylogenetic tree showing branch lengths. An unrooted tree with branch lengths derived from nucleotide substitution rates of anonymous aligned sequence in the greater CFTR region. Individual branch lengths are shown on each branch segment

Under the simplest scenario of identifying selective constraint, one is evaluating the likelihood that a segment of nucleotides has remained unchanged by chance, given an expected neutral rate of evolution D. For small values of D, we can use the Poisson distribution (e^{-D}) to approximate the probability that a neutrally evolving site will be unchanged (Cooper *et al.*, 2003; Eddy, 2005). For a man:mouse alignment with $D = 0.63$, there is a 53 per cent likelihood that a neutral site will be unchanged by chance.

In practice, a pairwise alignment between orthologous sequences cannot distinguish selective constraint from neutral evolution for a single nucleotide position. Rather, a region of consecutive nucleotides is evaluated collectively. The size of a region necessary to identify selective constraint scales inversely with the value of D for the analysis (Eddy, 2005). A simple way to increase the sensitivity of an analysis, to detect shorter or less conserved sequences, is to compare more distantly related sequences. Unfortunately, there are two important caveats to this strategy. First, the more diverged sequences are, the less accurate the alignments are between them (Pollard *et al.*, 2004), so constrained sequences may be missed at the alignment stage rather than in the analysis of the alignment. Second, is the issue of phylogenetic scope; diverged species are less likely to share biological functions or be subject to similar constraints.

An alternative approach for increasing total D of an analysis is to include more sequences through multiple alignment. Based on the branch-length values in Figure 6.1, a comparison of man and mouse has $D = 0.63$, but adding rat as a third species increases total D to 0.72. When calculating total D for an analysis, each unique section of branch is counted only once, so rat adds only $D = 0.086$ to the total analysis; considerably more power could be added by using dog instead of, or in addition to rat, as it would contribute $D = 0.244$ of unique branch length. A further advantage to increasing comparisons from pairwise to multiple sequences is that it allows the direction of mutational changes to be resolved, such as the discrimination of insertion from deletion and the ability to assign changes to a specific lineage.

Alignment of closely related pairs of sequences, such as man-chimpanzee or man-macaque orthologous regions ($D = 0.009$ and $D = 0.052$ respectively (Margulies *et al.*, 2003)), is of little use for phylogenetic footprinting studies (Section 6.3;

Figure 6.2). However, extending the approach described above to the alignment of many such similar sequences can in theory provide sufficient total D for useful detection of selective constraint (Cooper *et al.*, 2003; Eddy, 2005). As the sequences are closely related, their alignment should be highly accurate, covering most nucleotides (Pollard *et al.*, 2004), and the phylogenetic scope is narrow, so relatively little functional divergence is expected. This paradigm, known as phylogenetic shadowing (Boffelli *et al.*, 2003), represents an ideal combination of attributes for comparative genomic studies. Using phylogenetic shadowing, Boffelli *et al.* (2003) were able to demonstrate the identification of constrained sequences specific to primates, and showed that as few as four to eight well-chosen genomes could capture much of the information present in deeper alignments of up to 17 primate sequences. The principal limitation is the need for multiple closely related, orthologous sequences (Section 6.3).

6.4 Practicalities

6.4.1 Available genomic sequences

At the turn of the millennium, comparative genomic projects in vertebrates involved the laboratory-based identification of homologous regions and their sequencing (Davidson *et al.*, 2000), prior to any comparative analysis. This situation has changed markedly, with an extremely high-quality reference human genome sequence in hand (IHGSC, 2004) and high-quality draft sequences from mouse and rat (Waterston *et al.*, 2002; Gibbs *et al.*, 2004). The target for all three of these genomes is 'finished' sequence, highly accurate and completely contiguous. Finished sequence is the reference standard and the ideal for comparative analysis. Unfortunately, the production of finished vertebrate sequence currently demands considerable time and skill, and is correspondingly expensive. In contrast, a well-designed, whole-genome shotgun sequencing and assembly project (Weber and Myers, 1997) can be largely automated at every stage. As a result of these economics, most eukaryotic whole-genome sequencing projects now being undertaken have adopted a purely whole-genome shotgun strategy (Chapter 5), producing 'draft' assemblies with no finishing step planned for the foreseeable future.

Draft assemblies have been produced from multiple other vertebrates including chicken (*Gallus gallus* (Hillier *et al.*, 2004)), dog (*Canis familiaris* (Lindblad-Toh *et al.*, 2005)), zebrafish (*Danio rerio*), frog (*Xenopus tropicalis*), macaque (*Macaca mulatta*), chimpanzee (*Pan troglodytes*; Muzny), tiger pufferfish (*Takifugu rubripes* (Aparicio *et al.*, 2002)), domestic cattle (*Bos taurus*), rabbit (*Oryctolagus cuniculus*), armadillo (*Dasypus novemcinctus*), African elephant (*Loxodonta africana*), opossum (*Monodelphis domestica*), medaka (*Oryzias latipes*) and freshwater pufferfish (*Tetraodon nigroviridis* (Jaillon *et al.*, 2004)). This list is expanding at an accelerating rate, driven largely by the realization that sequence comparisons between multiple

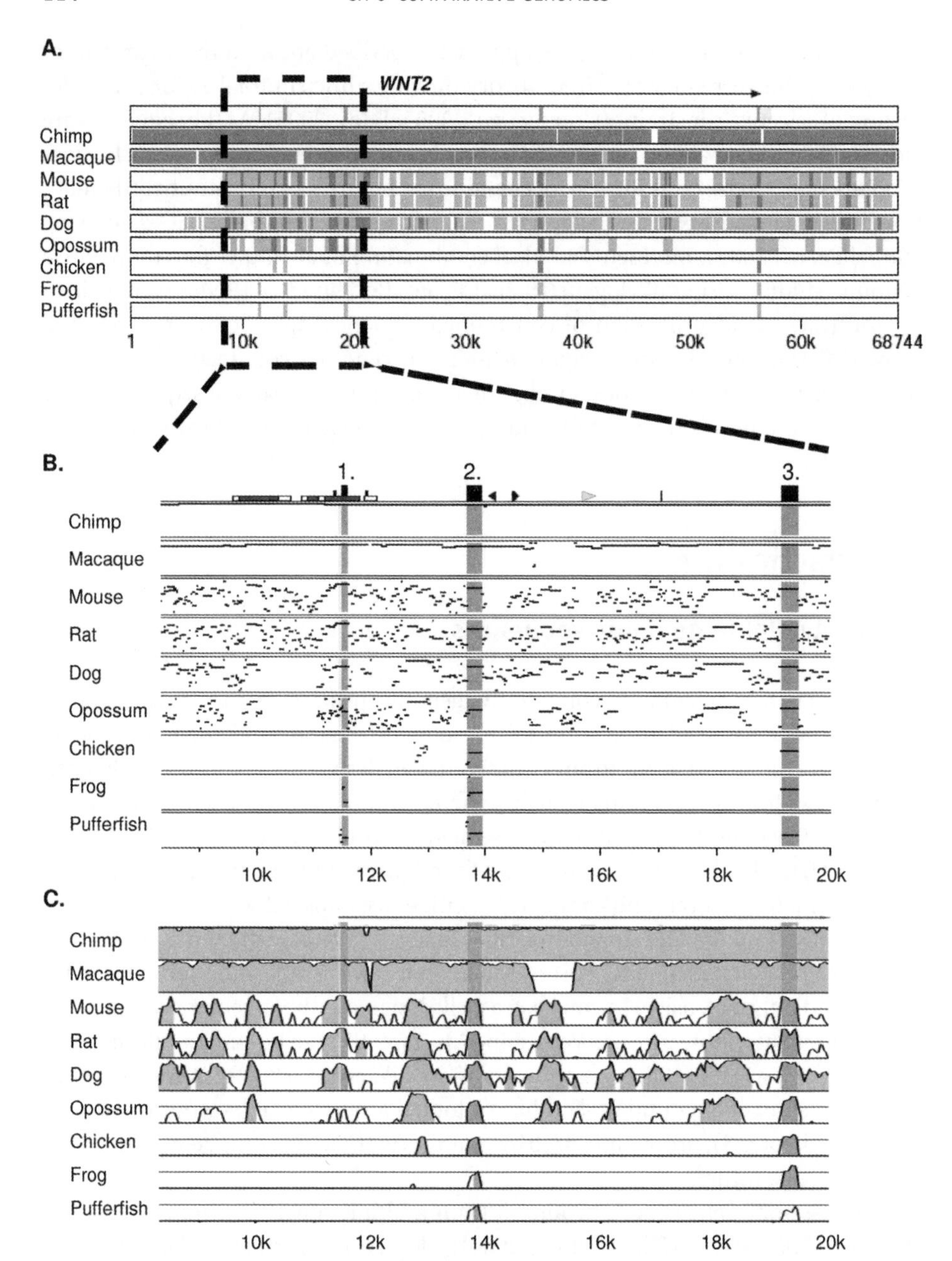

Figure 6.2 Visualization of genomic sequence alignments. The WNT2 locus was aligned between human and orthologous loci from nine other vertebrates for which at least a draft whole-genome shotgun sequence is available. Orthologous regions and extents were defined by the UCSC Nets. In each case, coordinates and annotation are shown for the human sequence and nucleotide identity from pairwise alignment. (A) Summary view from MultiPipMaker (Schwartz *et al.*, 2000) based on BlastZ alignments. The extent of the WNT2 transcript is shown above the alignment, protein-coding exons shaded in grey. Regions of local alignment are shown in light grey or dark

(*continued on following page*)

vertebrate genomes is crucial to understanding the structural and functional components encoded in the human genome (Collins *et al.*, 2003).

Whole-genome assemblies, rather than individual clone sequences, now provide the primary resource of genomic sequence for most comparative analyses in the vertebrate scope, and there is a similar situation for biologists focusing on prokaryotes, viruses, fungi, plants, nematodes and insects, with at least draft status sequence available for over 1000 genomes. However, the quality and completeness of sequences should be considered when undertaking an analysis. For a finished sequence, the accuracy is expected to be high; with less than one nucleotide error per 100 000 nucleotides and fewer than one insertion/deletion error per 200 000 nucleotides, the vast majority of which are located in tandemly repetitive sequence (IHGSC, 2004), and there should be no gaps in sequence coverage. The quality of draft sequences depends to a large degree on the depth of coverage. With eightfold (8×) coverage (every base sequenced on average eight times), a whole genome shotgun sequencing project can produce a high-quality sequence with good long-range ordering of sequences (Mullikin and Ning, 2003). As coverage is reduced, the rate of all types of error increase; in particular, there is a rapid reduction in sequence contiguity (Wendl and Yang, 2004).

Even in high-quality and 'finished' genomic sequences, there is still a chance of misassembly, especially in regions rich in repetitive elements. However, a more common issue is that of segmental duplication (Section 6.2), where very recently duplicated regions, which may encompass several genes, cannot be reliably discriminated during normal assembly procedures, resulting in the collapse of multiple duplications into a single sequence (She *et al.*, 2004). Efforts are currently being made to identify and resolve these problematic regions (Sharp *et al.*, 2005); however, it has become apparent that the copy number of high-identity (>97 per cent) segmental duplications is often polymorphic in the human population, diverges rapidly between species (Cheng *et al.*, 2005) and may be associated with disease susceptibility (Eichler, 2006). A further consideration is that the small number of differences between segmental duplicates will appear as polymorphisms in almost all assays,

Figure 6.2 (*Continued*) grey if a combined length and identity threshold is achieved (green or red respectively when viewed in colour). The region highlighted is shown in detail in panels B and C. (B) Detailed view of MultiPipMaker output, a percentage identity plot. Exons are denoted by black boxes above the plot and projected as grey shaded regions across it. Other features above the plot correspond to annotated repetitive elements (triangles and predicted CpG islands (grey and white boxes). (C) VISTA plot (Mayor *et al.*, 2000) summarizing mLagan (Brudno *et al.*, 2003b) global alignments of the sequences. Higher curves show greater conservation; regions meeting a threshold level of conservation are shaded (darker shading for protein-coding exons). Exons 2 and 3 are readily aligned in all cases, whereas the relatively short and poorly conserved exon 1 is not always aligned (panel C, frog and pufferfish). An additional complication when using draft sequences is the presence of assembly gaps; the apparent failure to detect exon 1 in chicken in this case coincides with a gap in the chicken assembly. (Figure generated by the authors using software from Mayor *et al.* and Brudno *et al.* Permission not required)

having potentially disruptive effects on genetic studies. Therefore, it is often prudent to check for indications of segmental duplication such as the 'WSSD' and 'Segmental Dups' tracks from the UCSC Genome Browser prior to investigating a new region.

Considerations of sequence quality and coverage are set to become more important, as the emphasis of genome sequencing continues its shift from high-accuracy sequencing to sampling more genomes but with lower individual coverage. As we have discussed above (Section 6.3), an optimal strategy for the identification of constrained sites is to analyse sequence from many closely related genomes to achieve a large total branch length. The cost of sequencing one genome to 8× is almost the same as eight genomes at 1×; there is, then, a trade-off between high-quality sequence and maximizing the number and diversity of sequenced genomes. Margulies *et al.* (2005) have explored this trade-off with both real and simulated data, demonstrating that as little as 2× shotgun, although insufficient to produce a good-quality assembly, can be useful in the identification of constrained sequences by directly aligning reads to more completely sequenced genomes.

The National Human Genome Research Institute (NHGRI; http://www.genome.gov/) has adopted this strategy of many genomes at low coverage and is currently coordinating the low-coverage sequencing of 16 additional mammalian genomes, selected to maximize total branch length for comparative analysis. The full list of organisms, target sequence coverage and progress in sequencing can be monitored online (http://www.genome.gov/10002154). Based on the equations of Eddy (2005) and simulations of Margulies *et al.* (2005), these genome sequences should provide resolution of selective constraint down to a segment length of eight nucleotides, approaching the same scale as individual transcription factor-binding sites. If successful, this strategy is likely to be applied to an even greater number of mammalian and other genomes (a fruitfly-based project is also currently under way; http://rana.lbl.gov/drosophila/multipleflies.html), the most exciting of which from the perspective of human biology is the proposal to sequence multiple primate genomes (http://www.genome.gov/12511814).

6.4.2 Defining and obtaining genomic sequences

When undertaking a comparative genomic study, it is necessary to delineate a locus or loci of interest and to obtain corresponding homologous, often orthologous, sequences. Typically, an approximate locus will be defined by either arbitrary distances from an identified feature of interest, the confidence intervals of a preceding genetic study, or the extent of a sequenced genomic fragment. It can be useful to extend a region of analysis slightly beyond the minimal extent so that the region is bounded by features that are well conserved between species, such as protein-coding exons, that serve as anchors for the analysis. A pair of well-conserved anchors provides confidence that the full extent of a locus has been isolated from each species under analysis.

Preassembled genomes are the most accessible source of defined genomic segments, as the problems of stitching together overlapping sequence fragments have already been tackled and the assemblies will have been subject to some degree of validation and quality control. Complete assemblies can be obtained from a number of disparate sites depending on the organism and assembly method. However, the UCSC Genome Browser (http://genome.ucsc.edu/), Ensembl (http://ensembl.org/) and the National Center for Biotechnology Information (NCBI) (http://www.ncbi.nlm.nih.gov/) all provide portals to the most current, and archived public assemblies. These sites also provide means of searching the assemblies, such as BLAST (Altschul *et al.*, 1997), BLAT (Kent, 2002) and SSAHA (Ning *et al.*, 2001) as well as precomputed annotation for the genome assemblies that can be readily incorporated into comparative genomic analyses.

There are several routes to identifying homologous loci in target genome sequences. An obvious approach is based on sequence similarity searches, but care must be taken to distinguish orthologous from paralogous loci. Processed pseudogenes, in particular, are common (Shemesh *et al.*, 2006); these are the reverse-transcribed copy of an mRNA that has integrated into the genome, but which does not code for a functional protein (Section 6.2). As processed pseudogenes lack introns, they can score better than an orthologous locus in a similarity search. Genome-wide, reciprocal best matches (Tatusov *et al.*, 2003) can be used to increase confidence that two loci are orthologous. Ensembl also provides precomputed assignments of gene orthology, currently based on reciprocal best matches for several genomes in the '*geneview*' pages and from the *EnsMart* data repository. Conservation of the order and orientation of genes in and neighbouring the locus can also provide additional support of the orthology of two loci.

Probably the simplest currently available route to identifying orthologous loci is with the *Net* alignments at UCSC. These genome-to-genome pairwise alignments show genome-wide best matches and local rearrangements within them. They provide a direct means of jumping between an orthologous location in two genomes and can be used directly to delineate an orthologous locus in a target genome. For example, with the genome browser showing a complete locus of interest in a human assembly, clicking on the human to dog *Net* will provide an option to open the dog genome browser in a corresponding window, from which the canine sequence and associated annotation can be obtained. An extension of this method is to use the genomic alignments to transfer annotation from one perhaps well-annotated genome to another that may have been recently assembled. The LiftOver tool at the UCSC Genome Browser (http://genome.ucsc.edu/cgi-bin/hgLiftOver) provides this facility for a limited set of genome pairs. This can provide a rapid way to get a baseline annotation, which can then be filtered and refined. The *Net* alignments are generally good quality, but problems do arise, in particular where segmental duplications and assembly gaps are involved.

If there is uncertainty in the assignment of paralogy or orthology between multiple sequences, it can often be resolved through rigorous phylogenetic analysis, of either whole genomic alignments or more discrete regions, such as protein-coding sequences, within them. This is often a problem with comparisons involving teleosts such as pufferfish and zebrafish, which may have been subject to a past whole-genome duplication (Hoegg *et al.*, 2004) with the subsequent loss of many genes.

6.5 Technology

There are three general challenges that are common to most comparative genome analysis: (i) the production of an alignment; (ii) visualization of the alignment; and (iii) detection of departures from neutral sequence evolution in the alignment. As alignments form the foundation of the comparative analysis, we spend some time discussing the different options available and the consequences for interpreting results. There are also several options available for the visualization of large-scale genomic alignments. We have already discussed the principles and general approaches taken for the detection of departures from neutrality (Section 6.4), in Section 6.5.3, we present the tools that are currently available to apply these methods.

6.5.1 Alignments

The starting point for the majority of comparative genomic analyses is an alignment between homologous sequences. Precomputed alignments are available between several whole genomes as well as tools (Table 6.1) for producing such alignments. To a large extent, the genomic alignment tools and precomputed alignments can be treated as 'black boxes'. It is not necessary to understand in fine detail the process of producing the alignment to address a biological question with it. However, knowing in general terms how an alignment was generated, and the parameters used, can be crucial to its meaningful interpretation, especially when considering the apparent absence of conservation. In this section, we present an overview of the genomic alignment problem, highlighting the limitations of available methods as well as recent advances in the field.

There are two general approaches to sequence alignment: local and global. When performing a local alignment, one is asking to be shown every similarity, scoring above a predefined threshold, between two sequences. The aligned subsequences (alignment segments) need not be in the same order or orientation in the parent sequences, and many-to-many matches are permitted. In contrast, in a global alignment, the entire length of one sequence is aligned with the entire length of the other through the insertion of gaps in both sequences. There is a maximum one-to-one correspondence between nucleotides and their order is constrained such that duplications, inversions and other rearrangements cannot be detected. Rather than competing and redundant, these approaches should be considered complementary, as they provide different insights into the relationship between two or more sequences.

Table 6.1 Summary of widely used and recommended genomic alignment tools. *G indicates global and L local alignment methods. **2 denotes pairwise and M multiple alignment tools. Note that several visualization tools, such as MultiPipMaker, emulate multiple alignment by stacking the percentage identity plots of multiple pairwise alignments without actually producing a character-based multiple alignment

Program	*	**	Reference	Comment
AVID	**G**	**2**	(Bray *et al.*, 2003)	http://genome.lbl.gov/vista/ http://genome.lbl.gov/vista/
BlastZ	**L**	**2**	(Schwartz *et al.*, 2000, 2003)	The most widely used local genomic alignment tool http://pipmaker.bx.psu.edu/pipmaker/
Blat	**L**	**2**	(Kent, 2002)	Efficient use of memory and rapid execution make this a good choice for defining approximate regions to align with more sensitive methods
CHAOS	**L**	**2**	(Brudno *et al.*, 2003a)	By itself lacks the heuristic refinements of BlastZ but is used by DIALIGN and Lagan to identify initial alignment matches
DIALIGN	**G**	**M**	(Morgenstern, 2004)	Only practical for alignment of large (>10 000 nucleotides) sequences when used in conjunction with CHAOS .(Brudno *et al.*, 2004) http://dialign.gobics.de/anchor/
GLASS	**G**	**2**	(Batzoglou *et al.*, 2000)	One of the first available tools, now superseded by AVID
Lagan	**G**	**2**	(Brudno *et al.*, 2003b)	http://genome.lbl.gov/vista/ http://genome.lbl.gov/vista/
MAVID	**G**	**M**	(Bray and Pachter, 2003)	http://genome.lbl.gov/vista/ http://genome.lbl.gov/vista/
mLagan	**G**	**M**	(Brudno *et al.*, 2003b)	http://genome.lbl.gov/vista/ http://genome.lbl.gov/vista/
MultiZ	**G**	**M**	(Blanchette *et al.*, 2004)	Based on BlastZ local alignments but with a tiling path of aligning segments chosen (chaining, see main text) and integrated into multiple sequence alignments. This is the method used to produce high-resolution alignments for the UCSC Genome Browser (http://genome.ucsc.edu/)
sLagan	**G**	**2**	(Brudno *et al.*, 2003c)	Also known as Shuffle-Lagan. Produces glocal alignments which have relaxed some constraints of global alignment so inversions, translocations and duplications can be detected http://genome.lbl.gov/vista/
TBA	**G**	**M**	(Blanchette *et al.*, 2004)	A prototype stand-alone tool to produce threaded blockset multiple sequence alignment, similar to the output of MultiZ
WABA	**G**	**2**	(Kent and Zahler, 2000)	Readily handles large gaps and can predict the protein-coding/non-coding status of a sequence region based in part on the periodicity of divergent sequences

Local genomic alignments

All of the local alignment methods commonly applied to genomic sequences (Table 6.1) employ an index-based search strategy based on the same principle as that employed in the original BLAST algorithm (Altschul *et al.*, 1997). Briefly, this approach produces an index of all k-length words (k-mers) in one of the input sequences and searches the other sequence for identical words. When a match is found, it is extended in both directions to define a maximally scoring segment of alignment. If that score is above the predefined threshold, the alignment is reported.

Three methods have been applied to genomic local alignment tools that elaborate this basic procedure to increase the sensitivity and specificity:

1. Requiring two matching words to be separated by a maximum distance from each other. This is a common approach used by BLASTN and most of the local alignment methods in Table 6.1. The principal exception is CHAOS, which identifies multiple matching words but does not perform alignment extension. Instead, the matching words are clustered (*chained*) if they lie in the same orientation and within a threshold distance of each other. It is this chain of words that is scored by CHAOS rather than BLAST-like extended initial matches.

2. Using degenerate k-mers, which can tolerate a mismatch in any position of the k-mer, is a strategy that adds considerably to the computational load in the initial search step but provides more flexibility in defining word matches. This method is used by CHAOS in conjunction with the novel chain-of-words approach.

3. Matching k-mers of non-consecutive positions, an idea introduced to the field by Ma *et al.* (2002). For example, a $k = 8$ word could be represented as 11111111, each number one denoting the position of an identity required for a match; a non-consecutive $k = 8$ could be represented as 11011011011. This is distinct from the degenerate k-mer approach, as a degenerate k-mer can tolerate a mismatch in any position, whereas the position of possible mismatches is constrained in the non-consecutive k-mer case.

Such patterns of matches can relate more directly to the underlying biology. The previous example could be useful to identify matches between coding sequence given the periodicity of codon conservation, due to the degeneracy of the genetic code. The non-consecutive k-mer also has a slight statistical advantage over the consecutive k-mer, as the failure to match overlapping non-consecutive k-mers is less strongly correlated between k-mers than the failure to match those which are overlapping and consecutive (Batzoglou, 2005).

Beyond the limits of sensitivity defined by the initial index search, there are many parameters that can be modified in the available tools to optimize them for a specific purpose or phylogenetic scope. For such insight, we direct the reader to the primary literature and associated web servers (Table 6.1). However, the program BlastZ is one of the most versatile and widely used in this class of program for comparative genomic studies and is the basis for a number of publicly available resources; as such, we consider its use in more detail here.

Developed principally by Scott Schwartz and Webb Miller (Schwartz *et al.*, 2000, 2003), BlastZ is based on the Gapped-BLAST algorithm (Altschul *et al.*, 1997). An alignment is seeded by a short perfect or defined imperfect match, extended by dynamic programming, initially without gaps and if score thresholds are achieved, and then with gaps. Sequence between anchoring alignments is again searched and alignments extended, but with a lower stringency than in the initial search, the stringency being determined by the separation distance between anchors. BlastZ employs heuristics to take sequence complexity into account, requiring low-complexity sequence to align better than high-complexity sequence, and dynamically to mask any regions with an unexpectedly large number of matches. As BlastZ is a local alignment tool, matches may overlap; they can be distributed between both strands and are unconstrained in their linear order. However, BlastZ has the option of constraining matches to be co-linear between input sequences (chaining) or to select only a best match to each region of a reference sequence (single coverage). Both of these options involve discarding data but can be useful in interpreting results and subsequent analysis.

Global genomic alignments

The prototypical global alignment method is that of Needleman and Wunsch (1970). However, this procedure does not scale well to the large alignments commonly required in comparative genomics. The approach employed by most of the genomic global alignment tools is to define a series of anchors, high-confidence matches between a pair of sequences that are constrained to be in the same order and orientation in both sequences. This is effectively the chaining method optionally employed by BlastZ, as discussed above. The portion of each sequence between adjacent anchors is then aligned with lower stringency, defining a new set of anchors, and the process is reiterated until all sequence is aligned. The strategy effectively breaks the large alignment down into a series of progressively smaller alignments, with two important consequences. First, the total search space is quickly reduced and continues to be refined with each iteration, allowing the alignments to be produced quickly and using little memory relative to the length of input sequences. Second, the chain-of-anchors approach is tolerant of large gaps, which are common in genomic sequence alignments, but poorly dealt with by gap penalties employed by the purely dynamic programming methods such as that of Needleman and Wunsch.

Table 6.1 summarizes the global sequence alignment tools that are often applied to genomic sequences. Of these, AVID (Bray *et al.*, 2003) and Lagan (Brudno *et al.*, 2003b) are the most widely used. AVID identifies maximal matches (identical runs of nucleotides) and from these selects a chain of non-overlapping alignment anchors using dynamic programming, iterating the process as described above until all bases are aligned, or there are no significant matches in the remaining subsequences. The full Needleman–Wunsch algorithm is applied if the remaining sequences are short (<4 kb); otherwise, a fully gapped alignment of these regions is returned.

A particularly useful feature of AVID is its ability to perform template-directed fragment assembly. Provided with a contiguous and a fragmented sequence, AVID will use high-confidence local matches to order and orient the fragmented sequence relative to the contiguous one, producing a 'merged draft', which is then used for pairwise alignment. Such a utility can be invaluable in the analysis of early-draft genomic sequences.

Lagan (Brudno *et al.*, 2003b) proceeds in a very similar iterative manner to that described for AVID, making use of the application CHAOS to produce local alignments from which the anchors are defined. In a further development, Brudno *et al.* (2003c) have generalized this approach by relaxing the criteria for co-linearity in the order of alignment anchors, instead requiring them to be sequentially ordered along only one of the input sequences, the designated reference sequence. This relaxation allows the detection of genomic rearrangements such as inversions, translocations and duplications relative to the reference sequence. This method is implemented as Shuffle–Lagan (Table 6.1). In recognition of similarities to both local and global methods, the authors have termed these 'glocal' alignments. The approach is innovative and has potential to be developed further, but there are two key drawbacks to the current implementation. First, because the two input sequences are treated differently, the resulting alignment depends on the order sequences are presented. The second limitation is our current lack of understanding in the frequency of genomic rearrangements to parameterize appropriately such alignments.

Multiple sequence alignments

The local and global sequence alignment methods we have discussed so far are able to produce only pairwise alignments. We have seen in Section 6.3, however, that the combined analysis of multiple sequences provides much greater insight, statistical power and resolution to comparative genomic studies. Unfortunately, the difficulties of producing pairwise genomic sequence alignments are exacerbated in the challenge of producing multiple alignments.

To perform a progressive multiple alignment in this manner, the phylogenetic relationship between sequences being aligned needs to be established. This either can be calculated from initial all-versus-all pairwise alignments of the sequences, or, for some programs, can be provided in the form of a previously established phylogenetic tree. If the multiple sequence alignment is between orthologous sequences, their relationship is often known in advance; for example, (((man,chimpanzee),(mouse,rat))dog). Provision of the tree in advance removes uncertainty in the order a program aligns the sequences, providing consistency between the alignment of multiple loci and expediting the alignment process, as well as ensuring that the correct phylogeny is used. After production of the initial multiple sequence alignment, the location of gaps can be optimized, making use of the greater information content of multiple sequences. There are multiple alignment versions of AVID and Lagan, denoted by the 'M' prefix to the name (MAVID and mLagan), both of which use the general method outlined above to produce the global multiple sequence alignments.

Although MAVID and mLagan produce true multiple sequence alignments, many visualization tools (Section 6.5.2) display conservation profiles relative to a chosen reference sequence. A similar approach is frequently used to integrate conservation measures between multiple pairwise alignments (Schwartz *et al.*, 2000) and to define multiply conserved sequences (Section 6.5). The reference sequence approach is an obvious choice if the objective is the annotation or investigation of a particular sequence. However, increasingly, comparative genomic studies intend to measure how a locus has evolved in multiple lineages and how selective forces have changed during that evolution, rather than just detecting regions of the reference sequence that are selectively constrained. For these analyses, the reference sequence approach has two major drawbacks. First, any regions conserved between a subset of aligned sequences, but not the reference, will not be detected. This problem can be overcome by generating several multiple sequence alignments, one with each of the sequences under study as the reference. This solution is time-consuming, raises the additional problem of integrating results between alignments, and exposes the second major drawback to the reference sequence approach; that is, the potential for inconsistencies when using alternate sequences as the reference.

A solution to the problems presented by reference sequence-based alignment and analysis has been proposed in the form of a 'threaded blockset' (Blanchette *et al.*, 2004). Under this proposition, a multiple sequence alignment is represented as a series of alignment blocks, termed 'blockset'. Within an individual block, each row corresponds exactly to an input sequence (or its reverse complement) if gap characters are ignored. That is, no sequence within a block has been rearranged. Additionally, an individual block need not involve every aligned sequence. From this blockset, multiple sequence alignments can be produced with any one of the aligned sequences as the reference sequence, simply by ordering and orienting the blocks according to the selected reference sequence, a process referred to as threading the blockset. This approach ensures consistency of alignment when alternate reference sequences are used and no portions of the alignment are discarded. The threaded blockset aligner (TBA) (Blanchette *et al.*, 2004) has been developed as a prototype tool to generate blocksets.

Many eukaryotic genomes are rich in repetitive sequences (Section 6.2); these can confuse alignment programs if not treated appropriately. The simplest treatment of interspersed repeats and low complexity regions is to mask the sequence prior to alignment, readily achieved with tools such as RepeatMasker (http://www.repeatmasker.org/) or available precomputed from the UCSC Genome

Browser for a wide range of whole genome assemblies. However, interspersed repeats can be interesting in their own right, and are a useful measure of mutation rate (Section 6.5). A more satisfactory treatment of repeats is to ignore them in the initial stages of alignment and then align through them if flanking non-repeat sequence has been aligned. This is often termed 'soft-masking', and is implemented in several alignment tools including LAGAN, Blat and AVID.

Assessing the quality of genomic alignment tools

Which alignment tool is the most accurate? This is an obvious question to ask when deciding which tool is the most appropriate to use. Unfortunately, this appears to be impossible to answer definitively. For protein-coding sequences, solved three-dimensional protein structures provide a reference standard against which alignment methods can be scored (Brenner *et al.*, 1998). No such equivalent exists for non-coding DNA. A possible solution is the *in silico* simulation of sequence divergence (Stoye *et al.*, 1998), which can provide a population of sequences related to a common ancestor by a precisely known sequence of mutational events, so that the true alignment is known.

There is a chance that an evaluation of alignment success based on simulated data is measuring the similarity of evolutionary models, rather than the sensitivity and specificity of the alignment methods themselves. Despite this limitation, Pollard *et al.* (2004) have performed such an analysis and produced some useful rules of thumb for genomic sequence alignment. All methods rapidly lost sensitivity with increasing divergence, with more than 50 per cent of nucleotides not accurately aligned by all methods with $D = 1.0$ (divergence, substitutions per site) in the most realistic simulations. Local aligners were successful at identifying constrained sites, but performed poorly on neutral sequence with $D > 1.0$. As would be expected from their mode of action, global aligners had the highest overall sensitivity to align orthologous sites accurately in both neutral and selectively constrained sequence. Lagan (Brudno *et al.*, 2003b) performed particularly well under almost all of the simulation scenarios. The simulations in this study did not include inversions and duplications, which would have been detected only by the local alignment methods considered.

Using whole-genome alignments

As we have seen, there is a good diversity of tools available to produce pairwise and multiple genomic sequence alignments. Although these tools are optimized for genomic sequence alignment, the alignment of whole eukaryotic genomes to each other is still a daunting and specialist task requiring considerable computational resources. Fortunately, several research groups that specialize in such large-scale genomic alignments have made their alignments publicly available (Table 6.2). The

Table 6.2 Precomputed whole eukaryotic genome alignment resources

Resource	URL	Reference
MultiZ at UCSC	http://genome.ucsc.edu/	(Blanchette *et al.*, 2004)
Berkeley Genome Pipeline	http://pipeline.lbl.gov/	(Couronne *et al.*, 2003)
GALA	http://gala.cse.psu.edu/	(Giardine *et al.*, 2003; Elnitski *et al.*, 2005)

utility of these alignments is not limited to whole-genome analyses, and they represent an excellent resource for investigations focused on defined loci.

These publicly available alignments have several significant advantages over proprietary alignments produced ad hoc to address specific questions. First, because they are a public resource, they are used by many members of the research community to address a multitude of questions; therefore, any systematic problems in their construction are likely to be highlighted, whereas in-house alignments are unlikely to be as rigorously vetted. Secondly, results based on the same alignments can be directly compared between research groups, as in the integration of findings in large collaborative projects (Waterston *et al.*, 2002; Gibbs *et al.*, 2004). Finally, it is faster and simpler than producing one's own alignments, especially in the cases where existing annotation has already been mapped to the alignments (http://pipeline.lbl.gov/) or can be readily mapped by easily accessible tools (http://pipmaker.bx.psu.edu/piphelper/).

However, there are, of course, limitations to the utility of precomputed alignments. The user is restricted by the predefined phylogenetic scope of the alignments; for example, at the time of writing, the human-based MultiZ alignments available from UCSC included alignments with chimpanzee, mouse, rat, dog, chicken, pufferfish and zebrafish; but assemblies for the genomes of opossum, rhesus macaque, cow and frog are also publicly available and could add considerably to the information content of the multiple sequence alignment. Moreover, for some analyses, a very specific set of alignment parameters or constraints are required (Keightley *et al.*, 2005), and these are unlikely to be met by off-the-shelf whole-genome alignments.

6.5.2 Visualizing genomic alignments

The visual representation of alignment-based data is an important aspect of comparative genomics, especially when the focus of the analysis is a locus of specific interest. One of the most intuitive and logical representations of a pairwise sequence alignment is a dotplot. Such a representation can summarize all regions of local similarity between two sequences, highlighting inversions, translocations, duplications and deletions. Plotting a sequence against itself is often an excellent first step in the comparative characterization of a locus, as it can highlight regions that are tandem repetitive and of low complexity, and that clearly show segmental duplications, all of which are potentially confusing to interpret when visualized with the other methods discussed below. For sequences of up to a few hundred kilobases, the Dotter software (Sonnhammer and Durbin, 1995) is able to produce a complete dotplot and incorporate arbitrary annotation. For sequences above this size, the computation of a complete dotplot is impractical, but tools such as PipMaker (Schwartz *et al.*, 2003) can produce dotplot style summaries of local alignments (Figure 6.2), which can be interpreted in essentially the same way.

The downside to dotplots is that they take up considerable space and are impractical when it comes to summarizing the similarity between multiple sequences. For these reasons, percentage identity plots (PIPs) were introduced (Hardison *et al.*, 2003) in which the x-axis represents the coordinates of a reference sequence and the y-axis shows percentage of identity (Figure 6.2). A horizontal bar within the plot then identifies a gap-free segment of local alignment, the horizontal position and extent of the bar defining the aligning section of the reference sequence. The position of the bar in the y-axis shows the percentage nucleotide identity for the ungapped local alignment. This is a versatile way of displaying pairwise sequence similarity, as it can be applied to both local and global alignments, and, through stacking of multiple such plots, can be adapted to show the conservation of a reference sequence aligned with any number of sequences.

Another intuitive and commonly used representation of nucleotide identity in sequence alignments is to plot a histogram of conservation (Figure 6.2). As with PIPs, identity is plotted against the coordinates of a chosen reference sequence. Rather than calculating the identity from an ungapped segment of alignment, however, it is calculated from a predefined range of nucleotides in the reference sequence. These can be discrete consecutive bins of, say, 10 alignment columns, or more commonly calculated as a sliding window. VISTA (Mayor *et al.*, 2000), for example, uses a window of 100 columns with sliding increments of 1, by default.

6.5.3 Detecting selection

Any significant departure from the neutral rate of sequence evolution can indicate the action of selection. If a collection of sequences that are thought a priori to be evolving in a neutral or nearly neutral manner can be defined, they can serve as a comparator for a set of test sequences. In a protein-coding sequence, this is often achieved through measuring the substitution rate at codon positions where a substitution would not result in an amino-acid change (synonymous sites) and comparing it to the rate at non-synonymous sites, where a substitution would change the amino acid (Kimura, 1977). In this case, the assumption is that selection is acting principally on the encoded amino-acid sequence. The ratio of non-synonymous (Ka) and synonymous (Ks) rates then provides a quantitative measure of net selection (these measures are also referred to as *dn* and *ds* respectively). $Ka/Ks > 1$ indicates positive selection, $Ka/Ks < 1$ is indicative of purifying selection, and a Ka/Ks not significantly different from 1 is consistent with neutral evolution. Outside the analysis of protein-coding sequence, Ks may not be the most appropriate category of sequence to estimate the neutral rate. Other categories of sequence used for this purpose include ancient repeats, anonymous sequence and pseudogenes. We consider the advantages and drawbacks of each of these below.

Fourfold degenerate sites

In the standard genetic code, there are eight instances where substituting the third codon position for any other nucleotide will not change the encoded amino acid (CTn=Leu; GTn=Val; TCn=Ser; CCn=Pro; ACn=Thr; GCn=Ala; CGn=Arg; GGn=Gly); these are synonymous or fourfold degenerate (4D) sites. 4D sites are readily identified from annotated or well-predicted coding sequences, and because they are embedded in generally well-conserved coding sequences, they can often be aligned between even highly divergent sequences with a high degree of confidence. For these reasons, 4D sites represent an excellent type of sequence from which to estimate the neutral rate. In general, such sites are readily identified as less conserved than other coding positions and non-4D third codon positions (Nei and Kumar, 2000). However, that is not to say that they are devoid of function or functional constraint – such sites may be involved in the regulation of splicing, translational efficiency, mRNA localization or transcript stability. 4D sites are generally considered to be good for the calibration of nucleotide substitution rates, and, as discussed above, they provide an excellent control sequence for the investigation of selection in a protein-coding sequence. However, they are of no use in measuring the neutral rate of insertion, deletion or rearrangements.

Ancient repeats

Interspersed repetitive elements (IREs) are widespread through most vertebrate genomes, and are thought to be free from selective constraint (Section 6.2). Unlike 4D sites, IREs are free to accumulate insertion, deletion and rearrangement as well as substitution changes (Petrov and Hartl, 1998). With the tool RepeatMasker (http://repeatmasker.org/), combined with an appropriate repeat database (Repbase; http://girinst.org/), IREs can be grouped into families and subfamilies based on sequence similarity. Each copy of an IRE subfamily is thought to have been almost identical at the time of insertion, as they were all produced from one, or a very small number of 'parent' elements in a brief period of activity before mutation robbed the parent element of its ability to transpose (Lander *et al.*, 2001). Therefore, an IRE that inserted into a genomic location in the common ancestor of a set of sequences being compared is expected to accumulate mutational changes independently in each of the diverging lineages, and those changes are likely to be invisible to selection.

This assumption of identity between IRE subfamily members at the time of insertion provides them with additional advantages over other categories of candidate neutral sequence. For example, if we use the IRE subfamily consensus sequence as an out-group, mutational changes can be assigned both a direction and a lineage from just pairwise comparisons, rather than requiring a minimum three aligned sequences.

For mammalian and other genomes rich in interspersed repeats, IREs appear to be the ideal means to measure the neutral rate of mutation. However, IREs are typically defined on the basis of their sequence similarity to previously defined repeats and to other sequences in the genome. This means that highly diverged members of a repeat family may not be detected, resulting in underestimation of the mutation rate. The distribution of IREs is non-random across a genome (Hardison *et al.*, 2003), some favouring A/T-rich insertion sites, and others showing preferential retention based on nucleotide composition. The non-random distribution may result a systematic bias in mutation rate estimation. The abundance of these elements in the genome may also lead to non-orthologous recombination between elements (Kazazian, 2004), resulting in a high frequency of gene conversion within the elements (Roy *et al.*, 2000).

Anonymous sequence

Another possibility is anonymous sequence. In genomes dominated by non-functional sequence, such as those of mammals (Section 6.2), the background mutation rate can be approximated by simply taking the average rate across the whole alignment. This estimate can be improved by specifically excluding annotated functional sites such as protein-coding exons and core promoters. The remaining unannotated (anonymous) regions of alignment will be enriched for selectively neutral sites. An interesting variation is to use sequences that align between closely related species but do not align with a more distant out-group, because the sequence has been inserted in one lineage or lost from the other (Cooper *et al.*, 2004). Again, it can be argued that the sequence is less likely to contain important functional elements and is thus enriched for selectively neutral sites.

Pseudogenes

Pseudogenes (Section 6.2) are particularly interesting for estimation of the neutral rate because their starting point is a functional gene, with all the associated sequence biases, periodicity and, in the case of non-processed pseudogenes, introns, splice junctions and regulatory sites. These are often the features we are most interested in identifying or investigating in comparative studies. If a gene pseudogenized before the common ancestor of compared sequences, we can see the effect of mutation and genetic drift free from the action of selection superimposed upon it. This is the ideal scenario. Unfortunately, non-processed pseudogenes are too rare – only 37 having been found in a systematic screen of the human genome (Lander *et al.*, 2001) – to be of general use in calculating background mutation rates. Processed pseudogenes have been useful for the investigation of protein-coding sequences (Ophir and Graur,

1997), but again their uneven distribution limits their use in estimating local mutation rates, and some sequences identified as pseudogenes may still have functional roles (Podlaha and Zhang, 2004).

Distribution of control and test sequences

For a comparative study that aims to identify highly constrained sequences that are evolving many times slower than the neutral rate, it might be adequate to estimate the neutral rate on a genome-wide basis. Such studies include the identification of protein-coding sequence with distant pairs of sequences, human versus pufferfish, for example (Davidson *et al.*, 2000, Taylor *et al.*, 2003), and the ultra-conserved elements (Bejerano *et al.*, 2004) discussed later (Section 6.6.2). For more sensitive studies, it is necessary to calculate the neutral rate in localized regions, as the rate of mutation has been found to vary spatially across genomes (Wolfe *et al.*, 1989; Hardison *et al.*, 2003; Taylor *et al.*, 2006) (Section 6.2).

We have seen that for the Ka/Ks measure in protein-coding sequence, both the test and neutral control sequences are interleaved. This is an ideal scenario, negating the confounding influence of regional variation in mutation rates. Regional estimates of the neutral rate can be calculated in a sliding window manner or by calculating it for an arbitrarily defined region of interest. The principal problem with this approach is that sites subject to selection cannot be assumed to be randomly distributed across the genome. For instance, anonymous sequence around the *PAX6* gene (Miles *et al.*, 1998) is highly enriched in functionally important conserved sites (Section 6.6.2). An estimate of the neutral rate based on anonymous sequence around this gene would give an artificially low estimate of the neutral mutation rate in the region. The larger the window used to estimate the regional neutral rate, the less likely it is to be dominated by non-neutral sites, but a larger window reduces the resolution for detecting regional variation in mutation rate. The optimum window size for neutral rate estimation will be a balance of these two opposing needs. Gaffney *et al.* (Gaffney and Keightley, 2005) found that within the rodent lineage, a window of 10 kb is likely to show a consistent neutral rate across its length, and even windows up to 1 Mb may have little variation in neutral rate across them. However, more recent findings suggest that these broad-interval analyses may mask considerable fine-grained variation in the mutation rate, particularly in the primate lineage (Taylor *et al.*, 2006).

Several studies have defined the extent of constrained regions on the basis of ungapped segments of alignment (Duret *et al.*, 1993; Dermitzakis *et al.*, 2002), a strategy that lends itself well to analyses based on local rather than global alignments. Often, these studies use precalibrated thresholds for significant constraint rather than calculating relative rates directly; for example, 70 per cent identity over 100 ungapped nucleotides is a commonly used parameter for man to rodent alignments (Dermitzakis *et al.*, 2002).

Sliding windows have been widely used to arbitrarily define the extent of sequences that are then evaluated for constraint (Mayor *et al.*, 2000; Waterston *et al.*, 2002). The approach can accommodate alignment gaps, generally treating them as nucleotide mismatches (Mayor *et al.*, 2000), but their sensitivity is crucially dependent on the size of the evaluation window and on how much the window is moved along the alignment for each evaluation. Analyses based on sliding windows have also been applied to phylogenetic shadowing (Boffelli *et al.*, 2003). In this case, rather than scoring conservation or substitution rate directly, the substitution rates for each alignment column were compared to the rates of sequences known to be evolving neutrally (e.g., ancient repeats) or subject to selection (e.g., exons), the final score being a likelihood ratio of neutral versus constrained evolution for each alignment column. A web server for phylogenetic shadowing analysis is available (http://bonaire.lbl.gov/shadower/).

An intuitive way of integrating measures of constraint across multiple aligned sequences is to define multiply conserved sequences (MCS). The common core of sequence that aligns in all (or most) sequences from a defined scope can then define the boundaries of the MCS (Margulies *et al.*, 2003; Thomas *et al.*, 2003). For instance, it is easy to see that exons 2 and 3 of *WNT2* can be considered MCS within vertebrates (Figure 6.2). The MCS definition is versatile, accommodating local or global alignments, and can tolerate missing sequence from incompletely sequenced genomes. This MCS paradigm can incorporate thresholds of alignment quality (identity and gap frequency), but, more commonly, a simple default of aligned or not-aligned is used, in which case the sensitivity of the alignment method becomes an arbitrary threshold score.

Two highly versatile tools, RankVISTA (Martin *et al.*, 2004) and phastCons (Siepel *et al.*, 2005), have recently been developed that quantify constraint and operate free of window size and identity thresholds. These tools are also noteworthy because they quantitatively measure constraint rather than the crude binary discrimination into constrained or unconstrained that is common to many of the methods discussed above. RankVISTA integrates pairwise relative rate scores across a multiple alignment, using a phylogenetic weighting scheme (conservation between distantly related species scores better than conservation between closely related species). The neutral rate estimates are derived from anonymous regions in the submitted alignment, and the final score is an easily interpretable probability of observing such conservation in a 10-kb fragment of neutrally evolving sequence. The optimal extent of constrained sequences is determined with a dynamic programming approach. This tool is available from the standard VISTA web server (http://genome.lbl.gov/vista/).

The tool phastCons (Siepel *et al.*, 2005) is one of the first practical implementations of a phylogenetic hidden Markov model (phylo-HMM (Felsenstein and Churchill, 1996)) to score conservation across genomic alignments, in effect scoring how well the observed pattern of substitution matches its internal model of a constrained site. The approach is also noteworthy because it takes into account the tendency for conservation levels to be similar at adjacent sites, and it is an extensible model that could be adapted to incorporate additional parameters. Regularly updated, precomputed phastCons results are available through the UCSC genome browser for multispecies whole-genome alignments. When interpreting phastCons results, it is worth remembering that they are based on genome-wide alignments. It is often the case that a discrete region is identified as highly conserved, but upon further investigation the aligning regions prove to be from non-syntenic loci. Many of these instances can be attributed to alignments involving process-pseudogenes. Consideration of the UCSC browser *Net* alignment track while interpreting phastCons results is a useful way of identifying these anomalous signals.

All of the methods described above focus on nucleotide substitution rates. Insertions and deletions (indels) have the potential to help detect constrained regions; however, estimation of their rate is more sensitive to alignment parameters than is the case of substitution rate calculations (Keightley and Johnson, 2004), and good stochastic models of insertion and deletion in non-coding DNA are not currently available. Alignment gaps are typically treated as either missing data (phastCons) or nucleotide substitutions (RankVISTA) when assessing selective constraint. Neither of these is a particularly satisfactory solution, and phastCons leads to artificially high scores over regions of gapped alignment. Recent work by Lunter *et al.* (2006) has shown that indel rates themselves can be a useful measure of selective constraint. Importantly, the analysis appears to be robust to a range of alignment parameters, suggesting that an accurate indel model may not be absolutely necessary to extract useful measures of selection. Indels to detect constrained regions have been used implicitly before in comparative genomics; for example, in a pip-plot (Figure 6.2, panel 2), exons clearly stand out as much for the length of the horizontal lines (indicating the absence of indels) as they do for the height of the lines on the y-axis (indicating nucleotide identity). However, the real advantage of indel-based measures of selection is that they can be used in conjunction with substitution rate measures, in the same sequence, allowing discordant selective pressures to be simultaneously measured – for example, positive selection driving amino-acid sequence diversity but purifying selection acting to constrain sequence length, or cases where the nucleotide sequence between two protein-binding sites is unconstrained but the spacing between elements is crucial.

6.5.4 Comparative genomics meets population genetics

We have seen that the comparison of sequences between species provides a powerful method to identify functional elements within genomic sequence. If within-species genetic variation (polymorphism) data are also available for any of the aligned sequences, an additional set of analyses becomes tractable, and this can provide independent tests of conclusions drawn from interspecies comparisons and open the door to new biological questions. The prototypical, combined intra- and interspecies analysis is the McDonald–Kreitman test (McDonald and Kreitman, 1991). The basic premise of this test is that mildly deleterious mutations will be present in a population as polymorphisms. However, as they are deleterious, they are unlikely to drift to fixation (frequency = 1.0). The vast majority of sequence differences between even closely related species are likely to be fixed differences; even the most famous example of overdominance maintaining polymorphism between man and chimpanzee has recently been shown to be convergent (Wooding *et al.*, 2006). Under neutral evolution, the Ka/Ks ratio derived from interspecies comparison should equal an analogous non-synonymous:synonymous (Π_a/Π_s) ratio derived from polymorphism data, an excess of interspecies amino-acid substitutions in this test indicating positive selection. In this test, it is not necessary to have complete ascertainment of variation, and control regions and the population frequency of polymorphisms used are unimportant. However, it is crucial that the ascertainment of polymorphism data be unbiased between test and control sequences. The McDonald–Kreitman test can be adapted to any pairwise comparison between test and nearly neutral control sequences, just like the Ka/Ks ratio test. If polymorphism data are available, especially if there are also allele frequency data, a number of measures can be used to reinforce conclusions drawn initially from interspecies comparative genomic studies and reveal the direction of recent selection (see .Hahn, in press, for an excellent review of this subject).

6.6 Applications

There have been a huge number of published studies that are either centred on comparative genomic analysis or utilize comparative genomics to address specific questions within a wider study. In the next few sections, we highlight a small number of examples that have given new insight into the general biology of vertebrate genomes and provide good examples of the application of methods described in this chapter.

6.6.1 How much of the human genome is constrained?

In Section 6.2, we provided a brief overview of the human genomic landscape. One of the most prominent features of that landscape was the apparent dearth of identified functional sequences, such as those encoding proteins, and an abundance of repeat sequences that presumably do not usefully contribute to the biology of the organism. With publication of both the draft human (Lander *et al.*, 2001) and mouse (Waterston *et al.*, 2002) genomes, it became possible to apply comparative genomic methodologies to the entire genome and test these presumptions. In particular, it became possible to estimate the total proportion of the human genome that is subject to selective constraint, and so estimate the proportion of the genome that has conserved function but is not protein coding. A conservation score was calculated for non-overlapping 50 nucleotide windows of human:mouse whole-genome pairwise alignments. Two sets of scores were derived, one for the complete alignment and a second only from aligned ancient repeats. As ancient repeats are thought to be unconstrained by selection (Section 6.2), the distribution of conservation scores should reflect the pattern expected under neutral evolution. The distribution of scores from the whole-genome alignments substantially overlaps those derived just from ancient repeats, although a significant shoulder specifically toward higher scores is evident (Waterston *et al.*, 2002). Subtracting the ancient repeat distribution from that of the whole genome suggests that approximately 5 per cent of 50 nucleotide windows are more highly constrained than expected under neutral evolution. Similar analyses based on human to rat comparisons have supported this conclusion (Gibbs *et al.*, 2004). These studies are not without their limitations. For example, isolated regions of constraint that are substantially shorter than the 50-nucleotide window size used will have gone undetected, suggesting that the estimate of 5 per cent may be a lower bound for the true value. Generally, similar fractions of 2.6–3.5 per cent of the human genome were found to show evidence of selective constraint by the indel-based method (Lunter *et al.*, 2006) described in the previous section. These studies have led to the important conclusion that much of the functionally constrained sequence in the human genome does not code for proteins.

If coding sequences are not the singularly dominant functional component of the genome, the question arises, what are the functions of non-coding sequence? Several types of non-coding elements are known, such as *cis*-regulators of transcription and splicing and RNA structures that influence transcript localization and stability, as well as transcripts whose functional product is RNA rather than protein (see Mattick (2004) for review). It is also likely that there are classes of functional elements that we have yet to discover. This potential naivety is well illustrated by the relatively recent realization that a major class of non-coding functional elements (microRNAs) has been almost entirety overlooked (Ambros, 2004; see Chapter 14).

It is one of the great strengths of comparative genomics that no prior assumption of the function is required to identify a sequence as functionally important. With the increasing depth of available genomes (Section 6.5) and the methods described above, we are rapidly approaching the stage where we can confidently identify short regions and possibly even single nucleotides as constrained. A remaining and significant challenge is to characterize the function of those sites. Again comparative genomics can help. We have already seen that there is a characteristic profile of conservation for protein-coding sequence (Section 6.5.1), and similar profiles may exist for other categories of functionally important sequence. Dermitzakis *et al.* (2004) found that conserved, non-genic sequences (CNGs) accumulated sequence changes in a manner that can be statistically distinguished from both protein-coding sequences and non-coding RNA genes. These patterns of sequence change most resembled clusters of protein-binding sites.

6.6.2 Ultra-conserved regions

The sequences studied by Dermitzakis *et al.* (2004) were selected, from chromosome 21, on the basis of a simple threshold identity in man to mouse alignment, and also on the ability to PCR amplify homologous sequences from 14 mammalian species.

Consequently, these sequences should represent the subset of CNGs that both have the highest nucleotide identity and are the most constrained through mammalian evolution. Ironically, a whole-genome analysis of non-coding conservation has since shown that human chromosome 21 is the only autosome devoid of so-called ultra-conserved elements (Bejerano *et al.*, 2004). These elements are also defined on the basis of simple and arbitrary length and identity thresholds, but, in this case, the very stringent criteria of 200-nucleotide ungapped alignment between human, mouse and rat, and 100 per cent nucleotide identity in all three species. In total, 481 of these incredibly well-conserved sequences were found.

Although defined initially on the basis of conservation between man and rodents, 97 per cent of the ultra-conserved elements could be identified in the chicken genome with, on average, over 95 per cent nucleotide identity, and more than 66 per cent of them could be aligned with a puffer fish genome (*Takifugu rubripes*). In contrast, only 5 per cent could be identified in any of the non-vertebrates *Ciona intestinalis* (sea squirt), *Drosophila melanogaster* (fruit fly) or *Caenorhabditis elegans* (nematode worm), and all of these were ultra-conserved elements that overlap protein-coding exons from known genes. It appears, then, that although these ultra-conserved elements have been highly constrained for 300–450 million years of vertebrate evolution (Bejerano *et al.*, 2004), they are largely confined to the vertebrates. A similar study making use of a recently available whole-genome sequence from multiple insects, has also identified ultra-conserved regions between fruit flies and the mosquito *Anopheles gambiae* (Glazov *et al.*, 2005). However, the majority of ultra-conserved elements identified in fruit flies were substantially shorter than the 200-nucleotide threshold used for the mammalian study, despite similar evolutionary distances, for some of the analyses, in both studies.

It has been noted in both mammals and fruit flies that ultra-conserved elements are often located in the introns of, or intergenic regions around, developmentally important genes (Bejerano *et al.*, 2004; Glazov *et al.*, 2005; Woolfe *et al.*, 2005). These developmental regulatory genes often encode DNA-binding transcription factors or RNA-binding proteins (Bejerano *et al.*, 2004) that are likely to be involved in the regulation of RNA processing and transport. These observations have invoked the notion of developmental master regulators: regions that integrate multiple signals coordinating the expression of genes that, in turn, regulate many more genes through transcription and RNA processing. Some experimental support for this idea has been provided by Woolfe *et al.* (2005) in a zebrafish experimental system. Of 25 non-coding sequence elements that are highly conserved between man and pufferfish, 23 showed significant transcriptional enhancer activity in one or more tissues during zebrafish development.

The idea that ultra-conserved elements act as developmental regulators fits well with the observation that they are highly conserved within phylogenetic clades that share similar developmental programs, but apparently are not conserved between more diverse groups. Could the ultra-conserved elements that are common to both man and pufferfish be the master regulators that define the basic vertebrate body plan:

skeletal structure, musculature and internal organs, and the developmental programs to orchestrate their construction? This is an attractive idea, but much more work is required to establish whether this is even close to accurate. In particular, some genes are known to be key regulators of developmental programs, and the orthologous genes in both man and fruit fly are apparently performing the same task in the same tissue. *PAX6*, for example, is crucial in the development of eyes in both man and fruit fly (van Heyningen and Williamson, 2002). The human *PAX6* locus is one of the richest in ultra-conserved elements (Bejerano *et al.*, 2004), and six out of seven tested elements show enhancer activity, four of which directed expression preferentially in the developing eye (Woolfe *et al.*, 2005). Despite the conserved role of *PAX6* in eye development between man and fruit fly and the demonstrated role of mammalian ultra-conserved elements in directing that expression, there is no identifiable sequence similarity between the ultra-conserved elements and the fruit-fly *PAX6* locus.

6.6.3 Specific locus studies

In this section, we focus on a small number of disease-related studies that have been substantially advanced through the application of comparative genomics. We make several references to Online Mendelian Inheritance in Man (OMIM), a key human curated resource that brings together published information relating human genetic diseases and disease genes. Full OMIM records can be obtained with their identifier number from the Entrez system (http://www.ncbi.nlm.nih.gov/entrez/).

Hirschsprung's disease is a congenital disorder characterized by intestinal abnormalities (OMIM:142623). The genetics of this disease have been well studied, but the pattern of inheritance is complex. Mutations have been found in eight loci that contribute to disease susceptibility (OMIM:142623 for review), but these account for only 30 per cent of cases (Emison *et al.*, 2005). Genetic evidence indicated that one of those eight loci, the *RET* gene, harboured additional, previously undetected mutations or variants that account for much of the remaining susceptibility (Gabriel *et al.*, 2002). All apparent protein-coding sequence of *RET* had already been screened for mutations, so the challenge was to identify additional functionally important non-coding sites within the locus or identify previously missed protein-coding sequence.

Emison *et al.* (2005) identified more than 30 regions of conserved non-coding sequence in 350 kb of genomic sequence centred on the *RET* gene. The analysis used the multiple conserved sequences paradigm (Section 6.5.3) based on alignment of 12 orthologous loci from vertebrates. Only five of the conserved non-coding regions were within the region maximally implicated by genetic evidence. The comparative analysis also indicated that a human single-nucleotide polymorphism (SNP) is located within one of the conserved regions, and not withstanding the polymorphism, the nucleotide has been highly conserved through vertebrate evolution, an obvious candidate for the functional variant. Emison *et al.* (2005) were able to show that this conserved element has enhancer activity and that the level of that activity is influenced by the SNP genotype. The comparative alignment allowed the ancestral and derived alleles to be discriminated, the lower enhancer activity and disease susceptibility being associated with the more recently derived allele. The non-coding SNP genotype was shown to account for much of the previously unaccounted for genetic susceptibility contributed by the *RET* locus.

The Hirschprung's disease *RET* locus is a good recent example of the utility of comparative genomics and its synergy with genetic studies. It also stands out for several of other reasons. The functional variant identified is common in the population, exceeding 50 per cent in some parts of East Asia, despite being disease-associated. The effect of the genotype is influenced by sex, demonstrating a form of epistasis, and the variant is regulatory rather than protein coding. All of these features are likely to be frequently encountered when searching for the genetic risk factors in common diseases (Marchini *et al.*, 2005) such as cancer, heart-disease, diabetes and stroke.

The *RPGR* gene has a similar story to the *RET* locus. *RPGR* is known to be a major locus for X-linked retinitis pigmentosa (OMIM:312610), a form of retinal degeneration. Several known disease-associated coding sequence mutations had been found, but it was apparent from genetic studies that many more cases of retinitis pigmentosa should be attributable to the locus than could be explained by the mutations in the coding sequence (Teague *et al.*, 1994; Vervoort *et al.*, 2000). Comparative genomics revealed a previously unknown, alternately spliced protein-coding exon that was specifically expressed in the retina and harboured the missing mutations (Vervoort *et al.*, 2000). In this case, all of the disease-associated mutations disrupted the encoded protein. It is likely that such missing mutations are common for even well-studied genes and that they are simply under-reported in the literature, because it is seldom practical to screen large genomic intervals for mutations, nor is it easy to demonstrate their causal role.

Our next example demonstrates over how wide an interval *cis*-regulatory sites can act, but also how, even when the region is large and complex, comparative genomics can allow functional sites to be identified and subsequently characterized. The mouse *Sasquatch* (*Ssq*) mutation was generated serendipitously in trying to insert a transgene into the genome. The transgene integration led to ectopic expression of the developmental signalling molecule *Sonic Hedgehog* (*SHH*) and resulted in preaxial polydactyly (extra digits) (Sharpe *et al.*, 1999). Intriguingly, genetic evidence demonstrated that the effect was specifically in *cis* (Sharpe *et al.*, 1999), but, as the integration site was over 1 Mb from *Shh* and located within the intron of an adjacent gene, identifying the functional regulatory element remained a challenge.

Multiple sequence alignment between orthologous regions from mouse, man, chicken and pufferfish identified a 0.8-kb stretch of sequence close to the transgene insertion site that has been highly conserved throughout vertebrate evolution (Lettice *et al.*, 2003). It has now been shown that the 0.8-kb element, known as the ZRS, is a limb bud-specific enhancer of *Shh* expression (Sagai *et al.*, 2004; Lettice and Hill, 2005) and that even the fish sequence can drive expression in the mouse limb bud. These studies of the *Shh* locus have shown that *cis*-regulatory elements can be located large distances, at least 1 Mb, along linear DNA from the genes they act to regulate. Not only can these elements be far from their targets, but they may also be closer to other genes on which they apparently have no regulatory role – the ZRS is located in the fifth intron of the *Lmbr1* gene, whose expression is unaffected by mutations in the ZRS.

Like the ultra-conserved sequences described above, the striking conservation of the ZRS throughout vertebrate development indicates that even single nucleotide substitutions in the region are likely to be detrimental and strongly selected against. Accordingly, point mutations in the ZRS have been found in four human families and two mouse lines, and in each case they lead to preaxial polydactyly (see Lettice and Hill, 2005, for review). In contrast to these point mutations, complete deletion of the ZRS in the mouse abolishes *Shh* expression in the limb bud and results in severely truncated limbs (Sagai *et al.*, 2005), a similar phenotype to human acheiropodia, which is also linked to the *Shh* locus (Ianakiev *et al.*, 2001). Several vertebrate lineages, such as snake, have substantially reduced or entirely lost limbs, although they were present in their ancestors. Sagai *et al.* (2004) have shown that for at least two of these cases, snakes and limbless newts, this has coincided with the loss of the ZRS, whereas it remains conserved in lizards and legged newts. Whether loss of the ZRS was a primary event in the morphological transition of either of these separate lineages, or whether it represents secondary losses, remains unclear; but it does illustrate two points rather well. First, the importance of selecting an appropriate phylogenetic scope for a comparative genomic study (Section 6.3); an analysis utilizing legless newts and snakes, rather than pufferfish and chickens, would not have revealed the ZRS in the first place. Second, it demonstrates the apparently modular nature of conserved non-coding sequence blocks in evolution. The ZRS can be lost without apparently disrupting the many other functions (OMIM:600725) of *Shh* during vertebrate development.

6.7 Challenges and future directions

There has been great progress in understanding the biology and functions encoded by the human genome since the first draft of a reference sequence was produced in 2001 (Lander *et al.*, 2001; (Venter *et al.*, 2001), and much of this insight has been gained by comparison both within and between genomes. However, as with many scientific endeavours, more questions arise with each increment in understanding. For example, we have now realized that much of the functionally constrained sequence in the human genome does not encode proteins, and our current understanding of these elements is poor. They are the dark matter of the genome. A major and current challenge is to identify each of these elements and to start dissecting their function. In particular, it is likely that they will harbour polymorphisms that affect human health, contributing to common disease susceptibility. The integration of comparative genomics with genetic variation data (IHC, 2005) to identify functional polymorphisms is likely to be a rapidly expanding field with the combined assets of multiple mammalian genome sequences and high-density confirmed polymorphism data available.

Sequence comparison alone may be able to identify all constrained sites, but it is unlikely to be able to establish their associated functions. Rather, it is the synergy of comparative studies with laboratory experiment that provides greatest insight. This approach is embodied by the Encyclopaedia of DNA elements (ENCODE) project, an international initiative with the ultimate aim of identifying all functional elements in the human genome (ENCODE Project Consortium (EPC), 2004), in effect to shed light on the dark matter of the genome. This is an ambitious and relatively long-term goal. As a first step, a pilot project has been undertaken to investigate 30 Mb of the human genome (approximately 1 per cent of the genome, selected primarily on the basis of gene density and evolutionary conservation) in great detail, applying a broad spectrum of experimental and computational methods to identify functionally important sites (http://genome.gov/10005107). These rigorously annotated regions will be important training and testing grounds for the development of methods in comparative genomics. The UCSC genome browser (http://genome.ucsc.edu/ENCODE/) provides a key portal to access the ENCODE pilot project data.

6.8 Conclusion

In the middle of 2000, credible estimates of the total number of human protein-coding genes plummeted from 80 000–100 000 to 30 000 or so (Roest Crollius *et al.*, 2000). These lower counts were essentially confirmed by the early analyses of the human genome (Lander *et al.*, 2001) and, if anything, the real numbers are likely to be smaller still (IHGSC, 2004). Although it is difficult, and perhaps even of little value, to interpret these results within the commonly perceived frameworks of organismal complexity, the fact remains that they have created a new impetus for looking beyond protein-coding genes to other classes of functional elements, such as non-coding RNAs and, in particular, the *cis*-acting elements regulating gene expression. At the same time, it is sobering to reflect on how unanticipated these downward revisions of gene count were, and accordingly to reserve judgement on exactly how many more functional elements of major relevance we may expect to find. The methods and early results presented in this review are merely the first steps on a long path to a broader understanding of the totality of information encoded in the genome.

Acknowledgements

This work is based, in part, on a paper published elsewhere (Taylor and Copley, in *Genomes to Therapies*, edited by Thomas Lengauer), and it benefited from editorial comments by Thomas Lengauer and an anonymous reviewer. M.S.T. and R.R.C. enjoy the financial support of the Wellcome Trust.

References

Altschul, S. F., Madden, T. L., Schaffer, A. A. *et al.* (1997). Gapped BLAST and PSI-BLAST: a new generation of protein database search programs. *Nucleic Acids Res* **25**, 3389–3402.

Ambros, V. (2004). The functions of animal microRNAs. *Nature* **431**, 350–355.

Aparicio, S., Chapman, J., Stupka, E. *et al.* (2002). Whole-genome shotgun assembly and analysis of the genome of *Fugu rubripes. Science* **297**, 1301–1310.

Batzoglou, S. (2005). The many faces of sequence alignment. *Brief Bioinform* **6**, 6–22.

Batzoglou, S., Pachter, L., Mesirov, J. P. *et al.* (2000). Human and mouse gene structure: comparative analysis and application to exon prediction. *Genome Res* **10**, 950–958.

Bejerano, G., Pheasant, M., Makunin, I. *et al.* (2004). Ultraconserved elements in the human genome. *Science* **304**, 1321–1325.

Bird, A. (2002). DNA methylation patterns and epigenetic memory. *Genes Dev* **16**, 6–21.

Bird, A., Taggart, M., Frommer, M. *et al.* (1985). A fraction of the mouse genome that is derived from islands of nonmethylated, CpG-rich DNA. *Cell* **40**, 91–99.

Blanchette, M., Kent, W. J., Riemer, C. *et al.* (2004). Aligning multiple genomic sequences with the threaded blockset aligner. *Genome Res* **14**, 708–715.

Boffelli, D., McAuliffe, J., Ovcharenko, D. *et al.* (2003). Phylogenetic shadowing of primate sequences to find functional regions of the human genome. *Science* **299**, 1391–1394.

Bray, N., Dubchak, I. and Pachter, L. (2003). AVID: a global alignment program. *Genome Res* **13**, 97–102.

Bray, N. and Pachter, L. (2003). MAVID multiple alignment server. *Nucleic Acids Res* **31**, 3525–3526.

Brenner, S. E., Chothia, C. and Hubbard, T. J. (1998). Assessing sequence comparison methods with reliable structurally identified distant evolutionary relationships. *Proc Natl Acad Sci U S A* **95**, 6073–6078.

Brudno, M., Chapman, M., Gottgens, B. *et al.* (2003a). Fast and sensitive multiple alignment of large genomic sequences. *BMC Bioinformatics* **4**, 66.

Brudno, M., Do, C. B., Cooper, G. M. *et al.* (2003b). LAGAN and Multi-LAGAN: efficient tools for large-scale multiple alignment of genomic DNA. *Genome Res* **13**, 721–731.

Brudno, M., Malde, S., Poliakov, A. *et al.* (2003c). Glocal alignment: finding rearrangements during alignment. *Bioinformatics* **19** Suppl 1, i54–62.

Brudno, M., Steinkamp, R. and Morgenstern, B. (2004). The CHAOS/DIALIGN WWW server for multiple alignment of genomic sequences. *Nucleic Acids Res* **32**, W41–44.

Carninci, P., Kasukawa, T., Katayama, S. *et al.* (2005). The transcriptional landscape of the mammalian genome. *Science* **309**, 1559–1563.

Cheng, Z., Ventura, M., She, X. *et al.* (2005). A genome-wide comparison of recent chimpanzee and human segmental duplications. *Nature* **437**, 88–93.

Collins, F. S., Green, E. D., Guttmacher, A. E. *et al.* (2003). A vision for the future of genomics research. *Nature* **422**, 835–847.

Cooper, G. M., Brudno, M., Green, E. D. *et al.* (2003). Quantitative estimates of sequence divergence for comparative analyses of mammalian genomes. *Genome Res* **13**, 813–820.

Cooper, G. M., Brudno, M., Stone, E. A. *et al.* (2004). Characterization of evolutionary rates and constraints in three mammalian genomes. *Genome Res* **14**, 539–548.

Couronne, O., Poliakov, A., Bray, N. *et al.* (2003). Strategies and tools for whole-genome alignments. *Genome Res* **13**, 73–80.

Darwin, C. (1859). *On the Origin of Species by Means of Natural Selection, or The Perservation of Favoured Races in the Struggle for Life.* London: Murray.

Davidson, H., Taylor, M. S., Doherty, A. *et al.* (2000). Genomic sequence analysis of *Fugu rubripes* CFTR and flanking genes in a 60 kb region conserving synteny with 800 kb of human chromosome 7. *Genome Res* **10**, 1194–1203.

Dermitzakis, E. T., Kirkness, E., Schwarz, S. *et al.* (2004). Comparison of human chromosome 21 conserved nongenic sequences (CNGs) with the mouse and dog genomes shows that their selective constraint is independent of their genic environment. *Genome Res* **14**, 852–859.

Dermitzakis, E. T., Reymond, A., Lyle, R. *et al.* (2002). Numerous potentially functional but non-genic conserved sequences on human chromosome 21. *Nature* **420**, 578–582.

Duret, L., Dorkeld, F. and Gautier, C. (1993). Strong conservation of non-coding sequences during vertebrate evolution: potential involvement in post-transcriptional regulation of gene expression. *Nucleic Acids Res* **21**, 2315–2322.

Eddy, S. R. (2005). A model of the statistical power of comparative genome sequence analysis. *PLoS Biol* **3**, e10.

Eichler, E. E. (2006). Widening the spectrum of human genetic variation. *Nat Genet* **38**, 9–11.

Elnitski, L., Giardine, B., Shah, P. *et al.* (2005). Improvements to GALA and dbERGE II: databases featuring genomic sequence alignment, annotation and experimental results. *Nucleic Acids Res* **33**, D466–470.

Emison, E. S., McCallion, A. S., Kashuk, C. S. *et al.* (2005). A common sex-dependent mutation in a RET enhancer underlies Hirschsprung disease risk. *Nature* **434**, 857–863.

EPC (2004). The ENCODE (ENCyclopedia Of DNA Elements) Project. *Science* **306**, 636–640.

Felsenstain, J. and Churchill, G. A. (1996). A Hidden Markov Model approach to variation among sites in rate of evolution. *Mol Biol Evol* **13**, 93–104.

Gabriel, S. B., Salomon, R., Pelet, A. *et al.* (2002). Segregation at three loci explains familial and population risk in Hirschsprung disease. *Nat Genet* **31**, 89–93.

Gaffney, D. J. and Keightley, P. D. (2005). The scale of mutational variation in the murid genome. *Genome Res* **15**, 1086–1094.

Giardine, B., Elnitski, L., Riemer, C. *et al.* (2003). GALA, a database for genomic sequence alignments and annotations. *Genome Res* **13**, 732–741.

Gibbs, R. A., Weinstock, G. M., Metzker, M. L. *et al.* (2004). Genome sequence of the Brown Norway rat yields insights into mammalian evolution. *Nature* **428**, 493–521.

Glazko, G. V., Koonin, E. V. and Rogozin, I. B. (2005). Molecular dating: ape bones agree with chicken entrails. *Trends Genet* **21**, 89–92.

Glazov, E. A., Pheasant, M., McGraw, E. A. *et al.* (2005). Ultraconserved elements in insect genomes: a highly conserved intronic sequence implicated in the control of homothorax mRNA splicing. *Genome Res* **15**, 800–808.

Gogarten, J. P. and Olendzenski, L. (1999). Orthologs, paralogs and genome comparisons. *Curr Opin Genet Dev* **9**, 630–636.

Hahn, M. (in press). Detecting natural selection on *cis*-regulatory DNA.

Hardison, R. C., Roskin, K. M., Yang, S. *et al.* (2003). Covariation in frequencies of substitution, deletion, transposition, and recombination during eutherian evolution. *Genome Res* **13**, 13–26.

Hillier, L. W., Miller, W., Birney, E. *et al.* (2004). Sequence and comparative analysis of the chicken genome provide unique perspectives on vertebrate evolution. *Nature* **432**, 695–716.

Hoegg, S., Brinkmann, H., Taylor, J. S. *et al.* (2004). Phylogenetic timing of the fish-specific genome duplication correlates with the diversification of teleost fish. *J Mol Evol* **59**, 190–203.

Ianakiev, P., Van Baren, M. J., Daly, M. J. *et al.* (2001). Acheiropodia is caused by a genomic deletion in C7orf2, the human orthologue of the Lmbr1 gene. *Am J Hum Genet* **68**, 38–45.

IHC (2005). A haplotype map of the human genome. *Nature* **437**, 1299–1320.

IHGSC (2004). Finishing the euchromatic sequence of the human genome. *Nature* **431**, 931–945.

Jaillon, O., Aury, J. M., Brunet, F. *et al.* (2004). Genome duplication in the teleost fish *Tetraodon nigroviridis* reveals the early vertebrate proto-karyotype. *Nature* **431**, 946–957.

Kapitonov, V. V. and Jurka, J. (2005). RAG1 core and V(D)J recombination signal sequences were derived from Transib transposons. *PLoS Biol* **3**, e181.

Kazazian, H. H., Jr (2004). Mobile elements: drivers of genome evolution. *Science* **303**, 1626–1632.

Keightley, P. D. and Johnson, T. (2004). MCALIGN: stochastic alignment of noncoding DNA sequences based on an evolutionary model of sequence evolution. *Genome Res* **14**, 442–450.

Keightley, P. D., Lercher, M. J. and Eyre-Walker, A. (2005). Evidence for widespread degradation of gene control regions in hominid genomes. *PLoS Biol* **3**, e42.

Kent, W. J. (2002). BLAT – the BLAST-like alignment tool. *Genome Res* **12**, 656–664.

Kent, W. J. and Zahler, A. M. (2000). Conservation, regulation, synteny, and introns in a large-scale *C. briggsae–C. elegans* genomic alignment. *Genome Res* **10**, 1115–1125.

Kimura, M. (1977). Preponderance of synonymous changes as evidence for the neutral theory of molecular evolution. *Nature* **267**, 275–276.

Kowalski, P. E., Freeman, J. D. and Mager D. L. (1999). Intergenic splicing between a HERV-H endogenous retrovirus and two adjacent human genes. *Genomics* **57**, 371–379.

Lander, E. S., Linton, L. M., Birren, B. *et al.* (2001). Initial sequencing and analysis of the human genome. *Nature* **409**, 860–921.

Law, S. Y., Fok, M., Cheng, S. W. *et al.* (1992). A comparison of outcome after resection for squamous cell carcinomas and adenocarcinomas of the esophagus and cardia. *Surg Gynecol Obstet* **175**, 107–112.

Lettice, L. A., Heaney, S. J., Purdie, L. A. *et al.* (2003). A long-range Shh enhancer regulates expression in the developing limb and fin and is associated with preaxial polydactyly. *Hum Mol Genet* **12**, 1725–1735.

Lettice, L. A. and Hill, R. E. (2005). Preaxial polydactyly: a model for defective long-range regulation in congenital abnormalities. *Curr Opin Genet Dev* **15**, 294–300.

Lim, L. P., Lau, N. C., Garrett-Engele, P. *et al.* (2005). Microarray analysis shows that some microRNAs downregulate large numbers of target mRNAs. *Nature* **433**, 769–773.

Linblad-Toh, K., Wade, C. M., Mikkelsen, T. S. *et al.* (2005). Genome sequence, comparative analysis and haplotype structure of the domestic dog. *Nature* **438**, 803–819.

Lunter, G., Ponting, C. P. and Hein, J. (2006). Genome-wide identification of human functional DNA using a neutral indel model. *PLoS Comput Biol* **2**, e5.

Lynch, M. (2006). The origins of eukaryotic gene structure. *Mol Biol Evol* **23**, 450–468.

Ma, B., Tromp, J. and Li, M. (2002). PatternHunter: faster and more sensitive homology search. *Bioinformatics* **18**, 440–445.

Maniatis, T. and Tasic, B. (2002). Alternative pre-mRNA splicing and proteome expansion in metazoans. *Nature* **418**, 236–243.

Marchini, J., Donnelly, P. and Cardon, L. R. (2005). Genome-wide strategies for detecting multiple loci that influence complex diseases. *Nat Genet* **37**, 413–417.

Margulies, E. H., Blanchette, M., Haussler, D. *et al.* 2003). Identification and characterization of multi-species conserved sequences. *Genome Res* **13**, 2507–2518.

Margulies, E. H., Vinson, J. P., Miller, W. *et al.* (2005). An initial strategy for the systematic identification of functional elements in the human genome by low-redundancy comparative sequencing. *Proc Natl Acad Sci U S A* **102**, 4795–4800.

Martin, J., Han, C., Gordon, L. A. *et al.* (2004). The sequence and analysis of duplication-rich human chromosome 16. *Nature* **432**, 988–994.

Mattick, J. S. (2004). RNA regulation: a new genetics? *Nat Rev Genet* **5**, 316–323.

Mayor, C., Brudno, M., Schwartz, J. R. *et al.* (2000). VISTA: visualizing global DNA sequence alignments of arbitrary length. *Bioinformatics* **16**, 1046–1047.

McDonald, J. H. and Kreitman, M. (1991). Adaptive protein evolution at the Adh locus in *Drosophila. Nature* **351**, 652–654.

Miles, C., Elgar, G., Coles, E. *et al.* (1998). Complete sequencing of the Fugu WAGR region from WT1 to PAX6: dramatic compaction and conservation of synteny with human chromosome 11p13. *Proc Natl Acad Sci U S A* **95**, 13068–13072.

Miller, W., Makova, K. D., Nekrutenko, A. *et al.* (2004). Comparative genomics. *Annu Rev Genomics Hum Genet* **5**, 15–56.

Morgenstern, B. (2004). DIALIGN: multiple DNA and protein sequence alignment at BiBiServ. *Nucleic Acids Res* **32**, W33–36.

Mullikin, J. C. and Ning, Z. (2003). The phusion assembler. *Genome Res* **13,** 81–90.

Needleman, S. B. and Wunsch, C. D. (1970). A general method applicable to the search for similarities in the amino acid sequence of two proteins. *J Mol Biol* **48**, 443–453.

Nei, M. and Kumar, S. (2000). *Molecular Evolution and Phylogenetics.* New York: Oxford University Press.

Nguyen, D. Q., Webber, C. and Ponting, C. P. (2006). Bias of selection on human copy-number variants. *PLoS Genet* **2**, e20.

Nielsen, R., Bustamante, C., Clark, A. G. *et al.* (2005). A scan for positively selected genes in the genomes of humans and chimpanzees. *PLoS Biol* **3**, e170.

Ning, Z., Cox, A. J. and Mullikin, J. C. (2001). SSAHA: a fast search method for large DNA databases. *Genome Res* **11**, 1725–1729.

Ohta, T. (1976). Simple model for treating evolution of multigene families. *Nature* **263**, 74–76.

Ophir, R. and Graur, D. (1997). Patterns and rates of indel evolution in processed pseudogenes from humans and murids. *Gene* **205**, 191–202.

Petrov, D. A. and Hartl, D. L. (1998). High rate of DNA loss in the *Drosophila melanogaster* and *Drosophila virilis* species groups. *Mol Biol Evol* **15**, 293–302.

Piganeau, G. and Eyre-Walker, A. (2003). Estimating the distribution of fitness effects from DNA sequence data: implications for the molecular clock. *Proc Natl Acad Sci U S A* **100**, 10335–10340.

Podlaha, O. and Zhang, J. (2004). Nonneutral evolution of the transcribed pseudogene Makorin1-p1 in mice. *Mol Biol Evol* **21**, 2202–2209.

Pollard, D. A., Bergman, C. M., Stoye, J. *et al.* (2004). Benchmarking tools for the alignment of functional noncoding DNA. *BMC Bioinformatics* **5**, 6.

Potter, S. C., Clarke, L., Curwen, V. *et al.* (2004). The Ensembl analysis pipeline. *Genome Res* **14**, 934–941.

Roest Crollius, H., Jaillon, O., Bernot, A. *et al.* (2000). Estimate of human gene number provided by genome-wide analysis using *Tetraodon nigroviridis* DNA sequence. *Nat Genet* **25**, 235–238.

Roy, A. M., Carroll, M. L., Nguyen, S. V. *et al.* (2000). Potential gene conversion and source genes for recently integrated Alu elements. *Genome Res* **10**, 1485–1495.

Sagai, T., Hosoya, M., Mizushina, Y. *et al.* (2005). Elimination of a long-range *cis*-regulatory module causes complete loss of limb-specific Shh expression and truncation of the mouse limb. *Development* **132**, 797–803.

Sagai, T., Masuya, H., Tamura, M. *et al.* (2004). Phylogenetic conservation of a limb-specific, *cis*-acting regulator of sonic hedgehog (Shh). *Mamm Genome* **15**, 23–34.

Schwartz, S., Kent, W. J., Smit, A. *et al.* (2003). Human-mouse alignments with BLASTZ. *Genome Res* **13**, 103–107.

Schwartz, S., Zhang, Z., Frazer, K. A. *et al.* (2000). PipMaker – a web server for aligning two genomic DNA sequences. *Genome Res* **10**, 577–586.

Sharp, A. J., Locke, D. P., McGrath, S. D. *et al.* (2005). Segmental duplications and copy-number variation in the human genome. *Am J Hum Genet* **77**, 78–88.

Sharpe, J., Lettice, L., Hecksher-Sorensen, J. *et al.* (1999). Identification of sonic hedgehog as a candidate gene responsible for the polydactylous mouse mutant Sasquatch. *Curr Biol* **9**, 97–100.

She, X., Jiang, Z., Clark, R. A. *et al.* (2004). Shotgun sequence assembly and recent segmental duplications within the human genome. *Nature* **431**, 927–930.

Shemesh, R., Novik, A., Edelheit, S. *et al.* (2006). Genomic fossils as a snapshot of the human transcriptome. *Proc Natl Acad Sci U S A* **103**, 1364–1369.

Siepel, A., Bejerano, G., Pedersen, J. S. *et al.* (2005). Evolutionarily conserved elements in vertebrate, insect, worm, and yeast genomes. *Genome Res* **15**, 1034–1050.

Singh, N. D., Arndt, P. F. and Petrov, D. A. (2005). Genomic heterogeneity of background substitutional patterns in *Drosophila melanogaster*. *Genetics* **169**, 709–722.

Sonnhammer, E. L. and Durbin, R. (1995). A dot-matrix program with dynamic threshold control suited for genomic DNA and protein sequence analysis. *Gene* **167**, GC1-10.

Stoye, J., Evers, D. and Meyer, F. (1998). Rose: generating sequence families. *Bioinformatics* **14**, 157–163.

Sved, J. and Bird, A. (1990). The expected equilibrium of the CpG dinucleotide in vertebrate genomes under a mutation model. *Proc Natl Acad Sci U S A* **87**, 4692–4696.

Tagle, D. A., Koop, B. F., Goodman, M. *et al.* (1988). Embryonic epsilon and gamma globin genes of a prosimian primate (*Galago crassicaudatus*). Nucleotide and amino acid sequences, developmental regulation and phylogenetic footprints. *J Mol Biol* **203**, 439–455.

Tatusov, R. L., Fedorova, N. D., Jackson, J. D. *et al.* (2003). The COG database: an updated version includes eukaryotes. *BMC Bioinformatics* **4**, 41.

Taylor, M. S., Devon, R. S., Millar, J. K. *et al.* (2003). Evolutionary constraints on the disrupted in schizophrenia locus. *Genomics* **81**, 67–77.

Taylor, M. S., Kai, C., Kawai, J. *et al.* (2006). Heterotachy in mammalian promoter evolution. *PLoS Genet* **2**, e30.

Taylor, M. S., Ponting, C. P. and Copley, R. R. (2004). Occurrence and consequences of coding sequence insertions and deletions in mammalian genomes. *Genome Res* **14**, 555–566.

Teague, P. W., Aldred, M. A., Jay, M. *et al.* (1994). Heterogeneity analysis in 40 X-linked retinitis pigmentosa families. *Am J Hum Genet* **55**, 105–111.

Thomas, J. W., Touchman, J. W., Blakesley, R. W. *et al.* (2003). Comparative analyses of multi-species sequences from targeted genomic regions. *Nature* **424**, 788–793.

Van Heyningen, V. and Williamson, K. A. (2002). PAX6 in sensory development. *Hum Mol Genet* **11**, 1161–1167.

Venter, J. C., Adams, M. D., Myers, E. W. *et al.* (2001). The sequence of the human genome. *Science* **291**, 1304–1351.

Vervoort, R., Lennon, A., Bird, A. C. *et al.* (2000). Mutational hot spot within a new RPGR exon in X-linked retinitis pigmentosa. *Nat Genet* **25**, 462–466.

Waterston, R. H., Lindblad-Toh, K., Birney, E. *et al.* (2002). Initial sequencing and comparative analysis of the mouse genome. *Nature* **420**, 520–562.

Weber, J. L. and Myers, E. W. (1997). Human whole-genome shotgun sequencing. *Genome Res* **7**, 401–409.

Wendl, M. C. and Yang, S. P. (2004). Gap statistics for whole genome shotgun DNA sequencing projects. *Bioinformatics* **20**, 1527–1534.

Wolfe, K. H., Sharp, P. M. and Li, W. H. (1989). Mutation rates differ among regions of the mammalian genome. *Nature* **337**, 283–285.

Wooding, S., Bufe, B., Grassi, C. *et al.* (2006). Independent evolution of bitter-taste sensitivity in humans and chimpanzees. *Nature* **440**, 930–934.

Woolfe, A., Goodson, M., Goode, D. K. *et al.* (2005). Highly conserved non-coding sequences are associated with vertebrate development. *PLoS Biol* **3**, e7.

Zhang, Z., Carriero, N. and Gerstein, M. (2004). Comparative analysis of processed pseudogenes in the mouse and human genomes. *Trends Genet* **20**, 62–67.

Section III

Bioinformatics for Genetic Study Design and Analysis

7 Identifying Mutations in Single Gene Disorders

David P. Kelsell[1], Diana Blaydon[1] and Charles A. Mein[2]

[1] *Centre for Cutaneous Research, Institute of Cell and Molecular Science, Queen Mary's School of Medicine and Dentistry, Whitechapel, London, UK*

[2] *Genome Centre, Queen Mary's School of Medicine and Dentistry, Charterhouse Square, London, UK*

7.1 Introduction

Genetic disorders due to the inheritance of an abnormality in a single gene are termed 'monogenic'. These can be subclassified into autosomal or sex-linked disorders, each with either a dominant (one mutation) or recessive (two mutations in the same gene) inheritance pattern. Though the genetic inheritance of a single gene disorder is not complex, the mechanism of mutation can be. In this chapter, we will highlight useful bioinformatics tools for chromosomally mapping a monogenic disease, and analytical tools and technologies for identification of genetic mutations. Finally, we will discuss some of the approaches for better understanding of the mechanism of the disease mutation. Such information may be invaluable to the understanding of human disease and ultimately drug development (Cohen *et al.*, 2004; Brinkman *et al.*, 2006).

7.2 Clinical ascertainment

Prior to any laboratory study of a monogenic disease, a number of non-bioinformatic and non-molecular aspects need to be considered. Paramount among these are the

Bioinformatics for Geneticists, Second Edition. Edited by Michael R. Barnes
 ISBN 978-0-470-02619-9 (HB) ISBN 978-0-470-02620-5 (PB)

requirement to apply for ethical permission before the study commences, correct clinical ascertainment, and informed patient consent. Other factors may also influence the study design, such as the mode of inheritance. During clinical ascertainment, consideration should be given to the possibility of phenocopies; that is, disease presentation occurring in an affected family member that is due to an environmental agent, epigenetics and/or a sporadic (or 'new') mutation appearing in a family that resembles the genetically inherited disorder segregating in the family. Examples of phenocopies include certain common cancers such as sporadic breast cancer occurring in families in which BRCA1 or BRCA2 mutations are segregating. Another possibility to consider is variable disease presentation between affected family members due to modifying genetic or environmental factors. A class of skin diseases called palmoplantar keratoderma shows variable expressivity in families and between families in which the same disease mutation is segregating. In inherited cancer syndromes, variable penetrance is also a factor, mutation carriers either not presenting with the disease at all or with variable age of onset. These variable clinical presentations suggest environmental or genetic modifiers. Beyond these confounding issues, the estimated statistical power of a given study design is a very important consideration, but this is largely outside the scope of this chapter; therefore, we direct the reader to other specialist texts on this matter (Balding *et al.*, 2003).

7.3 Genome-wide mapping of monogenic diseases

Two types of polymorphisms are routinely used to facilitate the localization of a single gene disorder to a specific chromosomal region: microsatellites or SNPs. Microsatellites are multiallelic and obviously much more polymorphic than biallelic SNPs (Database of Sequence Tag Sites http://www.ncbi.nlm.nih.gov/entrez/query.fcgi?db=unists). The physical and genetic distances between microsatellites can now be reliably compared with the human genome sequence as a physical mapping framework. This information can be visualized by one of the many available genome viewers (see Chapter 4). Estimates of recombination hotspots can also be derived from comparison of physical and genetic distances or by viewing HapMap recombination data. The standard microsatellite mapping panels with a density of 5 or 10 cM can now be supplemented to a much higher density to provide informative markers for fine-mapping efforts. Map coverage can be evaluated alongside other features, such as recombination hotspots and genes, with a genome viewer to visualize a particular chromosomal region. Figure 7.1 shows an example of the visualization of such information by the UCSC human genome browser (http://genome.ucsc.edu/).

7.3.1 Microsatellite mapping approaches

There are commercially available custom panels of fluorescently tagged primer pairs for genome-wide microsatellite panels. The Applied Biosystem (Foster City, CA,

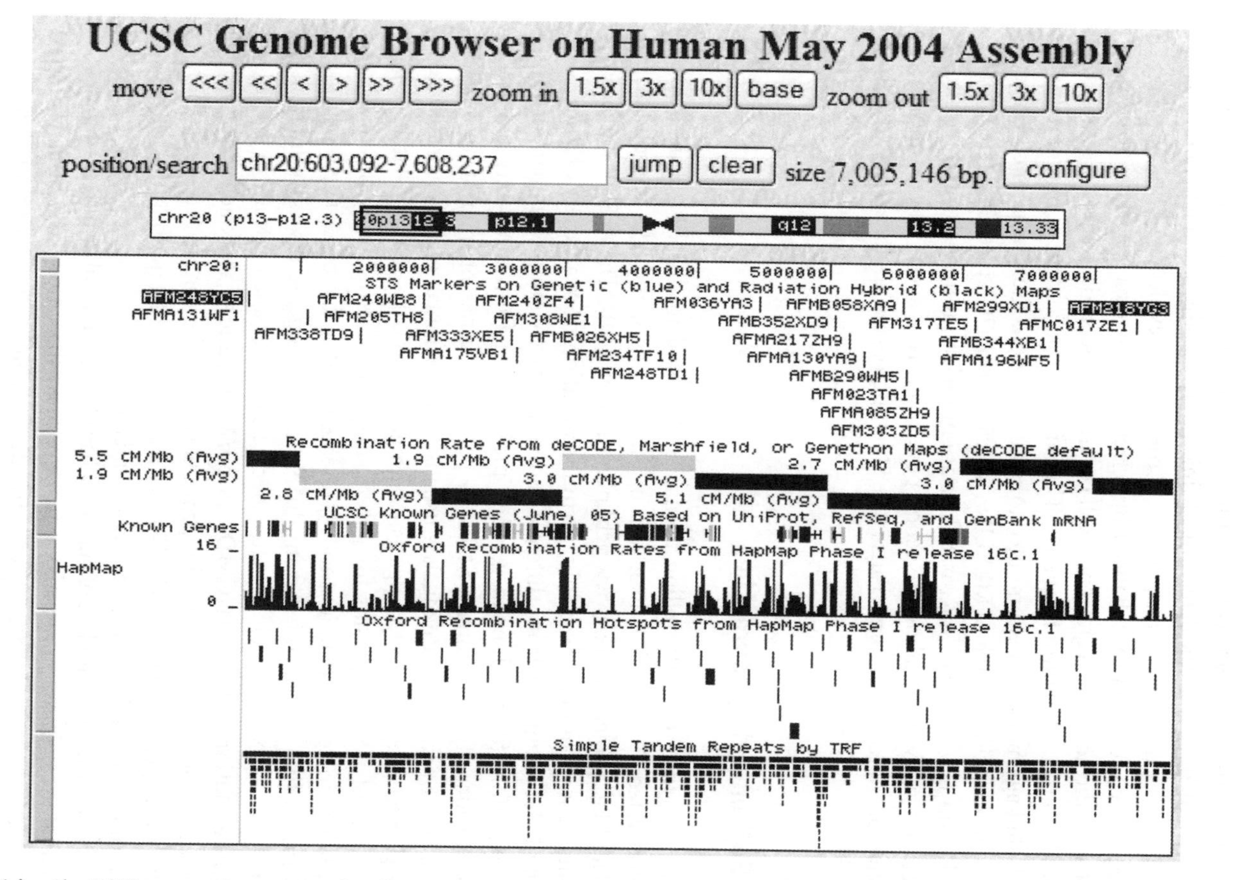

Figure 7.1 Using the UCSC genome browser to visualize genetic maps. The 7-Mb, D20S117–D20S115 region is displayed. The STS track has been modified to display only markers in the Genethon map. These are displayed alongside recombination rate data calculated from the DeCode, Marshfield and Genethon maps, known genes and recombination rates, and hotspots calculated from HapMap SNP data. The bottom track displays all tandem repeats across the region identified by the tandem repeat finder program. Generated using the UCSC Human Genome Browser, http://genome.ucsc.edu

USA) panel of 400 microsatellites with genome coverage of 10–15 cM is commonly used. The sizing and use of different fluorochromes allows pooling of PCR products from different microsatellites prior to size separation on capillary sequencers. However, the 'gaps' in genome coverage that these relatively sparse maps offer may inevitably result in very weak detection of linkage or possibly complete failure to detect linkage (Evans and Cardon, 2004). In these cases, evidence of linkage may be subsequently identified or enhanced by selecting further panels of microsatellites that span the genomic regions not well covered by the initial mapping set or in regions of high recombination frequency. The information content of microsatelite (and SNP) linkage maps can be computed to evaluate regions showing poor coverage with the '-information' switch in MERLIN (Abecasis *et al.*, 2002; see Chapter 10 for a worked example). However, microsatellites are more labour intensive and require more detailed analysis in calling the genotypes. In terms of speed and manpower, the method of choice for genome scans is the SNP microarray platform (Matsuzaki, *et al.*, 2004).

7.3.2 SNP-mapping approaches

There are currently over 10 million human SNPs listed in dbSNP – a development accelerated by the SNP consortium of the late 1990s, and later consolidated by the HapMap project (http://www.ncbi.nlm.nih.gov/SNP). The dbSNP database represents a compendium of (mainly) germline variation from a variety of sources, including SNPs identified from different individual DNA samples as part of the sequencing effort for the human genome project, and from more targeted approaches to identify all common SNPs, in particular genes of interest, by resequencing larger panels of individuals – usually with reference to particular disease areas (e.g., Seattle SNPs – http://pga.gs.washington.edu/). The dbSNP database is a rich source of polymorphisms for genetic mapping and the study of complex disease; however, certain data are not represented in this database. The most obvious variants that are excluded from dbSNP are monogenic disease mutations; these are represented separately in locus-specific databases and centrally in the Human Gene Mutation Database (HGMD) (http://www.hgmd.cf.ac.uk/ac/index.php). Somatic mutation data, generated by projects such as the cancer genome project (http://www.sanger.ac.uk/genetics/CGP), are also excluded from dbSNP – for coverage of somatic mutation resources, see Chapter 17.

The binary nature of this class of variation has allowed a number of technologies, in particular array-based platforms, to be developed that allow large scale typing of SNPs (Matsuzaki *et al.*, 2004). These provide more information at a much lower cost, in terms of both material and, more importantly, manpower; they have higher, more measurable accuracy than STR panels. A number of companies currently have off-the-shelf products that enable linkage to be assessed, including Affymetrix (Santa Clara, CA, USA) and Illumina (San Diego, CA, USA); arrays of 10 000–500 000 SNPs

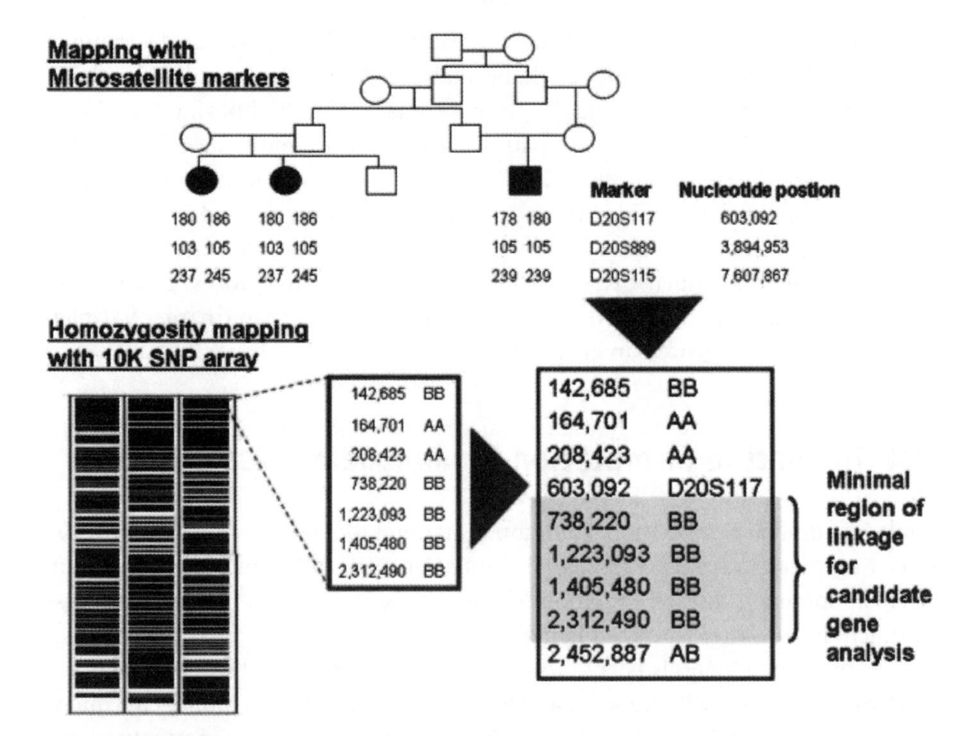

Figure 7.2 Linkage of anonychia to chromosome 20. No linkage was detected with microsatellite markers. Homozygosity mapping with Affymetrix 10K SNP array identified a region of homozygosity shared by all affected individuals, but no unaffected sibling (data not shown). Taken together, the data reveal a minimal region in which candidate genes may be analysed

are available (see Chapter 18 for details). The identities of these markers are easily available from company websites and are also listed on public websites, such as the UCSC genome browser, where their relative positions to other annotated features can easily be determined.

Certainly, for mapping recessive single gene disorders in consanguineous families, SNP genotyping has been the most frequently used method in very recent years, as it is rapid (results in 2–3 days), and the genome coverage of SNPs on the array (in most cases the 10K SNP array is sufficient) allows rapid disease mapping. In many cases, evidence of linkage can be identified by simply looking for shared homozygous blocks of SNPS between affected family members without the need to perform linkage analysis (Kelsell *et al.*, 2005). A number of studies have now reported detection of linkage with SNP arrays after failing to map the same disease gene with genome-wide microsatellite panels. Figure 7.2 shows an example of one of our experiences while trying to map the gene for autosomal recessive anonychia (no nail development) in consanguineous families. No linkage was detected with microsatellite markers. However, homozygosity mapping with the Affymetrix 10K SNP array identified a region of homozygosity shared by all affected individuals, but not an unaffected sibling (data not shown). Altogether, the data revealed a minimal region in which candidate genes could be analysed. Alternatively, additional genotyping may be required to refine the chromosomal region harbouring the disease gene further by identifying recombination events and/or identifying common disease haplotypes between families. Candidate gene mapping within this minimally defined region can be identified by any of the genome browsers and then analysed in order of probable functional candidacy, such as expression profile, homologue of gene mutated in similar disease or mechanistic roles. This process is similar in both complex and monogenic disease, so we direct the reader to Chapter 9, which discusses this whole process in great detail.

7.4 The nature of mutation in monogenic diseases

Table 7.1 summarizes the more common mutation types seen in monogenic disorders. In general, in recessive disease, the mutation leads to loss of protein function, while in dominant disease, mutations can act as either a dominant negative or loss of function (the latter is termed haploinsufficiency). Epigenetic effects, such as genomic imprinting, may also play a role, although, as these are not strictly mutations of DNA, we will deal with these separately below. Other common monogenic mutation mechanisms include mitochondrial mutations and expansion repeat mutations. An example of the latter is inherited dominant disease caused by trinucleotide repeat instability, including fragile X and Huntington's disease. The phenomenon of anticipation often occurs in these two disorders due to the number of repeats increasing through successive generations, leading to a more severe disease phenotype.

Another type of mutation are those that appear as gross chromosomal changes. It has been estimated that cytogenetically visible rearrangements are present in

Table 7.1 Summary of the more common mutation types seen in monogenic disorders

Mutation type	Subtype	Effect on protein
Single base substitution	Missense	One amino acid substituted for another
	Nonsense	One amino acid replaced with a stop codon
	Splice site	Create or destroy exon-intron splicing signals. Addition or deletion of amino acids and/or prematurely truncated protein
Deletion	In-frame	Deletion of one or more amino acids
	Frame shift	Prematurely truncated protein
Insertion	In-frame	Addition of one or more extra amino acids
	Frame shift	Prematurely truncated protein
Structural variation Copy number variation	Truncation or deletion or amplification	Protein truncated, entirely deleted or gene product amplified in the case of increases in gene copy number

~1 per cent of newborns. They are termed microdeletion syndromes. These chromosomal changes can cause a wide range of deleterious developmental effects, including mental retardation (Kriek *et al.*, 2004).

Heterozygous advantage is postulated for common recessive diseases which have a high mutation carrier frequency in certain populations; that is, being heterozygous for a recessive mutation confers some selective advantage on the individual. For example, the carrier frequency for the sickle cell gene is high in Africa, as it confers some protection against malaria. Heterozygote advantage is also seen in cystic fibrosis, where the invasion of *Salmonella typhi* into epithelial cells is restricted via the mutant CFTR chloride channel, thus providing protection against typhoid fever (Pier *et al.*, 1998). A more recent, but similar example is the association between a common deafness mutation in GJB2 and an improved skin barrier with a protective effect against bacterial invasion (Common *et al.*, 2004).

In view of the important role of natural selection and possible heterozygote advantage as a mechanism for the spread of monogenic diseases, some recent genome-wide studies of selection pressure may be of key importance to monogenic disease-gene hunting. Voight *et al.* (2006) published a map of signatures of positive selection that they detected in the human genome, using HapMap data. They made the results of their analysis available for viewing selection in the context of a gene, SNP or genomic region in Haplotter(http://hg-wen.uchicago.edu/selection/haplotter.htm).

7.4.1 Mutation detection by sequencing

Upon selection of genes for mutation analysis, the first method of choice for mutation screening is a combination of PCR of exons followed by sequencing. The steps involved in this method of mutation detection are shown in Figure 7.3. For initial mutation screening of a candidate gene, primers are designed to amplify each exon. Exons are first identified by mapping the mRNA sequence onto the genomic DNA sequence. At this stage, review of the gene in a genome browser, such as that of the UCSC, is important to ensure that all exons are identified, including novel exons present in EST data (see Chapter 5). Once all exons have been defined, amplified exon fragments are sequenced and analysed for nucleotide changes by comparison to the consensus of sequences reported in databases based on mRNA and genomic DNA data. A number of tools may then prove useful when determining whether the change identified is disease-causing or not (Chapters 11–14 review these methods in detail).

There are a number of mutation-screening strategies, which often involve PCR amplification of exons followed by conformational change analysis, including SSCP, heteroduplex and melt curve analysis. The reference standard, however, is re-sequencing. As sequencing technologies become more affordable and analysis packages improve, this is becoming the technology of choice for screening in many laboratories. A number of packages are available from commercial organizations, including Mutation

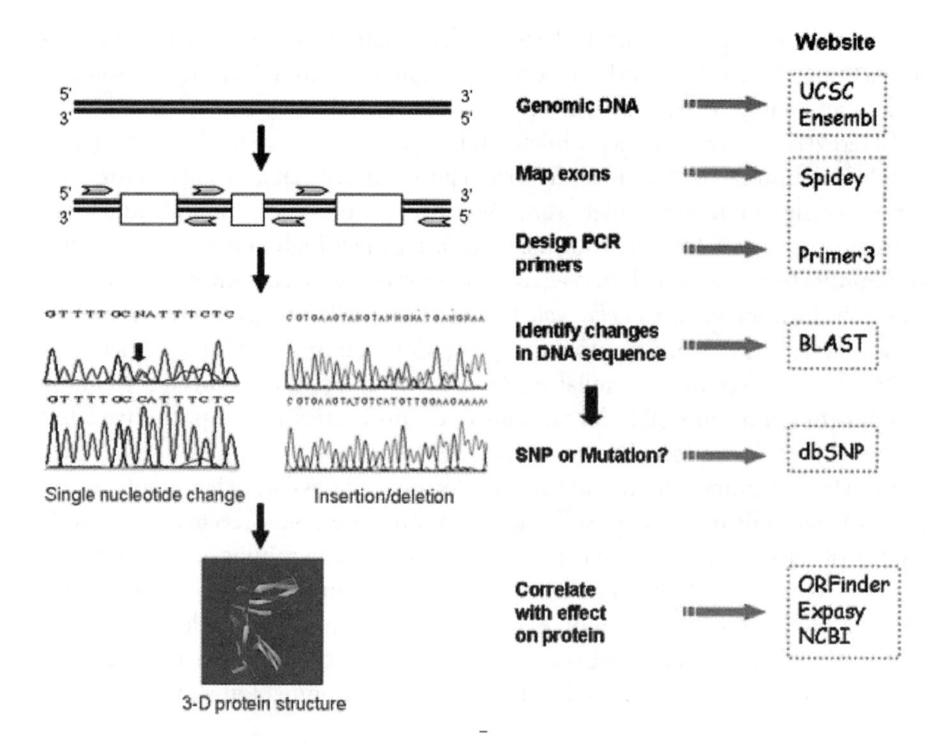

Figure 7.3 Mutation detection by sequencing

Surveyor (soft genetics http://www.softgenetics.com/ms/index.htm) and Seqscape (Applied Biosystems http://www.appliedbiosystems.com/), as well as, on an academic partnership basis, Polyphred (http://droog.gs.washington.edu/PolyPhred.html) (Stephens *et al.*, 2006), NovoSNP (Weckx *et al.*, 2005), SNPdetector (Zhang *et al.*, 2005) and InSNP (Manaster *et al.*, 2005). Each of these software packages allows comparison of resequencing with reference sequence files and gives quality scores for each inferred SNP. There is currently no objective survey comparing the efficacy of each method. Comparisons within original reports often favour the reported method over the comparator methods, false-positive and false-negative rates ranging from 5 per cent to 20 per cent (Manaster *et al.*, 2005; Weckx *et al.*, 2005; Zhang *et al.*, 2005; Stephens *et al.*, 2006). For identifying monogenic trait mutations, these platforms are likely to be an aid to target manual confirmation rather than being relied upon per se. Laboratories will need to investigate which package conforms to their specific requirements.

The sequencing approach will identify the majority of known mutation types including missense amino-acid substitutions, insertion of stop codons, small intragenic insertions or deletions, and splice-site mutations. However, there are other types of mutations that can be missed by this approach, so it can often be quite difficult to exclude a positional candidate gene purely by sequencing of exons and intron/exon junctions. In some cases, regions of low complexity or repeat sequences may be difficult to sequence, but perseverance is required, as these regions by their very nature may harbour insertion/deletions or other structural variants. This was the case during mutation analysis of the TNFRSF11A gene, which was almost excluded as a candidate for familial expansile osteolysis (Hughes *et al.*, 2000). Mutations in the gene candidate were excluded with the exception of a repetitive sequence in exon 1, which could not be sequenced. Polyacrylamide gel electrophoresis was eventually used to confirm an insertion in this region by the presence of an allele of increased size in affected individuals (Hughes *et al.*, 2000).

Other approaches may therefore need to be considered if all the genes have been excluded by sequencing. One point to be considered is that the databases have not identified all possible exonic sequences within the genomic DNA mapping in the minimal disease interval. Novel exons can be identified from spliced EST data, by genomic sequence conservation across vertebrates, or by the use of exon-prediction tools. Chapter 5 reviews some of the approaches that can be used to identify additional exons at a particular gene locus that may harbour the disease mutation. To maximize chances of successful mutation detection by sequencing, it can also help to have a number of unrelated families with the same disease that map to the same genomic region. It is likely that a subset of these families will have mutations that will be detectable by sequence analysis. However, this will not be possible when there is genetic heterogeneity, in which mutations in many genes cause the same monogenic disease. This is a particular problem for very rare diseases where only one or two families are diagnosed with the condition. An example of the latter is tylosis with oesophageal cancer (TOC). This is an autosomal dominant single gene disorder that occurs in three families, with two of the families related by disease haplotype analysis. The entire minimal region (34 kb) has been sequenced (except for highly repetitive regions) and no obvious disease-causing mutation has been identified, but a disease mechanism has been postulated (MacDonald *et al.*, 2006).

7.4.2 Other mutation detection approaches

Analysis of cDNA can reveal other mutations as well as indicating the effect of identified mutations on mRNA stability and confirming intron/exon splice site mutations. For example, a truncated message may indicate the presence of a large intragenic gene deletion. A larger than expected cDNA may indicate the presence of a larger intragenic insertion or the introduction of splicing mutations deep within the intronic sequence. Large single exon or multiple exon insertions or deletions can be difficult to detect unless they are in homozygous form, as they may be too large to be detectable by standard PCR and sequence analysis of exons, as the other allele (normal in dominant disease or carrying a different mutation in a recessive disease) is still being amplified in the PCR reaction (Figure 7.4). If cDNA is unavailable, there are a number of PCR-based genomic approaches available, including non-transmission

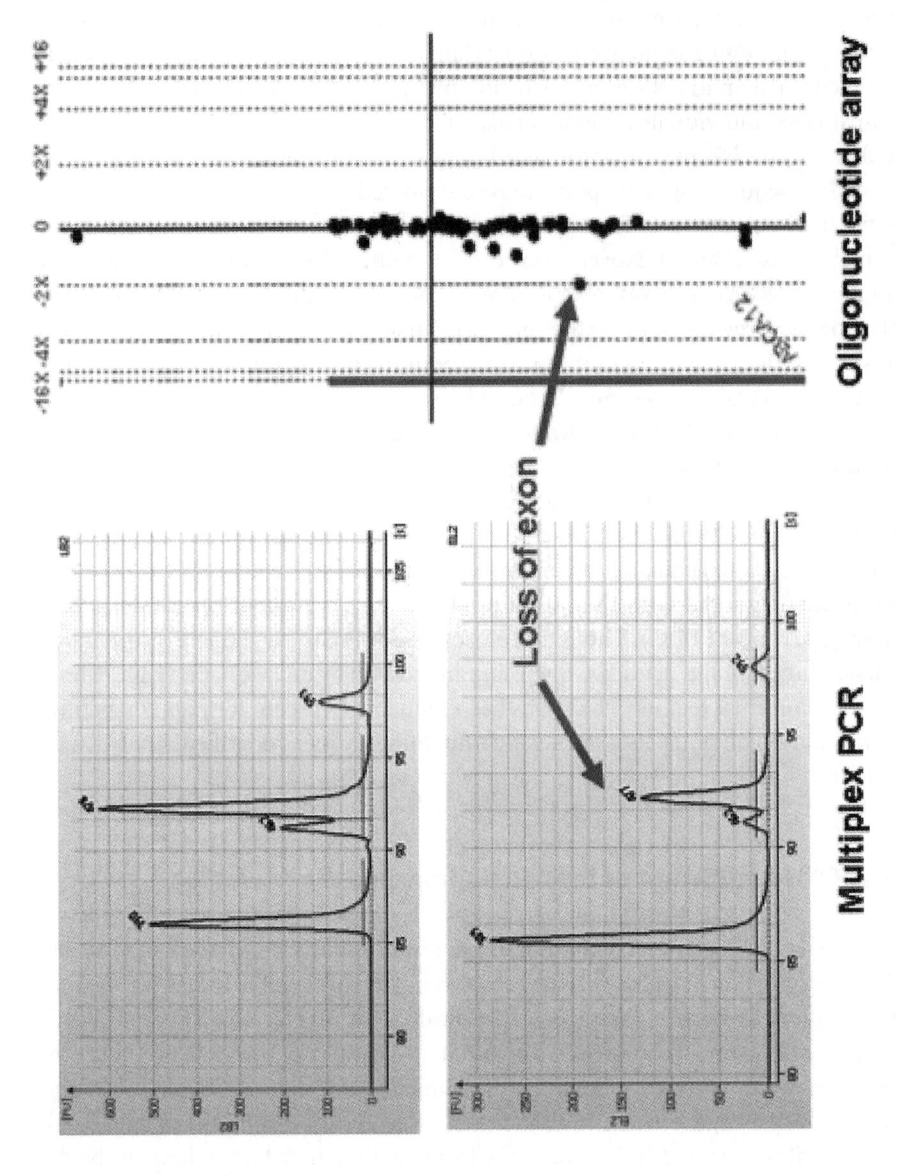

Figure 7.4 Heterozygous exon gene deletion identified by multiplex PCR or oligo array (thanks to Anna Thomas)

of SNPs to offspring (if SNP maps within deletion), multiple ligation (Hashimoto *et al.*, 2005), and multiplex analysis or copy number array analysis. The latter two examples appear in Figure 7.4, showing the detection of exon deletions in compound heterozygotes for ABCA12 mutations (Thomas *et al.*, 2006).

7.4.3 Arrays for detecting comparative genomic hybridization (CGH) and UPD

Other mutation mechanisms include large genomic deletions or duplications that encompass a number of genes. Examples include Williams syndrome and trisomy 21 in Down's syndrome. These can be easily detected by cytogenetic methods but could also be easily identified and mapped by CGH or copy number arrays. Another genetic mutation is uniparental disomy (UPD), which occurs when the chromosome copy number remains the same, but both homologues of a chromosome or part of a chromosome are identical to one another (reviewed by Kotzot and Utermann, 2005). Using SNP-based arrays, a number of studies have now revealed this as a major genetic mechanism in cancer (Figure 7.5). UPD also occurs via germline transmission, as in Prader–Willi syndrome (PWS) and Angelman syndrome (AS).

Copy number polymorphisms (CNPs) have been demonstrated to play a role in a number of monogenic diseases (such as DiGeorge syndrome and PWS) and, recently, complex diseases (Aitman *et al.*, 2006). However, they are increasingly acknowledged as part of normal genomic variation alongside SNPs and STRs (McCarroll *et al.*, 2006). CNPs must be taken into account when mapping monogenic traits to avoid confusing common variation with aetiological changes (see below). Encouragingly, initial reports suggest that CNPs may be tagged with adjacent SNPs in LD with the CNP, allowing relatively easy genotyping (Newman *et al.*, 2006). A number of approaches have been used to identify CNPs, including analysis of fosmid sequence data (Tuzun *et al.*, 2005) and CGH – either with BAC clones (de Vries *et al.*, 2005) or spotted (Lucito *et al.*, 2003; Barrett *et al.*, 2004; Sebat *et al.*, 2004) or synthesized (Bignell *et al.*, 2004; Hinds *et al.*, 2006) oligonucleotide arrays – or by inferring deletions from HapMap data (Conrad *et al.*, 2006; McCarroll *et al.*, 2006). There is currently no consensus on the number or size of CNP in a normal genome, and this reflects the technologies and sample sizes used to identify them. Estimates for the number of variant sites between any two individuals fall between 3 and 50. A model to assess the number of common CNPs greater than 5 kb gives figures of 900 in the Caucasian population and 1525 in the Yoruban population (Conrad *et al.*, 2006). This may underestimate the true value, as a number of the technologies and programs used to identify CNPs do not perform well in duplicate regions (Conrad *et al.*, 2006; McCarroll *et al.*, 2006), precisely those shown to harbour a high proportion of CNPs (Sebat *et al.*, 2004). However, they are likely to be several orders of magnitude less common than SNPs. Data from the majority of studies to date are represented in the Database of

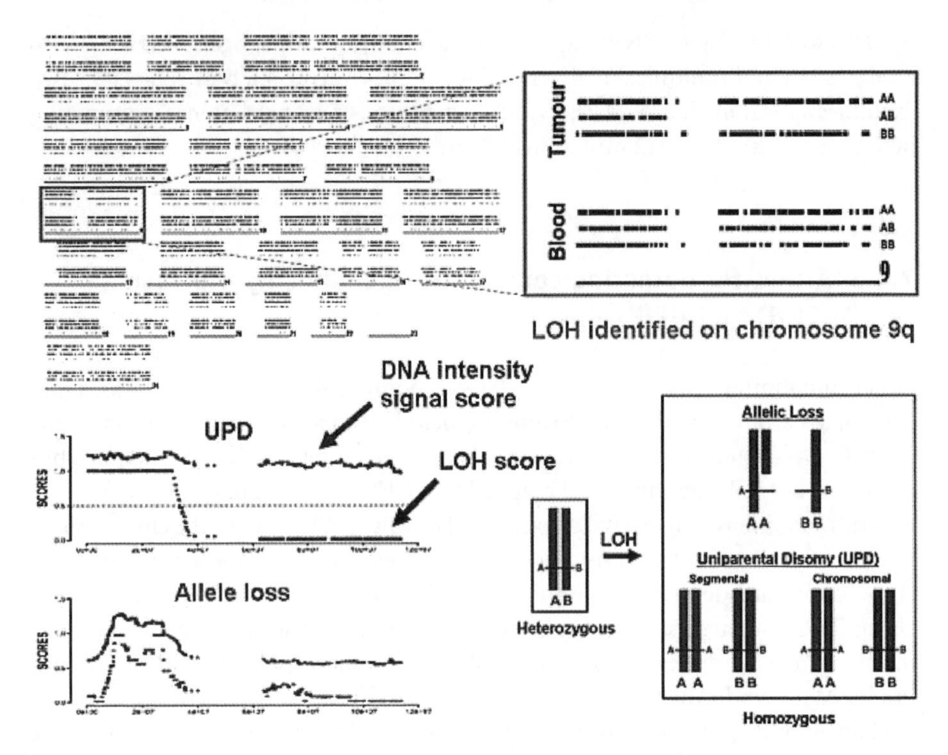

Figure 7.5 Identification of UPD in basal cell carcinoma. Affymetrix 10K SNP array analysis revealed loss of heterozygosity (LOH) on chromosome 9q when tumour DNA was compared to blood DNA. A corresponding decrease in DNA intensity signal score with LOH indicates loss of DNA. However, uniparental disomy (UPD) is signified when the DNA intensity signal remains the same in the presence of LOH

Genomic Variants http://projects.tcag.ca/variation/ and the HumanStructural Variation Database http://humanparalogy.gs.washington.edu/structuralvariation/. The data are most easily visualized at the latter site with a modified version of the UCSC browser. Figure 7.6 shows a view of the FCGR3B gene, using the Human Structural Variation Database. This is a good example of some of the data that can be interrogated by this tool. Aitman *et al.* (2006) reported a predisposition to systemic lupus erythematosus-related nephritis caused by CNPs. In a study of affected and unaffected individuals, the FCGR3B gene copy number was reduced in affected patients but generally increased in controls. In a 366-kb region in Figure 9.6, it is apparent that the entire FCGR3B region contains a substantial number of structural variations, including a gap in the genome assembly that may have been caused by the genome assembly problems that some of these structural variants have presented. If we look specifically at the FCGR3B gene, there appears to be evidence of a segmental duplication across the first three exons of the gene in the 'human WSSD' track. This track is based on data from the SSD database, which identifies putative segmental

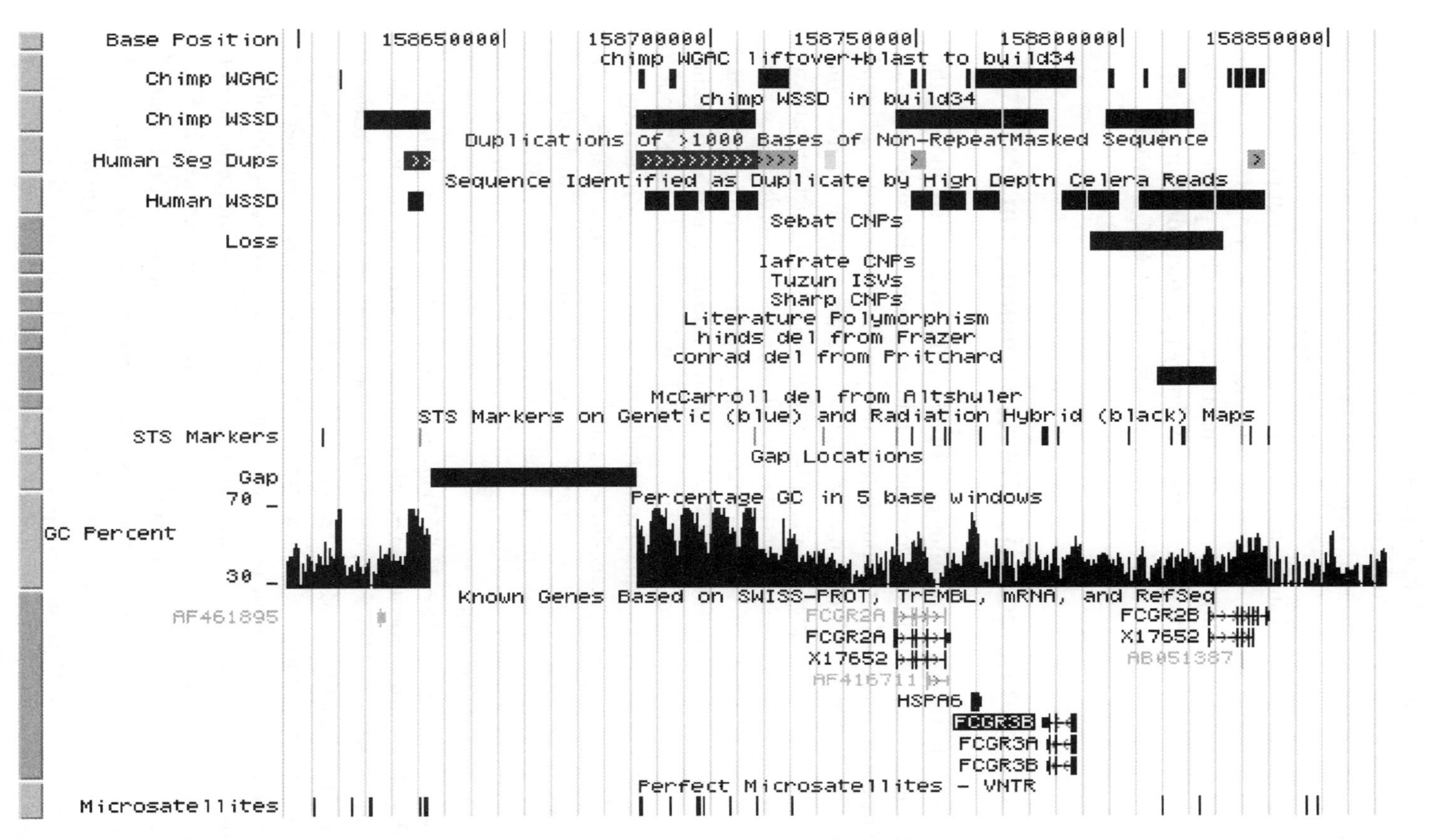

Figure 7.6 Using the human structural variation database. A 266-kb locus containing the FCGR3B gene viewed with an installation of the UCSC browser (http://humanparalogy.gs. washington.edu/structuralvariation/), displaying large-scale structural variation (LSV), copy number polymorphisms (CNPs) and intermediate-sized structural variation (ISV) as determined by array CGH (Iafrate *et al.*, 2004; Sharp *et al.*, 2005), representational oligonucleotide microarray analysis (Sebat *et al.*, 2004) and fosmid paired-end sequence analysis (Tuzun *et al.*, 2005)

duplications on the basis of Celera whole-genome shotgun read in depth across the genome assembly (Bailey *et al.*, 2001). Clicking on the WSSD region allows inspection of the data. Figure 7.7 is a detailed view of the AL590385 BAC clone, showing a higher than expected depth of Celera whole-genome shotgun reads across the left-hand region of the clone, and suggesting the presence of segmental duplications in this region. This seems to be in general concordance with the structural variation reported by Aitman *et al.* (2006). It is expected that these data will be presented in the Ensembl and UCSC genome browsers in the near future (personal communication). The 'WSSD duplication' track is already available in the main UCSC browser.

7.5 Considering epigenetic effects in mendelian traits

Epigenetics is concerned with the study of heritable changes other than those in the DNA sequence and encompasses two major modifications of DNA or chromatin: DNA methylation, the covalent modification of cytosine, and post-translational modification of histones, including methylation, acetylation, phosphorylation and sumoylation (Perini and Tupler, 2006). In terms of function, epigenetic modifications act to regulate gene expression and stabilize adjustments of gene dosage, as seen in X inactivation, tissue-specific gene silencing and genomic imprinting. There are a number of examples of epigenetic effects in human syndromes, such as PWS and AS, both of which show abnormalities of imprinted domains (Nicholls and Knepper, 2001), and specifically cancer-related syndromes, such as Werner syndrome (Agrelo *et al.*, 2006). Dysregulation of epigenetic mechanisms can also combine synergistically with genetic alterations in the development and progression of Mendelian disorders, such as Rett syndrome and facioscapulohumeral muscular dystrophy (FSHD), both of which result from altered gene silencing, but on the basis of very different mechanisms (Perini and Tupler, 2006).

Rett syndrome (RTT) (OMIM 312750) is a severe neurodevelopmental disorder, the second leading cause of mental retardation in females. Mutations in the methyl-CpG-binding protein 2 (MECP2) gene account for 75 per cent of RTT patients (Perini and Tupler, 2006). The MeCP2 protein is a ubiquitously expressed transcriptional repressor that participates in heterochromatin formation and gene silencing by direct binding of methylated CpG sequences and also by interactions with other transcriptional repressor complexes mediating repression through deacetylation of core histones. A very large number of MECP2 mutations have been seen in RTT, suggesting that a global alteration of transcriptional repression may be the basis of the disease. However, transcriptional studies in patients and MECP2 KO mice did not confirm this view. The targets of MeCP2 are still unclear, although it has been shown to bind brain-derived neurotrophic factor (BDNF), and it is also essential for the formation of a silent chromatin structure at Dlx5-Dlx6, a novel imprinted homeobox locus involved in brain patterning. Collectively, these studies emphasize

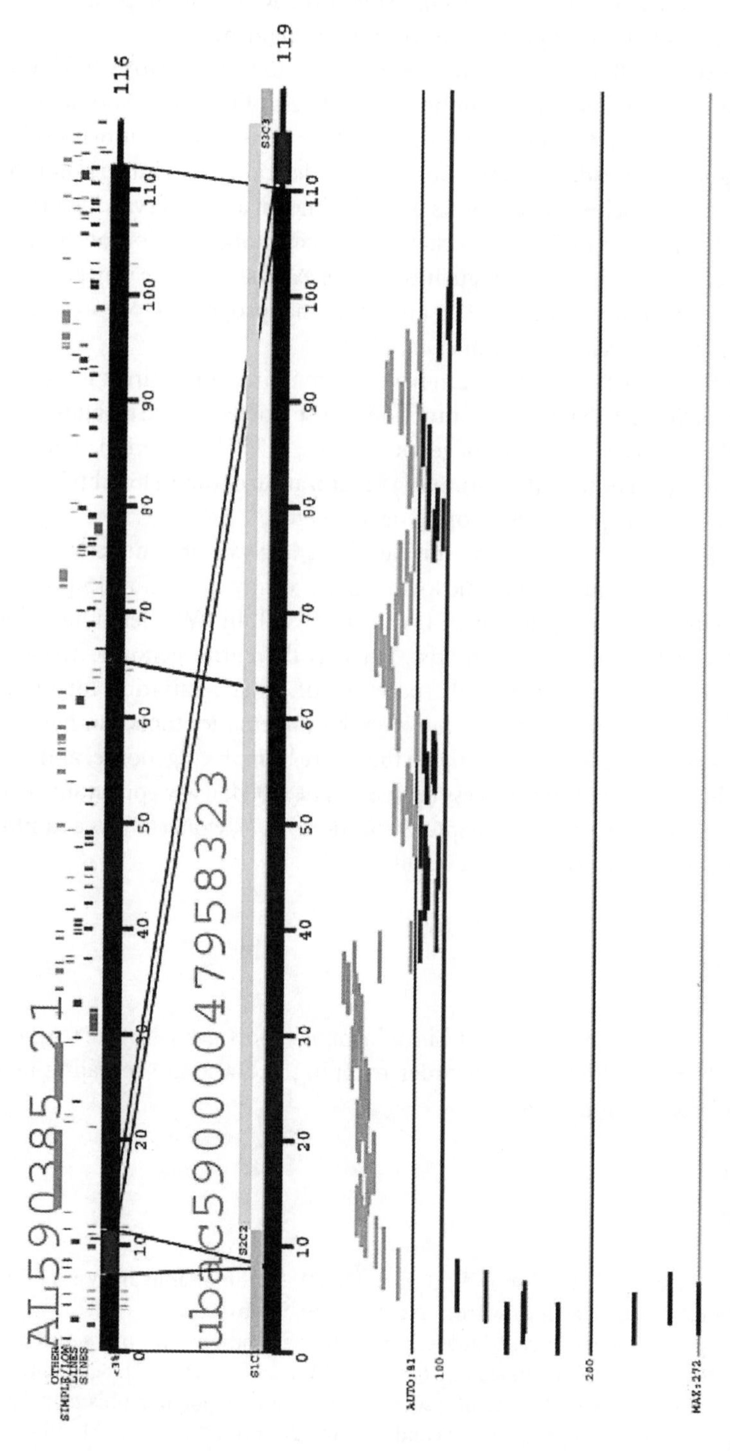

Figure 7.7 A detailed view of segmental duplication sequence identified by high-depth Celera whole-genome shotgun sequence reads. The AL590385 BAC clone shows a higher than expected depth of shotgun reads across certain regions, suggesting the presence of segmental duplications. For a description of the 'fuguization' detection method, see Bailey *et al.* (2001)

the functional plasticity of MeCP2 and suggest that the RTT phenotype may result from several different aberrant gene-transcription mechanisms.

FSHD (OMIM 158900) is an early-onset, autosomal-dominant myopathy characterized by progressive and variable atrophy and weakness of the facial, shoulder and upper-arm muscles. This disorder is not the result of a classical mutation within a protein-coding gene; instead, all FSHD patients carry deletions of an integral number of tandem 3.3-kb heterochromatic repeats on 4q35, known as D4Z4 (van Deutekom *et al.*, 1993). The deletion of D4Z4 affects the chromatin organization of the 4q35 subtelomeric region, leading to dysregulation of the expression of nearby genes. As a result, genes located upstream of D4Z4 are found inappropriately overexpressed, specifically in FSHD muscle (Gabellini *et al.*, 2002).

These examples show that genetics can have either a direct or indirect impact on epigenetic effects. In RETT syndrome, mutations cause a defect in the recognition of the correct DNA methylation pattern of genomic DNA. In FSHD, deletion of tandem heterochromatic repeats affects the chromatin organization of the 4q35 subtelomeric region and, consequently, the expression of nearby genes.

Epigenetic effects may also help to confound the analysis of genetic traits; for example, PWS can be found in a genetic form (e.g., a common deletion 15q11-q13) or in epigenetic forms (e.g., maternal UPD causing PWS). In PWS deletions, about four megabases of genomic DNA are lost, whereas the entire genomic sequence is normal for UPD cases. Evaluating the potential of genetic variation for impact on epigenetics is not a trivial exercise; however, bioinformatics tools can help this process. Efforts are under way to investigate the entire human epigenome, and tools are being developed to assist this process (Rakyan *et al.*, 2004; www.epigenome.org). Additionally, we refer the reader to Chapter 9 (Section 9.4.7), which covers a number of tools for epigenetic analysis in some detail.

7.6 Summary

The laboratory-based technology and bioinformatics tools available currently facilitate relatively rapid single gene disorder mapping and also are revealing novel mechanisms of disease mutation.

References

Abecasis, G. R., Cherny, S. S., Cookson, W. O. *et al.* (2002). Merlin – rapid analysis of dense genetic maps using sparse gene flow trees. *Nat Genet* **30**, 97–101.

Agrelo, R., Cheng, W. H., Setien, F. *et al.* (2006). Epigenetic inactivation of the premature aging Werner syndrome gene in human cancer. *Proc Natl Acad Sci USA* **103**(23), 8822–8827.

Aitman, T. J., Dong, R., Vyse, T. J. *et al.* (2006). Copy number polymorphism in Fcgr3 predisposes to glomerulonephritis in rats and humans. *Nature* **439**(7078), 851–855.

Bailey, J. A., Yavor, A. M., Massa, H. F. *et al.* (2001). Segmental duplications: organization and impact within the current human genome project assembly. *Genome Res* **11**(6), 1005–1017.

Balding, D. J., Bishop, M. and Cannings, C. (Eds) (2003). *Handbook of Statistical Genetics*, 2nd edn. Chichester: Wiley.

Barrett, M. T., Scheffer, A., Ben-Dor, A. *et al.* (2004). Comparative genomic hybridization using oligonucleotide microarrays and total genomic DNA. *Proc Natl Acad Sci U S A* **101**(51), 17765–17770.

Bignell, G. R., Huang, J., Greshock, J. *et al.* (2004). High-resolution analysis of DNA copy number using oligonucleotide microarrays. *Genome Res* **14**(2), 287–295.

Brinkman, R. R., Dube, M. P., Rouleau, G. A. *et al.* (2006). Human monogenic disorders – a source of novel drug targets. *Nat Rev Genet* **7**(4), 249–260.

Cohen, J. C., Kiss, R. S., Pertsemlidis, A. *et al.* (2004). Multiple rare alleles contribute to low plasma levels of HDL cholesterol. *Science* **305**(5685), 869–872.

Common, J. E., Di, W. L., Davies, D. *et al.* (2004). Further evidence for heterozygote advantage of GJB2 deafness mutations: a link with cell survival. *J Med Genet* **41**(7), 573–575.

Conrad, D. F., Andrews, T. D., Carter, N. P. *et al.* (2006). A high-resolution survey of deletion polymorphism in the human genome. *Nat Genet* **38**(1), 75–81.

de Bakker, P. I., Yelensky, R., Pe'er, I. *et al.* (2005). Efficiency and power in genetic association studies. *Nat Genet* **37**(11), 1217–1223.

de Vries, B. B., Pfundt, R., Leisink, M. *et al.* (2005). Diagnostic genome profiling in mental retardation. Am *J Hum Genet* **77**(4), 606–616.

Evans, D. M. and Cardon, L. R. (2004). Guidelines for genotyping in genomewide linkage studies: single-nucleotide-polymorphism maps versus microsatellite maps. *Am J Hum Genet* **75**(4), 687–692.

Gabellini, D., Green, M. R. and Tupler, R. (2002). Inappropriate gene activation in FSHD: a repressor complex binds a chromosomal repeat deleted in dystrophic muscle. *Cell* **110**, 339–348.

Gibbs, R. A., Belmont, J. W., Hardenbol, P. *et al.* (2003). The International HapMap Project. *Nature* **426**(6968), 789–796.

Hashimoto, M., Hupert, M. L., Murphy, M. C. *et al.* (2005). Ligase detection reaction/hybridization assays using three-dimensional microfluidic networks for the detection of low-abundant DNA point mutations. *Anal Chem* **77**(10), 3243–3255.

Hinds, D. A., Kloek, A. P., Jen, M. *et al.* (2006). Common deletions and SNPs are in linkage disequilibrium in the human genome. *Nat Genet* **38**(1), 82–85.

Hughes, A. E., Ralston, S. H., Marken, J. *et al.* (2000). Mutations in TNFRSF11A, affecting the signal peptide of RANK, cause familial expansile osteolysis. *Nat Genet* **24**(1), 45–48.

Kelsell, D. P., Norgett, E. E., Unsworth, H. *et al.* (2005). Mutations in ABCA12 underlie the severe congenital skin disease harlequin ichthyosis. *Am J Hum Genet* **76**(5), 794–803.

Klein, R. J., Zeiss, C., Chew, E. Y. *et al.* (2005). Complement factor H polymorphism in age-related macular degeneration. *Science* **308**(5720), 385–389.

Kotzot, D. and Utermann, G. (2005). Uniparental disomy (UPD) other than 15: phenotypes and bibliography updated. *Am J Med Genet* **136**(3), 287–305.

Kriek, M., White, S. J., Bouma, M. C. *et al.* (2004). Genomic imbalances in mental retardation. *J Med Genet* **41**(4), 249–255.

Lucito, R., Healy, J., Alexander, J. *et al.* (2003). Representational oligonucleotide microarray analysis: a high-resolution method to detect genome copy number variation. *Genome Res* **13**(10), 2291–2305.

Manaster, C., Zheng, W., Teuber, M. *et al.* (2005). InSNP: a tool for automated detection and visualization of SNPs and InDels. *Hum Mutat* **26**(1), 11–19.

Matsuzaki, H., Dong, S., Loi, H. *et al.* (2004). Genotyping over 100,000 SNPs on a pair of oligonucleotide arrays. *Nat Methods* **1**, 109–111.

McCarroll, S. A., Hadnott, T. N., Perry, G. H. *et al.* (2006b). International HapMap Consortium. Common deletion polymorphisms in the human genome. *Nat Genet* **38**(1), 86–92.

MacDonald, F. E., Liloglou, T., Xinarianos, G. *et al.* (2006). Down-regulation of the cytoglobin gene, located on 17q25, in tylosis with oesophageal cancer (TOC): evidence for trans-allele repression. *Hum Mol Genet* **15**(8), 1271–1277.

Nan, X., Meehan, R. R. and Bird, A. (1993). Dissection of the methyl-CpG binding domain from the chromosomal protein MeCP2. *Nucleic Acids Res* **21**, 4886–4892.

Newman, T. L., Rieder, M. J., Morrison, V. A. *et al.* (2006). High-throughput genotyping of intermediate-size structural variation. *Hum Mol Genet* **15**(7), 1159–1167.

Nicholls, R. D. and Knepper, J. L. (2001). Genome organization, function, and imprinting in Prader–Willi and Angelman syndromes. *Annu Rev Genomics Hum Genet* **2**, 153–175.

Perini, G. and Tupler, R. (2006). Altered gene silencing and human diseases. *Clin Genet* **69**(1), 1–7.

Pier, G. B., Grout, M., Zaidi, T. *et al.* (1998). *Salmonella typhi* uses CFTR to enter intestinal epithelial cells. *Nature* **393**, 79–82.

Rakyan, V. K., Hildmann, T., Novik, K. L. *et al.* (2004). DNA methylation profiling of the human major histocompatibility complex: a pilot study for the human epigenome project. *PLoS Biol* **2**(12), e405.

Sebat, J., Lakshmi, B., Trog, J. *et al.* (2004). Large-scale copy number polymorphism in the human genome. *Science* **305**(5683), 525–528.

Stephens, M., Sloan, J. S., Robertson, P. D. *et al.* (2006). Automating sequence-based detection and genotyping of SNPs from diploid samples. *Nat Genet* **38**(3), 375–381.

Thomas, A. C. , Cullup, T., Norgett, E. E. *et al.* (2006). ABCA12 is the major harlequin ichthyosis gene. *J Invest Dermatol* 10 Aug [Epub ahead of print].

Tuzun, E., Sharp, A. J., Bailey, J. A. *et al.* (2005). Fine-scale structural variation of the human genome. *Nat Genet* **37**(7), 727–732.

van Deutekom, J. C., Wijmenga, C., van Tienhoven, E. A. *et al.* (1993). FSHD associated DNA rearrangements are due to deletions of integral copies of a 3.2 kb tandemly repeated unit. *Hum Mol Genet* **2**, 2037–2042.

Voight, B. F., Kudaravalli, S., Wen, X. *et al.* (2006). A map of recent positive selection in the human genome. *PLoS Biol* **4**(3), e72.

Weckx, S., Del-Favero, J., Rademakers, R. *et al.* (2005). NovoSNP, a novel computational tool for sequence variation discovery. *Genome Res* **15**(3), 436–442.

Zhang, J., Wheeler, D. A., Yakub, I. *et al.* (2005). SNPdetector: a software tool for sensitive and accurate SNP detection. *PLoS Comput Biol* **1**(5), e53.

8

From Genome Scan to Culprit Gene

Refining Loci Implicated by Genome Scans

Ian C. Gray

Paradigm Therapeutics (S) Pte Ltd, Singapore

8.1 Introduction

Linkage analysis of complex traits with family-based samples typically results in a number of broad, ill-defined linkage peaks that represent several megabases of DNA (see, for example, Grettarsdottir *et al.*, 2002); beneath the expanse of each peak, there may lie a gene (or genes) associated with the disease in question. Historically, simple tandem repeat markers (STRs) have been used for genome-wide linkage scans (see Section 8.2.3 below); however, STR marker panels are gradually being replaced by single-nucleotide polymorphism (SNP) marker sets, which offer several advantages, including increased marker density and ease of use (Evans and Cardon, 2004). Under the prior assumption that a preliminary linkage analysis has been completed with SNPs, STRs or a combination of both, the goal of this chapter is to take the investigator through the process of characterizing and narrowing a linkage region by a population-based approach, with the ultimate aim of identifying candidate genes and testing them directly for association with the disease or trait in question. This is usually achieved by testing markers in the genomic interval of interest (the critical interval) for differences in allele frequency between case and control cohorts, where the cohorts consist of unrelated individuals (although methods employing family structure, based on the difference in frequency of allele transmission in a large number of small pedigrees

Bioinformatics for Geneticists, Second Edition. Edited by Michael R. Barnes
 ISBN 978-0-470-02619-9 (HB) ISBN 978-0-470-02620-5 (PB)

are also used – see below; this is covered in more detail in Chapter 10). In general, population-based methods for complex trait analysis offer large increases in both power and resolution over linkage-based approaches (McGinnis, 2000; Risch, 2000; discussed in Chapter 10) and are well suited to the follow-up of preliminary (and often equivocal) linkage results.

The extension of population-based genetic association methodology to cover the entire genome offers great promise for the rapid identification of genes involved in common disease and other complex traits, but requires a far larger number of markers than a linkage scan, as the increase in power to detect an effect is generally offset by a far shorter detection range for each marker. It is estimated that a panel of 300 000–600 000 SNPs is required to capture the majority of common variation across the genome in most human populations (International HapMap Consortium, 2005). Some progress toward routine use of marker sets of this magnitude has been made with the advent of Affymetrix microrrays that allow simultaneous typing of 100 000 SNPs (Nicolae *et al.*, 2006). The techniques described in this chapter are equally applicable to the follow-up of loci highlighted by whole-genome association studies using dense marker sets, although any loci thus identified are likely to be far shorter and more manageable than those identified by linkage. However, although the first whole-genome association studies are imminent at the time of writing (Thomas *et al.*, 2005), large numbers of loci with putative links to complex traits continue to be identified by linkage, and linkage methods are likely to run in parallel with association studies for the foreseeable future. Examples of the successful application of a two-step linkage-association approach include identification of the involvement of *ApoE* in Alzheimer's disease (Strittmatter *et al.*, 1993) and *NOD2* in Crohn's disease (Hugot *et al.*, 2001; Ogura *et al.*, 2001). In other cases, a convincing association between a genomic region and the trait under study has been uncovered, but the culprit gene has not yet been identified; for example, the recent discovery of a common genetic variant associated with prostate cancer in European and African populations (Amundadottir *et al.*, 2006). Although seemingly intractable at present, such anonymous associations will, no doubt, yield an increasing number of disease-related genes, as our understanding of gene and genome function and regulation improves. The first part of this chapter focuses on theoretical and practical considerations for good study design, while the second part covers a systematic approach to identification of the disease-associated gene, with emphasis on the application of methods, software tools and databases.

8.2 Theoretical and practical considerations

8.2.1 Choice of study population

Wherever possible, the study population selected for follow-up analysis of loci identified by genome scans should be derived from the same geographic area as the families or individuals used for the original scan. As the genetic components contributing to complex disease are likely to be varied, there is no guarantee that the predisposing genetic factors in one population will be the same in a second. If we use the term 'study population' in the broadest sense as applied to genetic association studies, a variety of study population structures may be considered. Three of the most common configurations are the case-control cohort, the discordant sib-pair cohort (i.e., one affected and one unaffected sib) and the parent-offspring triad (affected offspring with both parents) cohort. Each of these structures has advantages and disadvantages (for an evaluation of each, see Risch, 2000; Cardon and Bell, 2001).

Case-control cohorts simply consist of one group of individuals (cases) with the disease state and a second group without the disease (controls). Case-control cohorts have the advantage of being more straightforward to collect than the other two structures described above and generally provide more statistical power than similarly sized discordant sib or other nuclear family-based cohorts (McGinnis, 2000; Risch, 2000). However, case-control cohorts are prone to 'population stratification' (or substructure) effects. Population stratification occurs when the cohort under study contains a mix of individuals that can be separated on grounds other than the phenotype under study (most commonly on the basis of geographic origin). This can lead to allele frequency differences in cases and controls that are due to circumstances unrelated to the phenotypic difference under investigation, resulting in erroneous conclusions regarding association between the marker under test and the disease phenotype. Geneticists and statisticians often refer to such spurious associations as 'type I errors'; that is, rejection of the null hypothesis of no association when the null hypothesis is in fact correct and there is no true association. Conversely, failure to reject a false null hypothesis is referred to as a 'type II error'. Population stratification is a major potential source of type I errors in genetic association studies. Careful selection of individuals for inclusion in disease and control cohorts is necessary to ensure as homogeneous a background as possible and therefore avoid stratification. If stratification is suspected, it is possible to test for it by using randomly selected genetic markers (Pritchard and Rosenberg, 1999; Devlin and Roeder, 1999). It is also important to match the cohorts for phenotypic or environmental variables that may otherwise confound any genetic analysis; for example, hormone replacement therapy (HRT) has a large impact on bone mineral density (BMD), and it would be necessary to account for this in a search for genetic factors influencing BMD in a cohort of post-menopausal women (Giraudeau *et al.*, 2004).

Although population homogeneity and well-matched cases and controls are preferred, it may be possible to use a cohort even if stratification is present; Jonathan Pritchard and colleagues have developed a method for testing for genetic association in the presence of population stratification, by using unlinked markers to make inferences about population substructure and employing this information to test for associations within the identified subpopulations (Pritchard *et al.*, 2000; Falush *et al.*, 2003). STRUCTURE and STRAT, software tools for the detection of stratification and testing for genetic association in the presence of stratification can be downloaded from http://pritch.bsd.uchicago.edu/software.html. An alternative approach to correction for population stratification, termed 'genomic control', measures the degree of variability and magnitude of the test statistics observed at random loci and uses this information to adjust the critical value for significance tests at candidate loci by the appropriate degree (Devlin and Roeder, 1999; Clayton *et al.*, 2005). However, it should be noted that correction for stratification cannot completely remove the possibility of increased false-positive results under all circumstances (Cardon and Bell, 2001; Devlin *et al.*, 2001; Pritchard and Donnelly, 2001; Marchini *et al.*, 2004), and stratification should be avoided where possible.

The main advantage of using study populations that incorporate elements of family structure (such as discordant sibs or trios) is that, unlike case-control cohorts, they are immune to population-stratification effects. However, as mentioned above, family-based samples are typically more difficult to collect than case-control samples (particularly for late-onset diseases, that is, those that manifest in middle age or later) and generally offer less statistical power than the equivalent sized case-control cohort (McGinnis, 2000; Risch, 2000). The remainder of this chapter will focus predominantly on case-control methodology where reference to population structure is necessary; statistical methods for analysing family-based cohorts, such as the transmission disequilibrium (TDT) and sib transmission disequilibrium (S-TDT) tests, together with tools for the analysis of quantitative traits, are covered in Chapter 10.

Estimation of required cohort size for a genetic study depends on a number of factors, including the size of the effect of the locus under test, the frequency of the disease-risk-conferring allele, and the genetic nature of this 'risk allele', that is, whether recessive, dominant, additive, etc. If the causal variant is not being tested directly, the distance between the causal variant and the surrogate marker under test (see Section 8.2.3) is also relevant. Most of these factors are unknown prior to the start of the study, and the minimum required population size is usually based on assumptions concerning these factors (McGinnis, 2000; Risch, 2000). In reality, pragmatism typically dictates the available sample size; investigators use the largest obtainable cohort, with the caveat that the available sample may not provide sufficient statistical power to detect effects below a certain magnitude. To detect genetic factors of fairly small or moderate effect, cohorts of several hundred to a few thousand individuals may be required (McGinnis, 2000; Risch, 2000).

8.2.2 Sequence characterization at the locus under investigation

After a whole-genome linkage scan, the investigator is typically faced with several genetic loci of potential involvement in the disease process, the limits of each defined by two genetic markers spanning several centimorgans (cM). As 1 cM equates to roughly one megabase (Mb) on average, and each megabase contains an estimated average of nine genes (based on 25 000 genes in the entire 2900-Mb genome; International Human Genome Sequencing Consortium, 2004), this may represent several thousand kilobases of DNA and over 100 genes per locus. Whole-genome association scans are likely to yield far shorter intervals, as markers generally have a much smaller range for detection of a genetic effect in tests of association than linkage. The first task is to define the locus in the context of the human genome, in order to gain a comprehensive knowledge of genes and further genetic markers in the interval. Until recently, this involved the laborious laboratory process of identifying and ordering genomic clones into contigs and using those contigs as a framework for gene and marker identification. Fortunately, locus characterization has become far more straightforward in the wake of the human genome-sequencing project, with a number of web-based tools now available for exploiting this sequence. These tools are described very briefly in Section 8.3.1 and their practical application is covered in detail in Chapters 4 and 9.

8.2.3 SNPs, linkage disequilibrium, haplotypes and STRs

Introduction

This section provides a simple introduction to the underlying principles of the detection of genetic association with a population (i.e., non-family)-based approach. The majority of studies of this nature are undertaken with SNP markers (see Chapter 3). Biallelic SNPs are the marker of choice due to their abundance in the human genome and because they are amenable to high-throughput genotyping approaches. The other marker system commonly used for genetic studies is the multiallelic STR. The paragraphs below on linkage disequilibrium and haplotypes refer mainly to SNPs. The use of STRs for population-based association studies is discussed at the end of this section.

Linkage disequilibrium

A polymorphism associated with a disease state (in the true, rather than statistical, sense) may either directly contribute to the disease process, or be a surrogate marker co-inherited with an adjacent functional variant that contributes to the disease state. This co-inheritance of the surrogate marker with the disease allele, which can occur to varying degrees, is termed 'linkage disequilibrium' (LD). By strict definition, LD is said to be present if co-occurrence of the two polymorphisms happens with a frequency greater than would be expected by chance. A number of measures of LD are used, two of the most commonly employed being r^2 and D'. Both measures are based on the difference between the observed and expected (assuming independence) number of haplotypes (see below) bearing specified alleles of two markers (see Mueller, 2004, or Devlin and Risch, 1995, for a discussion of D', r^2 and other measures of LD). Although, by the strict definition given above, LD can occur between unlinked variants, for example, in the presence of recent population admixture, in the following paragraphs any reference to LD is specifically to LD between two linked markers.

Clearly, the greater the extent of LD between two polymorphisms, the larger the chances of detecting the phenotypic influence of one by genotyping the other in a case-control experiment. The degree of LD is dependent on the history of the two adjacent markers and is influenced by the relative times of appearance of the two polymorphisms in the population and the degree of recombination between them. An extreme example would be two polymorphisms that appeared simultaneously on the same chromosome through spontaneous mutation and between which no recombination events have occurred over 2000 generations. During this period, these two linked polymorphisms have attained a population frequency of 20 per cent through chance (random genetic drift) and are in absolute LD. Imagine an alternative scenario in which a new polymorphism arises adjacent to an ancient polymorphism that has already attained a frequency of 20 per cent over the previous 1000 generations; over the subsequent 1000 generations, there is a high degree of recombination between the markers, eroding the LD (Figure 8.1). Clearly, the former case would be more favourable for using one of the markers as a surrogate to detect the phenotypic influence of the other.

Haplotypes

A haplotype is a string of co-inherited alleles of different markers that are arranged in a successive fashion along a given stretch of DNA; hence, each haplotype represents a linear section of DNA rather than the single point corresponding to a single marker. The extent of discernible haplotype length varies widely for different regions of the genome; well-defined haplotypes (characterized by moderate or high LD) are punctuated by regions of extremely low LD, suggesting that the recombination processes, selective pressures and other factors that dictate the degree of LD vary widely in an abrupt fashion across the genome. However, recent evidence shows that most of the genome can be categorized into 'blocks' (contiguous LD intervals) of SNPs. These blocks vary in different populations according to population history. For example, in Caucasians, almost 90 per cent of genomic regions show block-like LD structures with an average of four common haplotypes per block and an average block length of 7 kb; in the African Yoruban population, less than 70 per cent of the genome can be defined in terms of LD blocks, and where blocks can be defined, the average number of common haplotypes per block is 5.6 and average block length is 16 kb (International HapMap Consortium, 2005).

In certain circumstances, statistical analysis of haplotypes for the identification of genes associated with the trait of interest is more powerful than single SNP analysis. This is because a SNP usually has only two allelic states, whereas a stretch of DNA can typically be represented by several different haplotypes; the chance that one of the many haplotypes shows strong association with a functional variant (i.e., a variant that influences the phenotype) is higher than the odds of a strong, pure correlation with one of only two possible alleles for a single SNP. In this sense, a

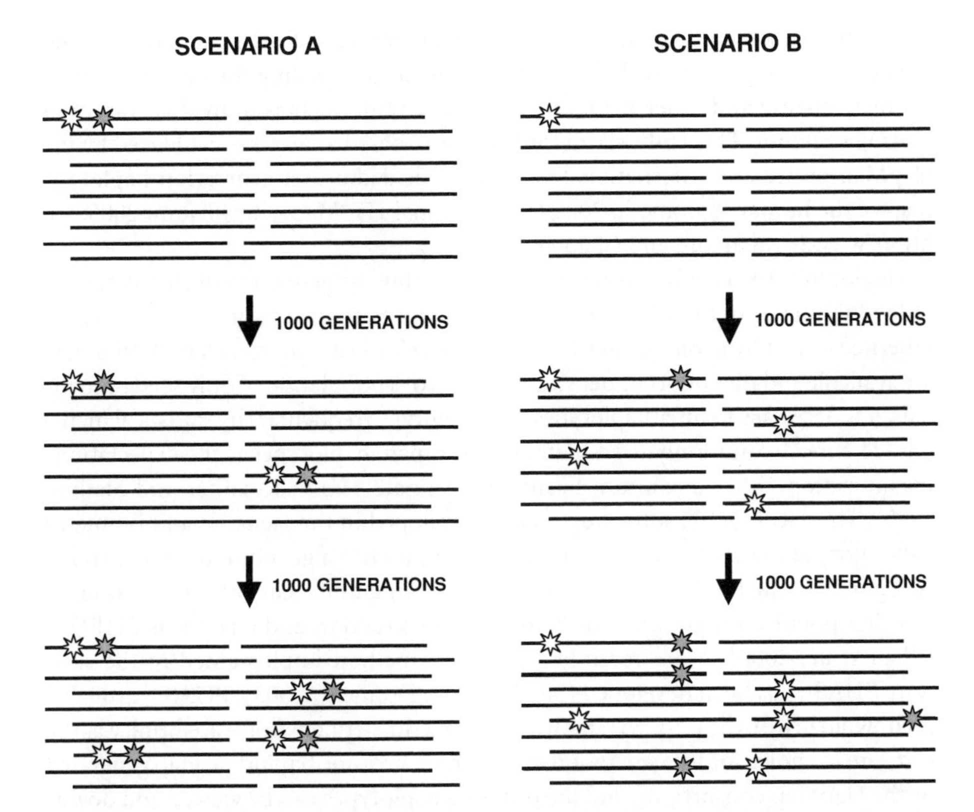

Figure 8.1 Alternative hypothetical scenarios depicting the evolution of a relationship between two SNPs. Identical stretches of DNA within a population are represented by black lines. In scenario A, two adjacent polymorphisms, represented by a white star and a grey star, arise simultaneously and by random drift achieve a population frequency of 0.1 after 1000 generations, increasing to 0.2 after 2000 generations, at which time they are still co-segregating as a tightly linked unit. In scenario B, a lone polymorphism (white star) reaches a frequency of 0.2 after 1000 generations, at which point a new polymorphism (grey star) arises spontaneously, some distance away. Note that although the grey polymorphism occurs only on a background bearing the white polymorphism, the association is less clear-cut than scenario A due to the chromosomes bearing the white polymorphism in the absence of the grey polymorphism. During the subsequent 1000 generations, association between the two polymorphisms is further clouded by recombination between the two SNPs and divergence through random drift. Unfortunately for the genetics investigator, scenario A is idealized and scenario B is more typical

series of haplotypes is analogous to a multiallelic STR marker (although regarded as more stable – see below). Clearly, if the functional variant itself is under test, or a polymorphism that shows perfect co-segregation with the functional variant, haplotypic analysis offers no advantage. It should also be noted that haplotype analysis is a double-edged sword and in addition to increasing statistical power it has the potential to reduce it, by introducing multiple testing and possibly by diluting an association signal due to undetected recombination within the haplotypes. These caveats notwithstanding, at the time of writing there is great excitement concerning the potential of genome-wide haplotype maps for unravelling the genetic basis of complex disease and other medically important traits such as individual variation in drug response. This enthusiasm for haplotype-based analyses is evidenced by the HapMap project, an international effort that yielded a first-generation haplotype map of the human genome in 2005 (International HapMap Consortium, 2005; see http://www.hapmap.org and Chapter 4).

Haplotypes are usually constructed by comparing the genotypes of closely related individuals at two or more linked markers and identifying groups of alleles that are co-inherited as a set from one generation to the next. However, where no family members are available and the cohort under study consists of a population of unrelated individuals, it is necessary to infer haplotypes and haplotype frequencies by statistical methods. The most commonly used method to estimate haplotypes is the expectation-maximization (EM) maximum likelihood estimate (MLE; Excoffier and Slatkin, 1995). The ARLEQUIN software package, developed in the Genetics and Biometry Laboratory at the University of Geneva, contains an EM algorithm for this purpose. ARLEQUIN can be downloaded from http://cmpg.unibe.ch/software/arlequin3/. Another popular programme for haplotype construction and analysis is EHPLUS (Zhao *et al.*, 2000). More recently, a Bayesian method implemented in the software PHASE has been developed (Stephens and Donnelly, 2003). PHASE, available for download from http://www.stat.washington.edu/stephens/software.html, was the software of choice for the generation of the whole-genome haplotype map generated by the HapMap consortium, and the resulting haplotypes can be viewed and downloaded from the HapMap Genome browser (http://www.hapmap.org). PHASE and other haplotype software packages are discussed in more detail in Chapter 10.

Note that haplotype construction using family inheritance patterns, although more robust than population-based statistical approaches, also typically requires a degree of inference, and the resulting haplotypes may be probable rather than actual (Hodge *et al.*, 1999). For absolute definition of all haplotypes, it is necessary to separate physically the two copies of each stretch of DNA under analysis, that is, reduction from a diploid to a haploid state, to allow unmixed analysis of a single haplotype. For very short stretches of DNA (up to approximately 10 kb), this can be achieved by allele-specific PCR (Michalatos-Beloin *et al.*, 1996); for large-scale haplotype construction, it is necessary to separate entire chromosomes. This strategy has been successfully employed by the California-based company Perlegen Sciences, Inc., who have used a rodent-human somatic cell hybrid technique to separate physically the two copies of human chromosome 21 for haplotype elucidation (Patil *et al.*, 2001). However, most investigators employ the less laborious population or family-based inference methods for haplotype construction and accept a certain degree of error or loss of power.

In addition to potentially providing greater power than single markers in subsequent statistical analyses, knowledge of the haplotypes representing the locus under study is extremely valuable for maximizing efficiency in study design. For example,

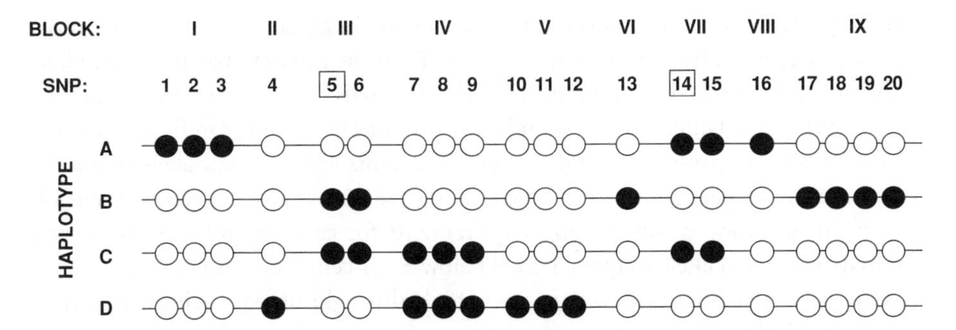

Figure 8.2 Using haplotypic information to maximise efficiency in genotyping study design. Twenty SNPs spanning four haplotypes are shown. Each SNP is represented by a circle; the circle is black or white, depending on the allelic state of the SNP. The SNPs can be grouped into nine blocks – each block contains a group of SNPs with an identical allelic pattern in the four haplotypes. Genotyping all SNPs in any given block is unnecessary, as the genotype of one SNP per block allows the genotypes of the other SNPs in the block to be inferred; for example, genotyping SNP 1 allows the genotypes of SNPs 2 and 3 to be predicted. Moreover, in this simplified example, all four haplotypes can be unambiguously identified by genotyping just two SNPs, 5 and 14 (boxed), yielding a 90 per cent reduction in genotyping compared to a 'blind' strategy (i.e., no knowledge of haplotypic structure)

two markers which always co-segregate (as in Figure 8.1, scenario A) will provide the same information, regardless of which of the two is genotyped; therefore typing both markers is inefficient, as the genotype of one can be inferred from the other. Consequently, detailed knowledge of the haplotypes across the interval theoretically allows a minimum marker set to be identified that will permit the extraction of all haplotypic information (Figure 8.2; see Johnson *et al.*, 2001; Patil *et al.*, 2001). The value of this knowledge in the design of association studies was a major impetus behind the establishment of the HapMap project (International HapMap Consortium, 2005). Several software algorithms are available to aid the selection of optimum marker sets based on haplotypic information (reviewed in Ke *et al.*, 2005). The SNPs that comprise these optimized marker sets are often referred to as 'tag SNPs' or 'htSNPs' ('haplotype tagging' SNPs). Tag SNP selection tools include Tagger (de Bakker *et al.*, 2005), which can be implemented through the Haploview interface at http://www.broad.mit.edu/mpg/haploview/ (Barrett *et al.*, 2005).

In order to select an optimized marker set for an association study spanning the genomic region of interest, it is necessary to identify all common SNPs within the area under study and select tag SNPs based on knowledge of LD and haplotypes across the region. At the time of writing, a significant amount of the raw information required for such a study (genome sequence, SNP data, haplotype and LD data and tag SNP sets) is publicly available, most notably as a result of the HapMap project (http://www.hapmap.org). It is possible that a significant proportion of the common variants contributing to complex phenotypes will be identified by HapMap SNP sets alone in coming years. Where detection of rarer functional variants is necessary,

HapMap SNPs alone are unlikely to be sufficient (Zeggini *et al.*, 2005) and more thorough identification of the majority of SNPs in the interval may be required, by sequence analysis of a significant number of individuals from the relevant population across the region of interest. For example, sequencing DNA from 24 individuals would give a 95 per cent probability of detecting all variants with a minor allele frequency of greater than 5 per cent (Nickerson and Kruglyak 2001). Five per cent is a sensible lower cut-off point, as sample size requirements for case-control studies increase dramatically when allele frequencies fall below 5 per cent (Johnson *et al.*, 2001).

Other than in laboratories with access to high-throughput sequencing capabilities, it is impractical to sequence a region covering several megabases (the typical outcome of a linkage scan) in 24 individuals. A more realistic approach is the identification of all genes in the interval and sequencing of the coding sequence plus flanking splice sites, together with 1–2 kb of putative promoter (that is, the region immediately upstream of the transcription start site) and any other known regulatory elements. Although not comprehensive, as unidentified regulatory elements can be intronic or tens to hundreds of kilobases away from the genes under their influence (Blackwood, 1998; Stranger *et al.*, 2005), this approach offers a good compromise between exhaustive coverage of the locus and practicality. For SNP identification purposes, it may be preferable to use individuals derived from the disease, rather than control, population. This will give a greater chance of detecting rare functional variants (mutations) that have a higher frequency in the disease population. For example, *NOD2* mutations predisposing to Crohn's disease were found to be at a frequency of 6–12 per cent in cases, but under 5 per cent in controls (Hugot *et al.*, 2001; Ogura *et al.*, 2001).

Having identified the majority of coding and regulatory sequence SNPs with a frequency of greater than 5 per cent, redundant SNPs can be removed by tag SNP software such as Tagger (de Bakker *et al.*, 2005), or more simply by pairwise comparison of LD between SNPs (Zeggini *et al.*, 2005). A subset of 96 individuals from the population under study should be sufficient to detect the majority of haplotypes with a frequency of greater than 5 per cent (B-Rao, 2001). These haplotypes can then be used as a basis for selecting a minimal SNP set for the full association study. It should be noted, however, that SNPs which suggest a strong possibility of functional consequence (such as those that alter residues that are conserved between a number of species, or result in non-conservative amino-acid changes; see Chapters 11–14) should not be excluded from analysis, and inclusion of such markers can be forced in most available tag SNP selection tools, including Tagger.

Simple tandem repeat markers (STRs)

STRs (also known as microsatellites) were the mainstay of monogenic trait linkage analysis during the 1990s, but are now frequently overlooked following the explosion of interest in SNPs for population-based studies. STRs are out of favour for two main reasons: (i) they are less amenable to cheap, high-throughput genotyping methodology than SNPs; (ii) STRs typically have a much higher mutation rate than SNPs (up to 10^{-3} per meiosis compared with an average of 10^{-9} for SNPs; Ellegren, 2000). It has been suggested that this extreme mutation rate, although rendering STRs highly informative for linkage, might confound genetic association studies, as a single allele may represent an excessive number of haplotypes, having independently arisen on the different haplotypic backgrounds through mutation events (Moffatt *et al.*, 2000). This may prevent the detection of association between the STR allele and an adjacent polymorphism associated with disease. However, comparison of the entire STR allele frequency distribution profiles for cases and controls may highlight differences reflecting a difference in the frequency of an adjacent disease-associated SNP, due to divergence of the STR profiles associated with SNP allele 1 and SNP allele 2, as a result of frequent STR mutation (Koch *et al.*, 2000; Abecasis *et al.*, 2001, Tanaka *et al.*, 2005). There is also some evidence to suggest that LD can be detected over greater distances with STRs than with SNPs, possibly ten times as far (Koch *et al.*, 2000, Horowitz *et al.*, 2005), perhaps because in some circumstances STR mutation significantly outstrips recombination at flanking sites. There is no doubt that STRs are a potentially useful tool in association studies; for example, a convincing association between an STR and prostate cancer was recently described by Amundadottir *et al.* (2006). However, due to the perceived cost and throughput advantages of SNPs, it is likely that STRs will be displaced by SNPs for the majority of future linkage and association studies.

8.2.4 Statistical analysis

Methods and software for the statistical analysis of both single marker and haplotype data in both case-control and family-based cohort scenarios are described in detail in Chapter 10. Briefly, a chi-square analysis may be used to test for departure between observed and expected allele frequencies for a single biallelic marker in a case-control cohort, while multiallelic systems, such as haplotypes, may be tested with software such as PHASE (Stephens and Donnelly, 2003). In addition to the haplotype construction capability described in Section 8.2.3 above, PHASE includes a permutation-based likelihood ratio test for comparing haplotype frequencies between case and control cohorts (for details of how the test of association employed in PHASE works, see Matthew Stephens' website: http://www.stat.washington.edu/stephens/phasefaq.html). Family-based samples such as parent-offspring trios and discordant sibs can be analysed by the transmission disequilibrium test (TDT) and associated methods (Spielman *et al.*, 1993; discussed in depth in Chapter 10), although the TDT was originally developed for biallelic markers, an extension of the TDT has been developed for testing multiallelic markers and haplotypes (Sham and Curtis, 1995). Many TDT-related tests are implemented through the software suite FBAT (Family Based Association Tests), which can be downloaded from http://www.biostat.harvard.edu/~fbat/fbat.htm (Laird and Lange, 2006).

8.3 A stepwise approach to locus refinement and candidate gene identification

Figure 8.3 gives an overview of the practical process of locus refinement, candidate gene selection and testing for phenotype–genotype association by a case-control approach. Each step is described in more detail in the following sections.

8.3.1 Sequence characterization

The most popular Web tools for the purpose of human genome sequence characterization are the human genome browser hosted by the National Center for Biotechnology Information (NCBI) at http://www.ncbi.nlm.nih.gov/, the Golden Path genome browser hosted by the University of California, Santa Cruz, at http://genome.ucsc.edu/ and the Ensembl human genome browser maintained by the European Bioinformatics Institute (EBI) and the Wellcome Trust Sanger Institute at http://www.ensembl.org/Homo_sapiens/. These browsers are described and reviewed in detail in Chapter 4, and the reader is referred to Chapter 9 for a comprehensive description of methods for defining a locus between two genetic markers at the sequence level with these three tools.

8.3.2 Preliminary analysis using HapMap tag SNPs

As a first step following complete locus characterization, an attempt to identify regions of potential association within the critical interval using HapMap tag SNPs is suggested. Tag SNPs across any given genomic region can be downloaded directly from the HapMap browser (http://www.hapmap.org). A reasonable degree of tag SNP 'portability' between populations (that is, the utility of the same tag SNP set for identification of common haplotypes in different but related populations) has been demonstrated (Gonzalez-Neira *et al.*, 2006; Montpetit *et al.*, 2006), and the HapMap population that most closely matches the population under test should be selected for tag SNP selection. Following genotyping of HapMap tag SNPs in the case-control cohort of interest, haplotype construction and haplotype frequency comparison between cases and controls can be performed with PHASE (Stephens and Donnelly, 2003; http://www.stat.washington.edu/stephens/software.html). A test for Hardy-Weinberg equilibrium (HWE) is a useful prior check for ensuring that there is no (or little) population stratification and that each SNP is giving the expected genotype

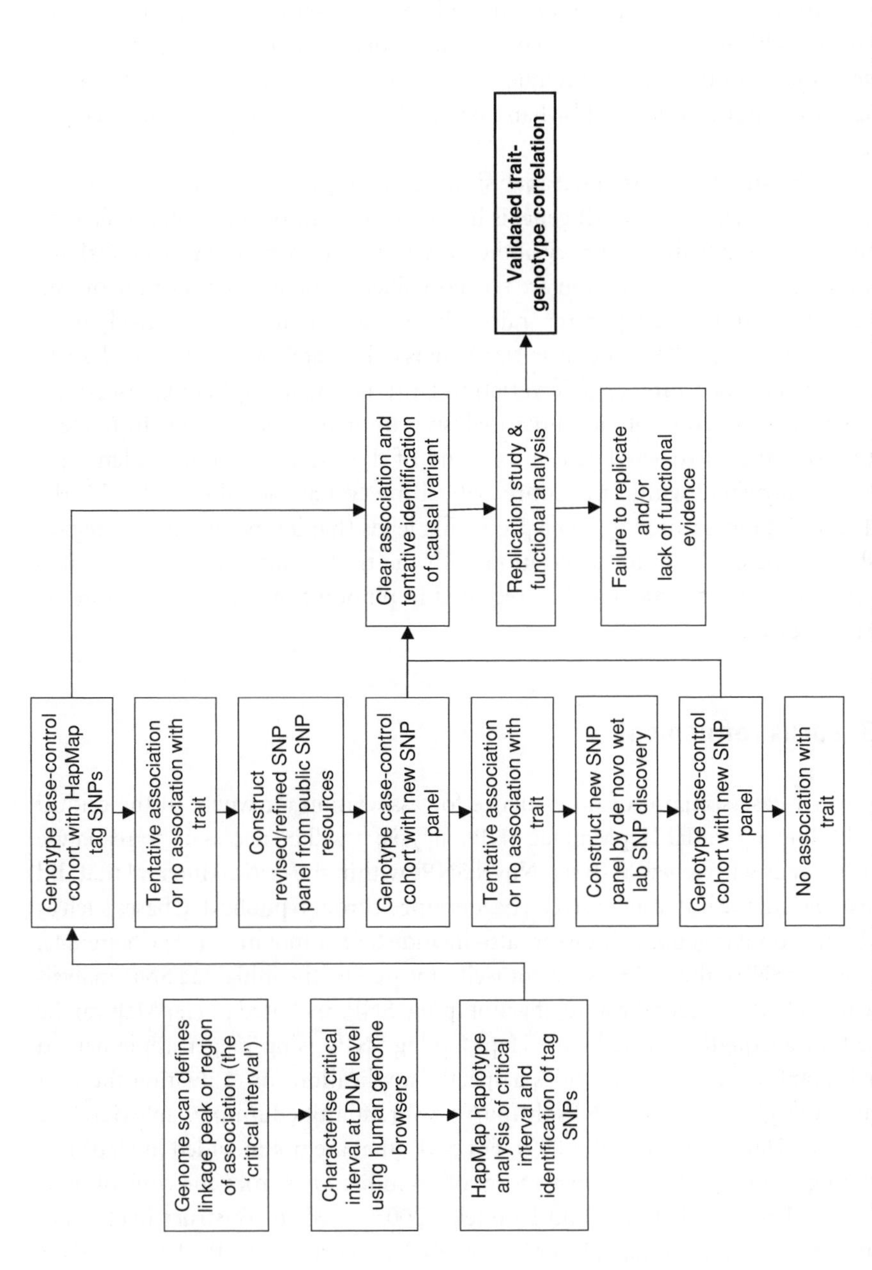

Figure 8.3 Flow diagram describing the logical steps in pinpointing and verifying a gene-phenotype association, using a case-control follow-up to a genome scan

distribution for the observed allele frequencies. Expected genotype frequencies are calculated from allele frequencies under the assumption $p^2 + q^2 + 2pq = 1$, where p and q are the allele frequencies and p^2, q^2 and 2pq correspond to the frequencies of the three possible genotypic states. The actual genotype frequencies are then tested for departure from the expected frequencies by the chi-square test. The calculation is simple and can be performed by hand or in a Microsoft Excel macro for biallelic markers.

The relative success of this initial analysis in narrowing the critical interval, or perhaps even identifying the culprit gene, is highly dependent on a number of factors, perhaps most notably the frequency of the causal variant. Common genetic variants that influence the trait in question are far more likely to be narrowed or uncovered with tag SNPs from HapMap; rarer trait-influencing variants are more likely to be missed (Zeggini *et al.*, 2005), because HapMap is deliberately focused on analysis of common, rather than rare, genetic variation (International HapMap Consortium, 2005). Thus, the outcome of this initial analysis is unpredictable, ranging from identification of a trait-associated gene at one end of the spectrum to no evidence for association anywhere within the entire critical interval at the other end, with the middle ground represented by some promising leads (haplotypes within the region that show some evidence for association with the trait of interest). However, it is likely that further refinement will be required to pinpoint the causal variant in the majority of cases.

8.3.3 Locus refinement

In the worst-case scenario of no evidence for association anywhere in the critical interval following initial HapMap tag SNP analysis, it will be necessary to reanalyse the entire locus with a new SNP set. Novel SNPs within the critical interval that did not form part of the initial analysis can be identified through public databases such as dbSNP; the popular genome browsers also include SNP annotation (see Chapter 4). Of these new SNPs, those that were not well captured by the initial tag SNP analysis can be included in a second round of genotyping. SNPs included in HapMap can be assessed for adequate capture by the HapMap tag SNPs using Tagger implemented through Haploview. Tagger includes a specific application for evaluating the performance of a given set of tag SNPs (http://www.broad.mit.edu/mpg/haploview/; de Bakker *et al.*, 2005; Barrett *et al.*, 2005). For SNPs that are not included in HapMap, genotyping can be performed on a subset of 96 individuals from the control population and PHASE (Stephens and Donnelly, 2003) used to construct haplotypes incorporating both the original HapMap tag SNPs and the new SNPs. The new SNPs can then be evaluated with regard to the amount of additional data that they are likely to capture prior to genotyping the entire case-control cohort and subsequent association analysis. If necessary, Tagger can be used to identify a further set of tag SNPs from the new SNPs to maximize genotyping efficiency. Where a single region or multiple regions within the critical interval show some evidence for association following initial HapMap tag SNP analysis, essentially the same strategy of novel SNP genotyping can be followed, but with a focus on the subregions highlighted in the initial analysis rather than reanalysis of the entire locus. If there are multiple, extensive tracts that warrant further analysis, it may be useful to arrange genes/regions in rank order for analysis, based on biological plausibility with respect to association with the disease or trait under study.

Ultimately, physical identification of novel SNPs in the population under study may be necessary for ultimate locus refinement and identification of the causal variant, particularly where the causal variant is rare. This can be achieved by sequencing the key interval in 24 individuals selected randomly from the disease population, as discussed in Section 8.2.3 above, followed by genotyping of all novel SNPs thus identified in 96 random individuals derived from the control population. The new SNP data can then be combined with previous SNP and haplotype information, to allow efficient SNP selection prior to full case-control genotyping and analysis, as before. If the regions for follow-up are extensive, the investigator may wish to consider prioritizing the identification and testing of SNPs in coding and known or putative regulatory regions in the first instance.

This iterative SNP testing, with increasing difficulty and cost depending on SNP source (public databases versus *de novo* SNP discovery in the investigator's own laboratory) and the availability of supporting resources, such as SNP allele frequency data and HapMap haplotypes, should allow the investigator to progress toward identification of the causal variant in the most efficient manner (Figure 8.3).

8.3.4 General comments on statistical analysis

Several statistical methods for the analysis of case-control data are available in addition to the method implemented in PHASE described in Section 8.2.4, including cladistic analysis methods and techniques that consider interactions between unlinked loci rather than taking a locus-by-locus approach (Durrant *et al.*, 2004; Marchini *et al.*, 2005). No doubt additional strategies will be developed over the coming years; investigators should carefully consider all available methods and employ the most appropriate for their data. Expert advice from a statistical geneticist at the early stages of study design is essential. Replication of positive findings in an independent cohort is generally considered to be key to confirming the validity of an association between polymorphism and trait, and investigators may wish to consider including scope for replication in the study design. For example, it may be beneficial to divide a cohort randomly into two groups for statistical analysis, to allow the possibility of replication of any positive association using the second subset. An acceptable P value threshold for declaring association between a marker and disease is the subject of considerable debate. Clearly, a nominal cut-off of $P = 0.05$ is inappropriate where multiple tests have been performed, as this value (or lower) may occur several times by chance. However, standard methods of correction for multiple testing, such as Bonferroni correction, are regarded as overly stringent (Cardon and Bell, 2001). The most pragmatic course of action is to avoid setting thresholds that are excessively rigorous and instead follow up promising leads, adding to the weight of evidence for involvement (or lack of involvement) in the disease process by additional means (see below).

8.3.5 The burden of proof – is an associated gene really involved in the disease process?

Detection of association between a gene and disease phenotype does not constitute definitive proof that the gene under test is involved in the disease process. Rather, it provides a single piece of evidence to suggest *possible* involvement in the disease process that requires further substantiation. Replication of the association in a second cohort considerably strengthens the argument for involvement; for example, the association between the insulin gene and type 1 diabetes has been reproduced a number of times (Bennett and Todd, 1996). However, even in the event of independent replication of results, one should consider the possibility that the replication is due to chance or that the apparent disease association is due to an adjacent gene in LD with the marker under test. If the polymorphism is in protein coding sequence and causes an amino-acid change, it may be possible to assess the possible impact on protein function by the nature of the change (conservative or non-conservative), the context in which it occurs (potential disruption of secondary or tertiary protein structure) and the degree of cross-species conservation and conservation within protein families. Conservation may also be used to gauge the potential impact of polymorphisms in putative regulatory elements. These areas are covered in detail in Chapters 11 and 13. However, it should also be remembered that polymorphisms that appear to be innocuous on cursory examination can have functional consequences; for example, synonymous coding changes that occur in exonic splicing enhancer (ESE) regions (Liu *et al.*, 2001).

Ultimately, it is likely that the investigator will wish to instigate additional laboratory-based experiments to judge the functional effect of the variant in question. These may include gene expression and cell-based reporter assays for putative promoter polymorphisms, functional enzyme or signal transduction assays for amino-acid changes and *in vivo* analysis in the mouse by gene knockout or polymorphism knock-in technology for studies in the context of the whole organism, to name but a small fraction of the available techniques.

8.4 Conclusion

This chapter provides a basic overview of the process of moving from a large genetic locus to the identification and screening of candidate genes for disease association.

More detailed information on all aspects of study design and data analysis can be gleaned from the references cited in the text and further review of the literature, and readers are strongly advised to broaden their knowledge beyond the limits of this chapter. Although a number of popular tools and techniques have been highlighted, several other equally valid approaches exist, and investigators are encouraged to seek out and develop further methods for comparison with those presented here. Continual development of new approaches and improvement of existing methodology are a dominant feature of this rapidly moving field; consequently there is a constant need for investigators to keep abreast of new developments to maximize the chances of success.

8.5 A list of the software tools and Web links mentioned in this chapter

- ARLEQUIN: http://cmpg.unibe.ch/software/arlequin3/
- Ensembl human genome browser: http://www.ensembl.org/Homo_sapiens/
- FBAT: http://www.biostat.harvard.edu/~fbat/fbat.htm
- Haploview: http://www.broad.mit.edu/mpg/haploview/
- HapMap: http://www.hapmap.org
- National Center for Biotechnology Information: http://www.ncbi.nlm.nih.gov/
- PHASE: http://www.stat.washington.edu/stephens/software.html
- STRUCTURE/STRAT: http://pritch.bsd.uchicago.edu/software.html
- Tagger: http://www.broad.mit.edu/mpg/tagger/
- UCSC Golden Path genome browser: http://genome.ucsc.edu/

Links to the software tools mentioned in this chapter and many others besides, via the North Shore LIJ Research Institute: http://www.nslij-genetics.org/soft/

Acknowledgements

Many thanks to Aruna Bansal for a critical reading of the original version of this chapter that appeared in the first edition, resulting in significant improvements.

Thanks also to Michael R. Barnes, for the insightful suggestions that led to further improvements and updates for the second edition.

References

Abecasis G. R., Noguchi, E., Heinzmann, A. *et al.* (2001). Extent and distribution of linkage disequilibrium in three genomic regions. *Am J Hum Genet* **68**, 191–197.

Amundadottir, L. T., Sulem, P., Gudmundsson, J. *et al.* (2006). A common variant associated with prostate cancer in European and African populations. *Nat Genet* **38**, 652–658.

Bennett, S. T. and Todd, J. A. (1996). Human type 1 diabetes and the insulin gene: principles of mapping polygenes. *Ann Rev Genet* **30**, 343–370.

Blackwood, E. M. and Kadonaga, J. T. (1998). Going the distance: a current view of enhancer action. *Science* **281**, 61–63.

B-Rao, C. (2001). Sample size considerations in genetic polymorphism studies. *Hum Hered* **52**, 191–200.

Cardon, L. R. and Bell, J. I. (2001). Association study designs for complex diseases. *Nat Rev Genet* **2**, 91–98.

Clayton, D. G., Walker, N. M., Smyth, D. J. *et al.* (2005). Population structure, differential bias, and genomic control in a large scale, case-control association study. *Nat Genet* **37**, 1243–1246.

de Bakker, P. I., Yelensky, R., Pe'er, I. *et al.* (2005). Efficiency and power in genetic association studies. *Nat Genet* **37**, 1217–1223.

Devlin, B. and Risch, N. (1995). A comparison of linkage disequilibrium measures for fine-scale mapping. *Genomics* **29**, 311–322.

Devlin, B. and Roeder, K. (1999) Genomic control for association studies. *Biometrics* **55**, 997–1004.

Devlin, B., Roeder, K. and Bacanu, S. A. (2001). Unbiased methods for population-based association studies. *Genet Epidemiol* **21**, 273–284.

Durrant, C., Zondervan, K. T., Cardon, L. R. *et al.* (2004). Linkage disequilibrium mapping via cladistic analysis of single-nucleotide polymorphism haplotypes. *Am J Hum Genet* **75**, 35–43.

Ellegren, H. (2000). Microsatellite mutations in the germline: implications for evolutionary inference. *Trends Genet* **16**, 551–558.

Excoffier, L. and Slatkin, M. (1995). Maximum-likelihood estimation of molecular haplotype frequencies in a diploid population. *Mol Biol Evol* **12**, 921–927.

Falush, D., Stephens, M. and Pritchard, J. K. (2003). Inference of population structure using multilocus genotype data: linked loci and correlated allele frequencies. *Genetics* **164**, 1567–1587.

Giraudeau, F. S., McGinnis, R. E., Gray, I. C. *et al.* (2004). Characterization of common genetic variants in cathepsin K and testing for association with bone mineral density in a large cohort of perimenopausal women from Scotland. *J Bone Miner Res* **19**, 31–41.

Gonzalez-Neira, A., Ke, X., Lao, O. *et al.* (2006). The portability of tagSNPs across populations: a worldwide survey. *Genome Res* **16**, 323–330.

Gretarsdottir, S., Sveinbjornsdottir, S., Jonsson, H. H. *et al.* (2002). Localization of a susceptibility gene for common forms of stroke to 5q12. *Am J Hum Genet* **70**, 593–603.

Hodge, S. E., Boehnke, M. and Spence, M. A. (1999). Loss of information due to ambiguous haplotyping of SNPs. *Nat Genet* **21**, 360–361.

Horowitz, A., Shifman, S., Rivlin, N. *et al.* (2005). Further tests of the association between schizophrenia and single nucleotide polymorphism markers at the catechol-*O*-methyltransferase locus in an Askenazi Jewish population using microsatellite markers. *Psychiatr Genet* **15**, 163–169.

Hugot, J. P., Chamaillard, M., Zouali, H. *et al.* (2001). Association of NOD2 leucine-rich repeat variants with susceptibility to Crohn's disease. *Nature* **411**, 599–603.

International HapMap Consortium (2005). A haplotype map of the human genome. *Nature* **437**, 1299–1320.

International Human Genome Sequencing Consortium (2004). Finishing the euchromatic sequence of the human genome. *Nature* **431**, 915–916.

Johnson, G. C., Esposito, L., Barratt, B. J. *et al.* (2001). Haplotype tagging for the identification of common disease genes. *Nat Genet* **29**, 233–237.

Koch, H. G., McClay, J., Loh, E. W. *et al.* (2000). Allele association studies with SSR and SNP markers at known physical distances within a 1 Mb region embracing the ALDH2 locus in the Japanese, demonstrates linkage disequilibrium extending up to 400 kb. *Hum Mol Genet* **9**, 2993–2999.

Kruglyak, L. and Nickerson, D. A. (2001). Variation is the spice of life. *Nat Genet* **27**, 234–236.

Laird, N. M. and Lange, C. (2006). Family-based designs in the age of large-scale gene-association studies. *Nat Rev Genet* **7**, 385–394.

Liu, H. X., Cartegni, L., Zhang, M. Q. *et al.* (2001). A mechanism for exon skipping caused by nonsense or missense mutations in BRCA1 and other genes. *Nat Genet* **27**, 55–58.

Marchini, J., Donnelly, P. and Cardon, L. R. (2005). Genome-wide strategies for detecting multiple loci that influence complex diseases. *Nat Genet* **37**, 413–417.

McGinnis, R. (2000). General equations for Pt, Ps, and the power of the TDT and the affected-sib-pair test. *Am J Hum Genet* **67**, 1340–1347.

Michalatos-Beloin, S., Tishkoff, S. A., Bentley, K. L. *et al.* (1996). Molecular haplotyping of genetic markers 10 kb apart by allele-specific long-range PCR. *Nucleic Acids Res* **24**, 4841–4843.

Moffatt, M. F., Traherne, J. A., Abecasis, G. R. *et al.* (2000). Single nucleotide polymorphism and linkage disequilibrium within the TCR alpha/delta locus. *Hum Mol Genet* **9**, 1011–1019.

Montpetit, A., Nelis, M., Laflamme, P. *et al.* (2006). An evaluation of the performance of tag SNPs derived from HapMap in a Caucasian population. *PLoS Genet* **2**, 282–290.

Mueller, J. C. (2004). Linkage disequilibrium for different scales and applications. *Brief Bioinform* **5**, 355–364.

Nicolae, D. L., Wen, X., Voight, B. F. *et al.* (2006). Coverage and characteristics of the Affymetrix GeneChip human mapping 100K SNP set. *PLoS Genet* **2**, 665–671.

Ogura, Y., Bonen, D. K., Inohara, N. *et al.* (2001). A frameshift mutation in NOD2 associated with susceptibility to Crohn's disease. *Nature* **411**, 603–604.

Patil, N., Berno, A. J., Hinds, D. A. *et al.* (2001). Blocks of limited haplotype diversity revealed by high-resolution scanning of human chromosome 21. *Science* **294**, 1719–1723.

Pritchard, J. K. and Donnelly, P. (2001). Case-control studies of association in structured or admixed populations. *Theor Popul Biol* **60**, 227–237.

Pritchard, J. K. and Rosenberg, N. A. (1999). Use of unlinked genetic markers to detect population stratification in association studies. *Am J Hum Genet* **65**, 220–228.

Pritchard, J. K., Stephens, M., Rosenberg, N. A. *et al.* (2000). Association mapping in structured populations. *Am J Hum Genet* **67**, 170–181.

Risch, N. J. (2000). Searching for genetic determinants in the new millennium. *Nature* **405**, 847–856.

Sham, P. C. and Curtis, D. (1995). An extended transmission/disequilibrium test (TDT) for multi-allele marker loci. *Ann Hum Genet* **59**, 323–336.

Spielman, R. S., McGinnis, R. E. and Ewens, W. J. (1993). Transmission test for linkage disequilibrium: the insulin gene region and insulin-dependent diabetes mellitus (IDDM). *Am J Hum Genet* **52**, 506–516.

Stephens, M. and Donnelly, P. (2003) A comparison of Bayesian methods for haplotype reconstruction from population genotype data. *Am J Hum Genet* **73**, 1162–1169.

Stranger, B. E., Forrest, M. S., Clark, A. G. *et al.* (2005) Genome-wide associations of gene expression variation in humans. *PLoS Genet* **1**, 695–704.

Strittmatter, W. J., Saunders, A. M., Schmechel, D. *et al.* (1993). Apolipoprotein E: high-avidity binding to beta-amyloid and increased frequency of type 4 allele in late-onset familial Alzheimer disease. *Proc Natl Acad Sci U S A* **90**, 1977–1981.

Tanaka, G., Matsushita, I., Ohashi, J. *et al.* (2005). Evaluation of microsatellite markers in association studies: a search for an immune-related susceptibility gene in sarcoidosis. *Immunogenetics* **56**, 861–870.

Thomas, D. C., Haile, R. W. and Duggan, D. (2005) Recent developments in genomewide association scans: a workshop summary and review. *Am J Hum Genet* **77**, 337–345.

Zeggini, E., Rayner, W., Morris, A. P. *et al.* (2005). An evaluation of HapMap sample size and tagging SNP performance in large-scale empirical and simulated data sets. *Nat Genet* **37**, 1320–1322.

Zhao, J. H., Curtis, D. and Sham, P. C. (2000). Model-free analysis and permutation tests for allelic associations. *Hum Hered* **50**, 133–139.

9

Integrating Genetics, Genomics and Epigenomics to Identify Disease Genes

Michael R. Barnes

Bioinformatics, GlaxoSmithKline Pharmaceuticals, Harlow, Essex, UK

9.1 Introduction

It has probably become apparent in the preceding chapters that definition of a genetic locus, gene or markers in the human genome sequence by a genome browser is one of the most fundamental bioinformatics processes that a geneticist needs to carry out. Defining a locus in the genome immediately places it in a wider context, with almost limitless options for further characterization. The wide range of possibilities that this action opens up, make it important to set clear objectives for further characterization. Firstly, it is important to define the genetic and physical structure of the region. This can be achieved in terms of HapMap LD, haplotypes and recombination rates. The physical structure of the locus can also directly influence the genetics, by influencing recombination, chromosome stability and epigenetics. A good understanding of the genetic and physical properties of the locus can also help to identify the likely extent of the locus beyond the immediately associated markers.

Once the genetic and physical terrain is evaluated, the next objective is to characterize the functional entities in the locus, these might include known and novel genes, regulatory RNAs and regulatory regions some of which may be remote to the genes undergoing regulation. With this information, a picture begins to emerge of the biology of the locus under study. The next objective is to place all this information

Bioinformatics for Geneticists, Second Edition. Edited by Michael R. Barnes
 ISBN 978-0-470-02619-9 (HB) ISBN 978-0-470-02620-5 (PB)

in the context of the trait being studied. What pathways are likely to be involved? What tissues are known to be involved in the trait? How might the functional entities (genes, regulatory elements, etc.) in the locus fit the biology of the trait under study? How might known variants in the locus affect the function of these entities? Naturally, the clarity of the answers to each of these questions is likely to vary greatly between traits, depending on the level of understanding of the trait and the quality of the available data. The final objective is to take all this information and, depending on the approach to follow-up, select markers for further analysis. This might involve identifying all potentially functional variants or markers which tag all known locus haplotypes or a combination of both approaches. There may also be a need to rank genes for follow-up by genetic and biological evidence of association. Most of this is possible *in silico* to a point, within certain limitations. These need to be understood and it is worth bearing in mind that in some situations stepping into the laboratory may be the only way forward.

9.2 Dealing with the (draft) human genome sequence

It is self-evident that the availability of complete genome sequences is a fundamental advance for genetics. The labours of the pre-genome sequence era have now been totally superseded by absolute genome localization to the nearest base pair. But the human genome sequence must be used with care; without some basic quality checks, genome sequences can create some distinct problems. Firstly, it is important to stress that, despite the announcement of the completion of the sequencing of the human genome in 2001, the sequence is still a *draft* assembly and is likely to stay so for the foreseeable future. In fact, there are no plans to sequence the heterochromatin and centromeric regions of the genome, so, arguably, it may never be complete. The order and orientation of DNA fragments is often not known from the sequencing process itself. In some cases, structural variations, such as copy number polymorphisms, exist (Feuk *et al.*, 2006); however, because of the nature of the genome assembly process, these will invariably be collapsed into a single contig that does not reflect the natural sequence. To address the technical challenges of whole-genome assembly, the human genome is released as defined 'builds' on a quarterly basis (Lander *et al.*, 2001; reviewed in Chapter 4). The increasing complexity of processes that map data to the genome implicitly involves some lag in availability of the most current sequence assembly. At the time of writing (May 2006), the May 2004 (NCBI35) human genome assembly was still in most frequent use by the majority of applications, while the HapMap was using the July 2003 (NCBI34) release. A March 2006 (NCBI36) release is just beginning to be incorporated into the UCSC and pre-Ensembl servers. This creates a potentially huge problem, of which the user must always be aware. When genomic coordinates of different data types are compared, it is *critical* to ensure that they are both based on the same NCBI genome build. Considering this, it is good practice to record the genome build with any data set containing genomic coordinates. Without this information, subsequent reference to this data, by either yourself or others, will be seriously compromised. Data mapped against different assemblies can be compared, although the process is rarely painless. The best tool for this purpose is probably the UCSC Batch coordinate conversion tool (http://genome.ucsc.edu/cgi-bin/hgLiftOver). On an individual basis, the UCSC browser is also very helpful; as it archives the four most recent NCBI genome builds, the user can view the last four genome assemblies and can convert a given region in one genome build to another (using the CONVERT function in the toolbar). In this way, one can view a region using NCBI34 coordinates and then convert the region to NCBI36 coordinates.

9.3 Progressing loci of interest with genomic information

In an ideal world, all genetic studies would progress smoothly from initial observations of linkage or association to alleles or haplotypes that explain molecular mechanisms of disease and other phenotypes of interest. Unfortunately, this linear path of discovery is rarely encountered. In the real world, the study of genetics is better characterized by false-positive associations, failures to replicate and analytical stalemates. Success in genetics was usually the result of painstaking study design, laborious sample collection, methodical analysis and a healthy portion of luck. Today this situation may be improving; with our knowledge of the genetic structure (HapMap) and biology of the genome, coupled with advances in genotyping technology, we may expect improved odds of success in genetics. However, increased genotyping capacities will create new problems, chiefly type I error (false-positive associations). With these new challenges and potentially huge rewards, mastery of data has never been so important. Effective bioinformatics is needed at every stage of genetic analysis for efficient refinement and integration of data, to build biological rationale and ascertain function. The step-by-step process of genetic analysis becomes an ever more challenging process of integrating new results with what is known already and drawing conclusions to make the next step. Figure 9.1 illustrates each step of this process of integration, interpretation and analysis, with genomic sequence as a common thread through every stage.

9.3.1 Defining a locus – dealing with synteny and orthology

A locus of interest for a particular trait may be identified by many routes and might relate to just a single genetic variant, a gene or a wider genomic region, containing many genes and genetic variants. Candidate loci for human traits and diseases might well be identified by synteny with loci identified in mammalian disease models. With the exception of the X chromosome, synteny (or evolutionarily conserved gene order) between human and mammalian genes extends over only limited regions. Billions of recombination events have tended to distribute orthologous genes across different chromosomes, making reconstruction of an equivalent locus between mouse and man quite a challenge. For example, a single locus in mouse may span multiple

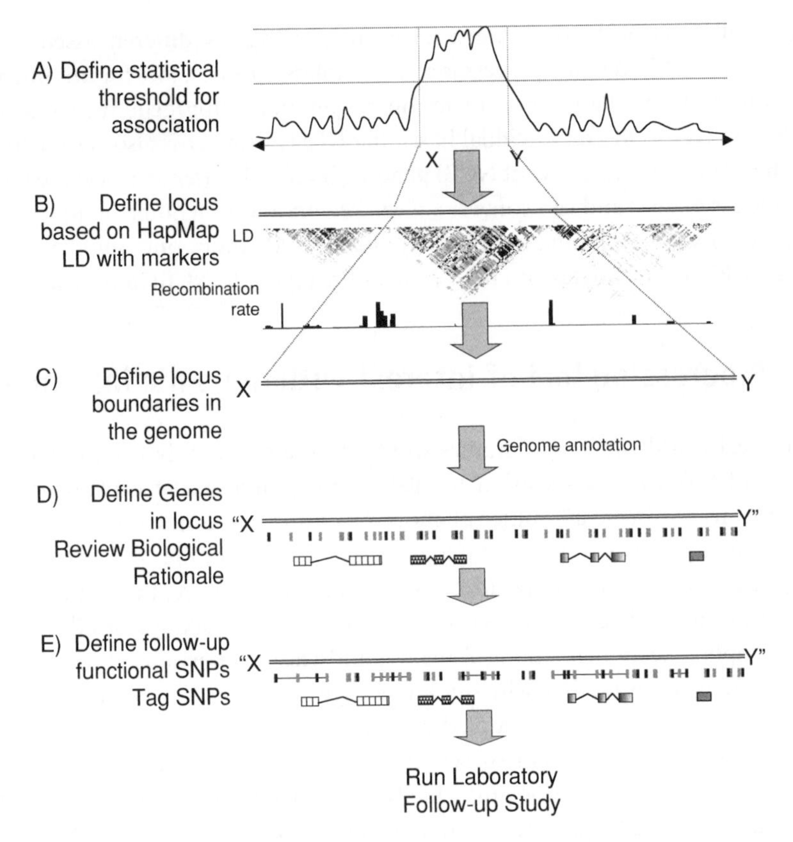

Figure 9.1 Using the genome and HapMap as a template for genetics. Key bioinformatic steps to take a genetic study from an initial linkage or association to laboratory genotyping are illustrated. The reader should note the role of genomic sequence as a common thread through every stage

regions in man (see Chapter 5). Similar issues also exist in the establishment of true orthology between genes in different species, where one is identified to play a role in a disease model. If two genes are truly orthologous, their evolution closely follows patterns of speciation (Fitch, 2000). The implicit assumption here is that the two genes might be expected to carry out a similar role in both organisms. However, these relationships are not always one to one, nor is similarity of function assured; for example, a gene identified in nematodes, with a role in apoptosis, might have more than one orthologue in man, or, worse still, it may have no identifiable orthologue (see Chapter 6 for a review of this area).

9.3.2 Defining a gene as a locus

Where regions of interest are identified directly in man, obviously the issues of locus definition for study are less complex, although pitfalls still exist. If a gene of interest

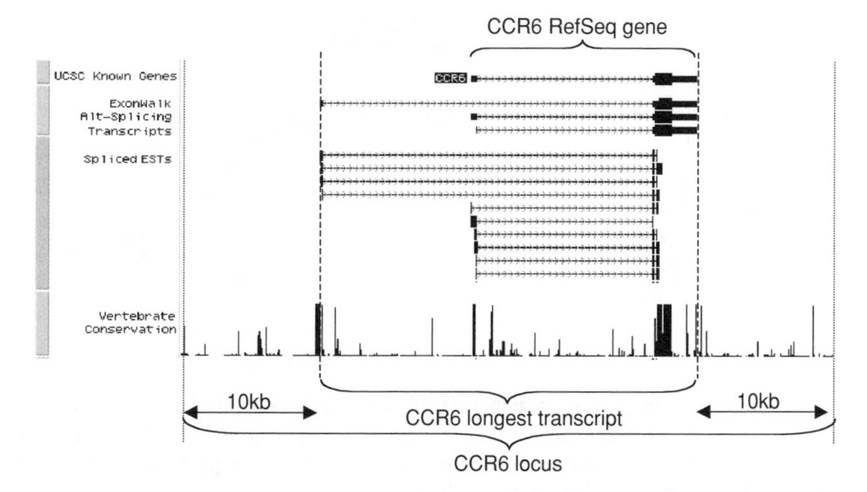

Figure 9.2 Definition of a gene locus. The CCR6 gene locus is defined by the most 5′ exon among all transcript variants and the most 3′ exon among all variants. 10 kb is added to either side of the locus to allow for regulatory regions. Note that known transcripts (e.g., RefSeq) may not always be the longest transcript, and a full review of other mRNA and EST evidence is recommended

is identified, perhaps by differential expression or biological rationale, the gene locus can be rapidly defined in a genome browser. The gene locus could reasonably be defined as the region encompassing all the exons and introns perhaps 10 kb upstream and downstream of the first and last exon, to allow for gene promoter and regulatory elements. In the example of the CCR6 gene, this can be achieved with the UCSC genome browser, for example, by entering the gene symbol and modifying the genome coordinates returned by the search by adding 10 kb upstream and downstream. However, even in this simple case, the locus should not be taken at face value. In the case of CCR6, most standard sources of gene information (e.g., RefSeq, Entrez-Gene) describe only one or two known CCR6 gene transcripts, both transcribed from the same first exon. However, review of UCSC spliced EST data identifies an additional transcript variant with an alternative first exon more than 11 kb upstream of exon 1 in the known gene (the structure of this variant is exemplified in the UCSC ExonWalk track). Clearly, this is important to investigate, as it suggests that CCR6 isoforms are driven by alternative promoters. Therefore, the appropriate locus would span 10 kb upstream of the most 5′ exon among all isoforms and 10 kb downstream of the most 3′ exon among all isoforms (Figure 9.2). The same rule of thumb would apply to genes that show complex alternative splice forms.

9.3.3 Defining genetic loci from linkage and association data

Disease susceptibility loci encompassing large genomic regions may be identified by family-based linkage, population-based linkage or association analysis. These

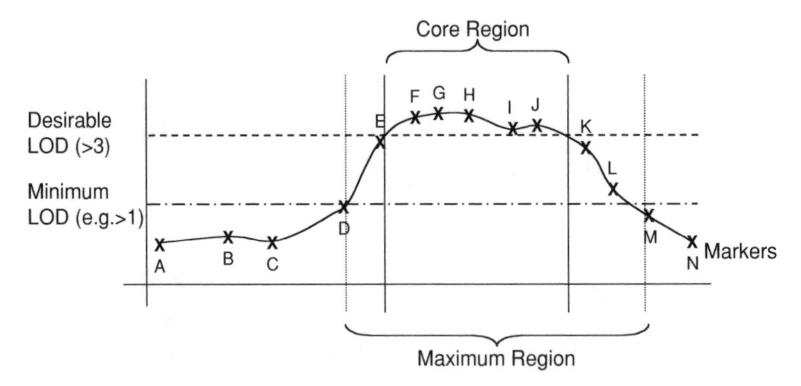

Figure 9.3 Definition of a linkage region by LOD score. In this example of a theoretical complex disease linkage peak, we define an acceptable 'core region' as any region with a LOD score of >3, with a 'maximum region' defined by markers with a LOD of >2 or perhaps >1 (respectively 10-fold and 100-fold drops in linkage probability)

different approaches produce results at very different resolutions, which in turn call for some distinct next steps to define the linked or associated region in the genome for further study (these differing approaches are discussed in Chapter 8). Taking population-based disease linkage as an example, the linkage region is likely to be very large (5–30 Mb) and may be defined by a broad peak, or by multiple peaks with a poorly defined apex. In such cases, without large numbers of family data, it is difficult to define a critical region; instead, the best approach may be to define a core region and a maximal region, defined by a 1-LOD drop in thresholds. Figure 9.3 shows an example of the approach based on a theoretical complex disease linkage peak. In this case, the 'core region' is defined by a LOD score of >3, while the acceptable 'maximal region' is defined by markers with a LOD of >2 or perhaps >1 (respectively 10-fold and 100-fold drops in linkage probability). If a large region is being prioritized for follow-up analysis, these graded regions can effectively be used as a method of weighting genes and markers for further analysis.

In the case of association analysis, particularly that based on a high-density map, such as an oligonucleotide array chip, the associated region may be very small indeed. Often an association may be with a single SNP (singleton association), but more often several SNPs, generally reflecting LD, which might range from 10 kb to as much as ~500 kb. High-density SNP association studies, and singleton-associated SNPs in particular, raise the vexed issue of type I error (false-positive associations). Briefly, 5 per cent of markers tested might be expected to show association by chance, so if 500 000 markers are tested, we can expect 25 000 associations by chance alone. Independent replication is the only practical solution to this problem, but the cost of sample ascertainment and further genotyping arguably makes detailed locus characterization and rigorous ranking of associated loci more important than ever. The general issues that high-density genome scans generate, including problems with multiple testing and locus prioritization, are reviewed in detail in Chapter 19.

Genetic associations obtained by candidate gene analysis have their own set of issues. Aside from the problems of multiple testing, a key issue is marker coverage – the limits of an associated candidate gene locus can be reliably defined only where sufficient markers have been typed on either side of the associated marker to see a decline in association. Without evidence of a lack of association in neighbouring regions, a marker in a candidate gene may simply be reflecting association with a neighbouring ungenotyped gene.

9.3.4 Defining a plan for follow-up – expedient or exhaustive?

However the locus is defined, the high-resolution data provided by the HapMap presents a real opportunity for exhaustive follow-up studies of genetic loci. In terms of laboratory-based studies, resource constraints are really the only limit to the scale of further studies. At this stage, one might also choose to evaluate candidate genes in a region for priority analysis. This approach is becoming increasingly attractive, as the rich range of available *in silico* data offers opportunities to identify genes with a strong rationale in the phenotype under study directly from very large loci. In such circumstances, it might be reasonable to test these genes immediately for association. However, it might be argued that there is a potential here for type I error; as available information continuously expands, so does the chance of identifying plausible candidate genes by chance.

In the case of high-resolution association analysis, the minimal region of association is defined by the extent of LD, and this can also create problems. LD is a phenomenon that can extend over several megabases, as in the human major histocompatibility complex (MHC) region (Horton *et al.*, 2004). However, on average, LD extends over 20–30 kb, so analysis can focus on in-depth dissection on a small number of candidate genes, and the impact of all known variation in the associated region may be assessed with relatively modest resources.

9.4 *In silico* characterization of the IBD5 locus – a case study

To illustrate the step-by-step process of genetic locus characterization by *in silico* methods, we will take the IBD5 Crohn's disease (CD) susceptibility locus as a case study. For more detail on IBD5, see the excellent review by Reinhard and Rioux (2006).

The genetics of the inflammatory bowel diseases have been among the few early successes in complex disease gene hunting, most notably with the discovery of the IBD1 locus by linkage and the subsequent identification of the CARD15 (NOD2) gene by association analysis in CD (Hugot *et al.*, 2001). The IBD5 locus was mapped around the same time as IBD1, to an 18-cM region on 5q31 by a genome-wide linkage scan. The locus was refined by association with a microsatellite-based transmission disequilibrium test to a region of around 500 kb. SNP genotyping produced a strikingly limited set of haplotypes, defining a ~250-kb risk haplotype that was replicated by association in six large studies, providing clear replication of IBD5 in CD. Given the robustness of this association, the challenge is to deconvolute the IBD5 risk haplotype, which contains at least five genes, in order to identify a causal gene or genes. In informatics terms, the first step in this process is to define the IBD5 haplotype in the genome.

9.4.1 Localizing markers in the genome

Accurate definition of locus boundaries in the genome should always be the first step in characterization. In Chapter 4, the three primary genome browsers, UCSC, Ensembl and MapView, are reviewed. These tools all offer the user an opportunity to localize markers to the human genome. Other specialist genetics tools are also available, such as the HapMap genome browser (see below), but these can be limited in scope for full genomic characterization. All genome browsers offer similar query capacities; usually, queries can be carried out by marker name, genome location or cytoband, or directly by sequence homology searches by BLAT (UCSC) or a similar tool. Searching by marker IDs, especially microsatellites, can be problematic, as no single tool contains a fully comprehensive index of genetic markers and their aliases. In such cases, the only way to locate an unindexed marker in the genome is by running a sequence similarity search with the marker sequence. The UCSC genome browser is highly recommended for genomic localization of markers. The browser is well indexed for SNPs and microsatellites; alternatively, one can use the fast BLAT service to localize a sequence (e.g., a marker sequence). To localize a marker or genomic locus, select from the home page the most current 'browser' from the top left-hand menu. (For this exercise, the May 2004 (ncbi35) version was used.) Type the marker names in the 'position' window. The UCSC interface is very flexible, accepting a wide range of formats, from a simple cytoband (e.g., 5q23.3) to a genome coordinate (e.g., chr5:131,651,169-131,829,974). In the original report of linkage, the IBD5 locus was defined betweenD5S1435 and D5S1480. To define a region in the genome between two markers, including SNPs or microsatellites, enter them separated by a semicolon. So, for example, in this case, enter 'D5S1435; D5S1480'. This returns a 27.1-Mb sequence interval, containing 314 distinct genes and transcripts. Note that in a microsatellite marker query, UCSC returns a larger interval with 100-kb flanking either side of the markers.

Table 9.1 IBD5 associated SNP alleles unique to the IBD5 risk haplotype

SNP ID	RefSNP ID	In HapMap	IBD5 *P* value
IGR2055a_1	rs2248116	Y	0.000019
IGR2060a_1	rs2522057	Y	0.000012
IGR2063b_1	–	N	0.000007
IGR2078a_1	rs4705950	Y	0.000063
IGR2096a_1	rs12521868	Y	0.000032
IGR2198a_1	rs11739135	Y	0.000048
IGR2230a_1	rs17622208	Y	0.000063
IGR2277a_1	rs4705938	N	0.000096
IGR3081a_1	–	N	0.000038
IGR3096a_1	rs7705189	Y	0.00004
IGR3236a_1	rs2301579	N	0.00023

Data from Rioux *et al.* (2001).

Using a dense SNP map, Rioux *et al.* (2001) narrowed the IBD5 locus down to a common risk haplotype spanning ~250 kb that shows strong association with CD. To define the IBD5 risk haplotype accurately in the genome, it is necessary to locate the SNPs described in the study. Unfortunately, the SNP IDs reported in this study are not standard dbSNP IDs, so it is not possible to query the UCSC or other tools by ID. There are several options available when the marker ID is proprietary or non-standard. Firstly, if the marker sequence is available, it is possible to use the UCSC BLAT tool to locate it by sequence homology searching. On this occasion, the sequences of the SNPs delineating the IBD5 risk haplotype, IGR2055a_1–IGR3236a_1 are available in the supplementary information from the original publication (http://www.broad.mit.edu/humgen/IBD5/5q31data.html). When only the primers for assaying the SNP are reported and these are too small for use in a homology searching tool such as BLAT, the alternative method to use is the UCSC *in silico* PCR tool. This tool is accessed on the UCSC toolbar by following the PCR link. The tool takes 5′ and 3′ primers as input and returns the genome coordinates for the PCR product. The entire locus can then be defined by recording the locations of the SNPs from the PCR result. In this case, using BLAT to find map locations for the SNPs in question, the locus is defined by a 251-kb region between chr5:131,581,239-131,832,246 (NCBI35); it also becomes apparent that the SNPs are represented in dbSNP, with IGR2055a_1 represented by rs2248116 and IGR3236a_1 represented by rs2301579 (Table 9.1). Conveniently, it is possible to save the view of any locus viewed in the UCSC by bookmarking the browser page, allowing the user to return later at any point. Figure 9.4 shows the IBD5 risk haplotype region in the UCSC browser; this clearly identifies seven known genes and a number of novel genes across the region, evidenced by human mRNAs and spliced ESTs. Methods for examining the biological rationale of each of these genes in IBD will be reviewed below.

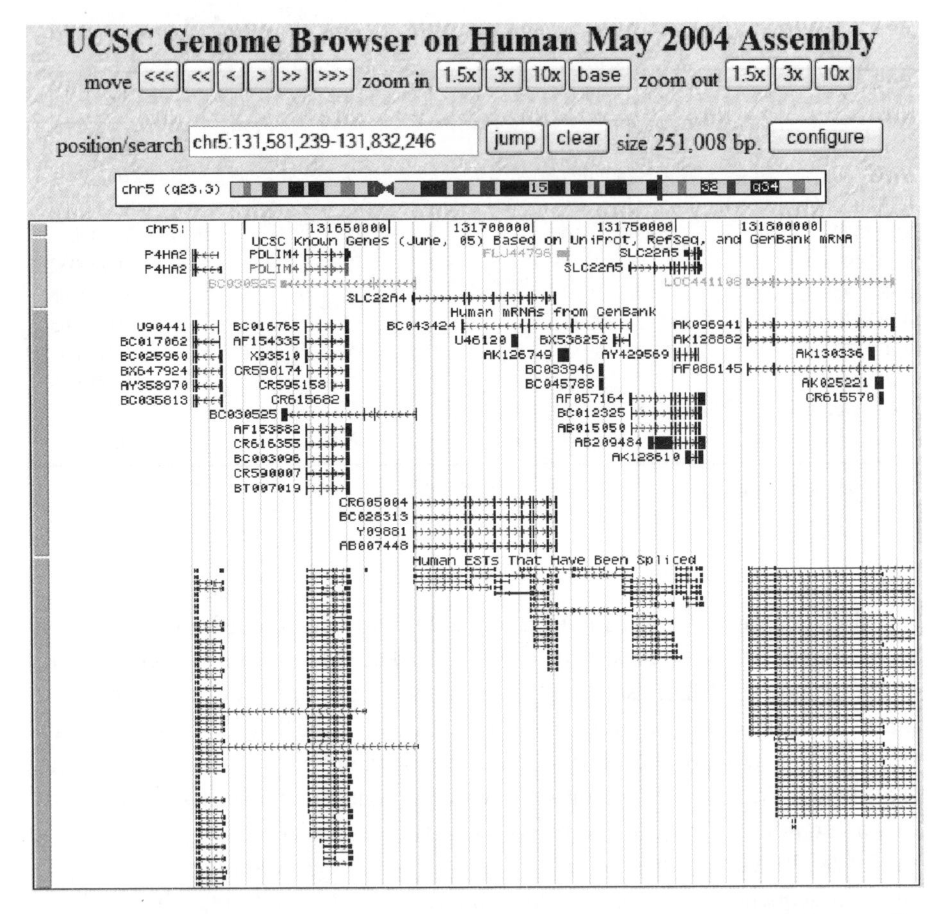

Figure 9.4 A view of genes across the IBD5 risk haplotype with the UCSC genome browser. The view immediately identifies seven known genes and evidence for a number of novel transcripts. Generated using the UCSC Human Genome Browser, http://genome.ucsc.edu

9.4.2 Extracting and annotating genomic sequence across the locus

At this point, with the locus defined in the genome, any number of approaches for further characterization is possible. If direct analysis of the genomic sequence across all or part of the locus is desired, the sequence can be extracted from the UCSC browser. To achieve this, select the 'DNA' link in the tool bar; this presents the user with a number of basic options to format the DNA sequence across the selected region. If the DNA sequence only is required, select 'all lower case' and press the submit button. Alternatively, select 'lower case repeats' to highlight repeats in the sequence, or mask them for primer design and other applications. There is also an option to reverse-complement the sequence. This is particularly useful to retrieve a sequence across a gene that is in the reverse orientation in the golden path. To receive full annotation of the sequence in terms of all the features reported by the UCSC browser, select 'extended case/color options' and press the submit button. This leads to a highly sophisticated annotation interface that allows annotation of almost every available UCSC feature on the sequence, with a combination of toggled case, underlining, boldface, italics and full-colour lettering, very useful for preparing figures for publication.

9.4.3 Defining the IBD5 locus in the HapMap

An important step in locus characterization before further genotyping or analysis is to review HapMap-related LD and haplotype information across the locus. The HapMap genome browser is the simplest access point to the data and can be used quite intuitively to view LD and haplotypes around a gene or region of interest, to select tagging SNPs, or to export genotypes or LD data in single or multiple populations. The browser, which can be reached by following the 'browse project data' link on the HapMap home page, can be searched by a gene, genomic region or SNP ID. The browser is fully configurable to display LD, haplotypes, recombination hotspots and tag SNPs alongside genes across the selected region. At the time of writing, the HapMap browser was transitioning data from NCBI34 to NCBI35 mapping coordinates, with only limited NCBI35 data availability. To view the NCBI35 region coordinates used in the UCSC genome browser in the NCBI34 version of the HapMap, it is necessary to convert the coordinates by the UCSC convert function (see Section 9.2). A region can be viewed by entering the same genomic coordinates used in the UCSC browser. The search format is similar, except the coordinates should be separated by '..' rather than '–'. So, in the case of the IBD5 risk haplotype, enter the NCBI34 converted coordinates 'chr5:131629556..131880563'.

The resulting region that is displayed shows fairly basic information across the locus, including HapMap genotyped SNPs and genes. However, the browser is configurable to display a range of valuable information. To view different features, select the 'Reports and Analysis' menu and select the desired feature. For example, to view haplotypes, select 'Annotate Phased Haplotype Display' and click the 'Configure' button. HapMap LD data can be viewed alongside a variety of other features, including Entrez genes from the NCBI, recombination hotspots, phased haplotypes and precomputed haplotype tagging SNPs. Figure 9.5 shows an example of output from the tool across the IBD5 region. From this view, a few points are clear. First, there is a recombination hotspot in the 5′ end of the IBD5 risk haplotype, and some evidence of variable recombination rates across the region. Rioux *et al.* (2001) reported that the IBD5 risk haplotype is broken into eight haplotype blocks separated by areas of elevated recombination. These reported blocks correlate reasonably well with the phased haplotypes reported by the HapMap. Finally, it is interesting to see that, based on the precomputed haplotype-tagging SNPs displayed across the region, 18 tagging SNPs (based on default Tagger parameters) would be sufficient

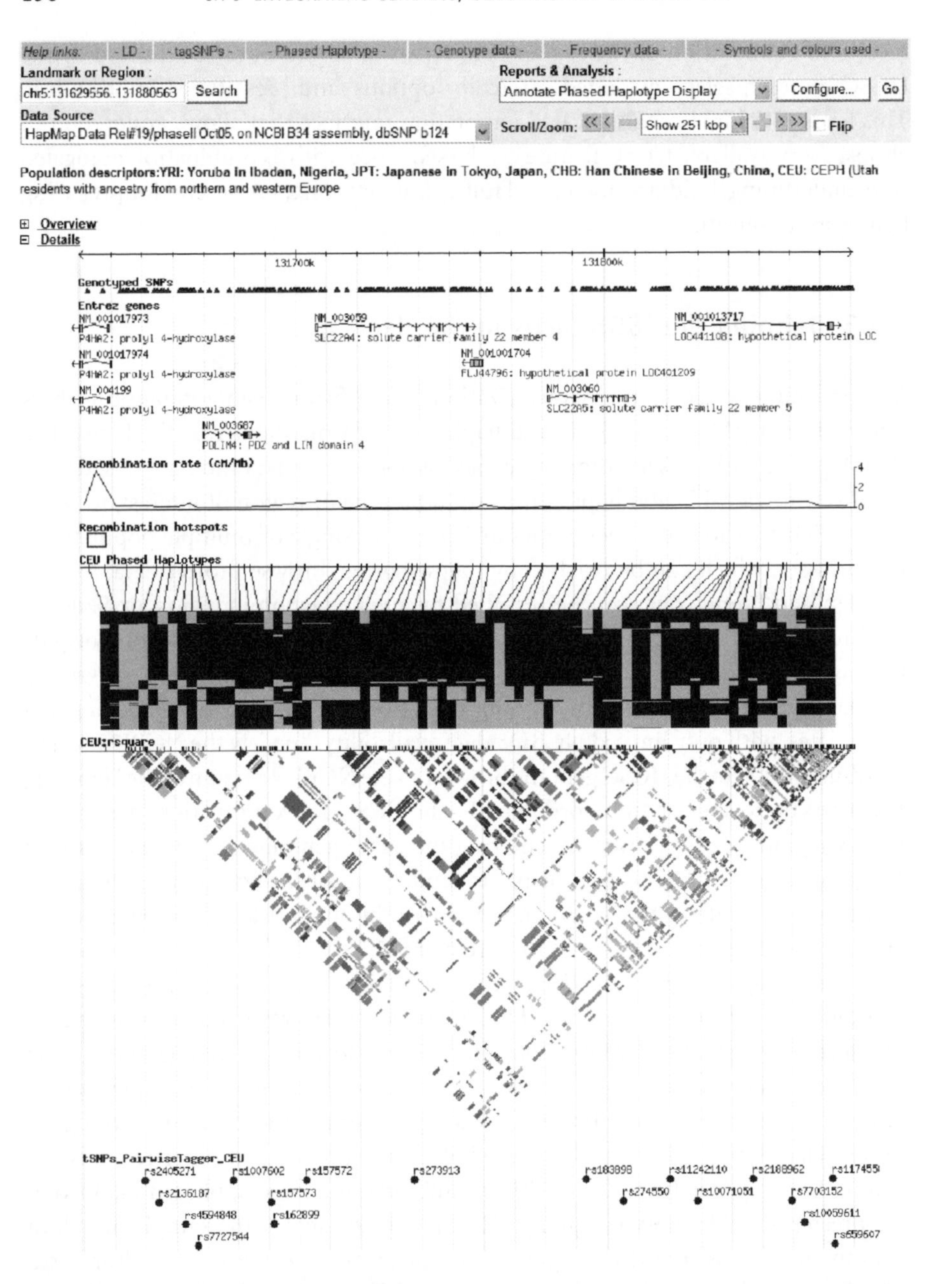

Figure 9.5 Viewing the IBD5 risk haplotype region in the HapMap genome browser. The view displays SNPs genotyped by HapMap, known genes, recombination rates, phased haplotypes and measures of LD in the CEU population group (Utah residents with European ancestry)

to capture information on the 375 SNPs genotyped by the HapMap across the region.

Aside from visualization functions, the HapMap genome browser also serves as an effective data-mining and analysis tool. It allows the user to export data on LD, SNP frequency and genotypes from the current window being visualized. The browser is closely integrated with the HaploView software package (Barrett *et al.*, 2005), allowing the user to pipeline genotype data from a region directly into HaploView, and giving more flexibility to recalculate LD and haplotypes and select tag SNPs. In a similar way, the user can also pipeline genotypes into Tagger (de Bakker *et al.*, 2005) to allow selection of tag SNPs which capture haplotype diversity across a region in a maximally efficient manner. This pipelining capability makes the HapMap browser a remarkably flexible tool, effectively providing the Web user with an LD and haplotype analysis capacity that would otherwise require hours of data loading, manipulation and database expertise.

While the HapMap genome browser is useful for exporting LD data across a defined region or gene, the HapMart tool, also accessed from the HapMap website, is one of the only practical Web tools for exporting LD data from specific populations across a range of chromosomes, by different filter criteria. These filters can include lists of query SNPs, a minor allele-frequency cut-off, a region, or a gene (Figure 9.6). Powerful queries can be effected with these filters. For example, filtering by SNP region, it is possible to retrieve all HapMap SNPs that are in LD with non-synonymous SNPs (nsSNPs). Alternatively, in the case of the IBD5 locus, it is possible to use the SNP filter function to retrieve all other SNPs that are in LD with a specified list of SNPs. In this case, we may be interested in retrieving SNPs in LD, with the SNPs showing association with CD in the IBD5 risk haplotype (Table 9.1). This enables functional analysis of all SNPs with evidence of LD with associated SNPs to identify putative causal SNPs. The output of such queries can be easily formatted into a custom track for display in tools such as the UCSC human genome browser. In Figure 9.7, we show an example of a custom track created from a HapMart query. In this case, the nine associated SNPs were entered into the HapMart filter, and all LD records with these SNPS were retrieved. All records with an $r^2 < 0.5$ were removed, and the remaining records were formatted as a UCSC custom track and loaded. The associated SNPs and the SNPs showing LD with the associated SNPs are displayed in the top two tracks of the browser in a full genomic context. This allows the user to zoom in and directly compare the location of the SNPs in relation to other genomic features to assist functional analysis (see Chapter 11 for a detailed overview of functional analysis methods).

HapMart is also useful for retrieving raw data for further analysis, such as genotypes for all SNPs that map to a submitted list of genes (based on HUGO IDs). These are powerful queries as they allow the user to extract specific data on genes of interest from the wider HapMap data set, which is otherwise quite intractable in the absence of a locally installed database environment.

9.4.4 Definition of known and novel genes across the IBD5 locus

Following clear definition of the IBD5 risk haplotype, it is important to identify all the known and novel genes in the region, so that they can be evaluated as candidates or to

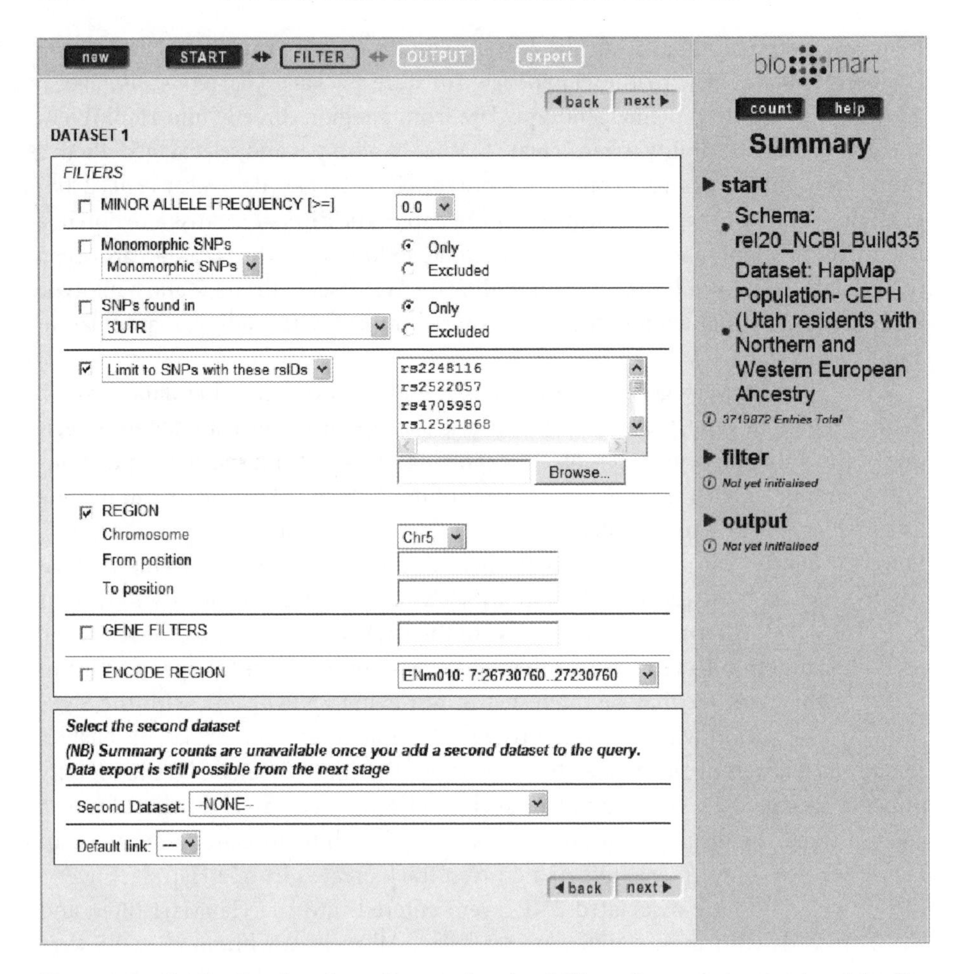

Figure 9.6 The HapMart interface. A range of optional filters allow complex queries of HapMap data, including genotypes, LD data and haplotypes

ensure that marker maps across the region have provided sufficient coverage to detect any genetic effect in genes or regulatory regions. The UCSC human genome browser and other tools such as Ensembl are valuable in this process. Both tools run the human genome sequence through sophisticated gene-prediction pipelines (Hubbard *et al.*, 2002). These analyses are coupled with a detailed view of supporting evidence for genes, such as ESTs, CpG islands and promoter predictions. Homology with other genomes is also presented; this is expanding constantly, but, at the time of writing (May 2006), 16 vertebrate genomes were mapped to human in the UCSC browser. This wealth of data probably makes further *de novo* gene prediction unnecessary in most cases. Improvement in the quality of annotation provided by Ensembl and the UCSC would require an in depth understanding of the intricacies of gene prediction, and this we cannot hope to convey within the scope of this book (see Rogic *et al.*,

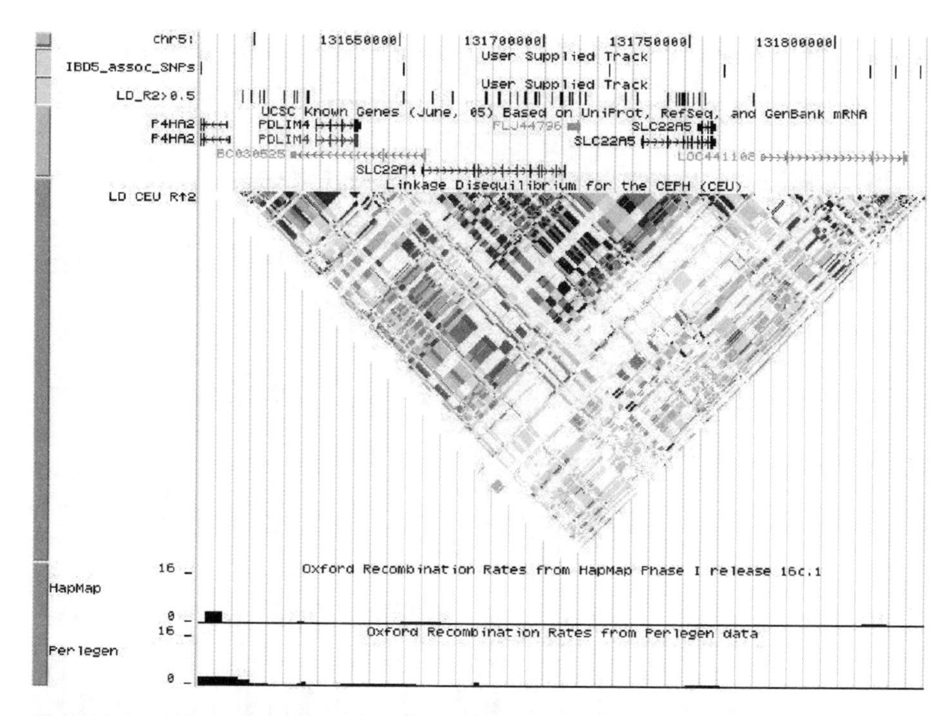

Figure 9.7 Using custom tracks to get a detailed view of genomic context in the UCSC genome browser. Custom tracks in the UCSC genome browser show the location of SNPs showing LD with IBD5-associated SNPs with $r^2 > 0.5$. An additional track describes known genes. HapMap LD information below is for the CEU individuals and suggests extended LD across the region, reflecting the extent of the IBD5 risk haplotype. Recombination rates independently calculated from HapMap and Perlegen data sets are displayed below the LD track. Generated using the UCSC Human Genome Browser, http://genome.ucsc.edu

2001, for an excellent review of this field). Instead we suggest a focus on the available data to build gene models based on existing annotation.

For the purposes of the study of the IBD5 locus, the objective is to identify all known and novel genes across the locus. This can be a painstaking process, as known and novel gene information across the genome can be overwhelming. To illustrate the process, Figure 9.8 focuses on an 83-kb segment of the IBD5 risk haplotype. This shows a UCSC view of the region between the end of the organic cation transporter, SLC22A4, and the beginning of the paralogous transporter, SLC22A5. For the purposes of this figure, the UCSC browser has been configured to show tracks directly applicable to the identification of genes in genomic sequence. These tracks include 'known genes', 'spliced ESTs', 'unspliced ESTs', 'CpG Islands' and 'vertebrate conservation'. The known genes across the region are clearly identified as SLC22A4, SLC22A5 and a hypothetical gene, FLJ44796. However, after review of all the data, especially focusing on the spliced expressed sequence tag (EST) information, it is clear that there are also some potentially novel transcripts across the region. Confidence in the

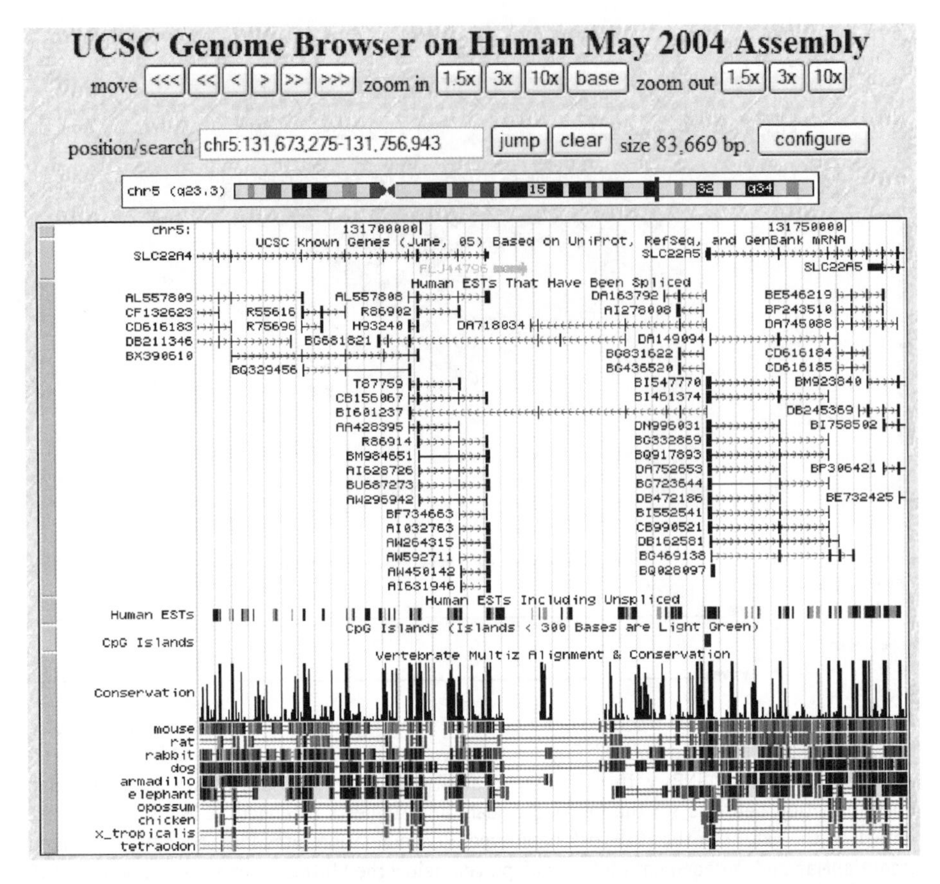

Figure 9.8 Using the UCSC human genome browser to investigate evidence for novel genes and transcripts. A range of evidence, including mRNAs, spliced ESTs and vertebrate homology, supports the existence of several novel transcripts in the intergenic region between SLC22A4 and SLC22A5. Generated using the UCSC Human Genome Browser, http://genome.ucsc.edu

identification of novel genes in genomic sequence is partly dependent on the range and nature of supporting evidence. The most convincing single item of evidence is a correctly spliced mRNA transcript, either an EST or whole transcript. In Figure 9.8, several human spliced ESTs (e.g., DA163792, DA718034 and BI601237) appear to be transcribed in the opposite direction from the SLC22A5 promoter region. EST data can be an important source of novel gene information; however, it can also be highly confusing. This is one of the reasons that EST data are divided into spliced and unspliced tracks. This is in recognition of the very high number of artefacts that are generated in EST libraries. In this case, these spliced ESTs could represent several things – without further laboratory characterization, it may be hard to determine what. First, in some cases, the ESTs overlap exons from SLC22A4, in which case they may represent read-through from the RNA polymerase during SLC22A4 transcription. Some of the ESTs also align to the first exon of SLC22A5, in which case they may represent uncharacterized upstream 5′ UTR exons from this gene. This seems unlikely, as there is a large CpG island in the region of the known first exon of SLC22A5. This is a strong hallmark of a promoter region, suggesting that the annotated first exon is real. In vertebrate conservation across the SLC22A4/SLC22A5 intergenic region, there are clear regions of conservation that correspond with many of the spliced ESTs in the intergenic region. This, along with the evidence of splicing in the ESTs, is strong evidence that these ESTs may actually be coding for a distinct gene in this region. It is not clear whether these ESTs encode a protein-coding transcript, or a non-coding RNA – it may be possible to identify protein sequence homology by using the NCBI BLASTX program (translated DNA v protein). Even if characterization of a novel transcript draws a blank, this is not necessarily a problem for genetic analysis. Simply acknowledging that a novel transcript exists, even if the function is unknown, makes it a candidate for further analysis. As an association is refined, if the signal still points to the novel transcript, further work (including laboratory characterization) would be worthwhile, but before this point it is probably premature to do more than record the transcript as a candidate.

Looking back at Figure 9.4, there is further evidence supporting novel genes in the IBD5 locus. A human cDNA sequence, BC043424, and several spliced ESTs offer strong supporting evidence of a gene that appears to be transcribed on the antisense strand of the SLC22A4 locus. Because mRNA is single stranded, the presence of a complementary antisense strand may alter transcription, elongation, processing, location stability and translation of mRNA. Functional antisense RNA has been widely implicated in gene regulation and differentiation in mammals (reviewed by Dolnick, 1997). Natural antisense transcripts may be coding or non-coding and usually arise via separate transcription initiation from the opposite DNA strand at the same genomic locus as the sense strand. Kiyosawa *et al.* (2003) carried out a genome-wide survey of 60 770 full-length enriched mouse cDNA sequences derived from various tissues at various developmental stages. They identified more than 2000 examples of sense–antisense transcript pairs, clearly indicating the prevalence of this mechanism and underlining its potential importance in mammalian gene regulation. There is some evidence linking SLC22A4 to CD (see below), so this also makes BC043424 a potentially significant gene with a possible role in CD susceptibility.

9.4.5 Building biological rationale around candidate genes

Once all the known and novel genes have been identified across the IBD5 risk haplotype, further genetic analysis could take two routes. One route would be to carry out further high-density association analysis in an attempt to define a subset of markers across the region that show increased evidence of association. Alternatively, specific candidate genes could be selected for follow-up studies. The route selected here depends on the size of the locus and the number of candidate genes it contains, but it might also be influenced by any compelling candidate genes with strong biological rationale. In the case of the IBD5 risk haplotype, there are seven known genes across the region and several novel genes evidenced by ESTs and cDNA sequences (Figure 9.4). With a relatively small number of genes to study, it would be quite reasonable to investigate each gene, but often a region contains a much larger number of genes, making follow-up of each gene an impractical approach. An alternative in such cases is to prioritize candidate genes by their biological rationale in the target phenotype or trait. Criteria for biological prioritization of candidate genes are discussed throughout this book. Genes can be prioritized by a known or putative role in the disease pathway, information from gene knockouts, expression in the disease tissue, functional polymorphism and many other criteria. In this exercise, we are looking for a gene with a possible role in irritable bowel diseases (IBD). Therefore, to prioritize the candidate genes, we might first review the literature to search for a link between each candidate gene in this region and IBD disease pathways. The aetiology of IBD, like many complex diseases, is poorly understood, making it difficult to establish clear biological rationale for genes in this disorder. Where biological rationale is found, it could range from convincing support, such as upregulation of the gene or a related gene or pathway component in a disease model or in a similar phenotype, to the most basic support, such as being expressed in a tissue affected by the disease – in this case, the intestinal tract.

Drawing together the complex strands of evidence in the literature is a skill that calls for a good background in biology and ideally a broad understanding of the disease under study. It is obviously not possible to gain an encyclopedic knowledge of biology and disease overnight; however, many tools such as OMIM at NCBI, offer very good encapsulated summaries of the underlying biology of genes and diseases, and so these are always a good place to start in literature searching. Appropriate tools for literature searching are PubMed at the NCBI, which searches journal abstracts. The new generation of full-text search engines is proving increasingly valuable for focused literature searching. Highwire (http://www.highwire.org) is highly recommended; this is particularly useful, as it reports the context of the search term in the sentence in which it occurs in the full-text journal article. Google Scholar (http://scholar.google.com/) is another member of this new generation of tools; it has the added advantage of searching the Web with a query, often turning up unexpectedly useful information.

It is perhaps a testament to the effectiveness of these tools that, with a little imagination, some form of biological rationale can usually be found for most genes in most phenotypes. However, reliance on literature-based evidence alone can run the risk of over-interpreting tenuous links between genes. This could be a particular problem in the case of poorly understood diseases, where a rationale for a novel pathway in a disease would largely fail to register. This issue is an argument to support a truly investigative approach to candidate gene identification. The candidate should be in the right place at the right time; beyond this, further assumptions may be misleading. Data presented by tools such as the UCSC can provide solid evidence to help to identify genes that are at least expressed in the tissues affected by the disease. Obviously, in the case of IBD, we are most interested in genes showing evidence of expression in the gut and intestines. This will inevitably include a large number of genes. In an analysis of the expression profiles of >33 000 genes, Su *et al.* (2004) found that, on average, any individual tissue expresses approximately 30–40 per cent of known genes. For candidate gene studies, this implies that 30–40 per cent of all genes are likely to be candidates for any given disease, based on expression in the disease tissue alone (assuming the disease affects only one tissue).

9.4.6 Analysis of gene expression with UCSC browser tools

A number of tracks in the UCSC browser provide information about the tissue-expression profiles of genes. The simplest level of information is provided by ESTs; each is implicitly a measure of gene expression, as each is derived from a specified tissue source. The number of ESTs represented in each tissue will also give a *very* rough idea of the expression levels of the gene, but it will not confirm the absence of a gene in a tissue. The UCSC provides an index page for all known genes and cDNAs (the page is reached by clicking on genes in the browser). This is a very useful link page to many expression-related resources. One of the best links provided is to the Stanford SOURCE website (http://source.stanford.edu). This provides useful summaries of gene function and approximated expression information, based on Unigene. The most comprehensive measure of gene expression at the UCSC is the GNF gene-expression atlas, contained in the 'GNF ratio' track (Su *et al.*, 2004). This track represents two replicates each of 79 human tissues run over Affymetrix U95 microarrays. In full display mode, all tissues are shown. In packed or dense mode, averages of related tissues are shown. The microarray data are displayed in a standard form, red indicating upregulation in the tissue, relative to the tissue-wide mean of experiments with the same gene-specific probe, and green indicating downregulation. The saturation of the colour corresponds to the magnitude of transcript variation. Black indicates an undetectable change in expression, and a white box indicates missing data. Su *et al.* (2004) offer a more detailed examination of this gene-expression visualization method. It is also possible to view and run queries on GNF gene expression data directly at the GNF SymAtlas website (http://symatlas.gnf.org/SymAtlas/).

UCSC gene expression data can also be viewed and interrogated more flexibly with the UCSC gene sorter, linked at the top of the genome browser. The UCSC gene sorter is a slightly idiosyncratic but powerful tool for mining most of the data contained within the UCSC site (Kent *et al.*, 2005). The tool aims at a gene-oriented view of the genome to complement the chromosome-oriented view of the genome browser. By default, the gene sorter sorts the displayed genes by their similarity in expression to the selected gene. This similarity is calculated as a weighted sum of differences in log expression ratio values. Genes can also be sorted by protein similarity, location in the genome, name and shared annotation terms. To view expression data for all genes

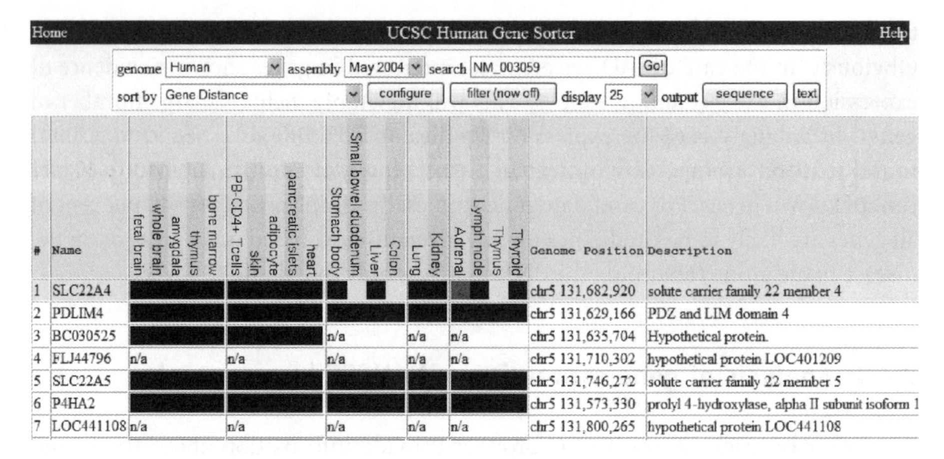

Figure 9.9 Using the UCSC human gene sorter to evaluate gene expression. The gene sorter shows expression information for genes contained within the IBD5 risk haplotype. Genes are sorted in order of genomic distance from the query gene, SLC22A4. The selected expression data from the GNF Atlas 2 shows that only SLC22A5 shows strong evidence of expression in the colon and small bowel duodenum. Generated using the UCSC Human Genome Browser, http://genome.ucsc.edu

in a locus, select 'sort by gene distance' and enter a gene in the centre of the locus; for the IBD5 locus, this is SLC22A4. The query returns all genes around SLC22A4 in order of distance from the query gene rather than physical genome order. The results of such a query are presented in Figure 9.9. Interestingly, only one gene in the IBD5 locus appears to show significant expression in gut tissue (based on colon, small bowel duodenum and stomach tissues) – SLC22A5 shows evidence of upregulation in small bowel duodenum and downregulation in colon. This perhaps serves to flag SLC22A5 as a potentially interesting candidate in this region, although it does not exclude the other genes.

9.4.7 Evaluating epigenomic and epigenetic effects

Epigenomics, the study of epigenetic modification on a genome-wide scale, is a newly emergent field of genomics that we should also consider in our analysis of the IBD5 locus. Epigenetics is concerned with the study of heritable changes other than those in the DNA sequence, and it encompasses two major modifications of DNA or chromatin: DNA methylation, the covalent modification of cytosine, and post-translational modification of histones, including methylation, acetylation, phosphorylation and sumoylation (Callinan and Feinberg, 2006). In terms of function, epigenetic modifications act to regulate gene expression and stabilize adjustments of gene dosage, as seen in X inactivation, gene silencing and genomic imprinting. Despite a number of examples of epigenetic effects in human syndromes and cancer (Agrelo

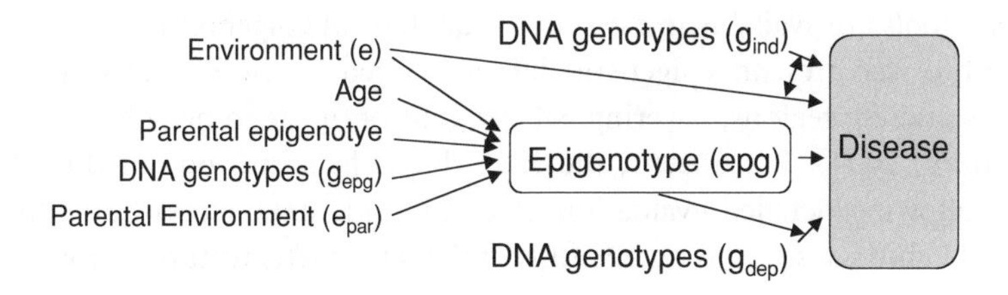

(Adapted from Bjornsson et al. 2004)

Figure 9.10 An integrated genetic and epigenetic approach to common disease. A schematic summary of how genetic and epigenetic (epg) factors might contribute to human disease and the factors that contribute to epigenetic variation. The sources of epg variation (genetic, environmental and stochastic) are also represented. For clarity, the subscript$_{ind}$ has been added to genes that affect disease independently of epigenetics and the subscript$_{epg}$ to genes that directly code for epg variation. g_{ind} might be epigenetically modified, but the epigenetic modification does not influence disease. The relative importance of g_{ind} is inversely proportional to the degree to which common disease is epigenetically determined, which is unknown at present. The modification of gene penetrance by epigenetic context is shown by the use of arrows, which point to relationships, rather than measured values (Reprinted from *Trends Genet* 20(8), Bjornsson, H. T., Fallin, M. D., Feinberg, A. P., pp. 350–358, Copyright 2004, with permission from Elsevier)

et al., 2006), to date, there are no robust examples of epigenetic effects in complex disease, although the prevailing view is that epigenetics may play a very important role in common diseases such as schizophrenia (Singh *et al.*, 2004).

Epigenetic effects are potentially a problem for genetic analysis methods, as a DNA sequence may be invariant between individuals, but their epigenetics may vary substantially. This variation also occurs within an individual in a tissue-specific manner. For example, Rakyan *et al.* (2004) showed that 10 per cent of all methylated sites in the HLA region display differential methylation between tissue types.

However, epigenetic analysis also presents opportunities for genetics; incorporating analysis of epigenetic variation into genetic studies may help to explain the late onset and progressive nature of common diseases and may help to accommodate the role of environment in disease development (see Figure 9.10 for a model of the interplay between genetics and epigenetics; Bjornsson *et al.*, 2004).

Incorporating epigenetics into genetic analysis can also enhance the predictive functional analysis of SNPs by highlighting regions of DNA that are accessible or inaccessible to protein binding by transcription factors and other regulatory proteins. SNPs may also lead to loss or gain of cytosine–guanine dinucleotide (CpG) methylation sites. Rakyan *et al.* (2004) suggested that such an event might affect the overall methylation profile of a locus and, consequently, promoter activity and gene expression. Alternatively, a non-CpG SNP located within an epigenetically sensitive regulatory element could also influence the epigenetic make-up of that region. Therefore, mutations in regulatory sequences could influence epigenetic profiles, resulting in altered phenotypes.

Several tools are available to assist in the analysis of epigenomic data. The UCSC genome browser presents epigenetic data generated by the ENCODE consortium across 44 discrete regions, covering ~1 per cent of the genome (ENCODE Project Consortium, 2004). Fortunately, the IBD5 locus falls into one of the ENCODE regions, allowing detailed evaluation of epigenomic data across the entire locus. Figure 9.11 shows a selected range of ENCODE epigenetic features across exon 1 of SLC22A4; many other features have been excluded for brevity. The features shown include chromatin immunoprecipitation (ChIP) data (Rodriguez and Huang, 2005), a DNA structural profile (Balasubramanian *et al.*, 1998), DNase I hypertensive sites

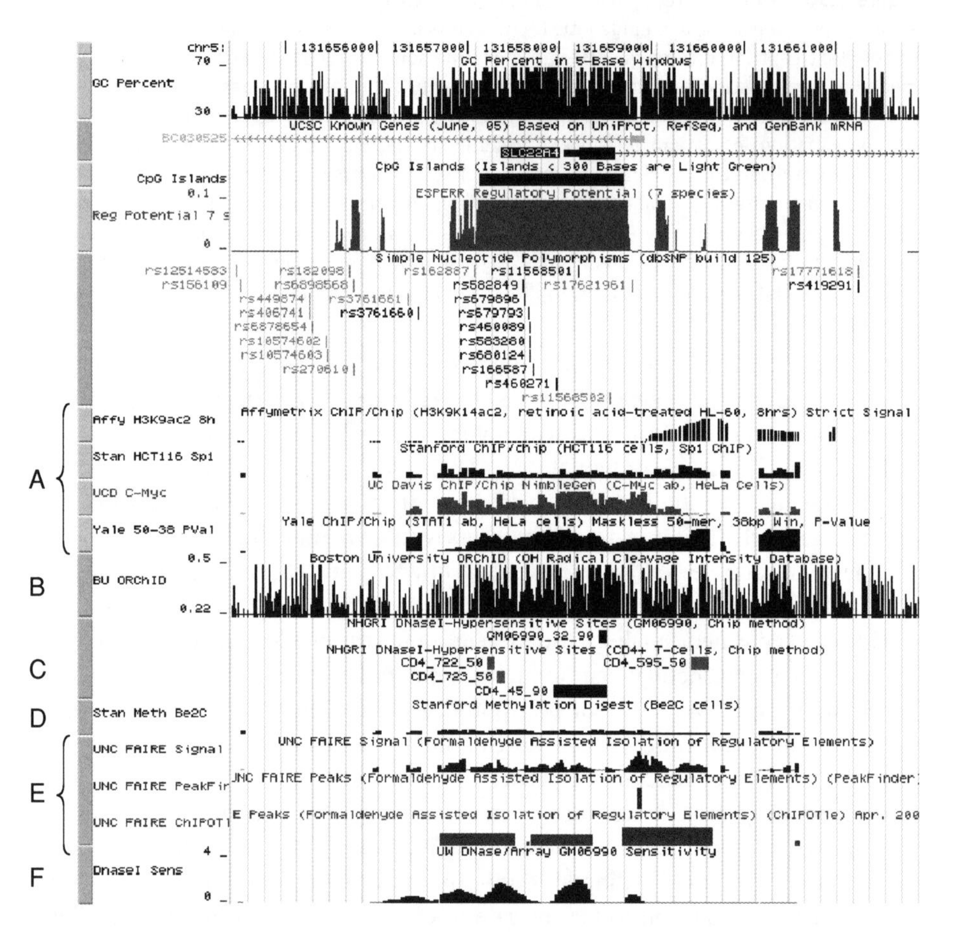

Figure 9.11 Evaluating epigenomic data across exon 1 of SLC22A4 with the UCSC genome browser. The figure shows a view of multiple epigenomic features across exon 1 of the SLC22A4 gene. The browser displays a range of epigenomic data generated by the ENCODE consortium. (A) ChIP data from four independent laboratories; (B) DNA structural profile based on predicted hydroxyl radical cleavage intensity on naked DNA; (C) DNase I hypersensitive sites; (D) DNA methylation; (E) formaldehyde-assisted isolation of regulatory elements (FAIRE), a procedure used to isolate chromatin resistant to the formation of protein-DNA cross-links; (F) DNase I sensitivity/hypersensitivity

(Crawford *et al.*, 2006), DNA methylation sites (Robertson and Wolffe, 2000), and formaldehyde-assisted isolation of regulatory elements (FAIRE) (Buck *et al.*, 2005) – a procedure used to isolate chromatin that is resistant to the formation of protein-DNA cross-links. It is not within the scope of this chapter to explain each of these data sets; however, in each case, a review of each data type is referenced. The UCSC also contains very detailed descriptions of the experimental methods used to generate each data track; this is accessible by clicking the grey button on the far left of the UCSC view. Essentially, these different data sources give similar information relating to the accessibility of DNA to binding by specific regulatory proteins. For example, the 'Affy H3K9K14ac2 8h' track shows regions that co-precipitate with antibodies against diacetylated H3 histones in retinoic acid-stimulated HL-60 cells harvested after 0, 2, 8, and 32 h. For brevity, only the 8-h time point is shown; data on four other DNA-binding proteins has also been excluded. ChIP is a valuable method for identification of regulatory elements. By this method, De Gobbi *et al.* (2006) were able to identify a SNP that created an inappropriate new transcriptional promoter element causing β-thalassaemia. A number of SNPs are co-located within these epigenetic features, although, notably, none are located in the H3 histone-binding region. These SNPs might have the potential to alter DNA binding by regulatory proteins, either directly by altering the DNA-binding motifs or indirectly by leading to a change in the epigenetic state of the DNA, as by altering methylation by removing or introducing a CpG dinucleotide. Further investigation of these SNPs is recommended.

The data generated across the ENCODE regions give a glimpse of the future complexity of epigenomic data on a genome-wide scale. One project is already well under way to investigate the entire epigenome. The Human Epigenome Project (HEP) aims to identify, catalogue, and interpret genome-wide DNA methylation phenomena (Rakyan *et al.*, 2004). Preliminary HEP data are already available and viewable in the MVP viewer(http://www.epigenome.org). Unfortunately, no HEP data are available across the IBD5 region; however, in Figure 9.12, we present HEP data in the MVP viewer across SLC22A1, a close homologue of SLC22A4. The figure shows a view of DNA methylation across CPG sites in the first exon of SLC22A1 gene. Below, SNPs and other genomic features from the UCSC genome browser are displayed. Reviewing the data in Figure 9.12, several points are apparent. First, methylation shows tissue-specific distribution; exon 1 of SLC22A1 is highly methylated in all tissues tested, with the exception of the liver, which shows lower levels of methylation. Second, methylation appears to show intersample variation, especially in the liver. Three SNPs are located in the methylated region – it may be worthwhile to evaluate these for impact on CpG residues. This can be carried out by submitting both alleles of the SNP to a CpG prediction tool, such as EMBOSS CpGplot. Using this tool on both alleles returns an interesting result. The non-synonymous T>C SNP, rs12208357, coding for a Cys61Arg change, also appears to alter the localized CpG profile, conceivably influencing the regulation of SLC22A1.

These are just a few examples of the kinds of epigenomic data that are available. As epigenomic technologies mature and genome-wide data sets become available, great advances may be made in understanding both epigenetics and gene regulation.

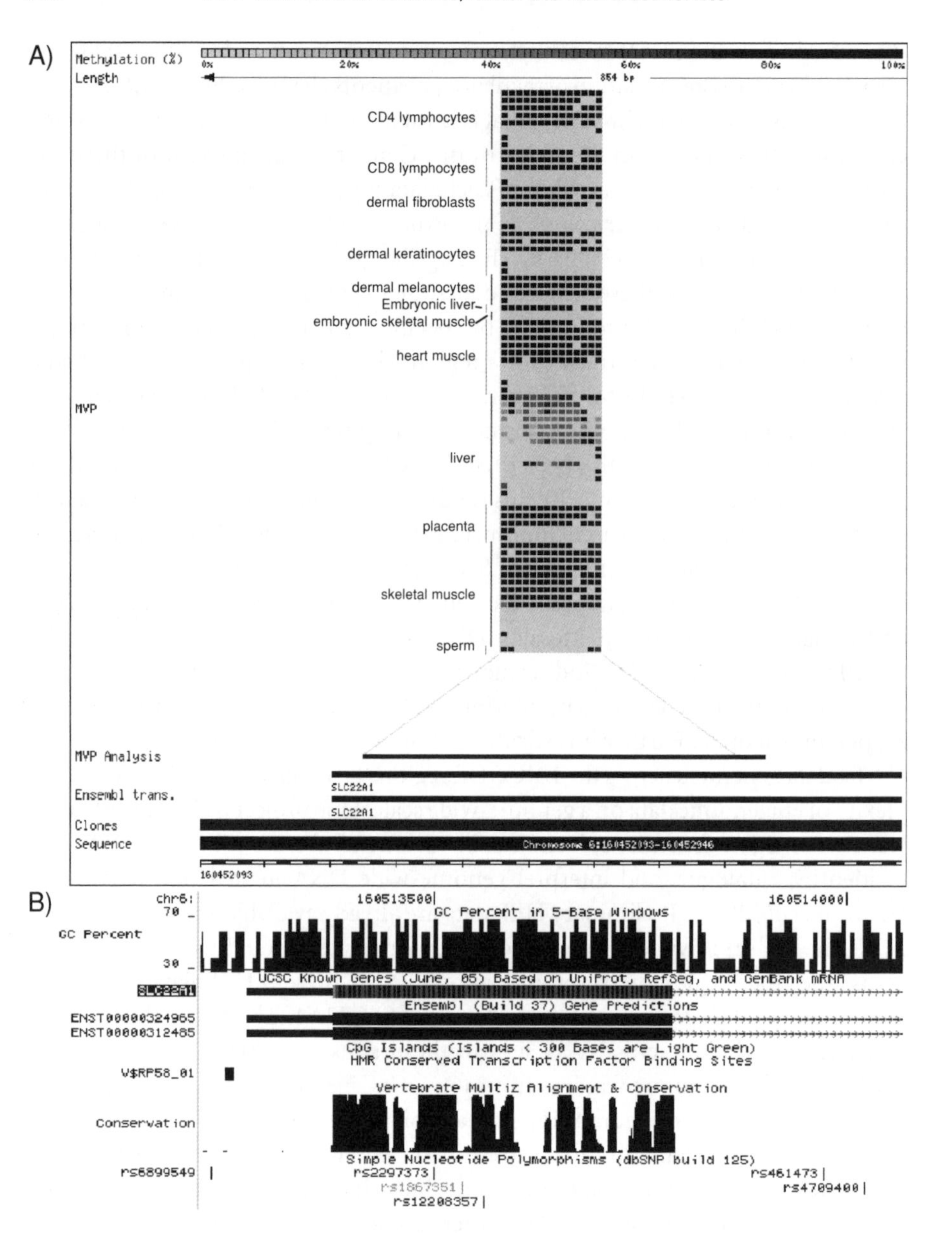

Figure 9.12 Evaluating epigenomic data across SLC22A1 with the MVP viewer and the UCSC genome browser. (A) View of DNA methylation across CPG sites in the first exon of the SLC22A1 gene, using the MVP viewer (http://www.epigenome.org). DNA methylation levels are displayed in the form of a matrix. Each shaded square within the matrix represents one CpG site. Shading represents graded levels of methylation, light grey representing 0 per cent and black representing 100 per cent. Clicking on a square reveals the tissue source of the sample (indicated in this panel to the left of the matrix) and the level of methylation observed at that particular CpG site. Samples are grouped by tissue type. Missing squares indicate CpG sites for which methylation levels could not be determined. (B) The equivalent region is displayed in the UCSC genome browser, showing percentage of GC ratio, conserved transcription factor-binding sites, known genes, vertebrate conservation and known SNPs

Progress is already being made in linking alterations in gene expression with genetic variation (Cheung *et al.*, 2005; see Chapter 16); however, making the link with regulatory elements is currently very difficult. The availability of epigenome-wide data sets like HEP is likely to revolutionize this area of biology.

9.4.8 Evaluating structural variation across a locus

Localized genomic rearrangements and large-scale copy number polymorphisms (CNPs) are a source of genetic variation that should be considered in any genetic analysis. Amplification and deletion of genomic regions can lead to differences in gene copy numbers (and expression levels) between individuals, contributing to the human phenotypic diversity (Feuk *et al.*, 2006).

Conrad *et al.* (2006) used SNP genotype data to identify polymorphic deletions in the genomes of HapMap CEU parent-offspring trios. They identified 586 distinct deletion polymorphisms spanning 267 genes in one or more of the families. Based on this analysis, they estimated that typical individuals are hemizygous for roughly 30–50 deletions larger than 5 kb, totaling around 550–750 kb of euchromatic sequence across their genomes. These CNPs may create problems for genotyping experiments, as they are likely to affect Mendelian inheritance patterns, Hardy–Weinberg equilibrium, and the general robustness, and reproducibility of genotypes (Wirtenberger *et al.*, 2006). However, studies by Newman *et al.* (2006) and McCarroll *et al.* (2006) have shown that SNPs flanking CNPs effectively tag them by LD. Thus, an association with a SNP immediately neighbouring a CNP could well be in LD with a CNP-bearing haplotype.

Several tools to study CNPs and other structural variation are available (Table 9.3). The UCSC genome browser has a track called 'WSSD duplication', which displays thousands of putative CNP regions identified by alignment of whole-genome shotgun sequences to the human genome (for an explanation of this method, see She *et al.*, 2004). There is also the Database of Genomic Variants, a resource dedicated to the curation of structural variants that have been reported in the literature (Iafrate *et al.*, 2004). A review of each of these resources fails to identify any form of structural variation across the IBD5 region, so on this occasion I think we can probably rule out this source of variation.

9.5 Drawing together biological rationale – hypothesis building

After review of all the evidence from the genomic annotations, expression and literature across the IBD5 locus, there is convincing evidence for at least eight genes, several of which have a clear rationale in inflammatory bowel diseases. A summary of

Table 9.2 Characteristics of IBD5 risk haplotype candidate genes

Gene	Name/function	Known expression	Rationale in Crohn's disease (CD)
P2HA2	Prolyl 4-hydroxylase, a key enzyme in collagen synthesis	High expression in colon (Unigene)	Increased activity leading to excessive collagen during wound healing, resulting in stricture in CD
PDLIM4	PDZ and LIM domain Protein 4, modulates actin turnover and association with actinin and actins	Low expression in colon (Unigene)	Alterations in actin could influence increase in intestinal permeability seen in CD
BC030525	Novel gene; unknown function	No evidence of expression in colon	None apparent
SLC22A4	Sodium-ion dependent, low affinity carnitine transporter	Highly expressed in intestinal cell types affected by CD including epithelial cells (Stanford SOURCE)	Impaired function may lead to decreased uptake of carnitine and increased uptake of pathological compounds, resulting in inappropriate inflammatory host response
FLJ44796	Novel gene; unknown function, contains reverse transcriptase domain	Low expression in colon (Unigene)	None apparent
SLC22A5	Functions both as organic cation transporter and sodium-dependent high-affinity carnitine transporter	Highly expressed in intestinal epithelial cells (Stanford SOURCE). Upregulated in small bowel duodenum, and downregulated in colon (GNF)	Impaired function may lead to decreased uptake of carnitine and increased uptake of pathological compounds, resulting in inappropriate inflammatory host response
BC043424	Novel gene; unknown function. Antisense transcript transcribed across SLC22A4 locus	Moderate expression in colon (Unigene)	Antisense transcription across the SLC22A4 locus may silence transcription of SLC22A4?
LOC441108	Similar to yeast, Rad50, a protein involved in DNA double-strand break repair	Moderate expression in colon (Unigene)	None apparent

this evidence is presented in Table 9.2. Perez-Iratxeta *et al.* (2005) have attempted to automate some of the decision-making processes leading to the prioritization of candidate genes, with some success. The Web tool G2D (Genes to Diseases) prioritizes genes across a user-entered chromosomal region according to their possible relation to an inherited disease by a combination of data mining of OMIM, PubMed MESH terms and Gene Ontology (GO) classification. The tool allows users to inspect any region of the human genome to find candidate genes related to a genetic disease or phenotype defined in OMIM. It does this by identifying GO terms that match MESH terms for an OMIM record. The GO terms are then compared to GO annotations of genes and transcripts in a query locus. Genes which show a significant match of GO terms are reported to the user. A query of G2D, using the IBD5 OMIM record (606348) and the locus for the IBD5 risk haplotype, identifies SLC22A4 (but, interestingly, not SLC22A5, which has similar GO annotation) and PDLIM4, principally on the basis of GO annotation similarities to the OMIM record.

Review of all the evidence linking the IBD5 risk haplotype genes to CD suggests that several genes could reasonably play a role in CD. Therefore, the next appropriate step would be to focus more closely on specific variants in each of the candidate genes that might help to provide a molecular basis for the association. Essentially, this is a process of hypothesis building and prioritization, ultimately leading to laboratory follow-up to test these hypotheses.

9.6 Identification of potentially functional polymorphisms

Aside from the ordered convenience that genome browsers bring to SNP data, with tools such as custom tracks at the UCSC, they also place SNPs that show evidence of association into a full and diverse genomic context (as seen in Figure 9.7), giving information on nearby genes, transcripts and promoters. At the simplest level, identification of potentially functional SNPs is a matter of identifying SNPs that overlap highly conserved regions or putative gene or regulatory features. The UCSC genome browser presents some detailed information on putative promoter regions, including CpG islands and conserved TFBS (transcription factor-binding sites) information. The UCSC browser also shows genome conservation among a wide range of vertebrate species. Genome conservation between vertebrates is generally restricted to genes (including undetected genes) and regulatory regions (Aparicio *et al.*, 1995). Hence, this is a simple but powerful method for identifying SNPs in regions that are potentially functionally conserved. Once a putative functional polymorphism is identified, the impact of different alleles can be evaluated by running the alleles through the tool originally used to predict the sequence feature. These could include tools for promoter prediction, splice site prediction or gene prediction. There is a great deal of coverage of SNP functional analysis in Chapters 11–14 in this book, so it is unnecessary to go into detail in this chapter. One final point on the functional characterization of SNPs is that although our knowledge of genome function is improving, it is still very limited, and so it is almost impossible to conclude that a SNP is *not* functional. For these reasons, conclusions on SNP function need to be balanced with evidence of association.

Reviewing the associated SNPs that make up the IBD5 risk haplotype and the HapMap SNPs showing LD with these SNPs, a few SNPs look potentially interesting in functional terms. One SNP, rs272893, codes for an Ile306Thr substitution in SLC22A4. Interestingly, a review of the UniProt entry for SLC22A4 shows that this SNP has already been functionally characterized and has been shown to have no impact on the physiological function of SLC22A4 (Kawasaki *et al.*, 2004). Peltekova *et al.* (2004) specifically evaluated the IBD5 candidate genes and presented evidence of a role of variants of both SLC22A4 and SLC22A5 in CD susceptibility. They identified two SNPs in LD that showed a stronger association with CD susceptibility than the IBD5 risk haplotype, forming a putative two-allele risk haplotype. One of the SNPs was a Leu503Phe substitution in SLC22A4 – a change from a medium-sized and hydrophobic residue to a large aromatic residue, which might be disruptive to protein structure. Physiological characterization of this variant showed a significantly reduced ability to transport carnitine, the natural substrate, or SLC22A4. The second SNP, −207G>C, was identified in the core promoter of SLC22A5. This was shown to fall within a heat-shock transcription factor (HSF)-binding element. Using gel shift assays to evaluate HSF binding to constructs bearing the two alleles, they found strong binding to the G allele and no binding to the C allele, demonstrating a possible regulatory impact *in vivo* in SLC22A5.

The data presented by Peltekova *et al.* (2004) quite convincingly point to both SLC22A4 and SLC22A5 as the most likely candidates for CD susceptibility; however, Reinhard and Rioux (2006) pointed out that this conclusion can be challenged. Peltekova *et al.* (2004) used the htSNP, IGR2078, to tag the IBD5 risk haplotype. This SNP is ~100 kb away from the SLC22A4/5 variants and separated by three intervals of recombination. Considering this, it would be preferable to compare the strength of association between the functional SNPs and a htSNP, uninterrupted by recombination intervals in the same haplotype block.

9.7 Conclusions

Although, clearly, there is still some uncertainty about the molecular basis of CD susceptibility in the IBD5 locus, it does seem likely that with some further molecular analysis, SLC22A4 and/or SLC22A5 may eventually be unequivocally associated with CD, as they both show expression in the appropriate tissues and have a strong rationale in the disease. But these conclusions are all graced with the benefit of hindsight (the most powerful analysis method!). Considered as a whole, the IBD5 locus contains four or five strong candidates for involvement in CD. This is not an atypical situation and clearly illustrates the challenge of complex disease genetics. To succeed in this paradigm, the geneticist will need to find the (true) associations, refine the region as much as possible, and master the *in silico* data to build a biological rationale around each candidate gene, finally taking in genetic, genomic and epigenomic information to define the most appropriate follow-up strategy to align the association

Table 9.3 Tools for genomic characterization of genetic loci

Tool	URL
Genome visualization	
UCSC genome browser	genome.ucsc.edu
ENSEMBL	http://www.ensembl.org
Ensembl Archives	archive.ensembl.org/
NCBI MapViewer	http://www.ncbi.nlm.nih.gov/mapview/map_search.cgi/
LD and haplotype data	
HapMap website	http://www.hapmap.org
HapMap Genome Browser	http://www.hapmap.org/cgi-perl/gbrowse/gbrowse/
HapMart	hapmart.hapmap.org/BioMart/martview
HaploView	http://www.broad.mit.edu/mpg/haploview/
Tagger	http://www.broad.mit.edu/mpg/tagger/
Structural genome analysis	
Database of Genomic Variants	projects.tcag.ca/variation/
Structural variation database	humanparalogy.gs.washington.edu/
Gene-expression analysis	
GNF Symatlas	symatlas.gnf.org/SymAtlas/
UCSC Gene Sorter	http://genome.ucsc.edu/cgi-bin/hgNear?
Epigenetic/epigenomic analysis	
MVP viewer	http://www.epigenome.org
CpG plot	http://www.ebi.ac.uk/emboss/cpgplot/
Methylation PCR design	http://www.urogene.org/methprimer/index1.html
Building biological rationale	
Stanford SOURCE	source.stanford.edu
DAVID	david.abcc.ncifcrf.gov/
UniProt	http://www.uniprot.org
Prospector	http://www.genetics.med.ed.ac.uk/prospectr/

to a molecular mechanism. This calls for some mastery of a dauntingly wide range of tools and data (Table 9.3). Hopefully this chapter helps to make these sources of information less daunting and clarifies how good bioinformatics can help to guide this process to a successful conclusion.

References

Adie, E. A., Adams, R. R., Evans, K. L. *et al.* (2005). Speeding disease gene discovery by sequence based candidate prioritization. *BMC Bioinformatics* **6**, 55.

Agrelo, R., Cheng, W. H., Setien, F. *et al.* (2006). Epigenetic inactivation of the premature aging Werner syndrome gene in human cancer. *Proc Natl Acad Sci USA* **103**(23), 8822–8827.

Aparicio, S., Morrison, A., Gould, A. *et al.* (1995). Detecting conserved regulatory elements with the model genome of the Japanese pufferfish, *Fugu rubripes*. *Proc Natl Acad Sci USA* **92**, 1684–1688.

Balasubramanian, B., Pogozelski, W. K. and Tullius, T. D. (1998). DNA strand breaking by the hydroxyl radical is governed by the accessible surface areas of the hydrogen atoms of the DNA backbone. *Proc Natl Acad Sci USA* **95**(17), 9738–9743.

Barrett, J. C., Fry, B., Maller, J. *et al.* (2005). Haploview: analysis and visualization of LD and haplotype maps. *Bioinformatics* **21**(2), 263–265.

Bjornsson, H. T., Fallin, M. D. and Feinberg, A. P. (2004). An integrated epigenetic and genetic approach to common human disease. *Trends Genet* **20**(8), 350–358.

Buck, M. J., Nobel, A. B. and Lieb, J. D. (2005). ChIPOTle: a user-friendly tool for the analysis of ChIP-chip data. *Genome Biol* **6**(11), R97.

de Bakker, P. I., Yelensky, R., Pe'er, I. *et al.* (2005). Efficiency and power in genetic association studies. *Nat Genet* **37**(11), 1217–1223.

Callinan, P. A. and Feinberg, A. P. (2006). The emerging science of epigenomics. *Hum Mol Genet* **15** Spec No 1, R95–101.

Conrad, D. F., Andrews, T. D., Carter, N. P. *et al.* (2006). A high-resolution survey of deletion polymorphism in the human genome. *Nat Genet* **38**(1), 75–81.

Crawford, G. E., Davis, S., Scacheri, P. C. *et al.* (2006). DNase-chip: a high-resolution method to identify DNase I hypersensitive sites using tiled microarrays. *Nat Methods* **3**(7), 503–509.

Dolnick, B. J. (1997). Naturally occurring antisense RNA. *Pharmacol Ther* **75**, 179–184.

ENCODE Project Consortium (2004). The ENCODE (ENCyclopedia Of DNA Elements) Project. *Science* **306**(5696), 636–640.

Feuk, L., Marshall, C. R., Wintle, R. F. *et al.* (2006). Structural variants: changing the landscape of chromosomes and design of disease studies. *Hum Mol Genet* **15** Suppl 1, R57–66.

Fitch, W. M. (2000). Homology: a personal view on some of the problems. *Trends Genet* **16**(5), 227–231.

Kawasaki, Y., Kato, Y., Sai, Y. *et al.* (2004). Functional characterization of human organic cation transporter OCTN1 single nucleotide polymorphisms in the Japanese population. *J Pharm Sci* **93**(12), 2920–2926.

Kent, W. J., Hsu, F., Karolchik, D. *et al.* (2005). Exploring relationships and mining data with the UCSC Gene Sorter. *Genome Res* **15**(5), 737–741.

Horton, R., Wilming, L., Rand, V. *et al.* (2004). Gene map of the extended human MHC. *Nat Rev Genet* **5**, 889–899.

Hubbard, T., Barker, D., Birney, E. *et al.* (2002). The Ensembl genome database project. *Nucleic Acids Res* **30**(1), 38–41.

Hugot, J. P., Chamaillard, M., Zouali, H. *et al.* (2001). Association of NOD2 leucine-rich repeat variants with susceptibility to Crohn disease. *Nature* **411**(6837), 599–603.

Iafrate, A. J., Feuk, L., Rivera, M. N. *et al.* (2004). Detection of large-scale variation in the human genome. *Nat Genet* **36**(9), 949–951.

Kiyosawa, H., Yamanaka, I., Osato, N. *et al.* RIKEN GER Group; GSL Members. (2003). Antisense transcripts with FANTOM2 clone set and their implications for gene regulation. *Genome Res* **13**(6B), 1324–1334.

Lander, E. S., Linton, L. M., Birren, B.*et al.* (2001). Initial sequencing and analysis of the human genome. *Nature* **409**, 860–921.

McCarroll, S. A., Hadnott, T. N., Perry, G. H. *et al.* (2006). Common deletion polymorphisms in the human genome. *Nat Genet* **38**(1), 86–92.

Newman, T. L., Rieder, M. J., Morrison, V. A. *et al.* (2006). High-throughput genotyping of intermediate-size structural variation. *Hum Mol Genet* **15**(7), 1159–1167.

Peltekova, V. D., Wintle, R. F., Rubin, L. A. *et al.* (2004). Functional variants of OCTN cation transporter genes are associated with Crohn disease. *Nat Genet* **36**(5), 471–475.

Perez-Iratxeta, C., Wjst, M., Bork, P. *et al.* (2005). G2D: a tool for mining genes associated with disease. *BMC Genet* **6**, 45.

Rakyan, V. K., Hildmann, T., Novik, K. L. *et al.* (2004). DNA methylation profiling of the human major histocompatibility complex: a pilot study for the Human Epigenome Project. *PLoS Biol* **2**(12), e405.

Reinhard, C. and Rioux, J. D. (2006). Role of the IBD5 susceptibility locus in the inflammatory bowel diseases. *Inflamm Bowel Dis* **12**(3), 227–238.

Rioux, J. D., Daly, M. J., Silverberg, M. S. *et al.* (2001). Genetic variation in the 5q31 cytokine gene cluster confers susceptibility to Crohn disease. *Nat Genet* **29**(2), 223–228.

Robertson, K. D. and Wolffe, A. P. (2000). DNA methylation in health and disease. *Nat Rev Genet* **1**(1), 11–19.

Rodriguez, B. A. and Huang, T. H. (2005). Tilling the chromatin landscape: emerging methods for the discovery and profiling of protein-DNA interactions. *Biochem Cell Biol* **83**(4), 525–534.

Rogic, S., Mackworth, A. K. and Ouellette, F. B. F. (2001). Evaluation of gene-finding programs on mammalian sequences. *Genome Res* **11**, 817–832.

She, X., Jiang, Z., Clark, R. A. *et al.* (2004). Shotgun sequence assembly and recent segmental duplications within the human genome. *Nature* **431**(7011), 927–930.

Singh, S. M., McDonald, P., Murphy, B. *et al.* (2004). Incidental neurodevelopmental episodes in the etiology of schizophrenia: an expanded model involving epigenetics and development. *Clin Genet* **65**(6), 435–440.

Su, A. I., Wiltshire, T., Batalov, S. *et al.* (2004). A gene atlas of the mouse and human protein-encoding transcriptomes. *Proc Natl Acad Sci* **101**, 6062–6067.

Wirtenberger, M., Hemminki, K. and Burwinkel, B. (2006). Identification of frequent chromosome copy-number polymorphisms by use of high-resolution single-nucleotide-polymorphism arrays. *Am J Hum Genet* **78**(3), 520–522.

10
Tools for Statistical Genetics

Aruna Bansal[1], Charlotte Vignal[1] and Ralph McGinnis[2]

[1] *Discovery and Pipeline Genetics, GlaxoSmithKline, Harlow, Essex, UK*

[2] *Wellcome Trust Sanger Institute, Hinxton, Cambridge, UK*

10.1 Introduction

A wealth of tools and methodology exists to aid in the identification of genetic variants that influence a trait of interest. The trait may be a biological measurement, possibly indicating risk of disease, or it may be the response to an environmental stimulus, such as a drug. This chapter sets out to do three things: introduce key methodology such as linkage analysis and association analysis, give a taste of available software and work through some examples. The majority of tools discussed may be downloaded, together with full documentation, by following links at http://linkage.rockefeller.edu. Web addresses for the few exceptions are provided in the text. Almost all are available free of charge.

10.2 Linkage analysis

Linkage analysis is applied in the early stages of searching for genes that cause a particular trait, and it is one means by which an initial, often broad, chromosomal interval of interest is defined. It is a process that uses family data to evaluate the correspondence between the inheritance pattern of genetic markers and the inheritance pattern of a disease or trait. Disease linkage manifests as a marker allele being inherited in diseased individuals more often than would be expected by chance.

Linkage analysis may be parametric, to test whether the inheritance pattern of the trait fits a specific model of inheritance, or it may be non-parametric (model-free).

Bioinformatics for Geneticists, Second Edition. Edited by Michael R. Barnes
 ISBN 978-0-470-02619-9 (HB) ISBN 978-0-470-02620-5 (PB)

The former is more powerful under a correctly specified model and is most informative for large, multiply affected pedigrees. The latter is more powerful when the mode of inheritance is unknown, as in complex trait analysis for which small pedigrees are often ascertained.

10.2.1 Parametric linkage analysis

By the parametric approach (and in certain non-parametric cases), evidence of linkage is measured by the LOD score (Morton, 1955). The term 'LOD score' stands for logarithm of the odds to the base 10, and its calculation proceeds by an assessment of the recombination fraction, often denoted by theta (θ). Theta is the probability of a recombination event between two loci on the same chromosome; as such, it is a function of distance. Two unlinked loci are given by $\theta = 0.5$, and the closer a pair of loci, the lower their recombination fraction. The LOD may be expressed as follows, using L to denote likelihood.

$$LOD = \log_{10} \frac{L(\theta = \hat{\theta})}{L(\theta = 0.5)}$$

The likelihood in the numerator is based upon the maximum likelihood estimate of the recombination fraction, derived from the data. It is compared to that calculated under the null hypothesis of no linkage ($\theta = 0.5$). A high LOD score is thus consistent with the presence of linkage. Due to the computational complexity of the likelihood calculation, software for exact parametric linkage analysis is constrained either by pedigree size or by the number of markers included in the calculation.

The software VITESSE (O'Connell and Weeks, 1995) allows rapid, exact parametric linkage analysis of extended pedigrees. At the expense of some speed, an alternative, FASTLINK (Cottingham *et al.*, 1993), allows the analysis of large pedigrees that also contain loops (marriages between related individuals). Both VITESSE and FASTLINK are based on an earlier program, LINKAGE (Lathrop *et al.*, 1984), and are available for UNIX, VMS and PC (DOS) systems. Using these pieces of software, analysis is typically conducted by means of a sliding window of one, two or four markers along the chromosome, although larger windows are also possible.

Parametric linkage analysis in more moderately sized pedigrees was greatly facilitated by the advent of GENEHUNTER (Kruglyak *et al.*, 1996). A major feature of this program is that it allows the rapid, simultaneous analysis of dozens of markers (often an entire chromosome) in a multipoint fashion, thereby providing increased power over single-marker analyses when map positions are known (Fulker and Cardon, 1994; Holmans and Clayton, 1995; Olson, 1995). In order to accommodate uncertainty in marker ordering, an option to perform single marker tests is also available.

Further advances were seen with the release of the software MERLIN (Abecasis *et al.*, 2002). The latter is a C++ program for UNIX or LINUX, again with a command-line interface. It offers further improvements in computational speed and reduction in memory constraints, making it more suited to very dense genetic maps. It has the attractive properties of incorporating error detection routines to improve power, and simulation routines to estimate *P*values. Graphical output is also provided.

10.2.2 Nonparametric (model-free) linkage analysis

Nonparametric linkage (NPL) analysis does not allow direct estimation of the recombination fraction, but one source of multiple testing – that derived from examining multiple models – is removed. The general principle is that relatives who share similar trait values exhibit increased sharing of alleles at markers that are linked to a trait locus. See Holmans (2001) for a review of the method.

Allele sharing may be defined as identical by state (IBS) or identical by descent (IBD). Two alleles are IBS if they have the same DNA sequence. They are IBD if, in addition to being IBS, they are descended from (and are copies of) the same ancestral allele (Sham, 1998). A statistical test is performed to compare the observed degree of sharing to that expected under the assumption that the marker and the trait are not linked. While the test statistic may take the form of a chi-square, normal or F statistic, often it is transformed to allow it to be expressed in LOD units.

NPL analysis often examines IBD or IBS allele sharing in sets of affected sib-pairs (ASPs) in which both siblings exhibit the trait of interest. In the absence of linkage, ASPs are expected to share 0, 1 or 2 alleles IBD, with probabilities 0.25, 0.5 and 0.25 respectively. The presence of linkage to a tested marker leads to a departure from these proportions that may be detected by the chi-square test (Cudworth and Woodrow, 1975). Another model-free test, the mean test, evaluates the null hypothesis that the proportion of IBD allele-sharing equals 0.5. The latter is implemented in the programs SAGE (1999) and SIBPAIR (Terwilliger, 1996), allowing for larger sibships and cases where IBD status cannot be determined unequivocally.

For dichotomous trait data measured on larger pedigrees, the degree of IBD sharing among affected pedigree members may be assessed, for example, by using the S_{all} or S_{pairs} scoring functions of Whittemore and Halpern (1994). These IBD-based scoring functions are then used to assign weights to conditional inheritance probabilities summed in the calculation of an asymptotically normally distributed NPL score. This form of NPL score is widely used and is implemented, for example, in GENEHUNTER-PLUS (Kong and Cox, 1997) and in MERLIN (Abecasis *et al.*, 2002).

For normally distributed quantitative traits (or those capable of being transformed to normality), variance-component analysis represents a powerful approach to the study of pedigrees of any size (Goldgar, 1990; Amos, 1994; Blangero and Almasy, 1996). The method is implemented in MERLIN (Abecasis *et al.*, 2002) and extensively in SOLAR (Blangero and Almasy, 1996), in which the size of each effect may be estimated and tested by a LR test. This is a powerful approach, and a major advantage is its scope for incorporating into models the effects of covariates, epistasis and gene–environment interaction. For highly complex problems, Markov Chain Monte Carlo methods are also available, as implemented, for example, in LOKI (Heath, 1997) and BLOCK (Jensen *et al.*, 1995). When the parameter set is large, however, the computational burden of these methods can be prohibitive.

10.2.3 Example: MERLIN (Abecasis *et al.*, 2002)

MERLIN (Multipoint Engine for Rapid Likelihood Inference) is a software package for single-point and multipoint linkage analyses of pedigree data (Abecasis *et al.*, 2002). MERLIN can be used for analysis of parametric linkage, non-parametric linkage (NPL) and quantitative trait linkage (QTL), incorporating variance-component analysis, IBD and kinship coefficient calculations, haplotyping, and genotype error detection. MERLIN was designed for the analysis of dense genetic maps of both biallelic and multiallelic markers. It uses sparse inheritance trees to represent gene flow in pedigrees and is one of the fastest packages for pedigree analysis. The website http://www.sph.umich.edu/csg/abecasis/Merlin offers an excellent tutorial.

Data import

The input files may be either in 'LINKAGE format', referring to the software in which it was first introduced (Lathrop *et al.*, 1984; Terwilliger and Ott, 1994), or in QTDT format (Abecasis *et al.*, 2000). In using QTDT format, MERLIN requires a pedigree file (*.ped*), a data file (*.dat*) and a map file (*.map*). It is worth noting that the *.ped* file looks the same for both LINKAGE and QTDT formats. Part of a coded pedigree file follows, where, for simplicity, three markers genotyped in two families are shown:

```
390  138    0    0  1  1  1  3  5  7  0  0
390  139    0    0  2  1  1  6  7  7  0  0
390  132  138  139  2  2  1  1  5  7  0  0
390  137  138  139  1  2  0  0  5  7  0  0
460  206  208  204  2  1  2  2  6  7  1  1
460  207  208  204  1  1  2  4  6  7  1  1
460  204    0    0  2  1  2  4  6  6  1  1
460  208    0    0  1  1  2  2  7  7  1  1
460  205  208  204  1  1  2  4  6  7  1  1
```

Each line corresponds to a single individual. The columns are as follows: kindred ID, individual ID, father's ID, mother's ID, sex (1 = male, 2 = female), affection status (U or 1 = unaffected; A or 2 = affected; X or 0 = missing) and marker genotypes. Genotypes are coded as pairs of space-delimited integers: one integer for each allele, with missing values coded as either X or 0. In practice, multiple paired columns of genotypes would be included in map order for each individual. For the

X chromosome, male genotypes are coded as pairs of identical alleles. Quantitative traits can also be added, again using X to denote missing values. This file therefore provides pedigree structure information, genotypes and phenotypes.

The data file *.dat* describes the contents of the *.ped* file, starting with column 6; the first five columns are standard. Each row corresponds to a data item. The first column in *.dat* indicates the data type: *A* for affection status, *T* for quantitative trait, *C* for covariate and *M* for marker. The second column indicates the label of each item. In our example, the first five rows of *.dat* take the following form:

```
A Affection
M 1
M 2
M 3
M 4
```

A summary of the pedigree and data files may be obtained by typing the following command. It provides a useful check that all the data have been formatted and loaded correctly.

```
prompt>pedstats –d npl.dat –p npl.ped
```

Lastly, the map file, *.map* lists the marker details in map order. The following example shows the first five rows of a map file, where each row gives the chromosome, the marker name and the map position in centimorgans (cM).

```
4  1  12.46
4  2  12.74267
4  3  13.95624
4  4  15.91956
4  5  17.96277
```

NPL analysis for a qualitative trait

At the command prompt, the user invokes a series of flags to indicate the files and options that are required for the analysis. NPL analysis can be executed using the following one-line command.

```
prompt>merlin –d npl.dat –p npl.ped -m npl.map --steps 4 --npl --markerNames --pdf >
npl_output
```

The user specifies an input data file ($-d$ parameter), pedigree file ($-p$ parameter) and map file ($-m$ parameter). MERLIN is then asked to perform a NPL analysis at four equally spaced points in each marker interval (*--steps* 4), using the Whittemore and Halpern $\mathrm{NPL_{all}}$ statistic to test for allele sharing among affected individuals

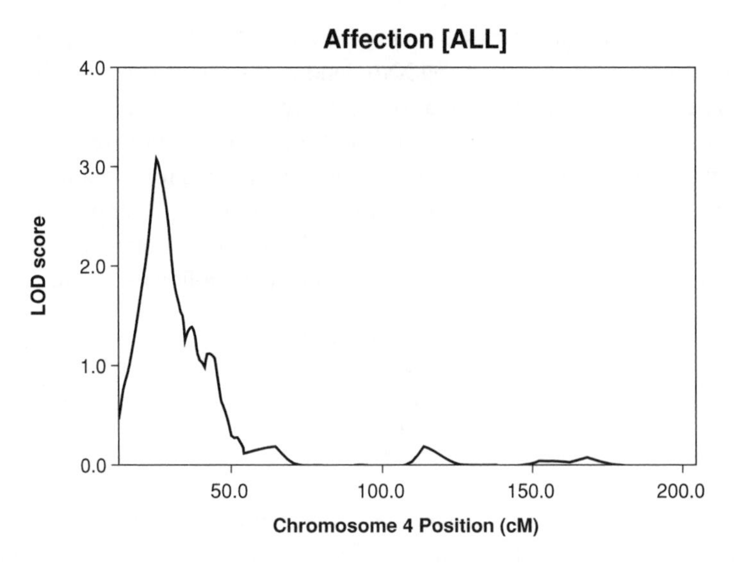

Figure 10.1 Graphical output from MERLIN after an NPL analysis of markers across chromosome 4

(*--npl*). The user has the option of showing marker labels in the output, instead of cM positions (*--markerNames*). NPL output can also be piped to an output file by using the > operator. Lastly, the user has the option of producing a graphical output that displays the LOD score across the analysed region (*--pdf*), as shown in Figure 10.1. The numerical output takes the following form:

Phenotype: Affection [ALL] (162 families)

Pos	Zmean	*p* value	delta	LOD	*p* value
min	−16.88	1.0	−0.271	−21.59	1.0
max	19.37	0.00000	0.622	46.33	0.00000
1	1.34	0.09	0.123	0.46	0.07
12.517	1.35	0.09	0.124	0.47	0.07
12.573	1.37	0.09	0.126	0.48	0.07
12.630	1.38	0.08	0.127	0.49	0.07
12.686	1.39	0.08	0.129	0.49	0.07
...					
13	3.39	0.0003	0.311	2.98	0.00011
26.443	3.39	0.0003	0.310	2.97	0.00011
26.510	3.39	0.0003	0.309	2.96	0.00011
26.577	3.39	0.0003	0.308	2.95	0.00011
26.643	3.40	0.0003	0.307	2.94	0.00012
14	3.40	0.0003	0.306	2.93	0.00012
27.267	3.29	0.0005	0.307	2.86	0.00014
27.825	3.19	0.0007	0.307	2.77	0.0002
28.382	3.09	0.0010	0.306	2.66	0.0002
28.939	3.00	0.0014	0.302	2.54	0.0003

The first two rows show the minimum and maximum possible scores given the data structure. These are followed by the NPL results at each location; in other words at each marker and at each of four steps between. In the current example, a linkage peak is shown close to marker 14 with a Z-score of 3.40 (*P* value of 0.0003), corresponding LOD score of 2.93 (*P* value of 0.00012).

As shown in Figure 10.1, a graphical representation of the results allows a rapid assessment of evidence for linkage and in this case evidence peaks close to marker 14. Localization cannot, however, be assumed to be precise, and separation of at least 10 cM may be seen between studies (Hauser and Boehnke, 1997). It is therefore usual to construct a support interval around a strong linkage signal (Conneally *et al.*, 1985). For example, having converted to LOD units, a one-unit support interval is the interval that includes all (possibly disjoint) map positions with LOD score less than one LOD unit below the peak score. A conservative approach is to adopt a 1.5 to 2 LOD support interval. All points within the support interval are considered to be of interest.

A determination of information content (the amount of IBD information extracted by the genotype data) is achieved in MERLIN with the – *information* switch. A graphical representation of information content plotted against map position is invaluable in interpreting the results of NPL analysis. Dips in the graph allow regions to be highlighted in which the typing of additional markers could be beneficial. Another useful MERLIN feature, the --*ibd* switch, provides a very rapid means of generating IBD probabilities, and its output may be used as input for other software such as QTDT (Abecasis *et al.*, 2000), to be discussed later. Another piece of software, SimWalk2 (Sobel and Lange, 1996), generates IBD probabilities for a wider range of family structures; but for small to moderate-sized pedigrees, MERLIN is faster.

10.3 Association analysis

Association analysis may be regarded as a test for the presence of a difference in allele frequency between cases and controls. A difference does not necessarily imply disease causality, as many factors, including population history and ethnic make-up, may yield this effect. In a well-designed study, however, evidence of association provides a flag for further study. In some instances, it is due to the marker being physically close to the causal variant.

Association testing for case-control or population data is often carried out with general (non-genetic) statistical software packages, such as SAS, R or S-PLUS. The chi-square test is applied to a contingency table, in which case/control status is tabulated by frequencies of either genotypes or alleles. The test takes the usual form,

$$\chi^2 = \sum \frac{(Obs - Exp)^2}{Exp}$$

where *Obs* and *Exp* are the observed and expected frequencies respectively, and the sum is taken over all cells in the table. The number of degrees of freedom is $(r-1)(c-1)$, where r is the number of rows, and c is the number of columns in the table. Equivalently, logistic regression can be applied, using disease status as the dependent variable and alleles or genotypes as the independent variables. See Clayton (2001) for a detailed review of the method. The remaining sections of this chapter all involve applications and extensions of the traditional association test.

10.3.1 Transmission disequilibrium tests

In recent years, there has been an upsurge in interest in family-based testing, owing to the concern that ethnic mismatching of non-family cases and controls (population stratification) can sometimes yield false-positive evidence of association. In particular, the transmission/disequilibrium test (TDT) (Spielman *et al.*, 1993) has gained prominence as a test of linkage in the presence of association, that does not give false evidence of linkage due to population stratification. The TDT is applied by counting alleles transmitted from heterozygous parents to one or more affected children in nuclear families. The alleles *not* transmitted to affected children may be regarded as control alleles, perfectly ethnically matched to the 'case' alleles seen in the affected children. The test takes the form of McNemar's test, which, under the null hypothesis of no linkage, follows a chi-squared distribution with one degree of freedom. The TDT is also a valid test for association, but only when applied to alleles transmitted from heterozygous parents to just one affected child per family.

Assuming a biallelic locus, let b denote the counts of heterozygous parent-to-offspring transmissions in which allele 1 goes to an affected child, while allele 2 is not transmitted. Let c denote the counts of transmissions the other way around, in which allele 2 is inherited in an affected child, while allele 1 is not transmitted. The test takes the following form:

$$\chi_1^2 = \frac{(b-c)^2}{(b+c)}$$

A number of groups have focused on generalizing the TDT to quantitative traits or to designs in which parental genotypes are not available. The sib-TDT, or S-TDT (Spielman and Ewens, 1998), does not use parental genotypes, and, like the original TDT, it is not prone to false positives due to population stratification. For association testing, the S-TDT requires that the data in each family consist of at least one affected and one unaffected sibling, each with different marker genotypes. This test and the original TDT are widely implemented; for example, in the Java-based program TDT/S-TDT (Spielman and Ewens 1996, 1998).

Multiallelic markers may be tested by ANALYZE (Terwilliger, 1995). This has the advantage of taking LINKAGE format files as input and so provides a natural follow-up to a genome scan. It does, however, require that LINKAGE (Lathrop *et al.*, 1984) be installed on one's system. Other software able to handle multiallelic markers include ETDT (Sham and Curtis, 1995) and GASSOC (Schaid, 1996).

For quantitative traits, a major development was the release of QTDT (Abacasis *et al.*, 2000), software, which allows TDT testing under a variance components framework. It is applicable to sibships with or without parental genotypes and incorporates a broad range of quantitative trait tests – those proposed by Rabinowitz (1997), Allison (1997), Monks *et al.* (1998), Fulker *et al.* (1999) and Abecasis *et al.* (2000). It is written in C++, to be run on UNIX and has a command-line interface. Its input files are based on LINKAGE format, but, in addition, one input file of IBD probabilities must be prepared in advance. QTDT assumes the IBD format generated by the programs SimWalk2 (Sobel and Lange, 1996) and MERLIN (Abecasis *et al.*, 2002). Covariates may also be modelled, but should be kept to a minimum in order to maintain performance.

10.3.2 Haplotype reconstruction

A haplotype is a string of consecutive alleles lying on the same chromosome. Each individual therefore has a pair of haplotypes for any chromosomal interval – one inherited from the paternal side and one inherited maternally. In statistical genetics, their importance lies in the fact that tests of association may be applied to haplotypes instead of single loci. This may yield increased power if the variant of interest is not being tested directly or if adjacent loci are contributing to a single effect (see Clark *et al.*, 1998; Nickerson *et al.*, 1998). Haplotypes may be inferred from the genotypes of parents or other family members (Weeks *et al.*, 1995) or by laboratory methods (Clark 1990; Nickerson *et al.*, 1998). Often, however, they are estimated by means of the expectation-maximization (E-M) algorithm (Dempster *et al.*, 1977; Little and Rubin, 1987; Excoffier and Slatkin, 1995; Hawley and Kidd, 1995; Long *et al.*, 1995).

The E-M algorithm is a method that aims to provide maximum likelihood parameter estimates in the presence of incomplete data. In the case of haplotype frequency estimation, it proceeds as follows (Schneider *et al.*, 2000):

1. An initial set of plausible haplotype frequencies is assigned – for example, the product of the relevant allele frequencies may be used.

2. The E-step: assuming Hardy–Weinberg equilibrium, the haplotype frequencies are used to estimate the expected frequencies of ordered genotypes.

3. The M-step: the expected genotype frequencies are used as weights to produce improved estimates of haplotype frequencies.

4. Steps 2 and 3 are repeated until the haplotype frequencies reach equilibrium.

Note that, as with other iterative techniques, it is wise to compare the results of multiple starting points, as the E-M algorithm may converge to a local, rather than global optimum. It is not always reasonable to assume that the maximum likelihood haplotype configuration has been reached.

Software written specifically for haplotype analysis includes EHPLUS (Zhao *et al.*, 2000), a reworked and extended version of the earlier program EH (Xie and Ott, 1993). It is written in C and is available in both UNIX and PC versions. EHPLUS can be applied to either case-control data or data assumed to come from a random-mating population. It accommodates large numbers of haplotypes and incorporates a companion program, PMPLUS, which reformats genotype data ready for use. Estimated haplotypes and their frequencies are output and may be subjected to association tests. A distinctive feature of these association tests is that PMPLUS provides P values for non-parametric tests or for parametric tests assuming a user-specified disease model or maximized over multiple disease models. Permutation features allow the calculation of empirical P values for these tests.

Further software for sophisticated haplotype analysis is available from ftp://ftp-gene.cimr.cam.ac.uk/software/clayton/. Resources include SNPHAP, a program that uses the E-M algorithm to estimate haplotype frequencies for large numbers of diallelic markers using genotype data. Another program, TDTHAP (Clayton and Jones, 1999), allows the TDT to be applied to extended haplotypes. STATA routines to aid SNP selection by haplotype tagging (Johnson *et al.*, 2001) are available at ftp://ftp-gene.cimr.cam.ac.uk/software/clayton/stata/htSNP/.

Haplotype reconstruction from family data can be achieved with SimWalk2 (Sobel and Lange, 1996). The derived haplotypes may then be imported to a pedigree-drawing package, such as Cyrillic (Chapman, 1990), for viewing recombinants in positional cloning. MERLIN (Abecasis *et al.*, 2002) and GENEHUNTER (Kruglyak *et al.*, 1996) also output haplotypes estimated from family data. Other software, such as TRANSMIT (Clayton, 1999) and FBAT (e.g., Laird, 2004), allows association testing of family-based haplotypes.

A newer piece of software, UNPHASED (Dudbridge, 2003), combines and builds upon many of the advantages of earlier code. It accommodates both single-marker and haplotype association testing with either quantitative or binary traits. The UNPHASED suite of programs allows one to test for association in unrelated individuals (QTPHASE, COCAPHASE), nuclear families (TDTPHASE), or extended pedigrees (PDTPHASE, QPDTPHASE). UNPHASED handles both biallelic and multiallelic markers, and provides the user with a wide range of command-line options, providing considerable flexibility in data analysis. For example, in performing haplotype analyses, the user can specify an analysis with a sliding window of a particular haplotype length (i.e., number of consecutive markers) or can specify that a particular set of non-consecutive markers define the haplotypes to be tested. The user can also specify various other options, such as dropping or pooling rare haplotypes, calculation of the pairwise linkage disequilibrium measures D' and r^2 (see later), or including data from only one affected sib per family. In addition to testing for association with phenotype, the E-M algorithm employed by UNPHASED can be used to estimate haplotype frequencies in subjects with phase unknown genotypes. The UNPHASED program suite runs on UNIX, LINUX and Windows platforms.

10.3.3 Example: UNPHASED (Dudbridge, 2003)

Data import

Each of the five programs comprising the UNPHASED suite requires a pedigree file to be input in standard LINKAGE format. This is the format of the *.ped* file described earlier and used as input to MERLIN. To recap, the file consists of one row per individual and tab or space-delimited columns laid out as follows: Kindred ID, Subject ID, Father ID, Mother ID, Sex, Affection Status, Allele1a, Allele1b, Allele2a, Allele2b ...

The Kindred ID can be alphanumeric, but all other columns must be numeric. For a binary trait, the Affection Status column would contain '2' for cases (affected) or '1' controls (unaffected); when association is being tested to a quantitative trait (QT), the Status column would contain each subject's numerical QT value. The UNPHASED documentation notes that an optional data file can be used to assign marker names and define input file formats other that shown above.

Here we illustrate Windows use of UNPHASED software by using QTPHASE to analyse data that was simulated to evaluate ability to detect association between haplotypes and bone mineral density (BMD), a quantitative trait indicative of bone strength. A study by Giraudeau *et al.* (2004) describes testing for association between BMD and haplotypes of the gene cathepsin K (CTSK), and the results described here are for a simulated, CTSK-like gene (sCTSK). Genotypes for four SNPs in sCTSK were evaluated in unrelated subjects; thus each subject had a different KindredID as well as specification of unknown ('0') for each FatherID and MotherID. Hence the first three lines of the input file (named *sCTSK_BMD.prn*) had the following space-delimited format:

```
Ped1  1  0  0  1  360.2  1  2  3  4  1  2  2  3
Ped2  2  0  0  1  328.5  2  2  3  3  1  2  2  2
Ped3  3  0  0  2  343.8  1  1  4  4  2  2  3  3
```

As a first step, *QTPHASE.exe* is commanded to conduct a global test of association (analogous to an omnibus ANOVA) to determine whether mean BMD associated with any sCTSK haplotype significantly differs from mean BMD associated with any other sCTSK haplotype(s). As shown in the first line of Figure 10.2, the issued command was:

```
QTPHASE sCTSK_BMD.prn –window 4 –EM
```

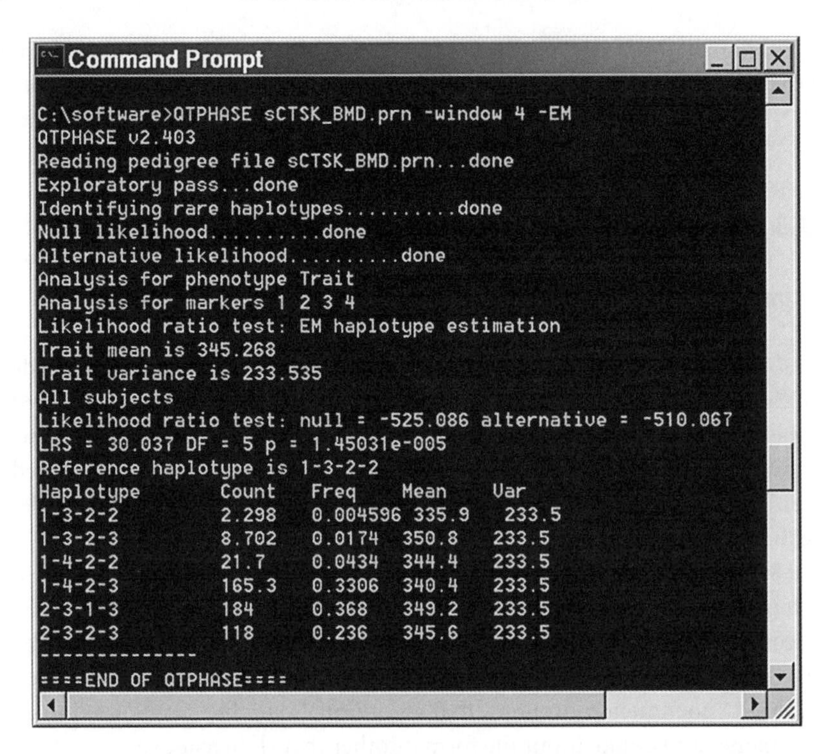

Command Prompt
C:\software>QTPHASE sCTSK_BMD.prn -window 4 -EM
QTPHASE v2.403
Reading pedigree file sCTSK_BMD.prn...done
Exploratory pass...done
Identifying rare haplotypes.........done
Null likelihood.........done
Alternative likelihood.........done
Analysis for phenotype Trait
Analysis for markers 1 2 3 4
Likelihood ratio test: EM haplotype estimation
Trait mean is 345.268
Trait variance is 233.535
All subjects
Likelihood ratio test: null = -525.086 alternative = -510.067
LRS = 30.037 DF = 5 p = 1.45031e-005
Reference haplotype is 1-3-2-2
Haplotype Count Freq Mean Var
1-3-2-2 2.298 0.004596 335.9 233.5
1-3-2-3 8.702 0.0174 350.8 233.5
1-4-2-2 21.7 0.0434 344.4 233.5
1-4-2-3 165.3 0.3306 340.4 233.5
2-3-1-3 184 0.368 349.2 233.5
2-3-2-3 118 0.236 345.6 233.5

====END OF QTPHASE====

Figure 10.2 Test of haplotype association with a quantitative trait using the QTPHASE component of the UNPHASED software suite

which instructs QTPHASE to estimate haplotype frequencies by the E-M algorithm and to use a sliding haplotype window of length 4 (note that as with MERLIN, a dash precedes each command line option). All subsequent lines in Figure 10.2 show QTPHASE output, which can be optionally directed to a file using the UNPHASED *-output* option). The output informs the user of a highly significant association between BMD and sCTSK haplotypes with a global P value $= 1.45031 \times 10^{-5}$. For each sCTSK haplotype, it also shows the estimated haplotype frequency, mean BMD value, and a single pooled error variance for the four estimated means.

To follow up the global *P* value test, QTPHASE also provides a test for a significant difference between the two means associated with any pair of haplotypes, thus enabling the user to identify pairwise differences that contribute to a global result. This is done by including – compare *haplotype1* – with *haplotype2* on the QTPHASE command line. For example, to compare the final two haplotypes shown in the output in Figure 10.2, we appended *-compare 2 3 1 3 – with 2 3 2 3* at the end of the command in first line of Figure 10.2. This generated output similar to that shown in Figure 10.2, except that a few additional lines specified the two haplotypes being compared and gave a P value of 0.0669 for the corresponding one degree of freedom test. It is important to note that the UNPHASED suite of programs enables users to verify asymptotic P values by calculating permutation-derived P values through shuffling the phenotype and genotype data. We also found that the haplotype frequencies estimated by UNPHASED are identical to frequencies estimated by the haplotype software EHPLUS (Xie and Ott, 1993; Zhao *et al.*, 2000).

10.4 Linkage disequilibrium

Linkage disequilibrium (LD) is a lack of independence, in the statistical sense, between the alleles at two loci. LD exists between two linked loci when particular alleles at these loci occur on the same haplotype more often than would be expected by chance alone. This phenomenon can provide valuable information in locating disease variants from marker data, as a marker in LD with the causal variant provides a flag for its location. LD information also provides a means by which the efficiency of high-density marker maps can be increased. If markers are in strong LD with each other, there is an argument for genotyping only a subset of them.

The extent of pairwise LD may be measured by the value D as follows (Lewontin, 1964). Assume two diallelic loci are linked and let p_{ij} be the proportion of chromosomes that have allele $i$at the first locus and allele j at the second locus. For example, p_{12} is the frequency of the haplotype with allele 1 at the first locus and allele 2 at the second locus. The disequilibrium coefficient D is the difference between the observed haplotype frequency p_{12} and the haplotype frequency expected under linkage equilibrium, the latter being the product of the two allele frequencies, say, p_{1+} and p_{+2}. It may be written as follows:

$$D = p_{12} - p_{1+}p_{+2}$$

A more commonly quoted measure of LD is D' (Lewontin, 1964). This is a normalized form, with numerator equal to D and denominator equal to the absolute maximum D that could be achieved given the allele frequencies at the two loci. D' can take values from -1 to $+1$ but, in general, its absolute value is presented and discussed. A value of 1 indicates absence of recombination event, whereas values less than 1 indicate that two loci have been separated through recombination. Intermediate values of D' may be difficult to interpret, as the measure tends to be inflated when sample size is small or allele frequencies are low.

The squared correlation coefficient, r^2 is sometimes preferred to quantify and compare the amount of LD between pairs of loci. r^2 is determined by dividing D' by the product of the four allele frequencies. When $r^2 = 1$, two markers provide identical information, not only having $D' = 1$ but also having equal allele frequencies. The main advantage of r^2 is its inverse relationship with the sample size required to detect genetic association between markers that are in complete LD (Pritchard and Przeworski, 2001). For instance, if cases and controls have only been genotyped for markers in the vicinity of a functional variant, the sample size should be increased by a factor $1/r^2$ in order to achieve the same power as would have been achieved by generating data at the susceptibility locus itself. r^2 is, however, more sensitive to allele frequencies than $|D'|$ and can be difficult to interpret when the two loci in question differ in allele frequencies.

P values obtained from pairwise significance tests of LD are also used to describe the pattern of LD. However, they should be used with care, as P values depend strongly on sample size. In large sample sizes, statistically significant P values can be obtained for low LD values. It is therefore not recommended to use P values to compare LD between studies with different sample sizes.

Other methods of estimating LD include the moment method, applicable to newly formed populations under certain assumptions concerning the evolutionary process (Hastabacka *et al.*, 1992; Lehesjoki *et al.*, 1993; Kaplan *et al.*, 1995). Maximum likelihood methods have also been explored (Hill and Weir, 1994; Kaplan *et al.*, 1995). Composite likelihood methods were proposed to evaluate the information from multiple pairs of loci simultaneously. Examples of software for the composite likelihood approach include DMAP (Devlin *et al.*, 1996) and ALLASS (Collins and Morton, 1998). The latter uses the Malecot isolation by distance equation and has the advantage of accommodating multiple founder mutations. Each method, however, relies upon population assumptions and may suffer reduced power when these are not met.

Understanding the pattern of LD has become of great interest in recent times, both for marker selection and for defining candidate regions. Many other valid measures of pairwise LD exist and have been reviewed elsewhere (Hedrick, 1987; Devlin and Risch, 1995). In all cases, their individual interpretation depends on the context and nature of the data at hand (Jorde, 2000; Ardlie *et al.*, 2002; Wall *et al.*, 2003).

Tests of various measures of LD can be achieved with software such as Arlequin (Schneider *et al.*, 2000). This is a C++ program available for the PC(Win), Linux and MacOS systems. The statistical significance of observed LD is estimated for phase-known (haplotype) data by Fisher's Exact Test. For phase-unknown data, a likelihood ratio test is applied. An alternative tool is GDA (Lewis and Zaykin, 2001), the PC(Win) companion program to the book, Genetic Data Analysis II (Weir, 1996). Both are well documented and perform a broad range of population genetic tests.

GOLD (Abecasis and Cookson, 2000), available for PC(Win), is another program to calculate D and D', and it is noteworthy in being one of the first to output them in graphical form. For each marker pair, the pairwise disequilibrium statistics are colour coded (bright red to dark blue) and plotted. The output is valuable for presentation purposes and provides a useful summary of the properties of dense maps. The software takes haplotype estimates as input, and in the case of family data, these must be reconstructed with software such as SimWalk2 (Sobel and Lange, 1996) prior to use. Case-control data are not well supported by GOLD, which relies for this purpose upon a limited interface to the software, EH (Xie and Ott, 1993). GOLDsurfer (Pettersson *et al.*, 2004) builds upon GOLD and has the attractive property of allowing simultaneous presentation of a variety of LD statistics and disease association statistics, in a three-dimensional plot. GOLDsurfer is embedded within a graphical software package available from http://www.umbio.com.

Haploview (Barrett *et al.*, 2005) is a recent software package for LD and haplotype analyses of biallelic markers. Haploview can analyse family- or non-family-based data. This software is designed to compute pairwise LD statistics, define haplotype blocks (genomic segments of high LD) and identify tagging SNPs for larger groups of markers. An accelerated E-M algorithm, similar to the method described in Qin *et al.* (2002), estimates haplotype frequencies. Single-marker and haplotype association tests can be performed for case/control data and family trios. A permutation test is also available to correct for multiple testing. An invaluable feature of Haploview is that it generates LD plots that display pairwise measures of LD, haplotype blocks and tagging SNPs. The user can choose among different LD measures, and three different algorithms are available for constructing haplotype blocks and selecting tagging SNPs. Haploview can be run on Windows, Mac OS X and Linux platforms and is designed to work on a Java Runtime Environment.

10.4.1 Example: HAPLOVIEW (Barrett *et al.*, 2005)

Data import

Haploview can analyse phase-known or phase-unknown data imported in the standard LINKAGE format. This format can be used for either family or non-family data (see earlier examples). Data can also be imported from the HapMap project website (http://www.hapmap.org). With the advent of the HapMap, it has become increasingly common for investigators to evaluate regional LD and select markers for experimentation based upon their performance within the HapMap. For the current example, we used the HapMap website to select 381 markers across a 1-Mb subregion spanning the MHC region (human chromosome 6, 29,001,919-29,996,718 [NCBI 34]). The HapMap data file was saved as *MHC.hmp*.

LD analysis of data from the HapMap website

1. Within Haploview, select *Load HapMap data* and load the HapMap file, *MHC.hmp*, as shown in Figure 10.3. In the current example, a value of zero has been entered to force all pairwise LD calculations to be performed (the default segment size is 500 kb). By default, subjects with more than 50 per cent missing genotypes are excluded from the subsequent analysis.

2. After loading the data set, Haploview displays the *Check Markers* window (Figure 10.4), which presents a series of data quality checks. These include, for each marker, the RS number, the map position, the observed heterozygosity, the

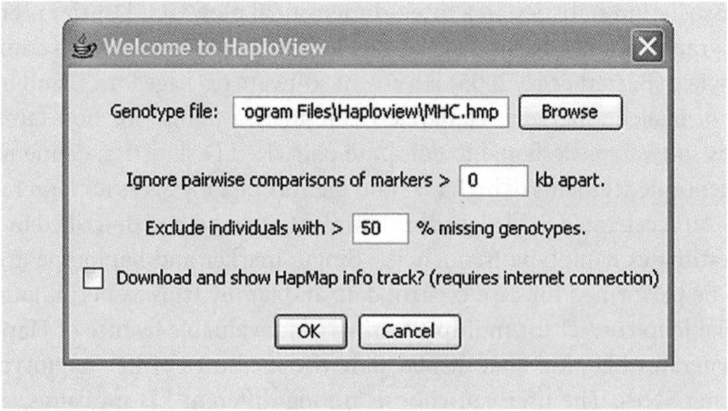

Figure 10.3 Haploview screen showing the loading of a chromosome 6 data file from HapMap

predicted heterozygosity, a check for Hardy–Weinberg equilibrium, the percentage of non-missing genotypes, the frequency of genotyped family trios, the Mendelian inheritance error rate, and the minor allele frequency. The *Rating* column filters out markers that fail quality control. The user can remove or add markers manually as necessary.

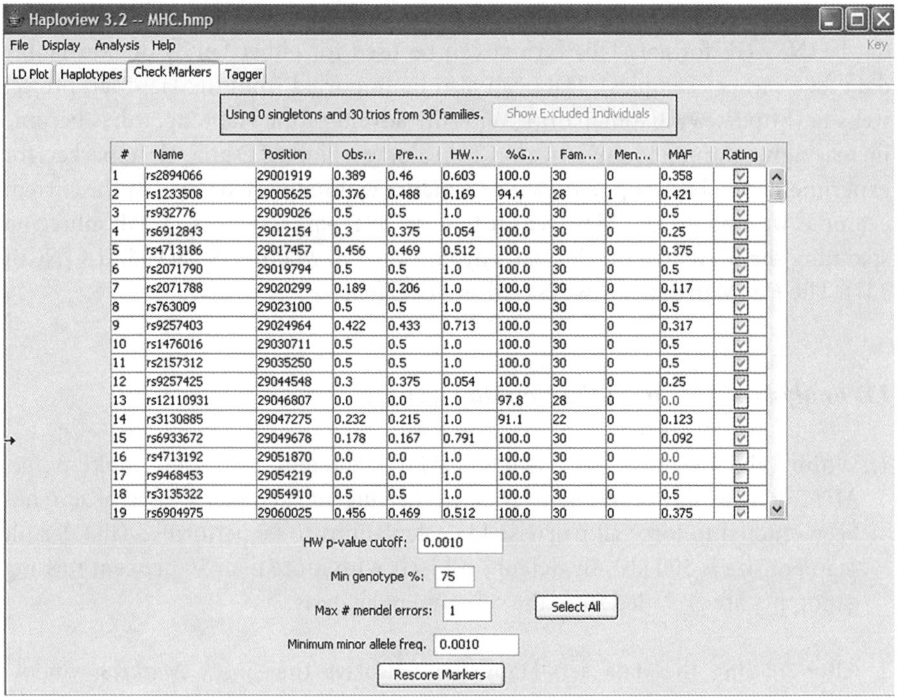

#	Name	Position	Obs...	Pre...	HW...	%G...	Fam...	Men...	MAF	Rating
1	rs2894066	29001919	0.389	0.46	0.603	100.0	30	0	0.358	☑
2	rs1233508	29005625	0.376	0.488	0.169	94.4	28	1	0.421	☑
3	rs932776	29009026	0.5	0.5	1.0	100.0	30	0	0.5	☑
4	rs6912843	29012154	0.3	0.375	0.054	100.0	30	0	0.25	☑
5	rs4713186	29017457	0.456	0.469	0.512	100.0	30	0	0.375	☑
6	rs2071790	29019794	0.5	0.5	1.0	100.0	30	0	0.5	☑
7	rs2071788	29020299	0.189	0.206	1.0	100.0	30	0	0.117	☑
8	rs763009	29023100	0.5	0.5	1.0	100.0	30	0	0.5	☑
9	rs9257403	29024964	0.422	0.433	0.713	100.0	30	0	0.317	☑
10	rs1476016	29030711	0.5	0.5	1.0	100.0	30	0	0.5	☑
11	rs2157312	29035250	0.5	0.5	1.0	100.0	30	0	0.5	☑
12	rs9257425	29044548	0.3	0.375	0.054	100.0	30	0	0.25	☑
13	rs12110931	29046807	0.0	0.0	1.0	97.8	28	0	0.0	☐
14	rs3130885	29047275	0.232	0.215	1.0	91.1	22	0	0.123	☑
15	rs6933672	29049678	0.178	0.167	0.791	100.0	30	0	0.092	☑
16	rs4713192	29051870	0.0	0.0	1.0	100.0	30	0	0.0	☐
17	rs9468453	29054219	0.0	0.0	1.0	100.0	30	0	0.0	☐
18	rs3135322	29054910	0.5	0.5	1.0	100.0	30	0	0.5	☑
19	rs6904975	29060025	0.456	0.469	0.512	100.0	30	0	0.375	☑

Figure 10.4 Haploview screen showing the marker quality check

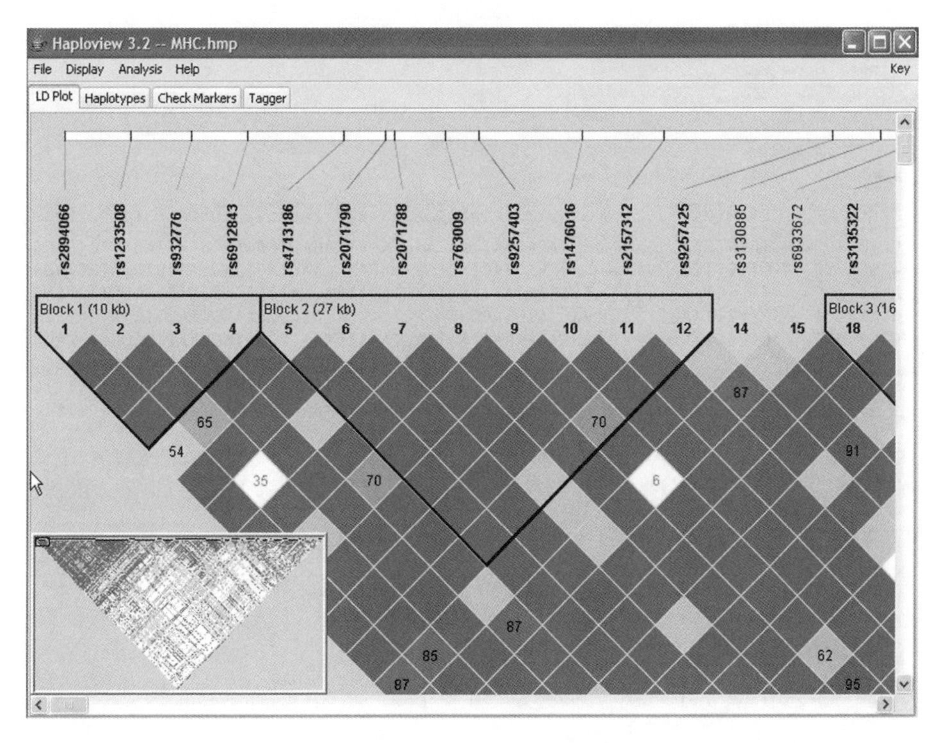

Figure 10.5 Haploview screen showing an LD plot of markers on chromosome 6 together with inferred haplotype blocks

3. Select *LD Plot* tab to display graphically pairwise measures of LD (Figure 10.5). Each square displays the amount of LD between a pair of markers. The strength of LD between two markers is given by the intensity of the colour of a box. By default, Haploview displays D' values, but the user can choose among several LD statistics. In Figure 10.5, thick black triangles depict haplotype blocks, which are genomic segments of high LD. In Haploview, blocks are assigned according to a user-specified definition. Definitions available are confidence intervals (Gabriel *et al.*, 2002), the four-gamete rule (Wang *et al.*, 2002) and the internally developed solid spine of LD, a block type in which the two end markers are in strong LD with intervening markers but intervening markers are not necessarily in LD with each other. By default, Haploview identifies blocks by the definition of Gabriel *et al.* (2002). Any marker can be removed or added into a block by clicking on its ID number. Groups of markers can also be manually selected to create new haplotype blocks.

4. Select the *Haplotype* tab to generate haplotypes within each block. As shown in the resultant Figure 10.6, each row represents a different haplotype with its estimated population frequency given on the right side of a block. By default, Haploview

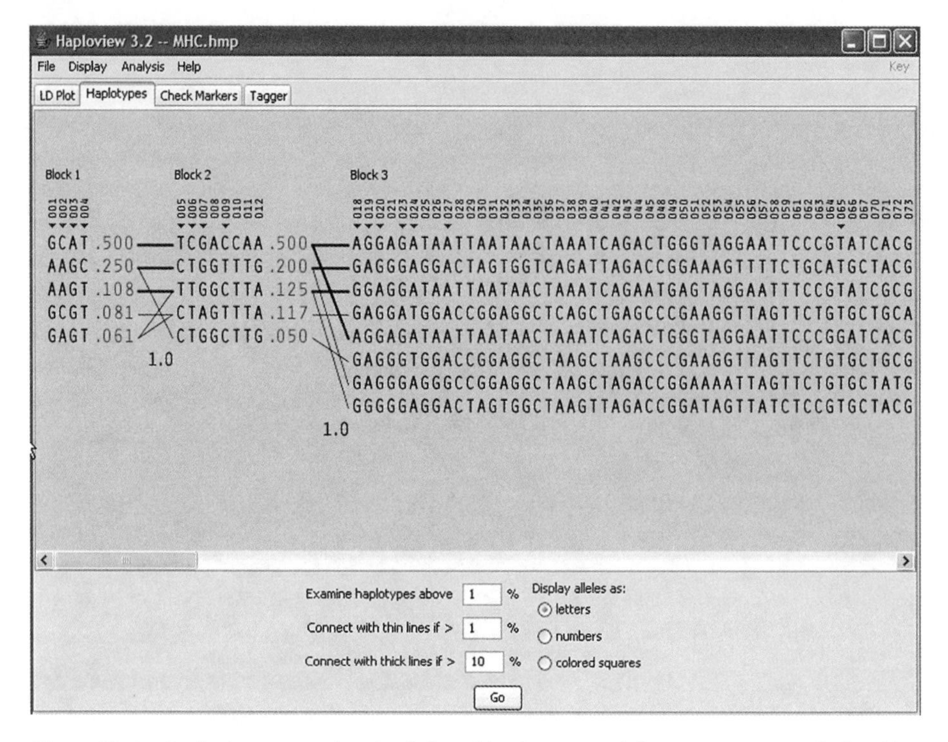

Figure 10.6 Haploview screen showing inferred haplotypes and the co-occurrence relationships among haplotypes in different inferred blocks

displays all haplotypes with a population frequency greater than 1 per cent. Markers within each block are shown above the block, and a small inverted triangle is given beneath IDs of tagging SNPs. Lines between two blocks indicate transitions from one block to the next, with thicker lines showing more frequent combinations. Thick lines indicate that more than 10 per cent of the chromosomes have the haplotype indicated on the left and the haplotype indicated on the right. Thin lines indicate that 1–10 per cent of the chromosomes have both haplotypes. In the crossing areas, a multiallelic D' value is given. This value corresponds to the level of recombination between two blocks, considering each haplotype within a block as an allele. A value of one indicates no evidence for historical recombination between two blocks, whereas a value close to zero indicates a great amount of historical recombination between two blocks.

Haploview is easy to use. It has a dynamic interface that provides graphical representations of LD relationships between markers in close proximity, haplotype blocks and tagging SNPs. At any time, the user can include or exclude markers and set up new thresholds for subsequent analyses. These modifications are instantly reproduced in the LD and haplotype analyses.

10.5 Quantitative trait locus (QTL) mapping in experimental crosses

In contrast to human studies, in which variances of phenotypic differences are used to establish the presence of linkage, QTL mapping in experimental crosses involves comparing means of progeny inheriting specific parental alleles. This is simpler and more powerful (Kruglyak and Lander, 1995). It can be achieved by any of a number of standard statistical methods, such as *t*-tests, analysis of variance (ANOVA), Wilcoxon rank-sum and regression techniques. Again, missing data can be accommodated by an application of the EM algorithm.

Of the very broad array of possible diploid crosses, the following are particularly common. They are derived from a pair of divergent inbred lines in which the genotypes at the majority of loci are homozygous and distinct, say, *aa* and *bb* for a particular locus in the two lines respectively. The filial F_1 generation results from crossing these two lines to produce individuals with heterozygous genotype *ab*. In the backcross (BC) design, F_1 is crossed with one of the parent strains. For example, in the case of a cross with the *aa* parent, half the offspring produced are *ab* and half are *aa*. In the filial F_2 design, the F_1 is selfed, or two F_1 individuals are crossed so that offspring are *aa*, *ab*, and *bb* in the ratio 1:2:1. Lastly, in the recombinant inbred line (RIL), each F_2 enters individually a single-seed, descent-inbreeding programme, so that all progeny are homozygous for the chosen allele.

The original analysis framework was based upon a marker-by-marker analysis. Of particular relevance to sparse maps, simple interval mapping (IM or SIM) allows the evaluation of any position within a marker interval. The maximum likelihood approach to IM proceeds by the calculation of a LOD score (Lander and Botstein, 1989). Similarly, and with lower computational burden, least-squares regression achieves the same goal (Haley and Knott, 1992; Martinez and Curnow, 1992). IM may be carried out by a range of software, including MAPMAKER/QTL (Lander *et al.*, 1987). This may appeal to regular users of GENEHUNTER, as the syntax is similar. It relies upon data preprocessing in MAPMAKER/EXP (Lander *et al.*, 1987), and provides simple graphical output.

Two newer and related methods are composite interval mapping (CIM) and multiple QTL mapping (MQM). Both involve performing a genome scan by moving stepwise along the chromosome and testing for the presence of the QTL with a predefined set of markers as cofactors (Jansen, 1992, 1993; Zeng, 1993, 1994; Jansen and Stam, 1994; Kao *et al.*, 1999; Zeng *et al.*, 1999). In other words, in the sparse map case, interval mapping is combined with multiple regression on markers. This approach allows one to control, to some extent, for effects of other QTLs. Software such as QTL Cartographer (Basten *et al.*, 1994, 1997) and PLABQTL (Utz and Melchinger, 1996) allow the selection of such cofactors by stepwise regression. These programs offer options that will automatically include or exclude background markers by user-defined criteria.

Lastly, Bayesian methods allow the consideration of multiple QTLs, QTL positions and QTL strengths (Jansen, 1996; Satagopan *et al.*, 1996; Uimari *et al.*, 1996; Sillanpaa and Arjas, 1998, Borevitz *et al.*, 2002). Multimapper (Sillanpaa, 1998), for example, allows the automatic building of models of multiple QTLs within the same linkage group. It is designed to work as a companion program to QTL Cartographer (Basten *et al.*, 1994, 1997) and allows a more detailed follow-up of regions of interest. As with other Markov Chain Monte Carlo methods, however, this approach is computer intensive and may suffer from problems of convergence to a local, rather than global optimum, or of lack of convergence if run for a short time.

Ten of the most prominent programs for QTL mapping are reviewed in greater detail by Manley and Olson (1999). The majority will perform IM and CIM for backcross, filial F_2 and recombinant inbred lines. Cordell (2002) provides worked examples of the use of three of them, MAPMAKER/QTL, QTL Cartographer and MapQTL (van Ooijen and Maliepaard, 1996a, 1996b).

In the future, genetically heterogeneous stocks may gain in prominence (Mott *et al.*, 2000). Talbot *et al.* (1999) were able to achieve a mapping resolution of less than 1 cM by the study of heterogeneous stocks from eight known inbred mouse progenitor strains that had been intercrossed over 30–60 generations. The group has released software called HAPPY (Mott *et al.*, 2000) that requires knowledge of the ancestral alleles in the inbred founders, together with the genotypes and phenotypes in the final generation. It applies variance component methods to test for linkage to the QTL.

10.5.1 Example: Map Manager QTX (Manley *et al.*, 2001)

Map Manager QTX is available for both MacOS and PC(Win). It has no licence fee and was selected here for the usefulness of its graphic user interface. It aids exploratory data analysis and mapping. It has both IM and CIM capability and can reformat data for use in other important software such as QTL Cartographer. Interval mapping is based on the Haley and Knott (1992) procedure, and CIM is achieved by adding background loci. Significance can be assessed by permutation (Churchill and Doerge, 1994).

The genotype data may derive from inbred or non-inbred stock, and options are provided for a variety of experimental designs. Extensive documentation can be downloaded in either pdf or Hypertext formats. The *Tutorial* is especially helpful, but readers should be aware that its files are somewhat inconspicuously tucked in with *Sample Data* files, rather than being included in the Map Manager QTX manual.

For the current example, genotype data were downloaded from the Mouse Genome Database (2001) (http://www.informatics.jax.org/). Specifically, it consists of mouse chromosome 1 genotypes from the Copeland–Jenkins backcross, and a selected subset of 10 markers spanning the entire ~100 cM length of the chromosome. Marker

En1 is located near the middle of the chromosome, between markers *Col6a3* and *D1Fcr15*, and it was used to simulate the quantitative trait (QT) for the 193 back-cross mice. Homozygotes (denoted as *b*) at *En1* received a QT value of 50 ± 20 (mean ±S.D.), while heterozygotes (*s*) at *En1* received a QT value of 100 ± 20. *En1* was then removed from the data set, and Map Manager QTX was used to analyse QT association with the remaining nine markers as shown below.

Data import

Map Manager QTX is launched by a mouse click on the Map Manager icon (*QTXb13.exe*), thus opening the main menu. The genotype data (alternatively termed 'Phenotype data' by Map Manager QTX) is imported by selecting *File* > *Import* > *Text.* The name of each marker and the genotypes (phenotypes) of the cross progeny are imported as a single line of text. The marker name is separated from the genotypes by a tab character, but the genotypes, each represented as above by a single letter, can be either given as an unbroken string of characters or space-separated. In our case, the first two lines of input therefore took the following form (with missing genotypes given by a hyphen):

Actn3<*tab*>sssbbbbbsbsbsbsssbbsbbsbbbbssbsbbsbsb-bbb-ssss<*CarriageReturn*>
Laf4<*tab*>-sbbbb--sb-------bb--bsbb-bbsb--s-bbbbbsbssb-bs<*CarriageReturn*>

Quantitative trait data are then read in from a second text file via *File* > *Import* > *Trait* Text. The format is almost identical, except that the name of the trait replaces marker name, and the trait value for each mouse must be separated from adjacent values by at least one space. Again, the name of the quantitative trait and all of the values for cross progeny must be in a single line of text.

Successful import of a text genotype file produces a small pop-up window (the *dataset* window), as shown in Figure 10.7, top left. Within it is a menu allowing selection of *Phen*, *Map*, *Stat* or *Ref.* Selecting one of these and double-clicking on a chromosome name in the *dataset* window produces the chosen window, as shown in Figure 10.7. The *Phenotype* window (top right) displays the marker names on the left side of the window, with one column for each member of the progeny. The body of the *Phenotype* window shows the genotype at each locus and also indicates locations of recombination events with an *X*. Pairs of question marks denote the possible locations of crossovers whose more precise location cannot be specified due to missing genotype data. The *Map* window (bottom left) shows a genetic *Map* with estimated distance (in cM) between markers, and the *Statistics* window (bottom right) summarizes useful numerical information, such as the number of recombination events between adjacent markers and LOD evidence for linkage.

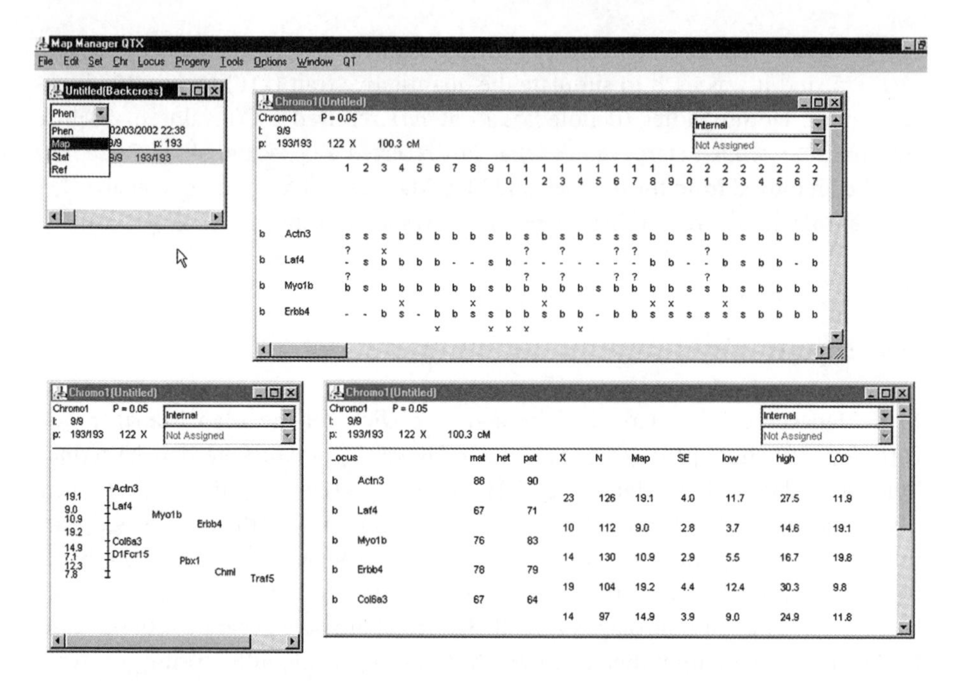

Figure 10.7 Screens in Map Manager QTX. The data-set window (upper left), the phenotype window (upper right), the map window (lower left), and the statistics window (lower right). Genotypes with permission from Mouse Genome Database (2001). Map Manager QTX, www.mapmanager.org/mmQTX.html. Described in Manly, K. F., Cudmore, Jr., R. H., Meer, J. M. (2001) Map Manager QTX, cross-platform software for genetic mapping. *Mammalian Genome* 12: 930–932

Single marker association

Testing for association between an individual marker and a quantitative trait is accomplished by first selecting a *P* value cut-off in the *Main* menu under *Options > Search&Linkage criteria*, and then choosing *QT > Links Report* in the *Main* menu. This produces a window allowing the user to select both the name of the quantitative trait to test and the background QTLs to be included in the analysis.

Figure 10.8 shows the table or *Links Report* that was produced by testing each of the nine markers in our panel for association with the simulated trait. Note that only eight markers appear in the table, as one marker did not meet the $P < 0.05$ criterion. Note also that marker *Col6a3* is highlighted as giving the strongest association and therefore as being the best marker to include as a background QTL in analyses of other chromosomal loci.

Simple interval mapping

Simple interval mapping of a QT across a series of markers is accomplished by choosing *QT > Interval Mapping* from the *Main* menu. This produces a window

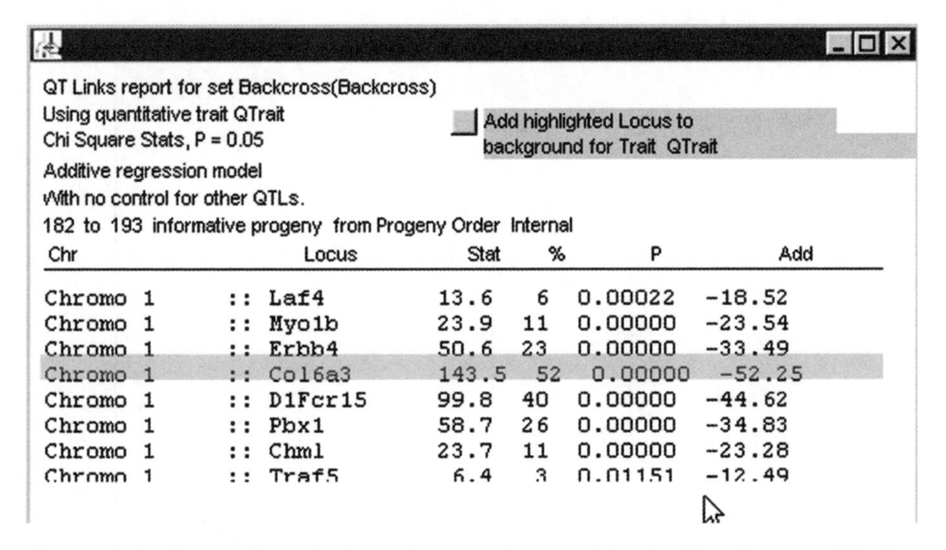

QT Links report for set Backcross(Backcross)
Using quantitative trait QTrait
Chi Square Stats, P = 0.05
Additive regression model
With no control for other QTLs.
182 to 193 informative progeny from Progeny Order Internal

Add highlighted Locus to background for Trait QTrait

Chr		Locus	Stat	%	P	Add
Chromo 1	::	Laf4	13.6	6	0.00022	-18.52
Chromo 1	::	Myo1b	23.9	11	0.00000	-23.54
Chromo 1	::	Erbb4	50.6	23	0.00000	-33.49
Chromo 1	::	Col6a3	143.5	52	0.00000	-52.25
Chromo 1	::	D1Fcr15	99.8	40	0.00000	-44.62
Chromo 1	::	Pbx1	58.7	26	0.00000	-34.83
Chromo 1	::	Chml	23.7	11	0.00000	-23.28
Chromo 1	::	Traf5	6.4	3	0.01151	-12.49

Figure 10.8 Output from single marker association testing in Map Manager QTX: The 'Links Report'. 'Add' denotes the additive regression coefficient for the association. Genotypes with permission from Mouse Genome Database (2001). Map Manager QTX, www.mapmanager.org/mmQTX.html. Described in Manly, K. F., Cudmore, Jr., R. H., Meer, J. M. (2001) Map Manager QTX, cross-platform software for genetic mapping. *Mammalian Genome* 12: 930–932

which again allows the user to specify the trait to be analysed and whether any background QTLs are to be included in the analysis. Once options in this window are specified, Map Manager QTX produces a table and a figure displaying the Interval Mapping results. Figure 10.9 shows the result of interval mapping our simulated trait across the nine markers on mouse chromosome 1. As indicated by the position of the cursor, the peak of the likelihood ratio statistic falls very close to the true location of the simulated QT locus, between markers *Col6a3* and *D1Fcr15*.

10.6 Closing remarks

This chapter provides a high-level overview of some key topics in statistical genetics with the emphasis on applications rather than theory. For the sake of brevity, some important topics have not been broached. For example, careful data management and error checking are prerequisites of the application of any of the methods described. Visualizing, summarizing and cleaning data at the outset save time and help to ensure appropriate interpretation later on. A great deal of software is available to assist in the preparation of human family-based data. A brief review is given by Almasy (2002).

The advent of the high-density SNP maps, the availability of HapMap data and advances in technology have revolutionized the way human geneticists work. Emphasis has shifted away from family-based studies and firmly toward case-control

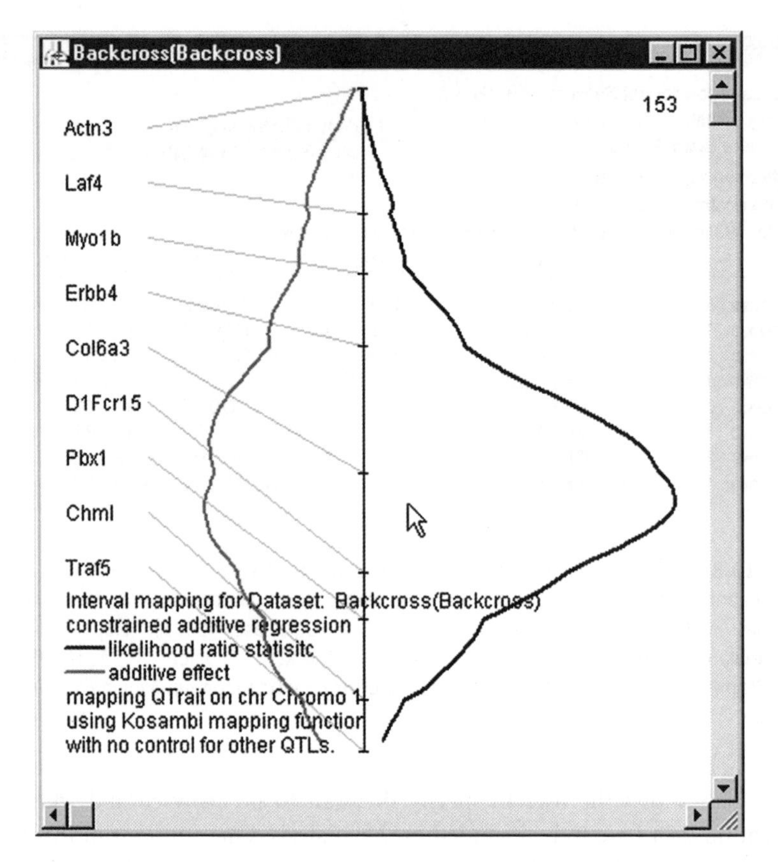

Figure 10.9 Output from Map Manager QTX. Results of interval mapping across nine markers. Genotypes with permission from Mouse Genome Database (2001). Map Manager QTX, www.mapmanager.org/mmQTX.html. Described in Manly, K. F., Cudmore, Jr., R. H., Meer, J. M. (2001) Map Manager QTX, cross-platform software for genetic mapping. *Mammalian Genome* 12: 930–932

association studies. Affymetrix, Illumina and Perlegen Sciences have broken new ground in allowing scientists to generate data on hundreds of thousands of genetic markers per subject. In response, research groups are completely re-evaluating their approach to loading, storing and analysing genetic data. Projects that used to involve thousands of data points, now involve billions of them. The importance of finding ways of distinguishing signal from noise has never been so great.

With the increasing number of markers comes an increasing number of statistical tests, and now more powerful data sets are needed. To be successful in determining the genetic basis of disease or drug response, research groups worldwide will need to collaborate openly and pool resources. The wholesale availability of the HapMap data has set a fine example, and already groups such as the Wellcome Trust are promising to generate and share further large data sets (http://www.wtccc.org.uk/). However, pooling of data should not be done blindly. Population substructure, if not properly recognized, can affect interpretation of data (e.g., Excoffier, 2001; Marchini *et al.*, 2004). In today's climate, this classical population genetics topic is gaining renewed importance, and tools to detect and address it are burgeoning (e.g., Devlin and Roeder, 1999; Pritchard *et al.*, 2000a, 2000b; Marchini *et al.*, 2005; Kohler and Bickeboller, 2006; Zheng *et al.*, 2006).

Acknowledgements

The authors wish to thank Pete Boyd, Heather Cordell, Dmitri Zaykin, Clive Bowman, Meg Ehm and Leonid Kruglyak for helpful discussions, advice and support during the writing of the first edition of this chapter.

References

Abecasis, G. R., Cardon, L. R. and Cookson, W. O. (2000). A general test of association for quantitative traits in nuclear families. *Am J Hum Genet* **66**, 279–292.

Abecasis, G. R., Cherny, S. S., Cookson, W. O. *et al.* (2002). Merlin – rapid analysis of dense genetic maps using sparse gene flow trees. *Nat Genet* **30**(1), 97–101.

Abecasis, G. R. and Cookson, W. O. C. (2000). GOLD – graphical overview of linkage disequilibrium. *Bioinformatics* **16**(2), 182–183.

Allison, D. B. (1997). Transmission-disequilibrium tests for quantitative traits. *Am J Hum Genet* **60**, 676–690.

Almasy, L. (2002). Software for genetic epidemiology. In *Biostatistical Genetics and Genetic Epidemiology*, pp. 726–736, R. Elston, J. M. Olson and L. Palmer (ed.). Chichester: Wiley.

Amos, C. I. (1994). Robust variance components approach for assessing genetic linkage in pedigrees. *Am J Hum Genet* **54**, 535–543.

Ardlie, K. G., Kruglyak, L. and Seielstad, M. (2002). Patterns of linkage disequilibrium in the human genome. *Nat Rev Genet* **3**(4), 299–309 (Erratum **3**(7), 566).

Blangero, J. and Almasy, L. (1996). SOLAR: Sequential Oligogenic Linkage Analysis Routines. Technical notes no. 6, Population Genetics Laboratory, Southwest Foundation for Biomedical Research, San Antonio, TX.

Barrett, J. C., Fry, B., Maller, J. *et al.* (2005). Haploview: analysis and visualization of LD and haplotype maps. *Bioinformatics* **21**(2), 263–265.

Basten, C., Weir, B. S. and Zeng, Z.-B. (1994). Zmap – a QTL cartographer. In *Proceedings of the 5th World Congress on Genetics Applied to Livestock Production: Computing Strategies and Software*, 22, pp. 65–66, C. Smith, J. S. Gavora, B. Benkel *et al.* (ed.).

Basten, C., Weir, B. S. and Zeng, Z.-B. (1997). *QTL Cartographer: A Reference Manual and Tutorial for QTL Mapping*. Raleigh, NC: Department of Statistics, North Carolina State University, http://statgen.ncsu.edu/qtlcart/.

Borevitz, J. O., Maloof, J. N., Lutes, J. *et al.* (2002). Quantitative trait loci controlling light and hormone response in two accessions of *Arabidopsis thaliana*. *Genetics* **160**(2), 683–696.

Chapman, C. J. (1990). A visual interface to computer programs for linkage analysis. *Am J Med Genet* **36**(2), 155–160.

Churchill, G. A. and Doerge, R. W. (1994). Empirical threshold values for quantitative trait mapping. *Genetics* **138**, 963–971.

Clark, A. G. (1990). Inference of haplotypes from PCR amplified samples of diploid populations. *Mol Biol Evol* **7**(2), 111–122.

Clark, A. G., Weiss, K. M., Nickerson, D. A. *et al.* (1998). Haplotype structure and population genetic inferences from nucleotide-sequence variation in human lipoprotein lipase. *Am J Hum Genet* **63**, 595–612.

Clayton, D. (1999). A generalization of the transmission/disequilibrium test for uncertain haplotype transmission. *Am J Hum Genet* **65**(4), 1170–1177.

Clayton, D. (2001). Population association. In *Handbook of Statistical Genetics*, pp. 519–540, D. J. Balding, M. Bishop and C. Cannings (ed.). Chichester: Wiley.

Clayton, D. and Jones, H. B. (1999). Transmission/disequilibrium tests for extended marker haplotypes. *Am J Hum Genet* **65**, 1161–1169.

Collins, A. and Morton, N. E. (1998). Mapping a disease locus by allelic association. *Proc Natl Acad Sci USA* **95**, 1741–1745.

Conneally, P. M., Edwards, J. H., Kidd, K. K. *et al.* (1985). Report of the committee on methods of linkage analysis and reporting. *Cytogenet Cell Genet* **40**, 356–359.

Cordell, H. (2002). Diabetes in the NOD mouse. In*Quantitative Trait Loci: Methods and Protocols*, pp. 165–198, N. J. Camp and A. Cox (ed.). Totowa, NJ: Humana Press.

Cottingham, R. W., Idury, R. M. and Schaffer, A. A. (1993). Fast sequential genetic linkage computation. *Am J Hum Genet* **53**, 252–263.

Cudworth, A. G. and Woodrow, J. C. (1975). Evidence for HLA-linked genes in 'juvenile' diabetes mellitus. *Br Med J* 3(5976), 133–135.

Davis, S., Schroeder., M., Goldin, L. R. *et al.* (1996). Nonparametric simulation-based statistics for detecting linkage in general pedigrees. *Am J Hum Genet* **58**, 867–880.

Dempster, A. P., Laird, N. M. and Rubin, D. B. (1977). Maximum likelihood from incomplete data via the EM algorithm. *J R Stat Soc* **B39**, 1–38.

Devlin, B. and Risch, N. (1995). A comparison of linkage disequilibrium measures for fine-scale mapping. *Genomics* **29**, 311–322.

Devlin, B., Risch, N. and Roeder, K. (1996). Disequilibrium mapping: composite likelihood for pairwise disequilibrium. *Genomics* **36**(1), 1–16.

Devlin, B. and Roeder, K. (1999). Genomic control for association studies. *Biometrics* **55**(4), 997–1004.

Dudbridge, F. (2003). Pedigree disequilibrium tests for multilocus haplotypes. *Genet Epidemiol* **25**, 115–121.

Excoffier, L. (2001). Analysis of population subdivision. In *Handbook of Statistical Genetics*, pp. 271–307, D. J. Balding, M. Bishop and C. Cannings (ed.). Chichester: Wiley.

Excoffier, L. and Slatkin, M. (1995). Maximum-likelihood estimation of molecular haplotype frequencies in a diploid population. *Mol Biol Evol* **12**(5), 921–927.

Fulker, D. W. and Cardon, L. R. (1994). A sib-pair approach to interval mapping of quantitative trait loci. *Am J Hum Genet* **54**, 1092–1103.

Fulker, D. W., Cherny, S. S., Sham, P. C. *et al.* (1999). Combined linkage and association sib-pair analysis for quantitative traits. *Am J Hum Genet* **64**, 259–267.

Gabriel, S. B., Schaffner, S. F., Nguyen, H. *et al.* (2002). The structure of haplotype blocks in the human genome. *Science* **296**, 2225–2229.

Giraudeau, F., McGinnis, R., Gray, I. *et al.* (2004). Charaacterization of common genetic variants in cathepsin K and testing for association with bone mineral density in a large cohort of perimenopausal women from Scotland. *J Bone Miner Res* **19**, 31–41.

Goldgar, D. E. (1990). Multipoint analysis of human quantitative genetic variation. *Am J Hum Genet* **47**, 957–967.

Haley, C. S. and Knott, S. A. (1992). A simple regression method for mapping quantitative trait loci in line crosses using flanking markers. *J Hered* **69**, 315–324.

Haseman, J. K. and Elston, R. C. (1972). The investigation of linkage between a quantitative trait and a marker locus. *Behav Genet* **2**, 3–19.

Hastabacka, J., de la Chapelle, A., Kaitila, I. *et al.* (1992). Linkage disequilibrium mapping in isolated founder populations: diastrophic dysplasia in Finland. *Nat* Genet **2**(3), 204–211.

Hauser, E. R. and Boehnke, M. (1997). Confirmation of linkage results in affected-sib-pair linkage analysis for complex genetic traits. *Am J Hum Genet* **61**, A278.

Hawley, M. E. and Kidd, K. K. (1995). HAPLO: a program using the EM algorithm to estimate the frequencies of multi-site haplotypes. *J Hered* **86**, 409–411.

Heath, S. (1997). Markov chain segregation and linkage analysis for oligogenic models. *Am J Hum Genet* **61**, 748–760.

Hedrick, P. W. (1987). Gametic disequilibrium measures: proceed with caution. Genetics **117**, 331–341.

Hill, W. G. and Weir, B. S. (1994). Maximum-likelihood estimation of gene location by linkage disequilibrium. *Am J Hum Genet* **54**, 705–714.

Hinds, D. and Risch, N. (1996). The ASPEX package: affected sib-pair mapping. ftp://lahmed.stanford.edu/pub/aspex.

Holmans, P. (2001). Nonparametric linkage. In *Handbook of Statistical Genetics*, pp. 487–505, D. J. Balding, M. Bishop and C. Cannings (ed.). Chichester: Wiley.

Holmans, P. and Clayton, D. (1995). Efficiency of typing unaffected relatives in an affected sib-pair linkage study with single locus and multiple tightly-linked markers. *Am J Hum Genet* **57**, 1221–1232.

Jansen, R. C. (1992). A general mixture model for mapping quantitative trait loci by using molecular markers. *Theor Appl Genet* **85**, 252–260.

Jansen, R. C. (1993). Interval mapping of multiple quantitative trait loci. *Genetics* **135**, 205–211.

Jansen, R. C. (1996). A general Monte Carlo method for mapping multiple quantitative trait loci. *Genetics* **142**, 305–311.

Jansen, R. C. and Stam, P. (1994). High resolution of quantitative traits into multiple loci via interval mapping. *Genetics* **136**, 1447–1455.

Jensen, C. S., Kong, A. and Kjaerulff, U. (1995). Blocking Gibbs sampling in very large probabilistic expert systems. *Int J Hum Comput Stud* **42**, 647–666.

Johnson, G. C., Esposito, L., Barratt, B. J. *et al.* (2001). Haplotype tagging for the identification of common disease genes. *Nat Genet* **29**(2), 233–237.

Jorde, L. B. (2000). Linkage disequilibrium and the search for complex disease genes. *Genome Res* **10**(10), 1435–1444.

Kao, C. H., Zeng, Z.-B. and Teasdale, R. D. (1999). Multiple interval mapping for quantitative trait loci. *Genetics* **152**(3), 1203–1216.

Kaplan, N., Hill, W. G. and Weir, B. S. (1995). Likelihood methods for locating disease genes in nonequilibrium populations. *Am J Hum Genet* **56**, 18–32.

Kohler, K. and Bickeboller, H. (2006). Case-control association tests correcting for population stratification. *Ann Hum Genet* **70**(1), 98–115.

Kong, A. and Cox, N. J. (1997). Allele sharing models: LOD scores and accurate linkage tests. *Am J Hum Genet* **61**, 1179–1188.

Kruglyak, L., Daly, M. J., Reeve-Daly, M. P. *et al.* (1996). Parametric and nonparametric linkage analysis: a unified multipoint approach. *Am J Hum Genet* **58**, 1347–1363.

Kruglyak, L. and Lander, E. S. (1995). Complete multipoint sib-pair analysis of qualitative and quantitative traits. *Am J Hum Genet* **57**, 439–454.

Laird, N. M. (2004). Family based tests for associating haplotypes with general phenotype data: application to asthma genetics. *Genet Epidemiol* **26**(1), 61–69.

Lander, E. S. and Botstein, D. (1989). Mapping mendelian factors underlying quantitative traits using RFLP linkage maps. *Genetics* **121**(1), 185–199.

Lander, E. S., Green, P., Abrahamson, J. *et al.* (1987). MAPMAKER: an interactive computer package for constructing primary genetic linkage maps of experimental and natural populations. *Genomics* **1**(2), 174–181.

Lathrop, G. M., Lalouel, J. M., Julier, C. *et al.* (1984). Strategies for multilocus linkage analysis in humans. *Proc Natl Acad Sci U S A* 81(11), 3443–3446.

Lehesjoki, A.-E., Koskiniemi, M., Norio, R. *et al.* (1993). Localization of the EPM1 gene for progressive myoclonus epilepsy on chromosome 21: linkage disequilibrium allows high resolution mapping. *Hum Mol Genet* **2**, 1229–1234.

Lewis, P. O. and Zaykin, D. (2001). Genetic Data Analysis: Computer Program for the Analysis of Allelic Data. Version 1.0 (d16c). Free program distributed by the authors over Internet: http://lewis.eeb.uconn.edu/lewishome/software.html.

Lewontin, R. C. (1964). The interaction of selection and linkage. I. General considerations; heterotic models. *Genetics* **49**, 49–67.

Little, R. J. A. and Rubin, D. B. (1987). *Statistical Analysis with Missing Data.* New York: Wiley.

Long, J. C., Williams, R. C. and Urbanek, M. (1995). An E-M algorithm and testing strategy for multiple locus haplotypes. *Am J Hum Genet* **56**, 799–810.

Manly, K. F., Cudmore, R. H. and Meer, J. M. (2001). Map Manager QTX, cross-platform software for genetic mapping. *Mamm Genome* **12**(12), 930–932.

Manly, K. F and Olson, J. M. (1999). Overview of QTL mapping software and introduction to map manager QT. *Mamm Genome* **10**(4), 327–334.

Marchini, J., Cardon, L. R., Philips, M. S. *et al.* (2004). The effects of human population structure on large scale genetic association studies. *Nat Genet* **36**(11), 1129–1130.

Marchini, J., Donnelly, P. and Cardon, L. R. (2005). Genome-wide strategies for detecting multiple loci that influence complex diseases. *Nat Genet* **37**(4), 337–338.

Martinez, O. and Curnow, R. N. (1992). Estimating the locations and the sizes of the effects of quantitative trait loci using flanking markers. *Theor Appl Genet* **85**, 480–488.

Monks, S. A., Kaplan, N. L. and Weir, B. S. (1998). A comparative study of sibship tests of linkage and/or association. *Am J Hum Genet* **63**(5), 1507–1516.

Morton, N. E. (1955). Sequential tests for the detection of linkage. *Am J Hum Genet* **7**, 277–318.

Mott, R., Talbot, C., Turri, M. *et al.* (2000). A new method for fine mapping quantitative trait loci in outbred animal stocks. *Proc Natl Acad Sci U S A* **97**, 12649–12654.

Mouse Genome Database (MGD) (2001). Mouse Genome Informatics Web Site, Jackson Laboratory, Bar Harbor, Maine. World Wide Web (URL: http://www.informatics.jax.org/).

Nickerson, D. A., Taylor, S. L., Weiss, K. M. *et al.* (1998). DNA sequence diversity in a 9.7-kb region of the human lipoprotein lipase gene. *Nat Genet* **19**, 233–240.

O'Connell, J. R. and Weeks, D. E. (1995). The VITESSE algorithm for rapid exact multilocus linkage analysis via genotype set-recoding and fuzzy inheritance. *Nat* Genet **11**, 402–408.

Olson, J. M. (1995). Multipoint linkage analysis using sib pairs: an interval mapping approach for dichotomous outcomes. *Am J Hum Genet* **56**, 788–798.

Pettersson, F., Jonsson, O. and Cardon, L. R. (2004). GOLDsurfer: three dimensional display of linkage disequilibrium. *Bioinformatics* **20**, 3241–3243.

Pritchard, J. K. and Przeworski, M. (2001). Linkage disequilibrium in humans: models and data. *Am J Hum Genet* **69**(1), 1–14.

Pritchard, J. K., Stephens, M. and Donnelly, P. (2000). Inference of population structure using multilocus genotype data. *Genetics* **155**, 945–959.

Pritchard, J. K., Stephens, M., Rosenberg, N. A. *et al.* (2000). Association mapping in structured populations. *Am J Hum Genet* **67**, 170–181.

Qin, Z. S., Niu, T. and Liu, J. S. (2002). Partition-ligation expectation-maximisation algorithm for haplotype inference with single nucleotide polymorphisms. *Am J Hum Genet* **71**, 1242–1247.

Rabinowitz, D. (1997). A transmission disequilibrium test for quantitative trait loci. *Hum Hered* **47**, 342–350.

Rubenstein, P., Walker, M., Carpenter, C. *et al.* (1981). Genetics of HLA disease associations. The use of the haplotype relative risk (HRR) and the 'haplo-delta' (Dh) estimates in juvenile diabetes from three racial groups. *Hum Immunol* **3**, 384 (abstr.).

Satagopan, J. M., Yandell, B. S., Newton, M. A. *et al.* (1996). A Bayesian approach to detect quantitative trait loci using Markov chain Monte Carlo. *Genetics* **144**, 805–816.

SAGE (1999). Statistical Analysis for Genetic Epidemiology, Release 4.0. Department of Epidemiology and Biostatistics, Rammelkamp Center for Education and Research, MetroHealth campus, Case Western Reserve University, Cleveland, OH.

Schaid, D. J. (1996). General score tests for associations of genetic markers with disease using cases and their parents. *Genet Epidemiol* **13**, 423–449.

Schneider, S., Roessli, D. and Excoffier, L. (2000). Arlequin ver. 2.000: a software for population genetics data analysis. Genetics and Biometry Laboratory, University of Geneva, Switzerland.

Sham, P. C. (1998). Statistics in Human Genetics. Arnold Publishers, London; John Wiley and Sons Inc, New York.

Sham, P. C. and Curtis, D. (1995). An extended transmission/disequilibrium test (TDT) for multi-allele marker loci. *Ann Hum Genet* **59**, 323–336.

Sillanpaa, M. J. (1998). *Multimapper Reference Manual.* http://www.RNL.Helsinki.F1/~mjs/.

Sillanpaa, M. J. and Arjas, E. (1998). Bayesian mapping of multiple quantitative trait loci from incomplete inbred line cross data. *Genetics* **148**, 1373–1388.

Slatkin, M. and Excoffier, L. (1996). Testing for linkage disequilibrium in genotypic data using the EM algorithm. *Heredity* **76**, 377–383.

Sobel, E. and Lange, K. (1996). Descent graphs in pedigree analysis: applications to haplotyping, location scores, and marker sharing statistics. *Am J Hum Genet* **58**, 1323–1337.

Spielman, R. S. and Ewens, W. J. (1996). The TDT and other family-based tests for linkage disequilibrium and association. *Am J Hum Genet* **59**, 983–989.

Spielman, R. S. and Ewens, W. J. (1998). A sibship test for linkage in the presence of association: the sib-transmission/disequilibrium test. *Am J Hum Genet* **62**, 450–458.

Spielman, R. S., McGinnis, R. E. and Ewens, W. J. (1993). Transmission test for linkage disequilibrium: the insulin gene region and insulin-dependent diabetes mellitus. *Am J Hum Genet* **52**, 506–516.

Talbot, C. J., Nicod, A., Cherny, S. S. *et al.* (1999). High-resolution mapping of quantitative trait loci in outbred mice. *Nat Genet* **21**, 305–308.

Terwilliger, J. D. (1995). A powerful likelihood method for the analysis of linkage disequilibrium between trait loci and one or more polymorphic marker loci. *Am J Hum Genet* **56**, 777–787.

Terwilliger, J. D. (1996). Program SIBPAIR – sib pair analysis on nuclear families. ftp://linkage.cpmc.columbia.edu.

Terwilliger, J. D. and Ott, J. (1994). *Handbook of Human Genetic Linkage.* Baltimore, MD: Johns Hopkins University Press.

Uimari, P., Thaller, G. and Hoeschele, I. (1996). The use of multiple markers in a Bayesian method for mapping quantitative trait loci. *Genetics* **143**, 1831–1842.

Utz, H. F. and Melchinger, A. E. (1996) PLABQTL: a program for composite interval mapping of QTL. *J Quant Trait Loci* **2**, http://probe.nalusda.gove:8000/otherdocs/jqtl/.

Van Ooijen, J. W. and Maliepaard, C. (1996a). MapQTL Version 3.0: software for the calculation of QTL positions on genetic maps. *Plant Genome* IV abstracts. http://probe.nalusda.gov:3000/otherdocs/pg/pg4/abstracts/p316.html.

Van Ooijen, J. W. and Maliepaard, C. (1996b). *MapQTL Version 3.0: Software for the Calculation of QTL Positions on Genetic Maps.* CPRO-DLO, Wageningen.

Wall, J. D. and Pritchard, J. K. (2003). Haplotype blocks and linkage disequilibrium in the human genome. *Nat Rev Genet* **4**(8), 587–597.

Weeks, D. E. and Lange, K. (1988). The affected-pedigree-member method of linkage analysis. *Am J Hum Genet* **42**, 315–326.

Weeks, D. E., Sobel, E., O'Connell, J. R. *et al.* (1995). Computer programs for multilocus haplotyping of general pedigrees. *Am J Hum Genet* **56**, 1506–1507.

Weir, B. S. (1996). *Genetic Data Analysis II.* Sunderland, MA: Sinauer Associates.

Whittemore, A. S. and Hapern, J. (1994). A class of tests of linkage using affected pedigree members. *Biometrics* **50**, 118–127.

Xie, X. and Ott, J. (1993). Testing linkage disequilibrium between a disease gene and marker loci. *Am J Hum Genet* **53**, 1107.

Zeng, Z.-B. (1993). Theoretical basis for separation of multiple linked gene effects in mapping quantitative trait loci. *Proc Natl Acad Sci USA* **90**, 10972–10976.

Zeng, Z.-B. (1994). Precision mapping of quantitative trait loci. *Genetics* **136**, 1457–1468.

Zeng, Z.-B., Kao, C. H. and Basten, C. J. (1999). Estimating the genetic architecture of quantitative traits. *Genet Res* **74**, 279–289.

Zhao, J. H., Curtis, D. and Sham, P. C. (2000). Model-free analysis and permutation tests for allelic associations. *Hum Hered* **50**, 133–139.

Zheng. G., Freidlin, B. and Gastwirth, J. L. (2006). Robust genomic control for association studies. *Am J Hum Genet* **78**(2), 350–356.

Section IV

Moving from Associated Genes to Disease Alleles

11 Predictive Functional Analysis of Polymorphisms: An Overview

Mary Plumpton[1] and Michael R. Barnes[2]

Bioinformatics, GlaxoSmithKline Pharmaceuticals, [1]Stevenage, Hertfordshire, UK, [2]Harlow, Essex, UK

11.1 Introduction

Human diseases with a strong genetic component are generally characterized by a profound range of phenotypic variability manifested in variable age of onset, severity, organ-specific pathology and response to drug therapy. The causes underlying this variability are likely to be diverse, influenced by differing levels of genetic and environmental modifiers. The vast majority of human genetic variants are likely to be neutral in effect, but some may cause or modify disease phenotypes. The challenge for bioinformatics is to identify the genetic variants that are most likely to show a non-neutral allelic effect. Geneticists studying complex disease are already seeking to identify these genetic determinants by genetic association of phenotypes with markers. The literature is now replete with reported associations, but moving from associated marker to disease allele is proving to be very difficult. So why are we so unsuccessful in making this transition? If we disregard false-positive associations (see Chapter 9), it may be that the diverse and subtle effects of genetic variation are helping disease alleles to elude us. Genetic variation can cause disease at any number of stages between promotion of gene transcription and post-translational modification of protein products. Many geneticists have chosen to focus their efforts on

Bioinformatics for Geneticists, Second Edition. Edited by Michael R. Barnes
 ISBN 978-0-470-02619-9 (HB) ISBN 978-0-470-02620-5 (PB)

the most obvious form of variation – non-synonymous coding variation in genes. While this category of variation is undoubtedly likely to contribute considerably to human disease, this may overlook many equally important categories of variation in the genome, namely, the effects of variation on gene transcription, temporal and spatial expression, transcript stability, and splicing.

Clearly, not all polymorphisms are equal. Analysis of polymorphism distribution across the human genome shows significant variations in polymorphism density and allele frequency distribution. Chakravarti (1999) showed an immediate difference between the density of SNPs in exonic regions and intragenic and intronic regions. SNPs occurred at average intervals of 1.2 kb in coding regions and 0.9 kb in intragenic and intronic regions. These differences point to different selection intensities in the genome, particularly in protein-coding regions, where SNPs may result in alteration of amino-acid sequences (non-synonymous SNPs (nsSNPs)) or the alteration of gene-regulatory sequences. These observations are intuitive – obviously, natural selection is likely to be strongest across gene regions, essentially encapsulating the objective of genetics – to identify non-neutral alleles with a role in disease.

So how should we go about identifying disease alleles? One approach used to identify disease mutations is direct screening of good candidate genes for mutations present in affected, but not unaffected, family members. This approach is very useful in the study of monogenic diseases and cancers, where transmission of the disease allele can generally be demonstrated to be restricted to affected individuals/tissues. But in complex diseases, the odds of identifying disease alleles by population screening of candidate genes would seem to be very high, and proving their role is problematic, as disease alleles are likely to be present in cases and controls. Instead, we detect common marker alleles in LD with rarer disease alleles. This methodical approach to disease gene hunting localizes disease alleles rather than identifying them directly, and the next step is to identify the disease allele from a range of alleles in LD with the associated marker. For conclusive identification of this allele, a functional mechanism for the allele in the disease needs to be identified.

11.1.1 Moving from associated genes to disease genes

Many potential associations have been reported between markers and disease phenotypes. Aside from the potential for false-positive association, magnitude of effect in complex disease is also a problem. There may be a few gene variants with major effects, but, generally, complex disease is very heterogeneous and polygenic; it therefore follows that studies of single gene variants will be inconclusive and inconsistent – this is just something we have to work with. We may also find a bewildering array of complex disease genes with somewhat indirect roles in disease, such as modifier genes and redundant genes that have many effects on phenotype. Understanding the mode of action of these associated alleles will help in determining how susceptibility genes may give rise to a multifactorial phenotype. Bioinformatics may be critical in this process. Follow-up studies need to be designed to ask the right questions, to get the right candidates tested, and to confirm the biological role of positive associations. It may also be necessary to attempt to characterize polymorphisms with a potential functional impact, to help to identify the molecular mechanisms by a combination of bioinformatics and laboratory follow-up. Many of these informatics approaches are similar to the approaches originally used to identify candidates, but, of necessity, these analyses benefit from a far more detailed approach as in depth analyses transfer to in-depth laboratory investigation.

Moving from an 'associated gene' to a 'disease gene' is not a purely academic objective. Genetics may sometimes be our only insight into the nature of a disease. Such insights may help to develop therapies that restore the normal function of disease genes in patients, or, better still, they may help prevent disease in the first place. Better diagnosis and treatments are also prospects afforded by better understanding of the pathology of disease. A validated 'disease gene' is one of the most tangible steps toward this end.

11.1.2 Candidate polymorphisms

To turn the arguments for association analysis on their head, there is also a theory that the direct identification of disease alleles may not be entirely futile. The common disease/common variant (CD/CV) hypothesis predicts that the genetic risk of common diseases is often due to disease-predisposing alleles with relatively high frequencies (Reich and Lander, 2001). There is not currently enough evidence to prove or disprove this hypothesis; however, several examples of common disease variants have been identified, some of which are listed in Table 11.1, and the allele frequency of these variants in the public databases is also listed.

The possibility that many disease alleles are common presents an intriguing challenge to genetics (and bioinformatics); if the CD/CV hypothesis prevails, a substantial number of disease alleles may already be present in polymorphism databases such as dbSNP, and they may be characterized in the HapMap. These might be termed 'candidate polymorphisms'. To extend this idea, just as genes with a putative biological role in disease are often prioritized for genetic association analysis, 'candidate polymorphisms' can be prioritized by predicted effect on the structure and function

Table 11.1 Disease alleles supporting the common disease/common variant hypothesis

Gene (allele)	Minor allele freq. (Caucasian)	Disease/trait association	OMIM review
APOE ε4	16 per cent (14 per cent)	Alzheimer's disease and cardiovascular disease	107741
Factor V^{leiden} R506Q	2–7 per cent (ND)	Deep vein thrombosis	227400
KCNJ11 E23K	14 per cent (25 per cent)	Type II diabetes	600937
COMT V158M	0.1–62 per cent (45 per cent)	Catechol drug pharmacogenetics	116990
PPARG Pro12Ala	88 per cent (Pro12)	Type II diabetes	601487
CARD15 3020C ins	4–8 per cent	Crohn's disease	605956

of regulatory regions, genes, transcripts or proteins. Thus, selection of candidate polymorphisms is an extension of the candidate gene selection process – but in this case, a link needs to be established between a predicted functional allelic effect and a target phenotype. As discussed earlier, DNA polymorphism can affect almost any biological process. Much of the literature in this area has focused on the most obvious form of variation – non-synonymous changes in coding regions of genes. Alterations in amino-acid sequences have accounted for a great number of Mendelian diseases. Coding variants may affect protein folding, active sites, protein–protein interactions, protein solubility or stability. But the effects of DNA polymorphism are by no means restricted to coding regions; variants in regulatory regions may alter the consensus of transcription factor-binding sites or promoter elements, variants in the untranslated regions (UTR) of mRNA may alter mRNA stability, and variants in the introns and silent variants in exons may alter splicing efficiency.

Approaches to evaluating the potential functional effects of DNA polymorphisms are almost limitless, but there are very few tools designed specifically for this task. Instead, almost any bioinformatics tool that makes a prediction based on a DNA or protein sequence can be commandeered to analyse polymorphisms – simply by analysing wild-type and mutant sequences and looking for an alteration in predicted outcome by the tool. Polymorphisms can also be evaluated at a simple level by looking at physical considerations of the properties of genes and proteins, or they can be evaluated in the context of a variant within a family of homologous or orthologous genes or proteins.

11.2 Principles of predictive functional analysis of polymorphisms

Faced with the diversity of disease phenotypes, analysis of polymorphism data calls for equally diverse methods, to assess functional effects that might cause these phenotypes. The complex arrangements that regulate gene transcription, translation and function are all potential mechanisms through which disease could act; therefore, analysis of potential disease alleles needs to evaluate almost every eventuality. Figure 11.1 illustrates the logical decision-making process that needs to be applied to the analysis of polymorphisms and mutations. The tools and approaches for the analysis of variation are completely dependent on the location of the variant within a gene or regulatory region. Many of these questions can be answered very quickly with genomic viewers such as Ensembl or the UCSC human genome browser (see Chapter 4 for a tutorial on these tools). Placing a polymorphism in full genomic context is useful to evaluate variants in terms of location within or near genes (exonic, coding, UTR, intronic, promoter region) and other functionally significant features, such as CpG islands, repeat regions or recombination hotspots. Once approximate localization is achieved, specific questions need to be asked to place the polymorphism in a specific genic or intergenic region. This will help to narrow down the potential range of functional effects attributable to a variant, and this will, in turn,

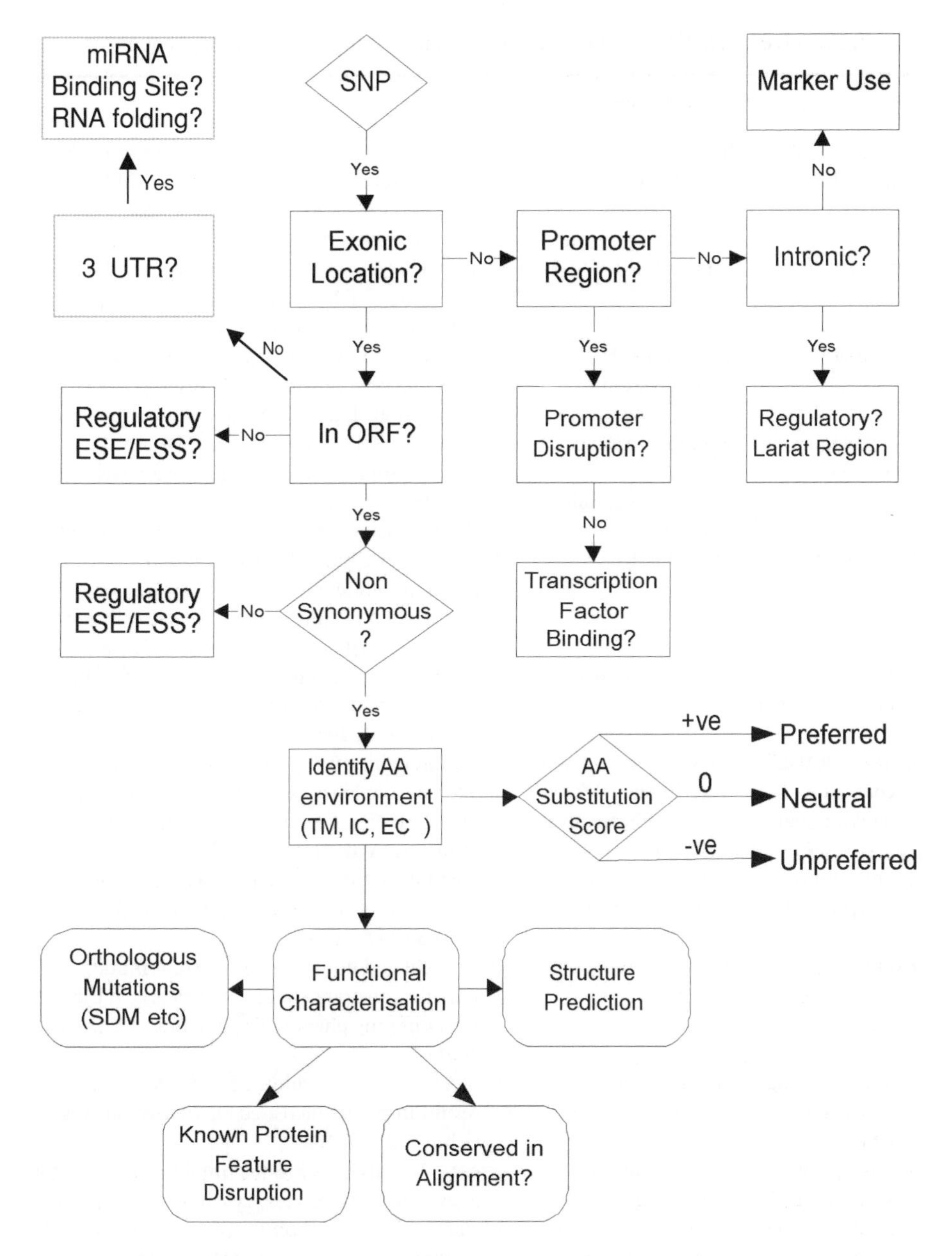

Figure 11.1 A decision tree for polymorphism analysis

help to identify the appropriate laboratory follow-up approach to evaluate function. Table 11.2 illustrates some carefully selected examples of non-coding polymorphisms in genes and transcripts; these publications were specifically selected, as each also includes a detailed laboratory based follow-up to evaluate each form of polymorphism. We refer the reader to these publications as a potential guide to assist in laboratory investigation.

Table 11.2 Functional polymorphisms in genes and gene regulatory sequences

Location	Gene/disease	Mechanism
Transcription factor binding	TNF in cerebral malaria	–376A SNP introduces OCT1 binding site altering TNF expression, associated with fourfold increased susceptibility to cerebral malaria (Knight *et al.*, 1999)
Promoter	CYP2D6	Common –48T > G substitution disrupts the TATA box of the CYP2D6 promoter, causing 50 per cent reduction in expression (Pitarque *et al.*, 2001)
Promoter	RANTES in HIV progression	–28G mutation increases transcription of the RANTES gene slowing HIV-1 disease progression (Liu *et al.*, 1999)
cis-regulatory element	Bruton's tyrosine kinasein X-linked agammaglobuli-naemia	+5G/A (intron1) shows reduced BTK transcriptional activity, suggesting a novel *cis*-acting element, involved in BTK downregulation, but not splicing (Jo *et al.*, 2001)
Lariat region	HNF-4alpha	NIDDM associated C/T substitution in polypyrimidine tract in intron 1b in an important *cis*-acting element directing intron removal (lariat region) (Sakurai *et al.*, 2000)
Splice donor/acceptor sites	ATP7A in Menke's disease	Mutation in donor splice site of exon 6 of ATP7A causes a lethal disorder of copper metabolism (Moller *et al.*, 2000)
Intronic 'RAGU' consensus flanking splice donor site	Neurofibrimin 1 gene (NF1) in neurofibromatosis type 1	Mutation in intron 3, position 5 G > C causes exon 3 to be completely skipped. Splice mutations identified with genomic DNA samples and a minigene assay (Barralle *et al.*, 2003)
Cryptic donor/ acceptor sites	β-Glucuronidase gene (GUSB) in MPS VII	A 2-bp intronic deletion creates a new donor splice site activating a cryptic exon in intron 8 (Vervoort *et al.*, 1998)
Exonic splicing enhancers (ESE)	BRCA1 in breast cancer	Both silent and nonsense exonic point mutations were demonstrated to disrupt splicing in BRCA1, with differing phenotypic penetrance (Liu *et al.*, 2001)
Intronic splicing enhancers (ISE)	α-Galactosidase in Fabry's disease	G > A transversion within 4 bp of splice acceptor results in greatly increased alternative splicing (Ishii *et al.*, 2002)
Exonic splicing silencers (ESS)	CD45 in multiple sclerosis	Silent C77G disrupts ESS that inhibits the use of the 5′ exon four splice sites (Lynch and Weiss, 2001)
Intronic splicing silencers (ISS)	TAU in dementia with parkinsonism	Mutations in TAU intron 11 ISS cause disease by altering Exon 10 splicing (D'Souza and Schellenberg, 2000)
Polyadenylation signal	FOXP3 in IPEX syndrome	A > G transition within the polyadenylation signal leads to unstable mRNA with 5.1 kb extra UTR (Bennett *et al.*, 2001)

11.2.1 A decision tree for polymorphism analysis

The first step in our decision tree for polymorphism analysis (Figure 11.1) is a simple question – is the polymorphism located in an exon? Answering this accurately may not always be simple or even possible with only *in silico* resources. As we have already seen in the previous section, delineation of genes is really the key step in all subsequent analyses; once we know the location of a gene, all other functional elements fall into place from their location in and around genes. As will be described in Section 11.3, the art of delineating genes must include methods for extending sequences to identify the true boundaries of a gene, not just its coding region. This activity may seem superfluous in the 'post-genome' era, but, in fact, we still know very little about the full diversity of genes, and the vast majority of genes are still incompletely characterized. Gene prediction and gene cloning have generally focused on the open reading frame – the protein-coding sequence (ORF/CDS) of genes. For the most part, UTR sequences have been neglected in the rush to find an open reading frame (ORF) and a protein. In the case of polymorphism analysis, these sequences should not be overlooked, as the extreme 5′ and 3′ limits of UTR sequence delineate the true boundaries of genes. This delineation of gene boundaries is illustrated in a canonical gene model in Figure 11.2. As the model shows, most of the known regulatory elements in genes are localized to specific regions by the location of the exons. For example, the promoter region is generally located in a 1–2-kb region immediately upstream of the 5′ UTR, and splice regulatory elements flank intron/exon boundaries.

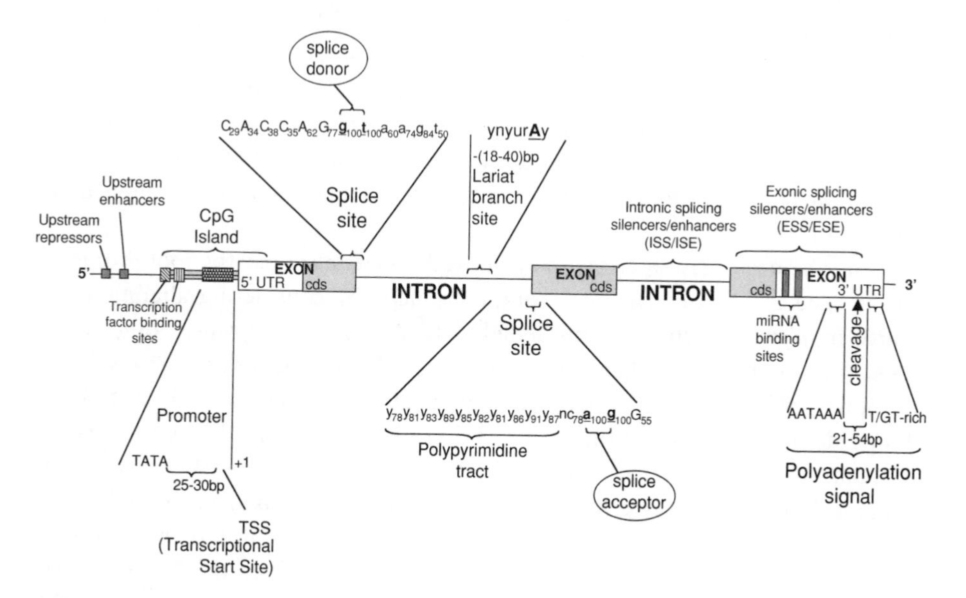

Figure 11.2 The anatomy of a gene. This figure illustrates some of the key regulatory regions that control the transcription, splicing and post-transcriptional processing of genes and transcripts. Polymorphisms in these regions should be investigated for functional effects

Many of these regulatory regions were first identified in Mendelian disorders, and now some are being identified in complex phenotypes. Table 11.2 lists some of the disease mutations and polymorphisms that have helped to shape our knowledge of this complex area.

11.3 The anatomy of promoter regions and regulatory elements

Prediction of eukaryotic promoters from genomic sequence remains one of the most challenging tasks for bioinformatics (Bajic *et al.*, 2004). The biggest problem is over-prediction; current methods, on average, predict promoter elements at 1-kb intervals across a given genomic sequence. This is in stark constrast to the estimated average 40–50-kb distance of functional promoters in the human genome (Reese *et al.*, 2000). Although it is possible that some of these predicted promoter elements may be used cryptically, the vast majority of predictions are likely to be false positives. To avoid false predictions, it is essential to provide promoter prediction tools with the appropriate sequence region, that is, the region located immediately upstream of the gene transcriptional start site (TSS). Accurate TSS definition has been hampered by the fact that the vast majority of cDNA-derived mRNAs are truncated at the 5′ end. However, through experimental enrichment of full-length 'capped' 5′ clones, the TSS(s) and corresponding promoter sequences for 8308 human genes are now available from the Database of Transcription Start Sites (DBTSS) (Table 11.3; Yamashita *et al.*, 2006). Smaller experimentally derived and verified TSS/promoter data sets are also now available. The Eukaryotic Promoter Database (EPD) provides sequences for ~2000 experimentally defined promoters (Schmid *et al.*, 2004). Trinklein *et al.* (2004) identified 1300 bidirectional promoters located between gene pairs arranged head to head on opposite strands and separated by less than 1000 bp; 90 per cent of the bidirectional promoters tested in a reporter assay showed significant activity over negative controls.

In the absence of an experimentally derived TSS data source, the tool Promoser can be used to identify TSS sites computationally by considering alignments of a large number of partial and full-length mRNA and EST sequences to genomic DNA, with provision for alternative promoters. The rewards of applying accurate promoter prediction to the functional analysis of SNPs is illustrated in a study by Hoogendoorn *et al.* (2003); one-third of SNPs located within the 500-bp region upstream of experimentally verified TSSs (from the EPD data set) affected transcription levels by 50 per cent or more in transient cell-based promoter assays. Functional deletion analysis of 45 experimentally validated promoters from ENCODE regions showed that the sequence –300 to –50 bp from the TSS positively contributed to core promoter activity (Cooper *et al.* 2006). In addition, putative negative elements were identified –1000 to –500 bp upstream of the TSS for 55 per cent of genes tested, suggesting that it may be important to analyse 2 kb or

Table 11.3 Tools for functional analysis of gene regulation and splicing

Experimentally determined TSS/promoter resources	
EPD	http://www.epd.isb-sib.ch/
DBTSS	http://dbtss.hgc.jp/
Transcription start site (TSS) / promoter prediction	
Promoser	http://biowulf.bu.edu/zlab/PromoSer/
First Exon Finder	http://rulai.cshl.org/tools/FirstEF/
Promoter 2.0	http://www.cbs.dtu.dk/services/Promoter/
NNPP	http://www.fruitfly.org/seq_tools/promoter.html
Transcription factor-binding site prediction	
ConSite	http://mordor.cgb.ki.se/cgi-bin/CONSITE/consite
TFSEARCH	http://www.cbrc.jp/research/db/TFSEARCH.html
Cis-regulatory module prediction	
PreMod	http://genomequebec.mcgill.ca/PReMod/welcome.do
Other DNA and mRNA regulatory elements	
UTRdb	http://bighost.area.ba.cnr.it/BIG/UTRHome/
ESE finder	http://exon.cshl.org/ESE/
Rescue ESE	http://genes.mit.edu/burgelab/rescue-ese/
miRBase	http://microrna.sanger.ac.uk/targets/v2/
Detection of novel regulatory elements and comparative genome analysis	
PipMaker	http://bio.cse.psu.edu/pipmaker/
TRES	http://bioportal.bic.nus.edu.sg/tres/
Regulatory Vista	http://www-gsd.lbl.gov/vista/rVistaInput.html
Integrated platforms for gene, promoter and splice site prediction	
TRED	http://rulai.cshl.edu/cgi-bin/TRED/tred.cgi?process=home
Webgene	http://www.itba.mi.cnr.it/webgene/
NNPP, SPLICE, Genie	http://www.fruitfly.org/seq_tools/

more upstream, particularly when the full extent of the 5′ UTR or TSS is not well defined.

Once a potential TSS has been identified, many tools can be applied to identify promoter elements and transcription factor-binding sites. The human genome browsers (UCSC and Ensembl) are the single most valuable resources for the analysis of promoters and regulatory elements. Specifically, Ensembl annotates putative promoter regions with the Eponine tool. The UCSC browser annotates transcription factor-binding sites that fall within human/mouse/rat conserved regions. This is a valuable confirmation of potential functional conservation; a binding site is considered to be conserved across the alignment if its score meets the threshold score for that binding site in all three species, thereby reducing the notorious risk of false-positive TFBS prediction. In higher eukaryotes, transcription factors rarely operate by themselves, but bind DNA in cooperation with other factors within a cluster known as a *cis*-regulatory module (CRM). Very recently, the first genome-wide map of predicted CRMs formatted for direct loading into the UCSC genome browser for intersection with SNPs has become available (Blanchette *et al.*, 2006) (PreMod, Table 11.3). This is a valuable data set that can increase confidence in the prediction of SNPs likely to alter regulation of gene expression.

These are very useful for rapid evaluation of the location of variants in relation to these features, although these data must be used with caution, as whole-genome analyses may over-predict or overlook evidence for alternative gene models. The approaches for promoter and transcription factor-binding site analysis are reviewed thoroughly in Chapter 12.

Characterization of gene promoters and regulatory regions is not only valuable for functional analysis of polymorphisms, but it can also provide important information about the regulatory cues that govern the expression of a gene. This may be valuable for pathway expansion to assist in the elucidation of the function of candidate genes and disease-associated genes.

11.4 The anatomy of genes

11.4.1 Gene splicing

Alternative splicing is an important mechanism for regulation of gene expression, expanding the coding capacity of a single gene to allow production of different protein isoforms, which can have very different functions. The completion of the human genome draft has given an interesting new insight into this form of gene regulation. Despite initial estimates of a human gene complement of >100 000 genes, direct analysis of the sequence suggests that man may only have 25 000–30 000 genes, which is only a two- to threefold gene increase over invertebrates. Indeed, a genome-wide survey of human alternative pre-mRNA splicing by exon-junction microarrays indicates that at least 74 per cent of human multiexon genes are alternatively spliced (Johnson *et al.*, 2003). This highlights the significance of post-transcriptional modifications, such as alternative splicing, as an alternative means to express the full phenotypic complexity of vertebrates without a very large number of genes.

11.4.2 Splicing mechanisms, human disease and functional analysis

Regulation of splicing is mediated by the spliceosome, a network of small nuclear ribonucleoprotein (snRNP) complexes and members of the serine/arginine-rich (SR) protein family. At its most basic level, pre-mRNA splicing involves precise removal of introns to form mature mRNA with an intact ORF. Correct splicing requires exon recognition with accurate cleavage and rejoining at the exon boundaries designated by the invariant intronic GT and AG dinucleotides, respectively known as the splice donor and splice acceptor sites (Figure 11.2). Other more variable consensus motifs have been identified in locations adjacent to the donor and acceptor sites. These include a weak exonic 'CACCAG' consensus flanking the splice donor site, an intronic 'RAGU' consensus 3′ of the donor site, an intronic polypyrimidine (Y: C or T)-rich tract flanking the splice acceptor site, and a weakly conserved intronic 'YNYURAY' consensus 18–40 bp from the acceptor site, which acts as a branch site for lariat formation (Figure 11.2). Other regulatory motifs are known to be involved in splicing, including exonic splicing enhancers (ESE) and intronic splicing enhancers (ISE), both of which promote exon recognition. Exonic and intronic splicing silencers (ESS and ISS respectively) have an opposite action, inhibiting the recognition of exons. DNA recognition motifs for splicing enhancers and silencers are generally quite degenerate. The degeneracy of these consensus recognition motifs points to quite promiscuous binding by SR proteins. These interactions can also explain the use of alternative and inefficient splice sites, which may be influenced by competitive binding of SR proteins and hnRNP determined by the relative ratio of hnRNP to SR proteins in the nucleus. A natural stimulus that influences the ratio of these proteins is genotoxic stress, which can lead to the often observed phenomenon of differential splicing in tumours and other disease states (Hastings and Krainer, 2001).

Mutations affecting mRNA splicing are a common cause of Mendelian disorders, and 10–15 per cent of Mendelian disease mutations affect pre-mRNA splicing (Human Gene Mutation Database, Cardiff, UK). These mutations can be divided into two subclasses, according to their position and effect on the splicing pattern. Subclass I (60 per cent of the splicing mutations) includes mutations in the invariant splice-site sequences, which completely abolish exon recognition. Subclass II includes mutations in the variant motifs, which can lead to both aberrantly and correctly spliced transcripts, by either weakening or strengthening exon-recognition motifs. Subclass II also includes intronic mutations, which generate cryptic donor or acceptor sites and can lead to partial inclusion of intronic sequences. These Mendelian disease mutations have helped to define our understanding of splicing mechanisms, and in view of the proven complexity of splicing in the human genome (Lander *et al.*, 2001), it seems reasonable to expect splicing abnormality to play a significant role in complex diseases, but examples are rare. This is explained in part by the power of family-based mutations, the inheritance of which can be traced between affected and unaffected relatives. It is difficult to determine similar causality for a population-based polymorphism.

11.4.3 Functional analysis of polymorphisms in putative splicing elements

If taken individually, many sequences within the human genome match the consensus motifs for splice sites, but most of them are not used. In order to function, splice sites need appropriately arranged positive (ESEs and ISEs) and negative (ESSs and ISSs) *cis*-acting sequence elements. These *cis*-acting arrangements of regulatory elements can be both activated and deactivated by DNA sequence polymorphisms (for a review, see Wang *et al.*, 2005). DNA polymorphisms at the invariant splice acceptor (AG) and donor (GT) sites are generally associated with severe diseases, and so are likely to be correspondingly rare (for example, only 429 out of 3.8 million HapMap SNPs are located in splice sites as defined by EnsEMBL genes (unpublished data)). But, as already described, recognition motifs for some of the elements that make up the larger splice site consensus are very variable, so splice site prediction from undefined genomic sequence is still imprecise at the best of times. Bioinformatics tools can fare rather better when applied to known genes with known intron/exon boundaries – this information can be used to carry out reasonably accurate evaluations of the impact of polymorphisms in putative splice regions. Several tools can predict the location of splice sites in a genomic sequence; all match and score the query sequence against a probability matrix built from known splice sites (Table 11.3). These tools can be used to evaluate the effect of splice region polymorphisms on the strength of splice site prediction by alternatively running wild-type and mutant alleles. As with any other bioinformatics prediction tool, it is always worth running predictions on other available tools to look for a consensus between different prediction methods. These tools can also evaluate the propensity of an exon to undergo alternative splicing. For example, an unusually low splice site score may indicate that aberrant splicing may be more likely at one exon than at exons with higher splice site scores. The phase of the donor and acceptor sites also needs to be taken into account in these calculations. Coding exons exist in three phases, 0, 1 and 2, based on the codon location of the splice sites. If alternative donor or acceptor sites are in unmatched phases, a frame-shift mutation will occur.

Splice site prediction tools will generally predict the functional impact of a polymorphism within close vicinity of a splice donor or acceptor site, although they will not predict the functional effect of polymorphisms in other elements such as lariat branch sites. Definition of consensus motifs for these elements (Figure 11.2) makes it reasonably easy to assess the potential functional impact of polymorphisms in these gene regions by simply inspecting the location of a polymorphism in relation to the consensus motif. As with all functional predictions, laboratory investigation is required to confirm the hypothesis.

ESEfinder (Cartegni *et al.*, 2003) and RESCUE-ESE (Fairbrother *et al.*, 2002) are Web tools that facilitate rapid analysis of exon sequences to identify putative ESEs (Table 11.3). The tools rely on different methodologies; ESEfinder matrices are derived by an *in vitro* functional 'SELEX' method, whereas RESCUE-ESE is based purely on computational prediction. As a result, ESE motifs recognized by ESEfinder and ESE-RESCUE do not significantly overlap (Wang *et al.*, 2005), and it is therefore advisable to use both tools to identify potential regulatory motifs in exons of interest.

Although a systematic approach to identify and analyse ESS sites has been developed (Sironi *et al.*, 2004), bioinformatics tools to predict the locations of these motifs and ISE and ISS sites are not currently available.

11.4.4 Polyadenylation signals

Polyadenylation of eukaryotic mRNA occurs in the nucleus after cleavage of the precursor RNA. Several signals are known to determine the site of cleavage and subsequent polyadenylation, of which the best known is a canonical hexanucleotide (AAUAAA) signal 20–50 bp from the 3′ end of the pre-RNA. This works with a downstream U/GU-rich element that is believed to regulate the complex of proteins necessary to complete 3′ processing (Pauws *et al.*, 2001). The specific site of cleavage of pre-RNA is located between these regulatory elements and is determined by the nucleotide composition of the cleavage region with the following nucleotide preference A>U>C>>G. In a study of 9625 known human genes, Pauws *et al.* (2001) found that 44 per cent of human genes regularly use more than one cleavage site, resulting in the generation of slightly different mRNA species.

Mutations in the canonical AAUAAA polyadenylation signal have been shown to disrupt normal generation of polyadenylated transcripts (Bennett *et al.*, 2001). This signal is needed for both cleavage and polyadenylation in eukaryotes, and failure to polyadenylate prevents maturation of mRNA from nuclear RNA (Wahle and Keller, 1992). The complete aggregate of elements that make up the polyadenylation signal, including the U/GU-rich region, may not be universally required for processing (Graber *et al.*, 1999). Single nucleotide variations in this region cannot be conclusively identified as functional, although any polymorphism in this region might be considered a candidate for further consideration.

11.4.5 Analysis of mRNA transcript polymorphism

The potential functional effects of genetic polymorphism can extend beyond a direct effect on the genomic organization and regulation of genes. Messenger RNA is far more than a simple coded message acting as an intermediary between genes and proteins. mRNA molecules have different fates related to structural features embedded in discrete regions of the molecule. The processing, localization, translation or degradation of a given mRNA may vary considerably, depending upon the environment in which it is expressed. Figure 11.3 illustrates a simplified model of an mRNA molecule, indicating the key features and regulatory motifs that could potentially be disrupted by polymorphism. At the most basic level, an mRNA molecule consists of a protein-coding ORF, flanked by 5′ and 3′ UTRs. Most polymorphism

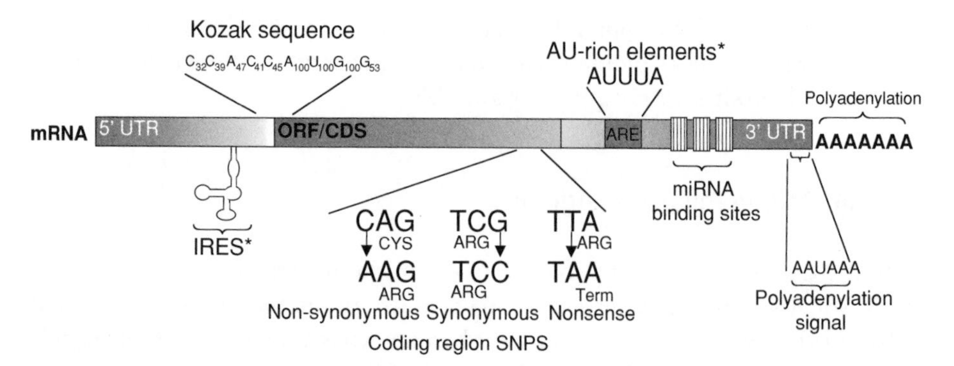

Figure 11.3 The anatomy of an mRNA transcript. This figure illustrates some of the key regulatory and structural elements that control the translation, stability and post-transcriptional processing of mRNA transcripts. Polymorphisms in these regions should be investigated for functional effects

analysis in the literature has tended to focus on the coding sequence of genes, but evidence suggests that UTR sequences also serve important roles in mRNA function and regulation. At the risk of generalizing, we may say that 5′ UTR sequences are important, as they are known to accommodate the translational machinery, while small non-coding RNAs have recently been shown to regulate gene expression by binding complementary sites in the 3′ UTR of target genes (Sontheimer and Carthew, 2005; see Chapter 13 for a review of the role of non-coding RNAs in gene regulation). In Table 11.4, we highlight some examples of polymorphisms which affect mRNA transcripts.

11.4.6 Initiation of translation

If a gene is known, the ORF will probably be well defined, but if a novel transcript is being studied, the ORF needs to be identified. We refer the reader to Chapter 5, which contains details on the extension of mRNA transcripts and ORF-finding procedures. The accepted convention is that the initiator codon will be the first in-frame AUG encoding the largest ORF in the transcript. There is evidence of a scanning mechanism for initiation of translation. The initiator codon generally conforms to a 'CCACCaugG' consensus motif known as the Kozak sequence (Kozak, 1996). However, Peri and Pandey (2001) and others have recently reappraised this convention, finding that more than 40 per cent of known transcripts contain in-frame AUG codons upstream of the actual initiator codon, some of which conform more closely to the Kozak motif than the authentic initiator codon. Their revised Kozak consensus '$C_{32}C_{39}A_{47}C_{41}C_{45}A_{100}U_{100}G_{100}G_{53}$' was much weaker. These observations have cast some doubt on the validity of the scanning mechanism for initiation of translation. Some have argued that the frequent occurrence of AUG codons upstream

Table 11.4 Functional non-coding polymorphisms in mRNA transcripts

Location	Gene/disease	Mechanism
miRNA binding site in 3′ UTR	SLITRK1 in Tourette's syndrome	A SNP in a miRNA-binding site of SLITRK1 replaces a G:U wobble with normal base pairing, leading to enhanced miRNA binding and increased gene expression (Abelson *et al.*, 2005)
Internal ribosome entry segment (IRES)	Proto-oncogene c-myc in multiple myeloma	C-T mutation in the c-myc-IRES causes aberrant translational regulation of c-myc, enhanced binding of protein factors, and enhanced initiation of translation, leading to oncogenesis (Chappell *et al.*, 2000)
Kozak initiation sequence	Platelet glycoprotein Ib-alpha (GP1BA) in ischaemic stroke	C/T polymorphism at the –5 position from the initiator ATG codon of the GP1BA gene is located within the 'Kozak' consensus nucleotide sequence. The presence of a C at this position significantly increases the efficiency of expression of the GPIb/V/IX complex (Afshar-Kharghan *et al.*, 1999).
Anti-termination mutation and 3′ UTR stability determinants	Alpha-globin in alpha-thalassaemia	UAA to CAA to anti-termination mutation allows translation to proceed into the 3′ UTR, masking stability determinants to decrease mRNA half-life substantially (Conne *et al.*, 2000)
UTR stability	Protein tyrosine phosphatase-1B (PTP1B)	1484insG in 3′ UTR causes PTP1B overexpression, leading to insulin resistance (Di Paola *et al.*, 2002)

of the putative initiator codon may indicate misassignment of the initiatior codon or cDNA library anomolies (Kozak, 2000); others point to the empirical increase in gene expression measured in the laboratory when initiator codons conforming to the Kozak consensus are compared to other sequences. This debate may never resolve conclusively, and it seems certain that the mechanism for translation initiation is still not fully understood.

Some polymorphisms in Kozak sequences appear to have a direct bearing on human disease. Kaski *et al.* (1996) reported a T > C SNP with an 8–17 per cent minor allele frequency at the –5 position from the initiator ATG codon of the GP1BA gene. This SNP is located within the most 5′ (and weakest) part of the Kozak consensus sequence. The cyotsine (C) allele at this position conforms more closely to the consensus, and subsequent studies of the SNP found that it was associated with increased expression of the receptor on the cell membrane, both in transfected cells and in the platelets of individuals carrying the allele. The polymorphism was also associated with cardiovascular disease susceptibility (Afshar-Kharghan *et al.*, 1999).

An alternative mechanism for translation initiation has been identified that does not obey the 'first AUG rule'; this involves cap-independent internal ribosome binding mediated by a Y-shaped secondary structure, denoted the 'internal ribosome entry site' (IRES), located in the 5′ UTR of 5–10 per cent of human mRNA molecules (see Le and Maizel, 1997, for a review of these elements). IRES elements are complex stem loop structures, and there is no reliable sequence consensus to allow prediction of the possible functional effects of polymorphisms in these elements. Instead this needs to be attempted by RNA secondary structure prediction tools such as MFOLD (see below). The sequences and links to published information for 50 IRESes located on eukaryotic transcripts are available from IRESdb (http://ifr31w3.toulouse.inserm.fr/IRESdatabase/) (Bonnal *et al.*, 2003).

11.4.7 mRNA secondary structure stability

While we have already established that nucleotide variants in mRNA can alter or create sequence elements directing splicing, processing, or translation of mRNA, variants may also influence mRNA synthesis, folding, maturation, transport and degradation. Many of these diverse biological processes are strongly dependent on mRNA secondary structure. Secondary structure is essentially determined by ribonucleotide sequence; therefore, folding of mRNA is also likely to be influenced by SNPs and other forms of variation at any location in a transcript. Shen *et al.* (1999) studied two common silent SNPs in the coding regions of two essential genes – a U1013C transition in human alanyl tRNA synthetase (AARS) and a U1674C transition in human replication protein A, a 70-kDa subunit (RPA70). Minor allele frequency was 0.49 for the AARS U allele and 0.15 for the RPA70 C allele. Using structural mapping and structure-based targeting strategies, they demonstrated that both SNPs had marked effects on the structural folds of the mRNAs, suggesting phenotypic consequences of SNPs in mRNA structural motifs. RNA stability is an intriguing disease mechanism; unfortunately, beyond this and a handful of other published studies (see Conne *et al.*, 2000, for a review), the true extent of detectable differences in mRNA folding caused by polymorphism is quite unknown, possibly reflecting the difficulties involved in studying such mutational effects *in vitro*.

Several tools can help to construct *in silico* secondary structure models of polymorphic mRNA alleles. One of the best is MFOLD (M. Zuker, Washington University, St Louis, MO, USA). This is maintained on the Zuker Laboratory home page, which also contains an excellent range of RNA secondary structure-related resources (http://bioinfo.math.rpi.edu/~zukerm/rna/). MFOLD will construct a number of possible models based on all structural permutations of a user-submitted mRNA sequence. Submission of mutant and wild-type mRNA alleles to this tool will give the user a fairly good indication of whether an allele can alter mRNA secondary structure. This can help to prioritize alleles for laboratory investigation in mRNA stability studies.

11.4.8 Regulatory control of mRNA processing and translation

Beyond splicing and promoter-based regulation, mRNAs are also tightly controlled by regulatory elements in their 5′ and 3′ untranslated regions (Figure 11.3). Proteins that bind to these sites are key players in controlling mRNA stability, localization and translational efficiency. Consensus motifs have been identified for many of these factors, usually corresponding to short oligonucleotide tracts, which generally fold in specific secondary structures that are protein-binding sites for various regulatory proteins. Some of these regulatory signals tend to be protein family specific, while others have a more general effect on diverse mRNAs. AU-rich elements (AREs) are the largest class of *cis*-acting 3′ UTR-located regulatory molecules that control the cytoplasmic half-life of a variety of mRNA molecules (reviewed by Barreau *et al.* 2006). One main class of these regulatory elements consists of pentanucleotide sequences (AUUUA) in the 3′ UTR of transcripts encoding oncoproteins, cytokines, and growth and transcription factors. Many RNA-binding proteins, mostly members of the highly conserved ELAV family, recognize and bind AREs (Chen and Shyu, 1995). Defective function of AREs can lead to the abnormal stabilization of mRNA; this forms the basis of several human diseases, including mantle cell lymphoma, neuroblastoma, and several immune and inflammatory diseases. Polymorphisms which disrupt AU-rich motifs in a 3′ UTR sequence may be worth evaluation as potentially functional polymorphisms. Some databases to assist in the identification of these motifs are described below.

11.4.9 Tools and databases to assist mRNA analysis

To assist in the analysis of diverse and often family-specific regulatory elements, such as ARE elements, Mignone *et al.* (2005) have developed UTRdb, a specialized non-redundant database of 5′ and 3′ untranslated sequences of eukaryotic mRNAs (http://bighost.area.ba.cnr.it/BIG/UTRHome/). In April 2006, UTRdb contained 83 184 non-redundant human entries; these are enriched with specialized information absent from primary databases, including the presence of RNA regulatory motifs with experimental proof of a functional role. It is possible to BLAST search the database for the presence of annotated functional motifs in a query sequence. Jacobs *et al.* (2002) have also developed Transterm, a curated database of mRNA elements that control translation (http://uther.otago.ac.nz/Transterm.html). This database examines the context of initiation codons for conformation with the Kozak consensus and also contains a range of mRNA regulatory elements from a broad range of species. Access is provided via a web browser in several different ways. A user-defined sequence can be searched against motifs in the database, or elements can be entered by the user to search specific sections of the database (e.g., coding regions, 3′ flanking regions or the 3′ UTRs) or the user's sequence. All elements defined in Transterm have associated biological descriptions with references.

11.5 Pseudogenes and regulatory mRNA

As a final word on the analysis of mRNA transcripts, it is important to be aware that not all mRNAs are intended to be translated. Some genes may produce transcripts that are truncated, retain an intron, or are otherwise configured in a way that precludes translation. It is difficult to clarify the role of some of these transcripts. If a transcript has multiple premature termination codons, it is likely to be a pseudogene; others may have no obvious ORFs and may be pseudogenes, or they may be regulatory mRNA molecules. Analysis of polymorphisms in these molecules is difficult, as they are very poorly defined in terms of functionality. Many non-coding RNA (ncRNA) molecules have been described that act as riboregulators directly influencing post-transcriptional regulation of gene expression. SNPs located in either the genes encoding regulatory mRNAs or in the target sites of genes regulated by these molecules have the potential to modify gene-expression levels. miRBase (Table 11.3) provides an automated pipeline for predicting miRNA-binding sites in target genes, and predicted sites are also displayed in the UCSC browser. (For a thorough review of the properties of regulatory mRNA and functional analysis of these genes and their target sites, see Chapter 14.)

11.6 Analysis of novel regulatory elements and motifs in nucleotide sequences

Geneticists are working at the vanguard of efforts to close the gap between our current understanding and the full complexity of human gene regulation. Genetics has already contributed greatly to the identification of new regulatory elements, often located far from gene promoters, by the identification of regulatory mutations and polymorphisms (Morley *et al.*, 2004).

In this chapter, we have reviewed a number of regulatory mechansims and motifs in DNA sequences, including motifs in promoter regions, splice sites, introns and transcripts. Functional analysis of polymorphisms located in the consensus sequences identified for some of these elements may be an important indicator of a potential functional effect. However, despite advances in bioinformatic tools, predictive functional analysis of sequence polymorphism is still difficult to validate without laboratory follow-up. Even with the benefit of laboratory verification, identification of deleterious alleles can be laborious, and the results of analyses do not always hold true between *in vitro* and *in vivo* environments. In a sense, evolution is an *in vivo* experiment on a grand scale; therefore, Sydney Brenner (2000) and others have proposed the concept of 'inverse genetics' to cover the use of information recovered from different genomes to clarify function. Brenner suggested comparing genomes to highlight conserved areas 'in a vast sea of randomness'. This is an elegant approach for the characterization of polymorphisms. Characterization by conventional genetics demands analysis of large sample numbers, complex *in vitro* analysis, or laborious transgenic approaches. In the case of inverse genetics, evolution and time have already done the work in a long-term 'experiment' that would be impossible to match in the laboratory.

Inverse genetics also has a wider application – analysis of a single promoter sequence will often identify many putative regulatory elements by chance alone. However, simultaneous analysis of many evolutionarily related but diverse promoter sequences will clearly identify known and novel conserved motifs that are likely to be functionally important to a particular family of genes. This approach, known as phylogenetic footprinting, has been used to elucidate many common regulatory modules (Gumucio *et al.*, 1996; Blanchette *et al.*, 2006) and conserved non-coding sequences corresponding to functional regulatory regions (see Nardone *et al.*, 2004, for a review). Kleiman *et al.* (1998) used a similar approach to identify a novel potential element in the polyadenylation regulatory apparatus, a TG deletion (deltaTG) in the 3′UTR of the HEXB gene, 7 bp upstream of the polyadenylation signal. The deltaTG HEXB allele, which occurred at 10 per cent frequency, showed 30 per cent lower enzymatic activities than wild-type individuals. Polyacrylamide gel electrophoresis analysis of the allele revealed that the 3′ UTR of the HEXB gene had an irregular structure. After studying a large range of eukaryotic mRNAs, including human, mouse and cat HEXB genes, they found that the TG dinucleotide was part of a conserved sequence (TGTTTT) immersed in an A/T-rich region observed in more than 40 per cent of mRNAs analysed. This study clearly illustrates how effective bioinformatic analysis of mRNA-processing signals may require more than sequence analysis of known regulatory motifs. Clearly, tools are needed to identify novel regulatory elements.

The UCSC genome browser conveniently displays a number of data sets relevant to the identification and subsequent refinement and analysis of conserved non-coding sequences (CNS). The 'conservation' track displays a measure of evolutionary conservation in 17 vertebrates, including mammalian, amphibian, bird and fish species. Comparisons of closely related species, such as man and chimpanzee, will identify regions where divergence is most readily tolerated by highlighting differences rather than similarities, whereas comparisons of distantly related organisms, such as mouse and chicken, will identify highly constrained sequences (Cooper *et al.*, 2003). Sequence comparison of moderately related species, such as human and mouse, is ideal for a survey analysis to begin to define CNS regions, and multiple species comparison can be used to increase the power of the technique and refine the regions further (Thomas *et al.*, 2003). Once a CNS region has been identified, other precomputed features displayed on the genome browser, such as conserved transcription factor-binding sites and DNase hypersensitive sites, can provide further evidence of a functional regulatory role.

11.6.1 TRES (http://bioportal.bic.nus.edu.sg/tres/)

TRES is a very flexible tool that can assist in the identification of novel elements from user-defined sequences. This approach is not just applicable to evolutionarily related sequences (see Chapter 6 for detailed coverage of this area); it can also be used to study unrelated sequences that may share similar regulatory cues, such as genes which show similar patterns of gene expression. TRES can compare as many as 20 nucleotide sequences. The tool is multifunctional, and it can identify either conserved sequence motifs between submitted sequences or known transcription factor-binding sites shared between sequences, using nucleotide frequency distribution matrices described in the TRANSFAC database (Heinemeyer *et al.*, 1999).

TRES also has a versatile search mode to detect palindromic motifs or inverted repeats shared between sequences. These have unique features of dyad symmetry that can form hairpins or loops to facilitate protein binding in homo- or heterodimer form. Many transcription factors have palindromic recognition sequences and bind as dimers; these motifs may be important to allow greater regulatory diversity from a limited number of transcription factors (Lamb and McKnight, 1991). Although TRES is generally focused on the identification of transcription factor-binding sites and promoter elements, its sequence motif identification facilities make it suitable also to identify other motifs in non-coding sequences, including UTR sequences and intronic sequences.

11.7 Functional analysis of non-synonymous coding polymorphisms

The impact of mutations and polymorphisms in protein sequences needs to be characterized on a case-by-case basis. To return to our decision tree for polymorphism analysis (Figure 11.1), the consequence of an amino-acid substitution is defined by three key areas:

a. the physicochemical environment in which the amino acid exists

b. the structural context of the amino acid

c. the functional context of the amino acid within the protein.

Investigation of amino-acid variants should take all of these areas into account to arrive at the most reliable hypothesis for the putative functional impact of a variant. At the time of writing the first edition of this book, this process was very challenging, with limited resources for performing these analyses and no curated data sets of preanalysed variants. Fortunately, the situation has improved markedly in the intervening years, and now there are rich resources available to carry out analysis of this nature, and there are also a number of precomputed analyses of amino-acid variants from which to obtain a second opinion or an initial pointer (see Section 11.7.5). In most cases, this will leave the geneticist with a choice – to rely on automated annotation of variants (a good stand-by) or to carry out ad hoc analysis (perhaps the reference standard). The following sections review the process of analysis of amino-acid variants. This can serve either as an explanation of automated data sets or as an outline of the steps required for ad hoc analysis.

11.7.1 Assessing the physicochemical environment of an amino-acid substitution

If we look first at the context of the amino acid, different cellular locations can have very different physicochemical environments, which can, in turn, have different effects on the properties of amino acids (these differences are addressed in detail in Chapter 13). The cellular location of proteins can be divided at the simplest level between intracellular, extracellular or transmembrane environments. The latter location is the most complex, as amino acids in transmembrane proteins can be exposed to all three cellular environments, depending upon the topology of the protein and the location of the particular amino acid within the known or predicted topology of the protein. In this case, the accuracy of protein topology prediction is particularly important; fortunately, tools such as UniProt (http://www.uniprot.org) offer detailed topological annotation for most known proteins. In the case of novel proteins, a number of tools can be used to predict topology (see Table 11.5). For predicting secondary structure in novel proteins, it is generally worth running several tools to try to obtain a consensus between tools.

Microenvironments around a specific amino acid may also differ in extracellular and intracellular proteins, depending on the location of the residue within the protein. Amino-acid residues may be buried in a protein core, or exposed on the protein surface. Once the environment of an amino-acid substitution has been defined, the properties of the different alleles (residues) can be assessed in the context of the physicochemical environment by a number of different tools (Table 11.5). For reference, we have provided four amino-acid substitution matrices in Appendix II of this book. These matrices can be used to evaluate amino-acid changes in extracellular, intracellular and transmembrane proteins; where the location of the protein or amino acid residue is unknown, a matrix for 'all proteins' is also available. Preferred (conservative) substitutions have positive scores, neutral substitutions have a zero score, and unpreferred (non-conservative) substitutions are scored negatively. These matrices are an example of 'inverse genetics' in action (Brenner, 2000), being constructed by observing the propensity during evolution for the exchange of one amino acid for another in the given amino-acid environment based on comparison of very large sets of related proteins (for more details, see Chapter 13 and http://www.russell.embl-heidelberg.de/aas).

Table 11.5 Tools for functional analysis of amino acid polymorphisms

Protein Secondary structure prediction	
TMPRED	http://www.ch.embnet.org/software/TMPRED_form.html
TMHMM	http://www.cbs.dtu.dk/services/TMHMM/
PREDICTPROTEIN	http://www.embl-heidelberg.de/predictprotein/
GPCRdb 7TM snake plots	http://www.gpcr.org/7tm/seq/snakes.html
Protein 3D structure analysis	
Protein data bank (PDB)	http://www.rcsb.org/pdb/Welcome.do
MODBASE	http://salilab.org/modbase
DeepView/Swiss PDB viewer	http://www.expasy.org/spdbv/
Cn3D	http://www.ncbi.nih.gov/Structure/CN3D/cn3d.shtml
STRAP	http://www.charite.de/bioinf/strap/
Identification of protein functional motifs	
INTERPRO	http://www.ebi.ac.uk/interpro/scan.html
PROSITE	http://www.ebi.ac.uk/searches/prosite.html
SIGNALP, NetPhos, NetOGlyc & NetNGlyc	http://www.cbs.dtu.dk/services/
Swissprot (Functional annotation)	http://www.expasy.ch/cgi-bin/sprot-search-ful
Evaluation of amino acid properties	
Properties of amino acids	http://www.russell.embl-heidelberg.de/aas/
PROWL	http://prowl.rockefeller.edu/aainfo/contents.htm
SNPPER AA properties	http://snpper.chip.org/bio/show-amino
Pre-computed multiple alignments	
DBAli	http://salilab.org/dbali
SPEED	http://bioinfobase.umkc.edu/speed/
Tools for prediction of non-synonymous SNP function	
LS-SNP	http://alto.compbio.ucsf.edu/LS-SNP/
SIFT	http://blocks.fhcrc.org/sift/SIFT.html
PolyPhen	http://genetics.bwh.harvard.edu/pph/index.html
SNPS3D	http://www.snps3d.org/
SNPeffect	http://snpeffect.vib.be/index.php
TopoSNP	http://gila-fw.bioengr.uic.edu/snp/toposnp/
MutDB	http://www.mutdb.org/
Pipeline tools for one stop Gene to SNP functional analysis	
UCSC	http://genome.ucsc.edu
pupaSNP	http://www.pupasnp.org/
FastSNP	http://fastsnp.ibms.sinica.edu.tw/

11.7.2 Defining the protein structural context of an amino-acid substitution

Defining the environment of an amino acid may be relatively straightforward if the protein is known, and good protein annotation exists. Analysis is even more effective if there is a known tertiary (3-D) structure (Skolnick *et al.*, 2000). There are a number of ways to access protein structural information; for example, it can be conveniently accessed from UniProt (http://www.uniprot.org). Sometimes the user may also need to consult the 'Niceprot View' linked at the top of the UniProt interface, as the available data seem to vary slightly between these two views on the SwissProt database. When the crystal structure of a protein has been determined, a link will be provided to the PDB database, the central repository of protein structural information. Alternatively, the NCBI protein search interface links to Blink, a useful source of precomputed protein annotation, including hits to known protein 3-D structures (e.g., the entry for ADRB3, http://www.ncbi.nlm.nih.gov/sutils/blink.cgi?pid=4557267).

The 3-D structures of proteins are determined by experimental methods, such as X-ray crystallography and NMR spectroscopy. At the time of writing (July 2006), 37 658 experimentally determined structures were in the Protein Data Bank (PDB) database, the international protein structure repository, and 7181 of these were derived from human proteins. This total figure is redundant, as a structure may be determined by multiple groups, in the same species or in species that are closely related on a protein level, such as mouse and human. The PDB provides a filter for highly similar structures to provide one or more representatives; if this is applied, there are ~14 000 non-redundant structures in the database.

Despite the best efforts of protein crystallographers, the protein structures of the majority of human proteins have not yet been directly determined (in fact, this may not be possible in the near future; see below). In many cases, however, the structure of an orthologue or close homologue of a human protein has been determined. In this case, it is possible to construct a reasonably accurate protein structural model based on an alignment between the protein with a known structure and the homologous protein. MODBASE (http://salilab.org/modbase; Pieper *et al.*, 2006) is a database that contains precomputed comparative protein structure models for all available protein sequences that can be matched to at least one known protein structure. The MODBASE database is updated regularly to capture new sequences and structures, and it currently contains reliable models for 908 507 sequences (July 2006). MODBASE also allows generation of new comparative models for proteins of interest with the automated modelling server MODWEB (http://salilab.org/modweb).

While tools such as MODBASE go part of the way toward providing comprehensive models of protein structure for all human proteins, there are some implicit limitations to the study of protein structure and hence our understanding of structure. Protein crystallography methods are practically limited to proteins that can be expressed *in vitro* in a soluble form. Hence, it is very difficult to determine the crystal structure of transmembrane proteins, as they are typically insoluble. G-protein-coupled receptors (GPCRs) are a good example of this; despite the size of the GPCR superfamily (>500 members in man) and their importance as drug targets, a crystal structure has been determined, to date, for only one member of the family, bovine rhodopsin. Therefore, all GPCR homology models in MODBASE are based on the structure of bovine rhodopsin, despite levels of similarity that can be as low as 20–30 per cent. In some cases, it may be possible individually to purify and determine the structure of a specific domain of a transmembrane protein, such as a kinase domain. Xie and Bourne (2005) carried out a survey of the current number of human genes that have some homology to a protein with a known protein structure or domain. They found that 37 per cent of the functional classes of proteins identified in the genome shared homology with at least one known structural domain, while 25 per cent shared homology with an entire protein structure. They predicted that even if structures were determined for all classes of human proteins that can be expressed and purified for crystallography, structural coverage of at least one domain would increase to only 69 per cent of functional classes, 44 per cent of functional classes sharing homology with an entire structure. From these figures, it is clear that protein structural information will be available for the evaluation of amino-acid variation only some of the time. When this information is unavailable, other sources of information, such as multiple alignments and secondary structural analysis and annotations, become much more important (see Section 11.7.4).

11.7.3 Reviewing the impact of amino-acid variation in protein structures

A wide range of tools are available for viewing and analysing protein structures (Table 11.5). From the point of view of genetic analysis, one of the most effective tools is DeepView, which is freely available to download (Schwede *et al.*, 2003). This tool has a number of features that make it very suitable for the analysis of amino-acid variants. Most notably, it allows the user to load a known structure and highlight residues, it allows the user to mutate residues, and, finally, it allows the user to identify other residues which might interact with a selected residue by identifying all residues within a user-defined 3-D radius around the selected residue. This makes it possible to identify residues that might interact with structural and functional protein features, such as active sites in 3-D-folded protein space. A detailed case study of exactly this kind of analysis involving the use of DeepView is presented in Chapter 19 (Section 19.3.2).

The use of protein sequence homology to build models of protein structure relies heavily on the quality of the alignment between the protein with the known structure and the protein being modelled by homology. Selected residues of the protein being analysed can be assigned to residues of the protein with the known structure, allowing structural and functional annotation. One issue that may confuse such analysis is the numbering of amino-acid residues. In the protein structure record, each residue is specified by a number; this numbering which may not necessarily start from one, and it may skip numbers. Consequently, it will differ from the residue numbering in annotation of the entire protein record, as in a SwissProt record. Therefore, it may be necessary to check the order of amino acids between the structural record and the protein sequence record to ensure that the same protein regions are being compared.

11.7.4 Defining the functional context of an amino-acid substitution

In taking into account the immediate physicochemical, structural context and annotated features of an amino-acid variant, the final step in characterization is to evaluate the conservation of the amino-acid position in an alignment of related proteins. Alignment of mutated amino-acid sequences with vertebrate and invertebrate orthologues and homologues in a protein family will indicate whether the residue is highly conserved throughout the gene family. Beyond evolutionary clues, there are many different sources of protein annotation (including precomputed protein family alignments) and tools to evaluate the impact of substitutions in known and predicted protein features; some of the best are listed in Table 11.5. The overriding principle of amino-acid variant analysis is to get to know a protein, first seek known annotation, then seek to annotate where annotation does not exist, and finally look at the impact of the variant in relation to all that you now know. Figure 11.4 shows an example of an evaluation of a mutation in Jagged1, a ligand for the Notch receptor family. Krantz *et al.* (1998) identified an Arg184Cys missense mutation in patients with Alagille syndrome (OMIM 118450). In terms of amino-acid substitutions, Arg > Cys is quite non-conservative (the extracellular substitution matrix score for this change is –5). Alignment of the mutated human amino-acid sequence with vertebrate and invertebrate orthologues and homologues in the Jagged family identifies the Arg184 residue as a highly conserved position throughout this protein family. A mutation to a cysteine at this position would be expected to lead to the aberrant formation of disulphide bonds with other cysteine residues in the Jagged protein. This is

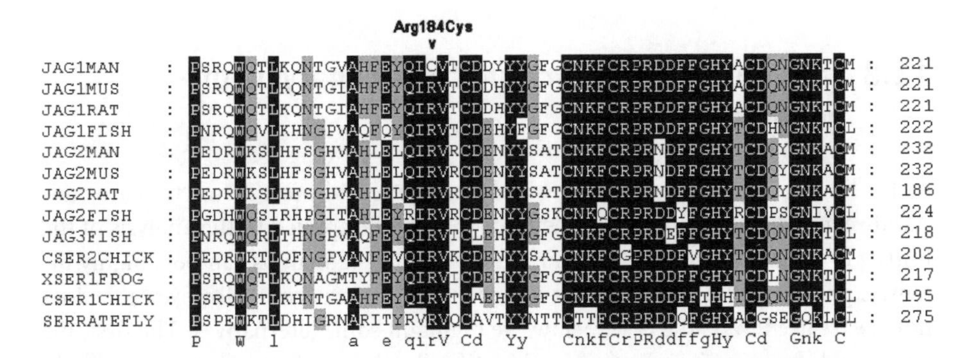

Figure 11.4 Functional evaluation of an Arg184Cys mutation in the Jagged protein family. Arg184Cys causes Alagille syndrome (OMIM 118450). Alignment of the mutated human amino-acid sequence with vertebrate and invertebrate orthologues and homologues in the Jagged family identifies the Arg184 residue in a highly conserved position throughout this gene family. A mutation to a cysteine at this position would be expected to lead to the aberrant formation of disulphide bonds with other cysteine residues in the Jagged protein; this is likely to have a disruptive effect on the structure of the Jagged1 protein

likely to have a disruptive effect on the structure of the Jagged1 protein, presumably leading to the Alagille syndrome phenotype. (See Chapter 13 for a description of the effects of inappropriate disulphide bond formation.)

11.7.5 Large-scale functional annotation of nsSNPs in the public domain

As mentioned at the start of this section, in the years following the first edition of this book, a number of public domain resources were developed that offer precomputed functional annotation of known mutations and polymorphisms, in both coding and non-coding regions. There are also other, less specialist resources covering both coding and non-coding SNPs; these are reviewed in Section 11.8. The quality of these annotations can vary, and it is important to check how often the analysis is updated. Some of the most widely used (and therefore best maintained) are LS-SNP (Karchin *et al.*, 2005), SIFT (Ng and Henikoff, 2006) and PolyPhen (Ramensky *et al.*, 2002), and some other resources are listed in Table 11.5. Several publications have compared different tools, usually finding different answers between tools for the same variant, highlighting the need to consult a range of tools (Letourneau *et al.*, 2005; Matyas *et al.*, 2006). How these tools are used is really up to the investigator. They can be used as a second opinion to add weight to the results of an analysis, or they can be used initially to identify SNPs with potential functional effects for further analysis. It is probably fair to say that the results and conclusions of these predominantly automated analysis approaches could always be enhanced by further detailed (manual) analysis, so generally they should be considered as a starting point rather than an end point.

11.8 Integrated tools for functional analysis of genetic variation

As the limits of our understanding of genome, gene and protein function are constantly expanding, so too are the methods that can be applied to the analysis of the impact of variation on these functions. Polymorphism analysis can be a highly intensive process, calling on the most skilled analysis approaches. However, this does not always need to be the case; below, we highlight some alternatives. The first, the UCSC table browser, is a great way to identify all polymorphisms that are located in regions with a potential for function, therefore making each polymorphism a candidate to be genotyped for an impact on this function. The second alternative is to use precomputed data on SNP function. A wealth of information is now available in this area, and there is a strong argument that SNPs should be genotyped from as many of these precomputed resources as possible. The final point to consider is that if we do not know about a function, we will not be able to predict an impact on this function! Genetic analysis, obviously after accounting for type I error, is arguably the ultimate test for impact on an unknown function.

11.8.1 UCSC table browser

The UCSC table browser is perhaps the single most powerful polymorphism analysis tool, although this was not the primary objective of its design. The rapidly evolving myriad of genomic feature annotations captured within the UCSC genome browser (Table 11.5) can be intersected with various sets of SNPs (including HapMap, dbSNP or user-defined sets entered as custom tracks) with the UCSC 'table browser' function (Karolchik *et al.*, 2004). By simply selecting the UCSC track of interest, such as 'TFBS conserved' and intersecting with a chosen SNP set, SNPs located within conserved TFBS sites are returned. This powerful tool is particularly well suited to analyse large numbers of SNPs, which can be rapidly assigned to exons (known and predicted), coding regions, potential regulatory regions, including predicted TFBS, microRNA-binding sites in 3′ UTR, and conserved elements. Assignment of a SNP to a functional region does not necessarily imply that the SNP will have a functional impact – further analysis is required to determine this. However, considering our limited understanding of many functional elements, it is reasonable to suggest that any SNP that co-locates with a functional element should be a candidate for genotyping. For this reason, the UCSC table browser is ideal for prioritizing SNPs for genotyping across genes and candidate regions.

11.8.2 PupaSNP and FastSNP

PupaSNP and FastSNP (Table 11.5) are two tools that can be used to semi-automate polymorphism analysis. Both are integrated platform applications that analyse all known (in the case of PupaSNP, also user-submitted) polymorphisms in a given gene or list of genes. This obviously offers great benefits to the user in terms of speed and convenience, and as a first-pass analysis, these tools both do a fine job. However, it is worth taking some time to explore fully all avenues of analysis to add to the output of these tools.

PupaSNP is for high-throughput analysis of SNPs with potential phenotypic effect. The tool takes a list of genes as an input and retrieves SNPs from evolutionarily conserved regions that could affect gene regulation and protein function. PupaSNP is quite comprehensive and uses a range of tools to investigate the impact of SNPs on splice boundaries, exonic splicing enhancers, transcription factor-binding sites (TFBS), and changes in amino acids. It also provides additional functional information from gene ontology (a descriptive hierarchy of gene function), OMIM and model organisms. FastSNP is also worth a mention (Table 11.5). This also provides a complete platform for SNP analysis, although the number of analyses performed is slightly reduced, focusing on TFBS, ESE prediction and amino-acid substitution analysis.

11.9 A note of caution on the prioritization of *in silico* predictions for further laboratory investigation

Just as the complexity of genes, transcripts and proteins is virtually limitless, so, too, are the possibilities for developing functional hypotheses. If every aspect of the analyses explored in this chapter was examined in any single polymorphism, it would probably be possible to assign a *potential* deleterious function to almost every one. But, clearly, the human genome does not contain millions of potentially deleterious mutations (thousands maybe, but not millions!), so it is important to treat *in silico* predictions with caution. If a polymorphism shows genetic association with a phenotype; it is important to consider first whether the polymorphism is causal or in LD with a causal mutation. Hypotheses need to be constructed and tested in the laboratory; for example, if a polymorphism is predicted to affect splicing, *in vitro* analysis methods need to be employed to investigate evidence of alternative transcripts.

11.10 Conclusions

In this chapter, we have presented an overview of some of the approaches to predictive functional analysis of polymorphisms in genes, proteins and regulatory regions. These methods can be applied at the candidate identification stage or at later stages to assist in the progression of associated genes to disease genes. The chapter has also examined the role of bioinformatics in the formulation of laboratory investigation for confirmation of functional predictions. As we have shown, functional prediction of the potential impact of variation requires a very good grasp of the full gamut of bioinformatics tools used to predict the properties and structure of genes, proteins and regulatory regions. This huge range of applications makes polymorphism analysis one of the most difficult bioinformatics activities to get right. The complexity of some analysis areas is worthy of special attention, particularly the analysis of polymorphisms in gene regulatory regions, non-coding RNA and protein sequences. The following three chapters specifically address some of these highly specialized analysis issues.

References

Abelson, J. F., Kwan, K. Y., O'Roak, B. J. *et al.* (2005). Sequence variants in *SLITRK1* are associated with Tourette's syndrome. *Science* **310**(5746), 317–320.

Afshar-Kharghan, V., Li, C. Q., Khoshnevis-Asl, M. *et al.* (1999). Kozak sequence polymorphism of the glycoprotein (GP) Ib-alpha gene is a major determinant of the plasma membrane levels of the platelet GP Ib-IX-V complex. *Blood* **94**, 186–191.

Bajic, V. B., Tan, S. L., Suzuki, Y. *et al.* (2004). Promoter prediction analysis on the whole genome. *Nat Biotechnol* **22**, 1467–1473.

Barallem, M., Barallem, D., De Contim, L. *et al.* (2003). Identification of a mutation that perturbs *NF1* agene splicing using genomic DNA samples and a minigene assay. *J Med Genet* **40**, 220–222.

Barreau, C., Paillard, L. and Osborne, B. H. (2006). AU-rich elements and associated factors: are there unifying principles? *Nucleic Acids Res* **33**(22), 7138–7150.

Bennett, C. L., Brunkow, M. E., Ramsdell, F. *et al.* (2001). A rare polyadenylation signal mutation of the FOXP3 gene (AAUAAA→AAUGAA) leads to the IPEX syndrome. *Immunogenetics* **53**(6), 435–439.

Blanchette,M., Bataille, A. R., Chen,X. *et al.* (2006). Genome-wide computational prediction of transcriptional regulatory modules reveals new insights into human gene expression. *Genome Res* **16**, 656–668.

Bonnal, S., Boutonnet, C., Prado-Lourenço, L. *et al.* (1993). IRESdb: the Internal Ribosome Entry Site database. *Nucleic Acids Res* **31**(1), 427–428.

Brenner, S. (2000). Inverse genetics. *Curr Biol* **10**(18), R649.

Cartegni, L., Wang, J., Zhu, Z. *et al.* (2003). ESEfinder: a Web resource to identify exonic splicing enhancers. *Nucleic Acids Res* **31**(13), 3568–3571.

Chakravarti, A. (1999). Population genetics – making sense out of sequence. *Nat Genet* **21** (Suppl), 56–60.

Chappell, S. A., LeQuesne, J. P., Paulin, F. E. *et al.* (2000). A mutation in the c-myc-IRES leads to enhanced internal ribosome entry in multiple myeloma: a novel mechanism of oncogene de-regulation. *Oncogene* **19**(38), 4437–4440.

Chen, C. Y. and Shyu, A. B. (1995). AU-rich elements: characterization and importance in mRNA degradation. *Trends Biochem Sci* **20**(11), 465–470.

Conne, B., Stutz, A. and Vassalli, J. D. (2000) The 3′ untranslated region of messenger RNA: a molecular 'hotspot' for pathology? *Nat Med* **6**, 637–641.

Cooper, G. M., Brudno, M., Green, E. D. *et al.* (2003). Quantitative estimates of sequence divergence for comparative analyses of mammalian genomes. *Genome Res* **13**, 813–820.

Cooper, S. J., Trinklein, N. D., Anton, E. D. *et al.* (2006). Comprehensive analysis of transcriptional promoter structure and function in 1 per cent of the human genome. *Genome Res* **16**, 1–10.

Di Paola, R., Frittitta, L., Miscio, G. *et al.* (2002). A variation in 3prime prime or minute UTR of hPTP1B increases specific gene expression and associates with insulin resistance. *Am J Hum Genet* **70**(3), 806–812.

D'Souza, I. and Schellenberg, G. D. (2000). Determinants of 4-repeat tau expression. Coordination between enhancing and inhibitory splicing sequences for exon 10 inclusion. *J Biol Chem* **275**(23), 17700–17709.

Fairbrother, W. G., Yeh, R. F., Sharp, P. A. *et al.* (2002). Predictive identification of exonic splicing enhancers in human genes. Science **297**(5583), 1007–1013.

Fontana, W., Konings, D. A. M., Stadler, P. F. *et al.* (1993). *Biopolymers* **33**, 1389–1404.

Fujimaru, M., Tanaka, A., Choeh, K. *et al.* (1998). Two mutations remote from an exon/intron junction in the beta-hexosaminidase beta-subunit gene affect 3′-splice site selection and cause Sandhoff disease. *Hum Genet* **103**(4), 462–469.

Graber, J. H., Cantor, C. R., Mohr, S. C. *et al.* (1999). *In silico* detection of control signals: mRNA 3′-end-processing sequences in diverse species. *Proc Natl Acad Sci U S A* **96**(24), 14055–14060.

Gumucio, D. L., Shelton, D. A., Zhu, W. *et al.* (1996). Evolutionary strategies for the elucidation of *cis* and *trans* factors that regulate the developmental switching programs of the beta-like globin genes. *Mol Phylogenet Evol* **5**, 18–32.

Hani, E. H., Boutin, P., Durand, E. *et al.* (1998). Missense mutations in the pancreatic islet beta cell inwardly rectifying K^+ channel gene (KIR6.2/BIR): a meta-analysis suggests a role in the polygenic basis of type II diabetes mellitus in Caucasians. *Diabetologia* **41**(12), 1511–1515.

Hastings, M. L. and Krainer, A. R. (2001) Pre-mRNA splicing in the new millennium. *Curr Opin Cell Biol* **13**(3), 302–309.

Heinemeyer, T., Chen, X., Karas, H. *et al.* (1999) Expanding the TRANSFAC database towards an expert system of regulatory molecular mechanisms. *Nucleic Acids Res* **27**, 318–322.

Hoogendoorn, B., Coleman, S. L., Guy, C. A. *et al.* (2003). Functional analysis of human promoter polymorphisms. *Hum Mol Genet* **12**(18), 2249–2254.

Ishii, S., Nakao, S., Minamikawa-Tachino, R. *et al.* (2002). Alternative splicing in the alpha-galactosidase A gene: increased exon inclusion results in the Fabry cardiac phenotype. *Am J Hum Genet* **70**(4), 994–1002.

Jacobs, G. H., Rackham, O., Stockwell, P. A. *et al.* (2002). Transterm: a database of mRNAs and translational control elements. *Nucleic Acids Res* **30**(1), 310–311.

Jo, E. K., Kanegane, H., Nonoyama, S. *et al.* (2001). Characterization of mutations, including a novel regulatory defect in the first intron, in Bruton's tyrosine kinase gene from seven Korean X-linked agammaglobulinemia families. *J Immunol* **167**(7), 4038–4045.

Johnson, J. M., Castle, J., Garrett-Engele, P. *et al.* (2003). Genome-wide survey of human alternative pre-mRNA splicing with exon junction microarrays. *Science* **302**(5653), 2141–2144.

Karchin, R., Diekhans, M., Kelly, L. *et al.* (2005). LS-SNP: large-scale annotation of coding non-synonymous SNPs based on multiple information sources. *Bioinformatics* **21**(12), 2814–2820.

Karolchik, D., Hinrichs, A. S., Furey, T. S. *et al.* (2004). The UCSC table browser data retrieval tool. *Nucleic Acids Res* **32**, D493–496.

Kaski, S., Kekomaki, R. and Partanen, J. (1996). Systemic screening for genetic polymorphism in human platelet glycoprotein Ib-alpha. *Immunogenetics* **44**, 170–176.

Kleiman, F. E., Ramirez, A. O., Dodelson de Kremer, R. *et al.* (1998). A frequent TG deletion near the polyadenylation signal of the human HEXB gene: occurrence of an irregular DNA structure and conserved nucleotide sequence motif in the 3′ untranslated region. *Hum Mutat* **12**(5), 320–329.

Knight, J. C., Udalova, I., Hill, A. V. *et al.* (1999). A polymorphism that affects OCT-1 binding to the TNF promoter region is associated with severe malaria. *Nat Genet* **22**(2), 145–150.

Kozak, M. (1996). Interpreting cDNA sequences: some insights from studies on translation. *Mamm Genome* **7**(8), 563–574.

Krantz, I. D., Colliton, R. P., Genin, A. *et al.* (1998) Spectrum and frequency of Jagged1 (JAG1) mutations in Alagille syndrome patients and their families. *Am J Hum Genet* **62**, 1361–1369.

Lamb, P. and McKnight, S. L. (1991). Diversity and specificity in transcription regulation: the benefits of heterotypic dimerization. *Trends Biochem Sci* **16**, 417–422.

Lander, E. S., Linton, L. M., Birren, B. *et al.* (2001) Initial sequencing and analysis of the human genome. *Nature* **409**, 860–921.

Le, S. Y. and Maizel, J. V. Jr. (1997). A common RNA structural motif involved in the internal initiation of translation of cellular mRNAs. *Nucleic Acids Res* **25**(2), 362–369.

Letourneau, I. J., Deeley, R. G. and Cole, S. P. (2005). Functional characterization of non-synonymous single nucleotide polymorphisms in the gene encoding human multidrug resistance protein 1 (MRP1/ABCC1). *Pharmacogenet Genomics* **15**(9), 647–657.

Liu, H, Chao, D., Nakayama, E. E. *et al.* (1999). Polymorphism in RANTES chemokine promoter affects HIV-1 disease progression. *Proc Natl Acad Sci U S A* **96**(8), 4581–4585.

Liu, H. X., Cartegni, L., Zhang, M. Q. *et al.* (2001). A mechanism for exon skipping caused by nonsense or missense mutations in BRCA1 and other genes. *Nat Genet* **27**(1), 55–58.

Lynch, K. W. and Weiss, A. (2001). A CD45 polymorphism associated with multiple sclerosis disrupts an exonic splicing silencer. *J Biol Chem* **276**(26), 24341–24347.

Matyas, G., Arnold, E., Carrel, T. *et al.* (2006). Identification and *in silico* analyses of novel TGFBR1 and TGFBR2 mutations in Marfan syndrome-related disorders. *Hum Mutat* **27**, 760–769.

Mignone, F., Grillo, G., Licciulli, F. *et al.* (2005). UTRdb and UTRsite: a collection of sequences and regulatory motifs of the untranslated regions of eukaryotic mRNAs. *Nucleic Acids Res* **33**(Database Issue), D141–146.

Moller, L. B. Tumer, Z., Lund, C. *et al.* (2000). Similar splice site mutations of the ATP7A gene lead to different phenotypes: classical Menkes disease or occipital horn syndrome. *Am J Hum Genet* **66**, 1211–1220.

Morley, M., Molony, C. M., Weber, T. M. *et al.* (2004) Genetic analysis of genome-wide variation in human gene expression. Nature **430**, 743–747.

Nardone, J., Le, D. U., Ansel, K. M. *et al.* (2004) Bioinformatics for the 'bench biologist': how to find regulatory regions in genomic DNA. *Nature* **5**(8), 768–774.

Ng, P. C. and Henikoff, S. (2006) Predicting the effects of amino acid substitutions on protein function. *Annu Rev Genomics Hum Genet* **7**, 61–80.

Pauws, E., van Kampen, A. H., van de Graaf, S. A. *et al.* (2001). Heterogeneity in polyadenylation cleavage sites in mammalian mRNA sequences: implications for SAGE analysis. *Nucleic Acids Res* **29**(8), 1690–1694.

Peri, S. and Pandey, A. (2001). A reassessment of the translation initiation codon in vertebrates. *Trends Genet* **17**(12), 685–687.

Pieper, U., Eswar, N., Davis, F. P. *et al.* (2006). MODBASE: a database of annotated comparative protein structure models and associated resources. *Nucleic Acids Res* **1**(34; Database Issue), D291–295.

Pitarque, M., von Richter, O., Oke, B. *et al.* (2001). Identification of a single nucleotide polymorphism in the TATA box of the CYP2A6 gene: impairment of its promoter activity. *Biochem Biophys Res Commun* **284**(2), 455–460.

Ramensky, V., Bork, P. and Sunyaev, S. (2002). Human non-synonymous SNPs: server and survey. *Nucleic Acids Res* **30**(17), 3894–3900.

Reese, M. G., Hartzell, G., Harris, N. L. *et al.* (2000). Genome annotation assessment in *Drosophila melanogaster. Genome Res* **10**(4), 483–501.

Reich, D. E. and Lander, E. S. (2001). On the allelic spectrum of human disease. *Trends Genet* **17**(9), 502–510.

Sakurai, K., Seki, N., Fujii, R. *et al.* (2000). Mutations in the hepatocyte nuclear factor-4alpha gene in Japanese with non-insulin-dependent diabetes: a nucleotide substitution in the polypyrimidine tract of intron 1b. *Horm Metab Res* **32**(8), 316–320.

Schmid, C. D., Praz, V., Delorenzi, M. *et al.* (2004). The Eukaryotic Promoter Database (EPD): the impact of *in silico* primer extension. *Nucleic Acids Res* **32**, D82–D85.

Schwede, T., Kopp, J., Guex, N. *et al.* (2003). SWISS-MODEL: an automated protein homology-modeling server. *Nucleic Acids Res* **31**(13), 3381–3385.

Shen, L. X., Basilion, J. P. and Stanton, V. P., Jr. (1999). Single-nucleotide polymorphisms can cause different structural folds of mRNA. *Proc Natl Acad Sci U S A* **96**(14), 7871–7876.

Sironi, M., Menozzi, G., Riva, L. *et al.* (2004). Silencer elements as possible inhibitors of pseudo-exon splicing. *Nucleic Acids Res* **32**(5), 1783–1791.

Skolnick, J., Fetrow, J. S. and Kolinski, A. (2000). Structural genomics and its importance for gene function analysis *Nat Biotechnol* **18**, 283–287.

Sontheimer, E. J. and Carthew, R. W. (2005). Silence from within: endogenous siRNAs and miRNAs. *Cell* **122**, 9–12.

Thomas, J. W., Touchman, J. W., Blakesley, R. W.*et al.* (2003). Comparative analyses of multi-species sequences from targeted genomic regions. *Nature* **424**, 788–793.

Trinklein, N. D., Aldred, S. F., Hartman, S. J. *et al.* (2004). An abundance of bidirectional promoters in the human genome. *Genome Res* **14**, 62–66.

Vervoort, R., Gitzelmann, R., Lissens, W. *et al.* (1998). A mutation (IVS8+0.6kbdelTC) creating a new donor splice site activates a cryptic exon in an Alu-element in intron 8 of the human beta-glucuronidase gene. *Hum Genet* **103**(6), 686–693.

Wahle, E. and Keller, W. (1992) The biochemistry of 3-end cleavage and polyadenylation of messenger RNA precursors. *Annu Rev Biochem* **61**, 419–440.

Wang, J., Smith, P. J., Krainer, A. R. *et al.* (2005). Distribution of SR protein exonic splice enhancer motifs in human protein encoding genes. *Nucleic Acids Res* **33**(16), 5053–5062.

Xie, L. and Bourne, P. E. (2005). Functional coverage of the human genome by existing structures, structural genomics targets, and homology models. *PLoS Comput Biol* **1**(3), e31.

Yamashita, R., Suzuki, Y., Wakaguri, H. *et al.* (2006). DBTSS: DataBase of Human Transcription Start Sites, progress report. *Nucleic Acids Res* **34**, D86–D89.

12

Functional *in Silico* Analysis of Gene Regulatory Polymorphism

Chaolin Zhang[1,2], Xiaoyue Zhao[1], Michael Q. Zhang[1]

[1] *Cold Spring Harbor Laboratory, Cold Spring Harbor, NY, USA*

[2] *Department of Biomedical Engineering, State University of New York at Stony Brook, NY, USA*

12.1 Introduction

Gene expression refers to the cellular processes that lead to functional products (primarily proteins) from the genetic information stored in the genomic sequences. Tightly regulated gene expression for specific cell types and developmental stages in response to different physiological conditions is driven by the orchestration of complex and multilayered gene regulatory networks (GRNs) (Maniatis and Reed, 2002). Inferring GRNs is of fundamental importance and a great challenge for molecular biologists and geneticists.

Mutations, including point mutations, insertions and deletions, translocations, and duplications, play critical roles in determining biological phenotypes and disease susceptibilities by perturbing the GRNs. Among them, single nucleotide polymorphisms (SNPs) generated by point mutations occur approximately one per 1000 bases and are the predominant variations in man. The interplay between the adaptive benefits introduced by mutations and natural selection shapes the genome into unique patterns of genetic variations in different regions. Therefore, investigating the functional roles of these genetic variations provides a great opportunity for understanding complex common diseases, such as cancer. The compilation of human

Bioinformatics for Geneticists, Second Edition. Edited by Michael R. Barnes
 ISBN 978-0-470-02619-9 (HB) ISBN 978-0-470-02620-5 (PB)

and other metazoan genome sequences (see Chapters 4 and 5) and the availability of genome-wide, high-resolution genotyping data (see Chapter 3) have provided extraordinary resources for this purpose.

By either family-based linkage analysis or population-based association studies, an increasing number of genes and genomic loci have been associated with disease traits, or more recently, gene expression quantitative traits (see Chapter 16). However, because of the linkage disequilibrium (LD), it remains extremely difficult to distinguish the real causative pathogenic loci from correlated markers, the key step to transform genetic findings into mechanistic understanding of GRNs and effective prevention, diagnosis and treatment of diseases. Nevertheless, this provides an important starting point to identify functional polymorphisms. Functional polymorphisms can be classified into two categories: *cis*-acting regulatory polymorphisms, which disrupt or create regulatory elements in DNA or RNA sequences, and *trans*-acting polymorphisms, which alter protein structures and activities, and potentially affect many target loci (Buckland, 2006). Methods to predict the impact of coding polymorphisms on protein structures will be discussed in Chapter 13. Here, we introduce the rapidly emerging and improved bioinformatics tools that can help the analysis of regulatory polymorphisms. We emphasize the principles underlying the leading algorithms in the field to help geneticists understand their advantages and disadvantages. We hope that this chapter will be a practical guide for geneticists to choose available tools and resources to facilitate their experimental studies.

Gene expression regulation can take place at any step during the path of expression, including transcription, mRNA splicing and processing, export and subcellular localization, translation and post-translational modifications. These steps are often coupled with each other (Maniatis and Reed, 2002). Currently, it is still too early to build comprehensive and accurate dynamic models for truly realistic GRNs. The majority of computational methods attempt to detect *cis-trans* relationships, the basic building blocks of GRNs, by modern statistical or machine learning approaches. In this chapter, we will focus on finding *cis*-regulatory elements or modules (multiple collaborative elements) at the transcriptional level (DNA) and the splicing (RNA) level, with emphasis on mammalian species. We choose these two fields because of the extensive research efforts in recent years and their representativeness. We first introduce methods for identifying regulatory regions, such as CpG islands and promoters. Then we describe tools to pinpoint specific regulatory elements, using the analysis of transcription factor-binding sites (TFBSs) as examples. Almost all of these methods can be employed to identify regulatory elements important for other regulation steps. Specific methods and resources for studying splicing regulatory elements are then given. Finally, we summarize steps of combining the genomic variation data and the prediction of regulatory elements, and give examples of how this approach can help inference from associated alleles to causative alleles. A selection of tools and resources is given in Table 12.1.

Table 12.1 Resources related to the analysis of gene regulatory sequences

Resource name	URL	References
Genome browsers and gene structure analysis		
UCSC genome browser	http://genome.ucsc.edu	(Hinrichs *et al.*, 2006)
Ensembl	http://www.ensembl.org	(Birney *et al.*, 2006)
Entrez gene	http://www.ncbi.nlm.nih.gov	(Maglott *et al.*, 2005)
Promoter databases and resources		
EPD	http://www.epd.isb-sib.ch	(Schmid *et al.*, 2006)
DBTSS	http://dbtss.hgc.jp	(Suzuki *et al.*, 2004)
CSHLmpd	http://rulai.cshl.edu/CSHLmpd2	(Xuan *et al.*, 2005)
CpGPlot	http://www.sanger.ac.uk/Software/EMBOSS	(Larsen *et al.*, 1992)
Eponine	http://www.sanger.ac.uk/Users/td2/eponine	(Down and Hubbard, 2002)
McPromoter	http://genes.mit.edu/McPromoter.html	(Ohler *et al.*, 2001)
Dragon PF and GSF	http://research.i2r.a-star.edu.sg/promoter	(Bajic *et al.*, 2002)
FirstEF	http://rulai.cshl.edu/tools/FirstEF	(Davuluri *et al.*, 2001)
Transcription factor binding site databases		
TRANSFAC®	http://www.biobase.de/pages/index.php?id=111	(Matys *et al.*, 2003)
JASPAR	http://jaspar.cgb.ki.se	(Sandelin *et al.*, 2004)
De novo*motif finding*		
CONSENSUS	http://bifrost.wustl.edu/consensus	(Hertz and Stormo, 1999)
Gibbs motif	sampler/http://bayesweb.wadsworth.org/gibbs/gibbs.html/	(Thompson *et al.*, 2003)
MEME	http://meme.sdsc.edu/meme	(Bailey and Elkan, 1994)
AlignACE	http://atlas.med.harvard.edu	(Roth *et al.*, 1998)
MDScan	http://ai.stanford.edu/~xsliu/MDscan	(Liu *et al.*, 2002)
DWE	http://rulai.cshl.edu/cgi-bin/TRED/tred.cgi?process=analysisMotifDWEForm	(Sumazin *et al.*, 2005)
DME	http://rulai.cshl.edu/software/index1.htm	(Smith *et al.*, 2005)
CisModule	http://www.people.fas.harvard.edu/~qingzhou/CisModScan/index.html	(Zhou and Wong, 2004)
Novel TFBS prediction		
MATCH™	http://www.biobase.de/pages/index.php?id=291	(Kel *et al.*, 2003)
Storm (in CREAD)	http://rulai.cshl.edu/cread	(Smith *et al.*, 2006)
MAST	http://meme.sdsc.edu/meme/mast.html	(Bailey and Gribskov, 1998)
CisModuleScan	http://www.people.fas.harvard.edu/~qingzhou/CisModScan/index.html	(Zhou and Wong, 2004)
Splice site prediction		
ASD	http://www.ebi.ac.uk/asd-srv/wb.cgi	(Thanaraj *et al.*, 2004)
MaxEntScan	http://genes.mit.edu/burgelab/maxent/Xmaxentscan_scoreseq.html	(Yeo and Burge, 2004)
Splice site prediction by neural network	http://www.fruitfly.org/seq_tools/splice.html	(Reese *et al.*, 1997)
Splicing enhancer and silencer prediction		
ESEfinder	http://rulai.cshl.edu/tools/ESE	(Cartegni *et al.*, 2003)
RESCUE-ESE	http://genes.mit.edu/burgelab/rescue-ese	(Fairbrother *et al.*, 2004)
PESX	http://cubweb.biology.columbia.edu/pesx	(Zhang and Chasin, 2004)
Regulatory SNP analysis		
PupaSNP Finder	http://pupasuite.bioinfo.cipf.es	(Conde *et al.*, 2004)
SNPselector	http://primer.duhs.duke.edu	(Xu *et al.*, 2005)

12.2 Predicting regulatory regions

The first step of studying regulatory polymorphisms is to determine whether the polymorphisms are located in a regulatory region or in a coding region. Different steps of regulation involve very different regulatory regions. For example, promoter is the most important regulatory region that controls and regulates the very first step of gene expression: mRNA transcription. The signal for splicing lies in splice sites at the boundaries of exons, as well as exonic and intronic sequences flanking the splice sites. The mRNA stability and localization are usually controlled by regulatory elements in 5′ UTRs and/or 3′ UTRs. For organisms like man or yeast, whose gene annotations are relatively complete, genome browsers are very useful for identifying gene structures and other related annotations (Birney *et al.*, 2006; Hinrichs *et al.*, 2006). These genome browsers include both transcript supported genes and computationally predicted genes. Many other resources, including promoter databases and computational methods for promoter predictions, are also available to characterize promoters more accurately.

12.2.1 An operational definition of promoter

Promoter is commonly referred to as the DNA region that is required for controlling and regulating the transcription initiation of the gene immediately downstream. A typical eukaryotic (Pol II or protein-coding) gene contains a core promoter about 100 bp centred at the transcriptional start site (TSS) and a proximal promoter about 500 bp immediately upstream of the core promoter. For most purposes, people use the region (–500, +100) with respect to a TSS as a specific definition.

The pre-initiation complex (PIC), which comprises of many general transcription factors (GTFs), assembles onto the core promoter by interacting with several core promoter elements, such as TATA-box, Inr, DPE, BRE and DCE. The core promoter can direct transcription mediated by purified GTFs and Pol II *in vitro* at the basal level. The functional form of the PIC *in vivo* must also contain coactivators/mediators and its interactions with other TFs, which recruit the complex to the core promoter and allow for response of the polymerase to the regulatory signals. During the development, genes are turned on and off in a pre-programmed fashion, a process orchestrated by TFs, whose binding sites aggregate in the promoters near their controlled genes. A combinatorial control is achieved via different combinations of ubiquitous and cell-specific regulatory factors. Moreover, genes can initiate transcription at multiple loci (alternative promoters), creating RNA isoforms with different 5′ regions. Alternative promoters are potentially important for gene-expression regulation or generating different protein products. Complex regulation *in vivo* can also involve many more features, such as enhancers, locus control regions (LCRs), and/or scaffold/matrix attachment regions (S/MARs). Enhancers are also referred to as the distal promoter elements, which can be either upstream of, downstream of, or within a gene and can be in any orientation. It should be noted that there is no real distinction between proximal and distal (enhancer) regulatory elements, as they often involve the same set of TF-binding sites. The cooperative binding of some TFs to enhancers and proximal promoters can lead to the assembly of nucleoprotein structures termed 'enhanceosomes'. For a comprehensive review on the related biology, see the excellent book by Carey and Smale (2000).

12.2.2 CpG islands

CpG island is an important signature of 5′ regions of more than 70 per cent mammalian genes, often overlapping with, or within 1000 bases downstream of the promoter (Ioshikhes and Zhang, 2000).

Vertebrate genomic DNA is known to be generally depleted of the dinucleotide CpG. In the human genome, for example, the occurrence of CpG dinucleotides is five times less than that statistically predicted from the nucleotide composition (Bird, 1980). CpG depletion is believed to result from methylation of Cs at 80 per cent CpG dinucleotides, leading to the mutation of the methylated C to T, and thus the conversion of the CpG dinucleotides to TpG. There are, however, genomic regions of high GC content, termed 'CpG islands', where the level of methylation is significantly lower than the overall genome. In these regions, the occurrence of CpGs is significantly higher, close to the expected frequency. As defined by Gardiner-Garden and Frommer (1987), CpG islands are greater than 200 bp in length, have more than 50 per cent GC content, and have a ratio of the CpG frequency to the product of the C and G frequencies above 0.6. The CpGPlot program in the EMBOSS package can be used to map CpG islands according to this definition (Larsen *et al.*, 1992). This information is also included in the UCSC genome browser (Hinrichs *et al.*, 2006).

12.2.3 Promoter databases and resources

One promoter resource with the best quality is the Eukaryotic Promoter Database (EPD), in which transcription start sites were determined experimentally (Schmid *et al.*, 2006). With high-throughput technologies, such as 5′ SAGE (Hashimoto *et al.*, 2004) or CAGE (Carninci *et al.*, 2005), emerging for mapping TSS, EPD starts collecting TSS from these databases with a built-in quality evaluation procedure. Currently, EPD (Release 86) contains 4809 promoters, including 2540 vertebrate promoters and 1871 human promoters. Database of Transcriptional Start Sites (DBTSS) is another useful source; it is based on full-length, oligo-capped cDNA sequences and provides alternative promoter annotations (Suzuki *et al.*, 2004). The current release of DBTSS (5.2.0) contains 30 964 human promoters and 425 117 corresponding TSS. By clustering TSSs, Suzuki *et al.* found that 8308 human genes and 4276 mouse genes have alternative promoters.

The Cold Spring Harbor Laboratory mammalian promoter database (CSHLmpd) (Xuan *et al.*, 2005) is a comprehensive promoter database for man, mouse and rat.

It used all known as well as predicted transcripts to construct gene sets. The corresponding promoter information was collected from multiple resources, including EPD, DBTSS, GenBank and also computational predictions. They are integrated with an internal quality quantitation and control system. It enables users to extract the sequences of their specified regions around TSS, with a specified quality. Promoters of orthologous genes can be compared to detect sequence conservations in those regions.

Recent advances in ChIP-chip technology (see Section 12.3.4) provide the opportunity to study the genome-wide map of active promoters in specific cell types. Using this technology, Kim *et al.* (2005) experimentally located the sites of PIC binding throughout the genome in human fibroblast cells. Databases based on 5′ SAGE, CAGE (fantom3) and oligo-capping (DBTSS) technology have also started to provide tissue information of each TSS. The accumulated tissue-specific mapping will be very useful for studying how genes are differentially expressed in different tissues.

12.2.4 Computational promoter prediction

Despite the availability of experimentally validated promoter resources, computational prediction algorithms are still of great importance in identification and characterization of novel genes, as well as large-scale annotations of many other species after genome sequencing. There have been extensive efforts to improve promoter predictions computationally (see Werner, 2003, and references therein). The primary goal of these programs is to identify TSS and/or core promoter elements for all (protein-coding) genes in a genome, in contrast to the programs identifying specific transcription factor-binding sites (TFBSs) that are shared by a particular set of co-regulated genes (see Section 12.3). The underlying principle of these programs is that promoter regions have some distinctive and characteristic features different from non-promoters. A classifier is trained on experimentally validated promoters/TSSs (obtained from databases such as EPD or DBTSS), and then used to scan novel genomic sequences. Different programs differ in the features and classification algorithms used.

Features important for computational promoter prediction programs include GC content, CpG ratio, TFBS density, word compositions and core promoter elements. These have been modelled by many programs. For example, PROMOTERSCAN (Prestridge, 1995) and AUTOGENE (Kondrakhin *et al.*, 1995) are two of the earliest programs utilizing different densities of TFBSs in promoters and non-promoter sequences, together with a TATA-matrix score. Due to the very limited number of TFs with known binding motifs (see Section 12.3.2), short sequences (words) more abundant in promoters than non-promoter regions have been employed for prediction. This idea, with some variations, has been implemented in PromFind (Hutchinson, 1996), CorePromoter (Zhang, 1998) and PromoterInspector (Scherf *et al.*, 2000). CpG_Promoter is an effective algorithm discriminating the promoter-associated CpG islands from the non-promoter-associated ones (Ioshikhes and Zhang, 2000). It uses three features to train a quadratic discriminant classifier: length, GC content and the CpGratio (observed/expected). This algorithm aims only for a promoter region, not the exact locations of TSSs. Other methods, especially those developed in recent years, try to integrate as much information as possible to improve the accuracy of promoter prediction. Here we introduce a few representative ones.

More comprehensive modelling of TFBSs

TSSG and TSSW (Solovyev and Salamov, 1997) both use LDA (linear discriminant analysis) to combine (a) a TATA score, (b) triplet preferences around TSS, (c) hexamer score in three non-overlapping windows of 100 bp upstream TSS, and (d) putative binding-site scores. The program Eponine (Down and Hubbard, 2002) models the preferential spacing between binding sites of TFs and TSS as well as the over-representation of the binding sites. Over-represented binding sites with conserved spacing receive high scores and are recovered *de novo* with a relevance vector machine. It was found that TATA box and the flanking region with GC enrichment are the most important signals. A linear combination of binding site scores is then used for prediction.

Physical properties

Regulatory regions often exhibit distinct physical properties such as DNA flexibility and GC content in their sequences. McPromoter integrates such structural features into a neural network in conjunction with the Markov modelling of the sequence information from different segments (upstream, core promoter and downstream) and is able to reduce false positives (Ohler *et al.*, 2001).

Cross-species conservation

It was observed that some major promoter components such as TSS, TATA and regulatory motifs are significantly more conserved than the sequences around them. A recent program PromH (Solovyev and Shahmuradov, 2003) uses linear discriminant functions that take into account the conservation features and nucleotide sequences of promoter regions in pairs of orthologous genes. To use PromH, othologous sequences must be provided. It should be noted that they are not always available due to the difficulty of aligning orthologous sequences, especially for distal species.

CpG related versus non-CpG related

It is computationally useful and biologically meaningful to treat CpG-related promoters and non-CpG-related ones separately. Non-CpG-related promoters are more heterogeneous and therefore more difficult for computational prediction. Two programs explicitly build different promoter models for these two classes. In FirstEF (Davuluri *et al.*, 2001), CpG-related promoters and non-CpG-related ones are modelled separately, each using three quadratic discriminant functions to recognize structural and compositional features of promoter regions, first exons and first splice-donor sites in conjunction. All these functions are then incorporated into a decision tree. The predictions of the first exons and promoter regions in the human genome are available in the UCSC genome browser. Another program, Dragon Promoter Finder (Dragon PF) (Bajic *et al.*, 2002), uses sensors for three functional regions: promoters, exons and introns, and then combines them via artificial neural networks (ANNs) for GC-rich and GC-poor sequences, respectively. Each sensor is based on the frequencies of pentamers at each position. Dragon Gene Start Finder (Dragon GSF) combines Dragon PF and the prediction of presence of CpG islands by ANN (Bajic and Seah, 2003).

Performance evaluation

Promoter prediction has been a difficult problem in gene finding and characterization. Choosing appropriate programs is very important, since the types of information built into different models are not completely the same. This is further complicated by the lack of benchmark data for training and evaluation during original publication. A most recent review compared eight programs for whole human genome predictions (Bajic *et al.*, 2004). According to its comparison, Dragon GSF and FirstEF might be the good choices to start with for general promoter predictions. Approximately, they can predict more than half of promoters correctly at the cost of one or a few false predictions for each correct one, at the resolution of several hundred to 2 kb. They are quite successful in locating the transcription start sites for CpG-related promoters, but the performance for non-CpG related ones is less satisfactory due to the diverse nature of vertebrate promoter sequences. Although they are improving, current programs are still insufficient to pinpoint TSSs; therefore, it is difficult to distinguish alternative promoters.

12.3 Modelling and predicting transcription factor-binding sites

Promoter prediction and TFBS identification are closely related. While promoter prediction is to locate the beginning and *cis*-regulatory regions of a gene, the focus of computational methods modelling and predicting TFBSs is to understand *cis-trans* interactions for transcription regulation. TFBSs are short (about 6-20 bp in length), usually degenerate, and often found in promoters. Both experimental and computational methods have been developed to identify TFBSs with different throughputs and at different resolutions. There are two general problems in the computational studies of TFBSs. First, with a set of sequences (e.g., promoters) believed to be co-regulated, statistical methods are used to identify the pattern of the binding sites (motif) for regulators. Second, given the motif of a specific TF, computational methods are used to scan for putative binding sites of that factor. Note that almost all methods described here are applicable to other types of protein-DNA/RNA Interactions, such as the regulation of mRNA splicing and stability.

12.3.1 Motif representation: consensus or matrix

The binding sequence of a TF allows a certain degree of variation, creating a spectrum of binding affinity. The variations of binding sites can be collected from known target genes, mutagenesis studies (Hallikas *et al.*, 2006), phylogenetic shadowing (orthologous binding sites in different species) (Ostrin *et al.*, 2006), and *in vitro* SELEX experiments (Liu and Stormo, 2005). Several recent technologies, such as SELEX-SAGE (Roulet *et al.*, 2002) and protein-binding microarray (PBM) (Mukherjee *et al.*, 2004), allow the determination of binding specificity in a high throughput manner.

The profile or motif of binding sites can then be described with a consensus sequence. By aligning the sites, the base(s) with the largest affinity (or the most frequent base among known binding sites) at each position is chosen as a representative. For example, the consensus of *E. coli* TATA-box can be written as TATAAT or TATRNT by the IUPAC code allowing degeneracy. This representation is straightforward and useful when the motif is relatively long and conserved. However, TF motifs of higher eukaryotes are generally degenerate. Consensus cannot quantitatively reflect the binding affinities of sites and thus is not optimal for predicting the occurrence of new sites. In most applications, a position weight matrix (PWM) is a better choice.

To maintain the binding affinity, point mutations inside binding sites must be constrained. This requirement, which connects energetic constraints and base frequencies, forms the foundation for statistical mechanic motif modelling (Berg and von Hippel, 1987). When the non-random base composition of the background or a control set (e.g., the whole genome) is taken into account, the relation can be expressed as follows:

$$s_{B,j} = \ln(p_{B,j}/p_{B,0}) \tag{12.1}$$

where $s_{B,j}$ is the score (PWM element) for a base B at position j ($j = 1, 2, \ldots, J$, J is the length of the motif), $p_{B,j}$ is the frequency of the base B at position j, and $p_{B,0}$ indicates the background base frequency which does not depend on position. Choosing $p_{B,0}$ appropriately (representing the correct background contrast) can be very important for searching new binding sites in a genome. With the assumption that the binding affinity of each position is independent and additive, the total score of the site is the sum of individual position scores at all J positions:

$$s = \sum_{j=1}^{J} s_{B_j, j}, \quad (12.2)$$

where B_j is the base at position j. A straightforward interpretation of the score is the log likelihood ratio of being a site to being a non-site. Similarly, the log likelihood ratio of observing K sites can be represented by

$$S = \sum_k s_k = K \sum_{j=1}^{J} \sum_{B=A}^{T} p_{B,j} \log\left(p_{B,j} / p_{B,0}\right). \quad (12.3)$$

Here

$$I = \sum_{j=1}^{J} \sum_{B=A}^{T} p_{B,j} \log\left(p_{B,j} / p_{B,0}\right)$$

is the information content of the motif and represents the level of degeneracy.

The PWM motif model can be visualized by a 'pictogram' or motif logo (Crooks *et al.*, 2004). In these visualizations, each position is a stack of letters, reflecting the frequency of observing each nucleotide. The total height of each position can be scaled according to the information content of that position.

The PWM representation can be generalized to more complex models, such as high-order Markov models (Zhang and Marr, 1993; Roulet *et al.*, 2002), the maximum entropy model (Yeo and Burge, 2004), Bayesian networks (Barash *et al.*, 2003), and generalizations of Bayesian networks (Ben-Gal *et al.*, 2005; Zhao *et al.*, 2005), when more binding sites are available. However, in most cases, the simpler PWM model is sufficient.

12.3.2 *De novo* motif finding

There are approximately 2000 TFs in man and probably also in other mammals (Messina *et al.*, 2004; Kummerfeld and Teichmann, 2006). TF motifs determined experimentally have been collected into databases such as SCPD (Zhu and Zhang, 1999), TRANSFAC (Matys *et al.*, 2003), and JASPAR (Sandelin *et al.*, 2004), which, however, contain limited data. For example, there are currently around 600 vertebrate motifs in TRANSFAC, many of which are redundant and/or derived from only a few known sites. In order to discover novel motifs, we must resort to *de novo* motif-finding algorithms. Given a set of related sequences, as described in Section 12.3.4, these algorithms attempt to find the most over-represented patterns of short sequences in a reasonable time. Numerous algorithms have been proposed in the past decade. These algorithms differ in motif representation, objective function and the procedure for optimization. More importantly, they incorporate different data or prior knowledge in the modelling and therefore can fit various situations. In the following, we introduce computational motif-finding algorithms according to these considerations, rather than technical details. Interested users are also referred to a recent review comparing 13 popular methods (Tompa *et al.*, 2005).

Finding the most over-represented motifs

Most earlier approaches attempted to identify motifs with the most over-represented binding sites. These algorithms achieve the goal by optimizing the local sequence alignment with Equation 12.3 or its variation as an objective function. Since neither the motif nor the binding sites are known, this optimization is a combinatorial problem, which needs heuristic searching strategies to get a reasonably good solution in a feasible time. Representative heuristic strategies include Greedy Search, implemented in CONSENSUS (Hertz and Stormo, 1999); expectation maximization (EM), implemented in MEME (Bailey and Elkan, 1994); and the Gibbs sampler (Lawrence *et al.*, 1993). In the last two approaches, the motif model and site locations are optimized iteratively by pretending that either the motif or the sites are known at the beginning. After the best motif is recovered, the sites are erased to identify the second best motifs and so on. A few other programs, such as MDScan (Liu *et al.*, 2002) and Weeder (Pavesi *et al.*, 2001), start from searching for over-represented consensus (allowing degeneracy), rather than from the matrix search directly.

Additional features included in these programs, as well as in their variations such as AlignACE (Roth *et al.*, 1998), Bioprospector (Liu *et al.*, 2001) and the Improbizer (http://www.cse.ucsc.edu/~kent/improbizer/), make them smarter. For example, MEME allows users to specify whether every sequence has one or multiple binding sites. MEME, CONSENSUS and Weeder can optimize motif length automatically. MEME and Bioprospector can limit the search to only two block motifs or palindromic motifs. Higher-order Markov models have been used in Biospector and MDScan to characterize background sequences more accurately, by which a significant improvement has been observed.

Note that repeat sequences should be masked before motif finding, and it might always be worth trying multiple (similar) programs to see whether a motif is detected consistently. It is reported that some programs are complementary to each other, and are thus able to improve specificity when used in combination (Tompa *et al.*, 2005). The identified matrices can be compared with matrices of known TFs to identify putative regulators (Schones *et al.*, 2005).

Finding discriminative motifs

In contrast to most motif-finding algorithms, discriminative motif finding attempts to identify the best motifs which discriminate two sets of sequences. This is extremely useful in studying gene-expression regulation, for example, to explain different responses of two groups of genes after a stimuli or to distinguish genes expressed in different tissues. The most discriminative motif is not necessarily the most abundant. Master regulators, such as $E2F$ family members and *myc*, have thousands of binding sites over the human genome (Cawley *et al.*, 2004), whereas some TFs may regulate only a handful of targets. Therefore, discriminative but subtle signals might be overwhelmed by common binding sites without careful modelling. The word-counting algorithms compare the relative enrichment of each word (may allow degeneracy) between the foreground sequences and background sequences. This approach is very effective in identifying short and less degenerate motifs, such as many typical TF sites in yeast (Zhang, 1999). Other programs of this category, with slightly different scoring functions, include WORDUP (Pesole *et al.*, 1992), DMOTIFS (Sinha, 2003) and DWE (Sumazin *et al.*, 2005). To detect more degenerate motifs, we must use the matrix model. The statistical framework used to find over-represented motifs (Liu *et al.*, 1995) can be easily extended to model the relative over-representation, as shown recently (Smith *et al.*, 2005). In addition, the implemented program, discriminative matrix enumerator (DME), 'exhaustively' searches the discretized space of matrices followed by a local optimization step. This method is very successful in identifying tissue-specific motifs, which can be highly degenerate (Smith *et al.*, 2005, 2006).

Finding conserved motifs

Sequence conservation across different species is an important indicator of functionality. Phylogenetic footprinting is referred to as the identification of functional regions by comparing orthologous genomic sequences between species (Fickett and Wasserman, 2000). With more sequenced genomes available, comparative analysis of noncoding regions has become an important approach in detecting promoters or regulatory regions in general (Bejerano *et al.*, 2004; Siepel *et al.*, 2005). Several earlier methods for detecting conserved blocks from a multiple alignment have been evaluated by Stojanovic *et al.* (1999). Programs designed for very long alignments of syntenic regions have also become available (see (Blanchette *et al.*, 2004, and references therein). A more detailed examination of comparative genomic approaches is presented in Chapter 6.

With the alignment of multiple orthologous sequences, it is possible to detect short TF motifs that are significantly more conserved than random. This idea has been applied to screen regulatory elements conserved in multiple yeast species (Kellis *et al.*, 2003) and recently in four mammalian species (Xie *et al.*, 2005). The motifs identified include many known ones as well as novel ones.

Given a set of presumably co-regulated sequences and their orthologues, it is also possible to incorporate both over-representation and conservation into motif-finding algorithms. A straightforward strategy is to use a two-step procedure: find conserved regions and then search for over-represented motifs only in those regions

(Wasserman *et al.*, 2000). The two steps can be applied in the opposite order: first over-represented motifs are identified separately in each species or in the pooled data, and then motifs without significant conservation are eliminated (GuhaThakurta *et al.* 2002; Pritsker *et al.* 2004; Li *et al.*, 2005). It was argued that these methods are somewhat ad hoc and may miss over-represented but divergent motifs or conserved motifs not very over-represented. Therefore, the two criteria can also be integrated into a single statistical framework for optimization (Prakash *et al.*, 2004; Li and Wong, 2005). However, since more parameters need to be estimated from an often small data set, these methods may also identify noisy motifs (Li *et al.*, 2005).

It should also be noted that the conservation of regulatory regions may vary widely. In principle, the regulatory programs that control early development in metazoan systems tend to be extremely complex, almost always involving distal enhancers and/or complicated locus control regions (LCRs). Subtle change of these programs can lead to dramatic effects. Therefore, lineage developmental master TFs and their binding sites are often more conserved. In contrast, in the terminally differentiated tissues, the regulatory program is often relatively simple; *cis*-regulatory regions in the promoters tend to be closer to TSS and many TFBS are less conserved among distant species.

Constructing cis-*regulatory modules*

Since genes are always regulated by multiple TFs and composite binding sites (*cis*-regulatory modules (CRMs)), simultaneous detection of CRMs rather than individual sites may provide a better specificity. A module can be composed of multiple sites of the same type (homotypic) or different types (heterotypic). Palindromic motifs can be regarded as a special type of CRMs and are common for TFs. However, most CRMs studies require that individual motifs be known. *De novo* CRM discovery is a much more difficult problem, which usually needs larger data sets (Bussemaker *et al.*, 2001; Zhou and Wong, 2004; Gupta and Liu, 2005). The Cis-Module algorithm has been applied to identify CRMs important for muscle-specific expression in *Ciona savignyi* (Johnson *et al.*, 2005). REDUCE (Bussemaker *et al.*, 2001), MotifRegressor (Conlon *et al.*, 2003), MARSMotif (Das *et al.*, 2004), and, more recently, MatrixREDUCE (Foat *et al.*, 2005) and MARSMotif-M (Das *et al.*, 2006) are regression-based algorithms that can maximize the explained variation of gene expression by a limited number of motifs in combination.

12.3.3 Predicting novel binding sites

Given a motif determined experimentally or computationally, an important task is to search new sequences for novel binding sites, using consensus matching or matrix scoring. However, one must first assess the quality of the motif and determine a threshold before using it for searching new sites. One way to do this is to perform a standard classification test by which both the threshold score and the motif length may be optimized by minimizing the classification (Bayesian) error. The MATCH program (Kel *et al.*, 2003) included in the TRANSFAC database uses precalculated thresholds with different stringencies. The storm program in the CREAD package is tailored to search a set of sequences for multiple motifs very quickly (Smith *et al.*, 2006).

Because of the short length and degeneracy of most TF motifs, the signal to noise ratio is quite low. Therefore, predicting novel functional binding sites of a known motif is by no means easier than *de novo* motif finding (Hu *et al.*, 2005). It was estimated that for a typical motif, without any other information except the matrix, the specificity of genomic search can be as low as 0.001, meaning one functional sites among 1000 predictions (Wasserman and Sandelin, 2004). These false predictions, which might bind TFs with high affinity *in vitro*, are never used *in vivo*, suggesting that important signals also reside outside the cognate binding sites to distinguish from decoy sites. These include CRMs, chromatin structure, and DNA stability and flexibility, which are important for determining the affinity and accessibility of the binding sites. Conservation information in other species is not accessible for the cellular machinery, but is effective for eliminating false positives by one order of magnitude (Wasserman and Sandelin, 2004).

One way to predict CRMs is by evaluating the significance of the co-occurrence of TFBSs within a certain distance. This approach requires the least prior knowledge. Claverie and Sauvaget (1985) published one of the earliest methods to detect two sites in a fixed distance and the same orientation in the heat-shock promoters. Alternatively, more subtle rules can be learned from known functional sites co-occurring in the same regions. Although still limited, several databases, such as COMPEL and TRRD, have started to collect experimentally validated CRMs (Heinemeyer *et al.*, 1998), and this will greatly facilitate advances in this field. An interesting example is the identification of regulatory modules that confer muscle-specific gene expression (Wasserman and Fickett, 1998), where logistic regression was used to combine matrix scores for multiple sites. The authors reported that focusing on CRMs rather than individual binding sites can reduce false positives by two orders of magnitude while retaining more than half of the true sites. In a recent study, not only co-occurrences, but also geometric constraints, were modelled quantitatively from known examples. The implemented program, called EEL, identified vertebrate enhancers successfully (Hallikas *et al.*, 2006). Therefore, the analysis of CRMs can improve prediction accuracy to a level that makes follow-up experimental investigations feasible.

Other functional annotations and co-localization information are also helpful. Computational approaches incorporating DNA mechanical properties and nucleosome structures are still rare, but represent important directions for the future.

12.3.4 Experimental approaches to identify co-regulated targets

Two types of high-throughput technologies, namely, microarrays and genomic occupancy assays, are highly effective to identify co-regulated genes, thereby narrowing down the putative regions of functional binding sites dramatically. These technologies are now routinely used to study gene-expression regulation.

Many studies have been based on expression microarrays. For the purpose of regulation studies, this tool becomes more powerful when data are collected under multiple conditions or at multiple time-points after the perturbation of the upstream TF (e.g., TF knockout, mutation in DNA binding domain or knock-down). The underlying assumption is that genes with similar expression profiles (co-expression) are likely to be regulated by the same factor(s) (co-regulation). Classical approaches are based on the clustering analysis to identify genes with correlated expression patterns, from which one could try to identify *cis*-elements enriched in their promoters (Spellman *et al.*, 1998; Hughes *et al.*, 2000). Some algorithms are specifically designed for motif finding by looking for 'tight clusters' of expression profiles (Tseng and Wong, 2005).

However, co-expression is not equal to co-regulation. When the perturbation is on some master regulators, it can also activate/repress many downstream TFs, as in the case of heat shock or other stress responses. Because multiple TFs are involved, responsive target genes would be a mixture of direct targets for different TFs. ChIP-chip (Lee *et al.*, 2002; Cawley *et al.*, 2004; Odom *et al.*, 2004; Carroll *et al.*, 2005) and ChIP-tag technologies (Impey *et al.*, 2004; Sabo *et al.*, 2004; Ng *et al.*, 2005; Wei *et al.*, 2006) allow for more direct detection of genomic regions occupied by endogenous transcription factors (see Chapter 9 for some examples of ChIP data). ChIP-chip cross-links binding proteins to chromatins *in vivo*. Immunoprecipitated DNA fragments are then hybridized to genomic DNA microarrays or sequenced by SAGE-tag technology. The power of these approaches has been demonstrated in many applications. Despite non-specific binding and cross-linking, it has been shown that highly enriched chip regions are very accurate in predicting *bona fide* targets. In a study of ER binding sites in chromosomes 21 and 22 with ChIP-chip data, all 57 predictions were validated to be real (Carroll *et al.*, 2005). Due to the current resolution of 500-2000 bp, computational analysis for motif finding and binding site prediction is indispensable. Almost all the motif-finding and TFBS prediction tools described in Sections 12.3.2 and 12.3.3 can be applied to chromatin occupancy data.

12.4 Predicting regulatory elements for splicing regulation

The next level of gene expression regulation is RNA processing, including capping, splicing, polyadenylation, editing, stability and transport. Many of these (in particular, the first three) steps are co-transcriptional and hence coupled to transcriptional regulation (Maniatis and Reed, 2002). In this section, we will focus on mRNA splicing, especially alternative splicing (AS), which is responsible for generating diverse protein isoforms from a single gene locus. Recent estimates of alternatively spliced genes are more than 60 per cent in man and probably other mammals (Lander *et al.*, 2001; Modrek and Lee, 2002; Johnson *et al.*, 2003). Alternative splicing plays critical roles in many regulatory pathways in metazoans, including those controlling cell growth, cell death, differentiation and development (Black, 2003). Aberrant splicing has been implicated in a large number of human diseases (Faustino and Cooper, 2003).

The boundaries of introns and exons are marked by splice sites. The canonical splice sites are composed of GU dinucleotide in the exon/intron boundary (5′ss or donor site), and AG dinucleotide in the intron/exon boundary (3′ss or acceptor site). Each dinucleotide is flanked by a larger, less conserved sequence. The branch site and polypyrimidine tract close to the 3′ss in the intron are also critical for splicing. Minor types of splice sites (e.g., AU/AC introns), although less than 0.1 per cent, also exist (Burset *et al.*, 2000). Although the key biochemical steps of splicing have been worked out, far less is known about the mechanism of accurate splicing regulation. In mammals, the signal carried by the splice sites is insufficient to drive specific exon recognition. The tight regulation of alternative splice site selection in response to different physiological conditions is mediated through the interactions of numerous *cis*-elements outside the splice sites, such as enhancers and repressors, and a very large protein/snRNA complex, the splicesome, which is composed of hundreds of proteins (Rappsilber *et al.*, 2002; Zhou *et al.*, 2002) and five critical snRNAs. Furthermore, the splicing studies rely heavily on *in vitro* systems (mini-genes). It is difficult to generate a mini-gene construct recapitulating the same splicing pattern *in vivo*. Previous computational analyses mainly focused on the detection of AS events and evolutionary properties based on cDNA/EST data. Methods for facilitating the understanding of the splicing regulation are emerging in recent years.

12.4.1 Statistical modelling and prediction of splice sites

The motif of splice sites can also be represented by consensus sequences or PWMs, as described in Section 12.3.1. However, this is perhaps the most appropriate place to test complex models, because a very large number of known splice sites are available by transcript-genome alignment. These methods include the higher-order Markov model (Zhang and Marr, 1993), the maximum entropy model (Yeo and Burge, 2004) and Bayesian networks (Chen *et al.*, 2005), all of which attempt to model the dependencies among different positions. It was claimed that integrating correlations between nucleotides helps discriminate authentic splice sites from pseudo splice sites.

12.4.2 Identification of splicing enhancers and silencers

Cis-elements for splicing regulation can be in exons or introns, and can be enhancers or silencers. They are important for both constitutive splicing and alternative splicing (Smith and Valcarcel, 2000). The best-characterized enhancers and silencers are recognized by one of two splicing factor classes: hnRNPs (heterogeneous nuclear ribonucleoprotein) and SR (serine/arginine rich) proteins. They are usually identified in tissue-specific exons or disease mutants with aberrant splicing. Many exonic splicing enhancers (ESEs) are purine rich. A well-studied example is the 73-nucleotide ESE in the alternative exon M2 of the mouse IgM gene. This ESE can stimulate splicing when inserted into a heterologous regulated intron of the *Drosophila* doublesex (dsx) gene (see references in Liu *et al.*, 1998). In another example, a single-nucleotide C/T silent transition causes the skipping of exon 7 in the human *SMN2* gene, which mediates the severity of spinal muscular atrophy (SMA) in the absence of the wild-type *SMN1* gene, a paralogue of *SMN2* (Lorson *et al.*, 1999). It was demonstrated that the transition disrupts an SF2/ASF-dependent ESE and creates an ESS bound by hnRNP A1 (Cartegni and Krainer, 2002; Kashima and Manley, 2003). Comprehensive lists of exonic and intronic splicing regulatory sequences reported in literature have been compiled (Ladd and Cooper, 2002; Zheng, 2004). Since elements recognized by a splicing factor seem to be very degenerate, it is difficult to derive motifs from these known examples.

SR-protein-binding sites and ESEfinder

The SR proteins are a family of highly conserved, serine/arginine-rich RNA-binding proteins. They are essential splicing factors with overlapping functions, involved in early steps of spliceosome assembly. They can regulate the selection of alternative splice sites in a concentration-dependent manner, in part by antagonizing the activity of hnRNP A1 (see references in Liu *et al.*, 1998). The binding sites of four SR proteins, including SF2/ASF, SC35, SRp40 and SRp55, have been determined by the Krainer Laboratory with a functional SELEX assay (Liu *et al.*, 1998, 2000). The ESE matrix for each SR protein was then derived from the winner sequences, as described in Section 12.3.1. These matrices have been included in a web-based resource called ESEfinder (Cartegni *et al.*, 2003), which can be used to predict and visualize novel ESEs of these SR proteins.

Exonic splicing silencers

A systematic screening for exonic splicing silencers has been performed by the Burge Laboratory (Wang *et al.*, 2004). The principle of the system is similar to that of SELEX except that (i) the screening is GFP-based, and (ii) random sequences are not selected for a specific splicing factor, but can be any element with repressive activities. This screening identified 141 ESS decamers. Most of these are probably repressive when introduced into heterogeneous gene contexts. These decamers can be clustered by sequence similarity to identify consensus motifs. Although some of them resemble the binding sites of known SFs, such as hnRNP A1 and hnRNP H, generally it is not clear which SFs can specifically recognize these sequences.

Enhancers and silencers derived in silico

Pure computational screening for enhancers and silencers has also been performed (Fairbrother *et al.*, 2002; Sironi *et al.*, 2004; Yeo *et al.*, 2004; Zhang and Chasin, 2004). As it is difficult to obtain a list of co-regulated AS events, these studies do not focus on specific splicing factors. Instead, they assume different densities of regulatory elements in different genic regions and attempt to identify general enhancing or repressive elements. In particular, in the RESCUE approach, Fairbrother *et al.* assumed that for an exon to be constitutively included, weak splice sites have to be complemented by a higher density of ESEs nearby. They also assumed that the density of ESEs in exonic regions is higher than that in intronic regions. All hexamers were scored by these two criteria of relative over-representation. As a result, 238 hexamers were identified as potential human ESEs (called RESCUE-ESEs), and then were clustered into 10 motifs by their sequence similarity. Some of these motifs resemble the binding sites of known SFs. When inserted into the test exon of a mini-gene construct, these hexamers can indeed enhance exon inclusion. This approach has been extended to mouse, zebrafish and fugu (Yeo *et al.*, 2004) for predicting intronic splicing enhancers (ISEs). Among hundreds of the predicted ISEs hexamers, the GGG motif is the most prevalent and is contained in majority of ISEs hexamers. The predicted ESEs were included in the RESCUE-ESE server and can be used to scan new sequences (Fairbrother *et al.*, 2004).

Since ESEs are largely imposed on coding constraints, Zhang and Chasin (2004) argued that the codon usage bias might complicate the ESE identification. Therefore, in their study, only constitutively spliced, internal, non-coding exons were used. They assumed that ESEs should have a relative enrichment in these spliced non-coding exons compared to 5′ UTRs of intronless genes and pseudoexons (intronic regions flanked by splice-site like sequences with a similar length as real exons). Using this approach, they identified 2069 octomers as putative ESEs (PESEs) and 974 octomers as putative ESSs (PESSs). An online tool is also available for identifying the occurrences of these PESEs and PESSs.

It should be noted that the *trans*-acting factors interacting with these enhancers and silencers predicted *in silico* are not obvious, although their splicing role has been demonstrated in *in vitro* splicing assays and endogenous genes (Zhang *et al.*, 2005). Since these elements were originally identified by constitutively spliced exons, although they are probably important for alternative splicing as well, it is not very clear how much bias might be introduced. The constitutive and alternative splicing may result from the different balances of the same positive and negative splicing signals. Moreover, at least for several known SFs, their specific elements are associated predominantly with tissue-specifically spliced exons. These elements are probably missed by these *in silico* predictions.

The enhancers and silencers derived by different approaches have also been compared for their ability to predict splicing alterations caused by point mutations. Interestingly, although these methods (SELEX, RESCUE ESEs, PESEs) are comparable in predictive power, the overlap is moderate (Wang *et al.*, 2005; Zhang *et al.*, 2005). This might imply that a number of ESEs, as well as ESSs, have not been identified.

12.4.3 Splicing microarrays

The catalogue of regulatory elements described above is only the first step towards understanding splicing regulation. A more challenging step is to understand how the interaction of these regulatory elements and splicing factors generates highly regulated splicing patterns in different tissues types or under different conditions in response to stimuli. The combinatorial interaction of multiple factors may contribute greatly to the subtle regulation. The variation of expression of splicing factors may also add another layer of complexity. Currently, the detailed mechanistic studies of splicing regulation are limited to only a few model systems using mini-gene constructs. One concern is whether the rules inferred from these models can reflect the regulation *in vivo* and whether they are general enough to extend to other genes. As in the study of transcriptional regulation, the high-throughput technologies measuring splicing activities and protein-RNA interactions under specific conditions can provide invaluable information.

The feasibility of using microarrays to study the regulation of RNA splicing was first demonstrated in yeast (Clark *et al.*, 2002). These splicing microarrays are designed to distinguish splicing variants by probes in exon bodies and exon junctions (Modrek and Lee, 2002). It was demonstrated that the loss of key mRNA-processing factors leads to dramatic splicing defects, which can be measured by microarrays. Since 40–60 per cent of the mammalian genes have introns (a typical gene has about eight introns) compared to 3.8 per cent intron-containing genes in yeast, detecting AS in a mammalian system with microarrays has become possible only very recently (Johnson *et al.*, 2003; Pan *et al.*, 2004; Li *et al.*, 2006; Sugnet *et al.*, 2006). The largest study so far measured AS in more than 50 human tissues and cell lines, using 36-nucleotide oligonucleotide probes tiled on every consecutive exon junctions of Refseq genes (Johnson *et al.*, 2003). Since probes are included even for ‘constitutive’ exon junctions, where there is no cDNA/EST evidence of alternative splicing, this platform can identify novel AS events in a more unbiased manner than in the EST sequencing approach. However, it should be noted that these data are noisy, since each AS event is represented by only one or two probes (in affymetrix arrays, more than 10 probes are used to summarize the mRNA abundance level). As there are very limited choices for probe positions in exon junctions to optimize hybridization efficiency and specificity, some probes might behave poorly and thus do not reflect the correct abundance of the splicing junctions. In the recent affymetrix exon microarrays, multiple probes are tiled on each exon to get more reliable signals of exon inclusion. However, there are no junction probes, and it is difficult to infer splicing patterns. In several other studies, arrays are designed for the AS events with transcript evidence (Pan *et al.*, 2004; Ule *et al.*, 2005; Sugnet *et al.*, 2006). In these designs, each AS event is represented by multiple probes in exons, introns or exon junctions. Therefore, they can measure AS more accurately and are suitable to infer rules of splicing regulation. For example, motifs have been discovered from the flanking intronic regions of brain/muscle specific AS exons (Sugnet *et al.*, 2006). Other microarray-based assays have also been developed. For example, in DASL, a high specificity is achieved by the ligation of a pair of oligos across the splice junction. The primer extension step before ligation makes the choice of probes more flexible (Fan *et al.*, 2004). This approach has been applied to screen a panel of prostate cancer tissues and normal tissues to identify signature splicing events (Li *et al.*, 2006) and compared with conventional microarrays (Zhang *et al.*, 2006). The combination of splicing microarrays and knock-down experiments of splicing factors have been demonstrated as a powerful tool to dissect important pathways regulated through tissue-specific splicing (Ule *et al.*, 2005). Obviously, splicing microarrays will be routinely used to study GRNs in the coming years.

Several technical difficulties should be noted. As transcription and splicing are intrinsically coupled, it is very difficult to separate their individual contributions to the steady-state levels of the transcripts; there have been several attempts (Johnson *et al.*, 2003; Cline *et al.*, 2005; Li *et al.*, 2006; Shai *et al.*, 2006). Moreover, the alterations detected by microarrays are individual AS events, which are local. Usually, the complete isoforms and protein products cannot be tracked unambiguously (Wang *et al.*, 2003). Another challenge, even combined with knock-out experiments, is to distinguish direct effects and indirect effects, as the direct measurement of protein–RNA interaction on a large scale is still in its infancy (Ule *et al.*, 2003).

12.5 Evaluating the functional importance of regulatory polymorphisms

The functional importance of polymorphisms is determined by how they can affect the regulation of gene expression. In the most common scenario for geneticists, a polymorphism, such as a SNP, is linked to a disease or gene expression trait, and one has to ask whether it is a causative allele or just an allele with LD. The first step is to determine the regulatory region it falls in, the basis to choose appropriate tools. For polymorphisms in promoters, it would be a strong indication of disrupting transcription if they overlap with a transcription factor-binding sites. Putative binding sites can be predicted, as described in Section 12.3. Although the false-positive rate of such *in silico* prediction for a single binding site is usually high, accuracy can be improved by incorporating information from different resources.

- If a binding site overlaps with the SNP, do the different alleles have different binding affinities (motif scores)? A study of 127 SNPs in the promoters of cell-cycle checkpoint genes found that a majority of them potentially affect binding affinity, as validated at a high success rate by Gel shift *in vitro* (Belanger *et al.*, 2005). A further step is to ask whether the identified binding site co-occurs with others; this might form a CRM.

- Is the SNP in the core promoter? According to the report assays that evaluated the effect of 674 haplotypes in 247 promoters for promoter activity, there is a strong indication that functional polymorphisms are close to the TSS (Buckland *et al.*, 2005). This is probably due to the fact that the density of regulatory elements is higher near the TSS. Subtle changes in the mechanical or geometric properties of DNA may also alter the efficiency of transcription (Buckland, 2006) even if the SNP does not change a binding site directly.

- Is the overlapped region conserved? The UCSC genome browser is an excellent resource for interactive analysis of cross-species conservation. See Bejerano *et al.* (2005) for a step-by-step tutorial on how to screen conserved regions for functional elements.

- Is there any *in vivo* binding evidence? Besides the known binding sites collected in different databases, it is also worth checking whether a genome-wide assay for chromatin occupancy has been performed for the TF under study.

- Is there any functional annotation (such as gene ontology, tissue-specific expression or protein interaction) for the associated gene? Is it involved in the pathway implicated in the disease?

These *in silico* analyses are economical and fast and often can provide good evidence about the potential importance of the gene. The candidates which pass these filtering steps may be validated by a reporter assay before further investigation.

If the SNP is located in the exonic region or flanking intronic region, it might alter mRNA splicing.

- Disruption of splice sites is very suggestive of aberrant splicing.

- Otherwise, we must determine whether it disrupts or creates splicing enhancers or silencers. Successful examples have been demonstrated to identify point mutations affecting ESEs with ESEfinder (Liu *et al.*, 2001; Cartegni and Krainer, 2002). It is easier to predict erroneous exon skipping than other splicing errors (e.g., cryptic splicing), as shown in these examples.

- Can splicing alteration lead to dramatic change of protein product? Skipping of an internal coding exon whose length is not a multiple of three or a non-sense mutation generally induces mRNA non-sense mediated decay (NMD) or a large truncation at the 3′ part of the protein. If the exon overlaps with an important domain, skipping of the exon can also have dramatic effects. Different isoforms can be virtually translated into different proteins, which can then be analysed for their protein structures (see Chapter 13).

Several tools have been developed for partial automation of these analyses (Conde *et al.*, 2004; Xu *et al.*, 2005).

Acknowledgements

MQZ Laboratory is supported by grants from NIH, NSF and CSHL Associations.

References

Bailey, T. and Elkan, C. (1994). Fitting a mixture model by expectation maximization to discover motifs. In *Biopolymers: The Second International Conference on Intelligent Systems for Molecular Biology*. Menlo Park, CA: AAAI Press.

Bailey, T. L. and Gribskov, M. (1998). Combining evidence using p-values: application to sequence homology searches. *Bioinformatics* **14**, 48–54.

Bajic, V. B. and Seah, S. H. (2003). Dragon Gene Start Finder: an advanced system for finding approximate locations of the start of gene transcriptional units. *Genome Res* **13**, 1923–1929.

Bajic, V. B., Seah, S. H., Chong, A. *et al.* (2002). Dragon Promoter Finder: recognition of vertebrate RNA polymerase II promoters. *Bioinformatics* **18**, 198–199.

Bajic, V. B., Tan, S. L., Suzuki, Y. *et al.* (2004). Promoter prediction analysis on the whole human genome. *Nat Biotech* **22**, 1467–1473.

Barash, Y., Kaplan, T., Friedman, N. *et al.* (2003). Modeling dependencies in protein-DNA binding sites. *Proceedings of the 7th International Conference on Research in Computational Molecular Biology (RECOMB)*, 28–37.

Bejerano, G., Pheasant, M., Makunin, I. *et al.* (2004). Ultraconserved elements in the human genome. *Science* **304**, 1321–1325.

Bejerano, G., Siepel, A. C., Kent, W. J. *et al.* (2005). Computational screening of conserved genomic DNA in search of functional noncoding elements. *Nat Methods* **2**, 535–545.

Belanger, H., Beaulieu, P., Moreau, C. *et al.* (2005). Functional promoter SNPs in cell cycle checkpoint genes. *Hum Mol Genet* **14**, 2641–2648.

Ben-Gal, I., Shani, A., Gohr, A. *et al.* (2005). Identification of transcription factor binding sites with variable-order Bayesian networks. *Bioinformatics* **21**, 2657–2666.

Berg, O. G. and Von Hippel, P. H. (1987). Selection of DNA binding sites by regulatory proteins: statistical-mechanical theory and application to operators and promoters. *J Mol Biol* **193**, 723–743.

Bird, A. P. (1980). DNA methylation and the frequency of CpG in animal DNA. *Nucleic Acids Res* **8**, 1499–1504.

Birney, E., Andrews, D., Caccamo, M. *et al.* (2006). Ensembl 2006. *Nucleic Acids Res* **34**, D556–561.

Black, D. L. (2003). Mechanisms of alternative pre-messenger RNA splicing. *Annu Rev Biochem* **72**, 291–336.

Blanchette, M., Kent, W. J., Riemer, C. *et al.* (2004). Aligning multiple genomic sequences with the threaded blockset aligner. *Genome Res* **14**, 708–715.

Buckland, P. R. (2006). The importance and identification of regulatory polymorphisms and their mechanisms of action. *Biochim Biophys Acta* **1762**, 17–28.

Buckland, P. R., Hoogendoorn, B., Coleman, S. L. *et al.* (2005). Strong bias in the location of functional promoter polymorphisms. *Hum Mutat* **26**, 214–223.

Burset, M., Seledtsov, I. A. and Solovyev, V. V. (2000). Analysis of canonical and non-canonical splice sites in mammalian genomes. *Nucleic Acids Res* **28**, 4364–4375.

Bussemaker, H. J., Li, H. and Siggia, E. D. (2001). Regulatory element detection using correlation with expression. *Nat Genet* **27**, 167–174.

Carey, M. and Smale, S. (2000). *Transcriptional Regulation in Eukaryotes: Concepts, Strategies, and Techniques.* Cold Spring Harbor, NY: Cold Spring Harbor Laboratory Press.

Carnici, P., Kasukawa, T., Katayama, S. *et al.* (2005). The transcriptional landscape of the mammalian genome. *Science* **309**, 1559–1563.

Carroll, J. S., Liu, X. S., Brodsky, A. S. *et al.* (2005). Chromosome-wide mapping of estrogen receptor binding reveals long-range regulation requiring the forkhead protein FoxA1. *Cell* **122**, 33–43.

Cartegni, L. and Krainer, A. R. (2002). Disruption of an SF2/ASF-dependent exonic splicing enhancer in SMN2 causes spinal muscular atrophy in the absence of SMN1. *Nat Genet* **4**, 377–384.

Cartegni, L., Wang, J., Zhu, Z. *et al.* (2003). ESEfinder: a web resource to identify exonic splicing enhancers. *Nucleic Acids Res* **31**, 3568–3571.

Cawley, S., Bekiranov, S., Ng, H. H. *et al.* (2004). Unbiased mapping of transcription factor binding sites along human chromosomes 21 and 22 points to widespread regulation of noncoding RNAs. *Cell* **116**, 499–509.

Chen, T.-M., Lu, C.-C. and Li, W.-H. (2005). Prediction of splice sites with dependency graphs and their expanded Bayesian networks. *Bioinformatics* **21**, 471–482.

Clark, T. A., Sugnet, C. W. and Ares, M., Jr. (2002). Genomewide analysis of mRNA processing in yeast using splicing-specific microarrays. *Science* **296**, 907–910.

Claverie, J.-M. and Sauvaget, I. (1985). Assessing the biological significance of primary structure consensus patterns using sequence databanks. I. Heat-shock and glucocorticoid control elements in eukaryotic promoters. *Comput Appl Biosci* **1**, 95–104.

Cline, M. S., Blume, J., Cawley, S. *et al.* (2005). ANOSVA: a statistical method for detecting splice variation from expression data. *Bioinformatics* **21**, i107–115.

Conde, L., Vaquerizas, J. M., Santoyo, J. *et al.* (2004). PupaSNP Finder: a web tool for finding SNPs with putative effect at transcriptional level. *Nucleic Acids Res* **32**, W242–248.

Conlon, E. M., Liu, X. S., Lieb, J. D. *et al.* (2003). Integrating regulatory motif discovery and genome-wide expression analysis. *Proc Natl Acad Sci U S A* **100**, 3339–3344.

Crooks, G. E., Hon, G., Chandonia, J.-M. *et al.* (2004). WebLogo: a sequence logo generator. *Genome Res* **14**, 1188–1190.

Das, D., Banerjee, N. and Zhang, M. Q. (2004). Interacting models of cooperative gene regulation. *Proc Natl Acad Sci U S A* **101**, 16234–16239.

Das, D., Nahle, Z. and Zhang, M. (2006). Adaptively inferring human transcriptional subnetworks. *Mol Syst Biol* (in press).

Davuluri, R. V., Grosse, I. and Zhang, M. Q. (2001). Computational identification of promoters and first exons in the human genome. *Nat Genet* **29**, 412–417.

Down, T. A. and Hubbard, T. J. P. (2002). Computational detection and location of transcription start sites in mammalian genomic DNA. *Genome Res* **12**, 458–461.

Fairbrother, W. G., Yeh, R.-F., Sharp, P. A. *et al.* (2002). Predictive identification of exonic splicing enhancers in human genes. *Science* **297**, 1007–1013.

Fairbrother, W. G., Yeo, G. W., Yeh, R. *et al.* (2004). RESCUE-ESE identifies candidate exonic splicing enhancers in vertebrate exons. *Nucleic Acids Res* **32**, W187–190.

Fan, J.-B., Yeakley, J. M., Bibikova, M. *et al.* (2004). A versatile assay for high-throughput gene expression profiling on universal array matrices. *Genome Res* **14**, 878–885.

Faustino, N. A. and Cooper, T. A. (2003). Pre-mRNA splicing and human disease. *Genes Dev* **17**, 419–437.

Fickett, J. W. and Wasserman, W. W. (2000). Discovery and modeling of transcriptional regulatory regions. *Curr Opin Biotech* **11**, 19–24.

Foat, B. C., Houshmandi, S. S., Olivas, W. M. *et al.* (2005). Profiling condition-specific, genome-wide regulation of mRNA stability in yeast. *Proc Natl Acad Sci U S A* **102**, 17675–17680.

Gardiner-Garden, M. and Frommer, M. (1987). CpG islands in vertebrate genomes. *J Mol Biol* **196**, 261–282.

Guhathakurta, D., Palomar, L., Stormo, G. D. *et al.* (2002). Identification of a novel *cis*-regulatory element involved in the heat shock response in *Caenorhabditis elegans* using microarray gene expression and computational methods. *Genome Res* **12**, 701–712.

Gupta, M. and Liu, J. S. (2005). De novo *cis*-regulatory module elicitation for eukaryotic genomes. *Proc Natl Acad Sci U S A* **102**, 7079–7084.

Hallikas, O., Palin, K., Sinjkushina, N. *et al.* (2006). Genome-wide prediction of mammalian enhancers based on analysis of transcription-factor binding affinity. *Cell* **124**, 47–59.

Hashimoto, S.-I., Suzuki, Y., Kasai, Y. *et al.* (2004). 5[prime]-end SAGE for the analysis of transcriptional start sites. *Nat Biotech* **22**, 1146–1149.

Heinemeyer, T., Wingender, E., Reuther, I. *et al.* (1998). Databases on transcriptional regulation: TRANSFAC, TRRD and COMPEL. *Nucleic Acids Res* **26**, 362–367.

Hertz, G. Z. and Stormo, G. D. (1999). Identifying DNA and protein patterns with statistically significant alignments of multiple sequences. *Bioinformatics* **15**, 563–577.

Hinrichs, A. S., Karolchik, D., Baertsch, R. *et al.* (2006). The UCSC Genome Browser Database: update 2006. *Nucleic Acids Res* **34**, D590–598.

Hu, J., Li, B. and Kihara, D. (2005). Limitations and potentials of current motif discovery algorithms. *Nucleic Acids Res* **33**, 4899–4913.

Hughes, J. D., Estep, P. W., Tavazoie, S. *et al.* (2000). Computational identification of *cis*-regulatory elements associated with groups of functionally related genes in *Saccharomyces cerevisiae*. *J Mol Biol* **296**, 1205–1214.

Hutchinson, G. B. (1996).The prediction of vertebrate promoter regions using differential hexamer frequency analysis. *Comput Appl Biosci* **12**, 391–398.

Impey, S., McCorkle, S. R., Cha-Molstad, H. *et al.* (2004). Defining the CREB regulon: a genome-wide analysis of transcription factor regulatory regions. *Cell* **119**, 1041–1054.

Ioshikhes, I. P. and Zhang, M. Q. (2000). Large-scale human promoter mapping using CpG islands. *Nat Genet* **26**, 61–63.

Johnson, D. S., Zhou, Q., Yagi, K. *et al.* (2005). *De novo* discovery of a tissue-specific gene regulatory module in a chordate. *Genome Res* **15**, 1315–1324.

Johnson, J. M., Castle, J., Garrett-Engele, P. *et al.* (2003). Genome-wide survey of human alternative pre-mRNA splicing with exon junction microarrays. *Science* **302**, 2141–2144.

Kashima, T. and Manley, J. L. (2003). A negative element in SMN2 exon 7 inhibits splicing in spinal muscular atrophy. *Nat Genet* **34**, 460–463.

Kel, A. E., Gossling, E., Reuter, I. *et al.* (2003). MATCHTM: a tool for searching transcription factor binding sites in DNA sequences. *Nucleic Acids Res* **31**, 3576–3579.

Kellis, M., Patterson, N., Endrizzi, M. *et al.* (2003). Sequencing and comparison of yeast species to identify genes and regulatory elements. *Nature* **423**, 241–254.

Kim, T. H., Barrera, L. O., Zheng, M. *et al.* (2005). A high-resolution map of active promoters in the human genome. *Nature* **436**, 876–880.

Kondrakhin, Y. V., Kel, A. E., Kolchanov, N. A. *et al.* (1995). Eukaryotic promoter recognition by binding sites for transcription factors. *Comput Appl Biosci* **11**, 477–488.

Kummerfeld, S. K. and Teichmann, S. A. (2006). DBD: a transcription factor prediction database. *Nucleic Acids Res* **34**, D74–81.

Ladd, A. and Cooper, T. (2002). Finding signals that regulate alternative splicing in the post-genomic era. *Genome Biol* **3**, reviews0008.1–reviews0008.16.

Lander, E., Linton, L., Birren, B. *et al.* (2001). Initial sequencing and analysis of the human genome. *Nature* **409**, 860–921.

Larsen, F., Gundersen, G., Lopez, R. *et al.* (1992). CpG islands as gene markers in the human genome. *Genomics* **13**, 1095–1107.

Lawrence, C. E., Altschul, S. F., Boguski, M. S. *et al.* (1993). Detecting subtle sequence signals: a Gibbs sampling strategy for multiple alignment. *Science* **262**, 208–214.

Lee, T. I., Rinaldi, N. J., Robert, F. *et al.* (2002). Transcriptional regulatory networks in *Saccharomyces cerevisiae. Science* **298**, 799–804.

Li, H.-R., Wang-Rodriguez, J., Nair, T. M. *et al.* (2006). Two-dimensional transcriptome profiling: identification of messenger RNA isoform signatures in prostate cancer from archived paraffin-embedded cancer specimens. *Cancer Res* **66**, 4079–4088.

Li, X. and Wong, W. H. (2005). Sampling motifs on phylogenetic trees. *Proc Natl Acad Sci USA* **102**, 9481–9486.

Li, X., Zhong, S. and Wong, W. H. (2005). Reliable prediction of transcription factor binding sites by phylogenetic verification. *Proc Natl Acad Sci U S A* **102**, 16945–16950.

Liu, H.-X., Chew, S. L., Cartegni, L. *et al.* (2000) Exonic splicing enhancer motif recognized by human SC35 under splicing conditions. *Mol Cell. Biol* **20**, 1063–1071.

Liu, H.-X., Zhang, M. and Krainer, A. R. (1998). Identification of functional exonic splicing enhancer motifs recognized by individual SR proteins. *Genes Dev* **12**, 1998–2012.

Liu, J. and Stormo, G. D. (2005). Combining SELEX with quantitative assays to rapidly obtain accurate models of protein–DNA interactions. *Nucleic Acids Res* **33**, e141.

Liu, J. S., Neuwald, A. F. and Lawrence, C. E. (1995). Bayesian models for multiple local sequence alignment and Gibbs sampling strategies. *J Am Stat Assoc* **90**, 1156–1170.

Liu, X., Brutlag, D. and Liu, J. (2001). BioProspector: discovering conserved DNA motifs in upstream regulatory regions of co-expressed genes. *Pac Symp Biocomput* **6**, 127–138.

Liu, X. S., Brutlag, D. L. and Liu, J. S. (2002). An algorithm for finding protein-DNA binding sites with applications to chromatin-immunoprecipitation microarray experiments. *Nat Biotech* **20**, 835–839.

Lorson, C. L., Hahnen, E., Androphy, E. J. *et al.* (1999). A single nucleotide in the SMN gene regulates splicing and is responsible for spinal muscular atrophy. *Proc Natl Acad Sci U S A* **96**, 6307–6311.

Maglott, D., Ostell, J., Pruitt, K. D. *et al.* (2005). Entrez Gene: gene-centered information at NCBI. *Nucleic Acids Res* **33**, D54–58.

Manitis, T. and Reed, R. (2002). An extensive network of coupling among gene expression machines. *Nature* **416**, 499–506.

Matys, V., Fricke, E., Geffers, R. *et al.* (2003). TRANSFAC(R): transcriptional regulation, from patterns to profiles. *Nucleic Acids Res* **31**, 374–378.

Messina, D. N., Glasscock, J., Gish, W. *et al.* (2004). An ORFeome-based analysis of human transcription factor genes and the construction of a microarray to interrogate their expression. *Genome Res* **14**, 2041–2047.

Modrek, B. and Lee, C. (2002). A genomic view of alternative splicing. *Nat Genet* **30**, 13–19.

Mukherjee, S., Berger, M. F., Jona, G. *et al.* (2004). Rapid analysis of the DNA-binding specificities of transcription factors with DNA microarrays. *Nat Genet* **36**, 1331–1339.

Ng, P., Wei, C.-L., Sung, W.-K. *et al.* (2005). Gene identification signature (GIS) analysis for transcriptome characterization and genome annotation. *Nat Methods* **2**, 105–111.

Odom, D. T., Zizsperger, N., Gordon, D. B. *et al.* (2004). Control of pancreas and liver gene expression by HNF transcription factors. *Science* **303**, 1378–1381.

Ohler, U., Niemann, H., Liao, G.-C. *et al.* (2001). Joint modeling of DNA sequence and physical properties to improve eukaryotic promoter recognition. *Bioinformatics* **17**, S199–206.

Ostrin, E. J., Li, Y., Hoffman, K. *et al.* (2006). Genome-wide identification of direct targets of the *Drosophila* retinal determination protein Eyeless. *Genome Res* **16**, 466–476.

Pan, Q., Shai, O., Misquitta, C. *et al.* (2004). Revealing global regulatory features of mammalian alternative splicing using a quantitative microarray platform. *Mol Cell* **16**, 929–941.

Pavesi, G., Mauri, G. and Pesole, G. (2001). An algorithm for finding signals of unknown length in DNA sequences. *Bioinformatics* **17**, S207–214.

Pesole, G., Prunella, N., Liuni, S. *et al.* (1992). WORDUP: an efficient algorithm for discovering statistically significant patterns in DNA sequences. *Nucleic Acids Res* **20**, 2871–2875.

Prakash, A., Blanchette, M., Sinha, S. *et al.* (2004) Motif discovery in heterogeneous sequence data. *Pac Symp Biocomput* **9**, 348–359.

Prestridge, D. S. (1995). Predicting Pol II promoter sequences using transcription factor binding sites. *J Mol Biol* **249**, 923–932.

Pritsker, M., Liu, Y.-C., Beer, M. A. *et al.* (2004). Whole-genome discovery of transcription factor binding sites by network-level conservation. *Genome Res* **14**, 99–108.

Rappsilber, J., Ryder, U., Lamond, A. I. *et al.* (2002). Large-scale proteomic analysis of the human spliceosome. *Genome Res* **12**, 1231–1245.

Reese, M., Eckman, F., Kulp, D. *et al.* (1997). Improved splice site detection in Genie. *J Comput Biol* **4**, 311–323.

Roth, F. P., Hughes, J. D., Estep, P. W. *et al.* (1998). Finding DNA regulatory motifs within unaligned noncoding sequences clustered by whole-genome mRNA quantitation. *Nat Biotech* **16**, 939–945.

Roulet, E., Busso, S., Camargo, A. A. *et al.* (2002). High-throughput SELEX-SAGE method for quantitative modeling of transcription-factor binding sites. *Nat Biotech* **20**, 831–835.

Sabo, P. J., Hawrylycz, M., Wallace, J. C. *et al.* (2004). Discovery of functional noncoding elements by digital analysis of chromatin structure. *Proc Natl Acad Sci U S A* **101**, 16837–16842.

Sandelin, A., Alkema, W., Engstrom, P. *et al.* (2004). JASPAR: an open-access database for eukaryotic transcription factor binding profiles. *Nucleic Acids Res* **32**, D91–94.

Scherf, M., Klingenhoff, A. and Werner, T. (2000). Highly specific localization of promoter regions in large genomic sequences by PromoterInspector: a novel context analysis approach. *J Mol Biol* **297**, 599–606.

Schmid, C. D., Perier, R., Praz, V. *et al.* (2006). EPD in its twentieth year: towards complete promoter coverage of selected model organisms. *Nucleic Acids Res* **34**, D82–85.

Schones, D. E., Sumazin, P. and Zhang, M. Q. (2005). Similarity of position frequency matrices for transcription factor binding sites. *Bioinformatics* **21**, 307–313.

Shai, O., Morris, Q. D., Blencowe, B. J. *et al.* (2006). Inferring global levels of alternative splicing isoforms using a generative model of microarray data. *Bioinformatics* **22**(5), 606–613.

Siepel, A., Bejerano, G., Pedersen, J. S. *et al.* (2005). Evolutionarily conserved elements in vertebrate, insect, worm, and yeast genomes. *Genome Res* **15**, 1034–1050.

Sinha, S. (2003). Discriminative motifs. *J Comput Biol* **10**, 599–615.

Sironi, M., Menozzi, G., Riva, L. *et al.* (2004). Silencer elements as possible inhibitors of pseudoexon splicing. *Nucleic Acids Res* **32**, 1783–1791.

Smith, A. D., Sumazin, P., Xuan, Z. *et al.* (2006). DNA motifs in human and mouse proximal promoters predict tissue-specific expression. *Proc Natl Acad Sci U S A* **103**, 6275–6280.

Smith, A. D., Sumazin, P. and Zhang, M. Q. (2005). Identifying tissue-selective transcription factor binding sites in vertebrate promoters. *Proc Natl Acad Sci U S A* **102**, 1560–1565.

Smith, C. W. J. and Valcarcel, J. (2000). Alternative pre-mRNA splicing: the logic of combinatorial control. *Trends Biochem Sci* **25**, 381–388.

Solovyev, V. and Salamov, A. (1997). The Gene-Finder computer tools for analysis of human and model organisms genome sequences. *Proc Int Conf Intell Syst Mol Biol* **5**, 294–302.

Solovyev, V. and Shahmuradov, I. (2003). PromH: promoters identification using orthologous genomic sequences. *Nucleic Acids Res* **31**, 3540–3545.

Spellman, P. T., Sherlock, G., Zhang, M. Q. *et al.* (1998). Comprehensive identification of cell cycle-regulated genes of the yeast *Saccharomyces cerevisiae* by microarray hybridization. *Mol Biol Cell* **9**, 3273–3297.

Stojanovic, N., Florea, L., Riemer, C. *et al.* (1999). Comparison of five methods for finding conserved sequences in multiple alignments of gene regulatory regions. *Nucleic Acids Res* **27**, 3899–3910.

Sugnet, C. W., Srinivasan, K., Clark, T. A. *et al.* (2006). Unusual intron conservation near tissue-regulated exons found by splicing microarrays. *PLoS Comput Biol* **2**, e4.

Sumazin, P., Chen, G., Hata, N. *et al.* (2005). DWE: discriminating word enumerator. *Bioinformatics* **21**, 31–38.

Suzuki, Y., Yamashita, R., Sugano, S. *et al.* (2004). DBTSS, DataBase of transcriptional start sites: progress report 2004. *Nucleic Acids Res* **32**, D78–81.

Thanaraj, T. A., Stamm, S., Clark, F. *et al.* (2004). ASD: the alternative splicing database. *Nucleic Acids Res* **32**, D64–69.

Tompa, M., Li, N., Bailey, T. L. *et al.* (2005). Assessing computational tools for the discovery of transcription factor binding sites. *Nat Biotech* **23**, 137–144.

Thompson, W., Rouchka, E. C. and Lawrence, C. E. (2003). Gibbs Recursive Sampler: finding transcription factor binding sites. *Nucl Acids Res* **31**, 3580–3585.

Tseng, G. C. and Wong, W. H. (2005). Tight clustering: a resampling-based approach for identifying stable and tight patterns in data. *Biometrics* **61**, 10–16.

Ule, J., Jensen, K. B., Ruggiu, M. *et al.* (2003). CLIP identifies Nova-regulated RNA networks in the brain. *Science* **302**, 1212–1215.

Ule, J., Ule, A., Spencer, J. *et al.* (2005). Nova regulates brain-specific splicing to shape the synapse. *Nat Genet* **37**, 844–852.

Wang, H., Hubbell, E., Hu, J.-S. *et al.* (2003). Gene structure-based splice variant deconvolution using a microarry platform. *Bioinformatics* **19**, i315–322.

Wang, J., Smith, P. J., Krainer, A. R. *et al.* (2005). Distribution of SR protein exonic splicing enhancer motifs in human protein-coding genes. *Nucleic Acids Res* **33**, 5053–5062.

Wang, Z. F., Rolish, M. E., Yeo, G. *et al.* (2004). Systematic identification and analysis of exonic splicing silencers. *Cell* **119**, 831–845.

Wasserman, W. W. and Fickett, J. W. (1998). Identification of regulatory regions which confer muscle-specific gene expression. *J Mol Biol* **278**, 167–181.

Wasserman, W. W., Palumbo, M., Thompson, W. *et al.* (2000). Human-mouse genome comparisons to locate regulatory sites. *Nat Genet* **26**, 225–228.

Wasserman, W. W. and Sandelin, A. (2004). Applied bioinformatics for the identification of regulatory elements. *Nat Rev Genet* **5**, 276–287.

Wei, C.-L., Wu, Q., Vega, V. B. *et al.* (2006). A global map of p53 transcription-factor binding sites in the human genome. *Cell* **124**, 207–219.

Werner, T. (2003) The state of the art of mammalian promoter recognition. *Brief Bioinform* **4**, 22–30.

Xie, X., Lu, J., Kulbokas, E. J. *et al.* (2005). Systematic discovery of regulatory motifs in human promoters and 3[prime] UTRs by comparison of several mammals. *Nature* **434**, 338–345.

Xu, H., Gregory, S. G., Hauser, E. R. *et al.* (2005). SNPselector: a web tool for selecting SNPs for genetic association studies. *Bioinformatics* **21**, 4181–4186.

Xuan, Z., Zhao, F., Wang, J. *et al.* (2005). Genome-wide promoter extraction and analysis in human, mouse, and rat. *Genome Biol* **6**, R72.

Yeo, G. and Burge, C. B. (2004). Maximum entropy modeling of short sequence motifs with applications to RNA splicing signals. *Comput Biol* **11**, 377–394.

Yeo, G., Hoon, S., Venkatesh, B. *et al.* (2004). Variation in sequence and organization of splicing regulatory elements in vertebrate genes. *Proc Natl Acad Sci U S A* **101**, 15700–15705.

Zhang, C., Li, H.-R., Fan, J.-B. *et al.* (2006). Profiling alternatively spliced mRNA isoforms for prostate cancer classification. *BMC Bioinformatics* **7**(1): 202.

Zhang, M. (1999). Promoter analysis of co-regulated genes in the yeast genome. *Comput Chem* **23**, 233–250.

Zhang, M. Q. (1998). Identification of human gene core promoters *in silico*. *Genome Res* **8**, 319–326.

Zhang, M. Q. and Marr, T. G. (1993). A weight array method for splicing signal analysis. *Comput Appl Biosci* **9**(5), 499–509.

Zhang, X. H.-F. and Chasin, L. A. (2004). Computational definition of sequence motifs governing constitutive exon splicing. *Genes Dev* **18**, 1241–1250.

Zhang, X. H.-F., Kangsamaksin, T., Chao, M. S. P. *et al.* (2005). Exon inclusion is dependent on predictable exonic splicing enhancers. *Mol Cell Biol* **25**, 7323–7332.

Zhao, X., Huang, H. and Speed, T. P. (2005). Finding short DNA motifs using permuted Markov models. *J Comput Biol* **12**, 894–906.

Zheng, Z.-M. (2004). Regulation of alternative RNA splicing by exon definition and exon sequences in viral and mammalian gene expression. *J Biomed Sci* **11**, 278–294.

Zhou, Q. and Wong, W. H. (2004). CisModule: *de novo* discovery of *cis*-regulatory modules by hierarchical mixture modeling. *Proc Natl Acad Sci U S A* **101**, 12114–12119.

Zhou, Z., Licklider, L. J., Gygi, S. P. *et al.* (2002). Comprehensive proteomic analysis of the human spliceosome. *Nature* **419**, 182–185.

Zhu, J. and Zhang, M. Q. (1999). SCPD: a promoter database of the yeast Saccharomyces cerevisiae. *Bioinformatics* **15**, 607–611.

13

Amino-Acid Properties and Consequences of Substitutions

Matthew J. Betts and Robert B. Russell

Structural and Computational Biology Programme, EMBL, Meyerhofstrasse 1, 69117 Heidelberg, Germany

13.1 Introduction

Since the earliest protein sequences and structures were determined, it has been clear that the positioning and properties of amino acids are key to understanding many biological processes (Pal *et al.*, 2006). For example, the first-determined protein structure, haemoglobin, provided a molecular explanation for the genetic disease sickle cell anaemia. A single nucleotide mutation leads to a substitution of glutamate in normal individuals with valine in those who suffer the disease. The substitution leads to a lower solubility of the deoxygenated form of haemoglobin, and it is thought that this causes the molecules to form long fibres within blood cells that lead to the unusual sickle-shaped cells that give the disease its name.

Haemoglobin is just one of many examples now known where single mutations can have drastic effects on protein structure, function and associated phenotype. The current availability of thousands or even millions of DNA and protein sequences means that we now have knowledge of many mutations, either naturally occurring or synthetic. Mutations can occur within one species, or between species at a wide variety of evolutionary distances. Whether mutations cause diseases or have subtle or drastic effects on protein function is often unknown; however, large-scale efforts are under way to quantify these effects (Cavallo and Martin, 2005; Karchin *et al.*, 2005).

Bioinformatics for Geneticists, Second Edition. Edited by Michael R. Barnes
 ISBN 978-0-470-02619-9 (HB) ISBN 978-0-470-02620-5 (PB)

The aim of this chapter is to give some guidance on how to interpret mutations that occur within genes that encode for proteins. Both of us have been approached previously by geneticists who want help interpreting mutations by protein sequence and structure information. This chapter attempts to summarize our thought processes when giving such help. Specifically, we discuss the nature of mutations, and the properties of amino acids in a variety of different protein contexts. The hope is that this discussion will help in anticipating or interpreting the effect that a particular amino-acid change will have on protein structure and function. We will first highlight features of proteins that are relevant to considering mutations: cellular environments, three-dimensional structure and evolution. Then we will discuss classifications of the amino acids by evolutionary, chemical or structural principles, and the role of amino acids of different classes in protein structure and function in different contexts. Last, we will review several studies of mutations, including naturally occurring variations, SNPs, site-directed mutations, mutations that allow adaptive evolution, and post-translational modification.

13.2 Protein features relevant to amino-acid behaviour

It is beyond the scope of this chapter to discuss the basic principles of proteins, since this can be gleaned from any introductory biochemistry textbook. However, a number of general principles of proteins are important to place any mutation in the correct context.

13.2.1 Protein environments

A feature of key importance is cellular location. Different parts of cells can have very different chemical environments with the consequence that many amino acids behave differently. The biggest difference is between *soluble* proteins and *membrane* proteins. Whereas soluble proteins tend to be surrounded by water molecules, membrane proteins are surrounded by lipids. Roughly speaking, this means that these two classes behave in an ‘inside-out’ fashion relative to each other. Soluble proteins tend to have polar or hydrophilic residues on their surfaces, whereas membrane proteins tend to have hydrophobic residues on the surface that interacts with the membrane.

Soluble proteins also come in several flavours. The biggest difference is between those that are *extracellular* and those that are *cytosolic* (or *intracellular*). The cytosol is quite different from the more aqueous environment outside the cell; the density of proteins and other molecules affects the behaviour of some amino acids quite drastically, especially cysteine. Outside the cell, cysteines in proximity to one another can be oxidized to form disulphide bonds, sulphur–sulphur covalent linkages that are important for protein folding and stability. However, the reducing environment inside the cell makes the formation of these bonds very difficult; in fact, they are so rare as to warrant special attention.

Cells also contain numerous compartments, the organelles, which can also have slightly different environments from each other. Proteins in the nucleus often interact with DNA, meaning they contain different preferences for amino acids on their surfaces (e.g., positive amino acids or those containing amides most suitable for interacting with the negatively charged phosphate backbone). Some organelles, such as mitochondria or chloroplasts, are quite similar to the cytosol, while others, such as lysosomes or Golgi apparati, are more akin to the extracellular environment. It is important to consider the likely cellular location of any protein before considering the consequences of amino-acid substitutions.

A detailed hierarchical description of cellular location is one of the three main branches of the classification provided by the Gene Ontology Consortium (Ashburner *et al.*, 2000), the others being 'molecular function' and 'biological process'. The widespread adoption of this vocabulary by sequence databases and others should enable more sophisticated investigation of the factors governing the various roles of proteins.

13.2.2 Protein structure

Proteins themselves also contain different microenvironments. For soluble proteins, the surface lies at the interface with water and thus tends to contain more polar or charged amino acids than one finds in the core of the protein, which is more likely to comprise hydrophobic amino acids. Proteins also contain regions that are directly involved in protein function, such as active sites or binding sites, in addition to regions that are less critical to the protein function, and where mutations are likely to have fewer consequences. We will discuss many specific roles for particular amino acids in protein structures in the sections below, but it is important to remember that the context of any amino acid can vary greatly depending on its location in the protein structure.

13.2.3 Protein evolution

Proteins are nearly always members of homologous families. Knowledge of the family a protein belongs in will generally give insights into the possible function, but several things should be considered. Two processes can give rise to homologous protein families: *speciation* or *duplication*. Proteins related by speciation only are referred to as 'orthologues', and, as the name suggests, these proteins have the same function in different species. Proteins related by duplications are referred to as 'paralogues'. Successive rounds of speciation and intragenomic duplication can lead to confusing situations where it becomes difficult to say whether paralogy or orthology applies.

To be maintained in a genome over time, paralogous proteins are likely to evolve different functions (or have a dominant negative phenotype, and so resist decay by point mutation (Gibson and Spring, 1998)). Differences in function can range from subtle differences in substrate (e.g., malate versus lactate dehydrogenases), to only weak similarities in molecular function (e.g., hydrolases), to complete differences in cellular location and function (e.g., an intracellular signalling domain homologous to a secreted growth factor (Schoorlemmer and Goldfarb, 2001)). At the other extreme, the molecular function may be identical, but the cellular function may be altered, as in the case of enzymes with differing tissue specificities.

Similarity in molecular function generally correlates with sequence identity. Mouse and human proteins with sequence identities in excess of 85 per cent are likely to be orthologues, provided there are no other proteins with higher sequence identity in either organism. Orthology between more distantly related species (e.g., man and yeast) is harder to assess, since the evolutionary distance between organisms can make it virtually impossible to distinguish orthologues form paralogues by simple measures of sequence similarity. An operational definition of orthology can sometimes be used, for example, if the two proteins are each other's best match in their respective genomes. However, there is no substitute for constructing a phylogenetic tree of the protein family to identify which sequences are related by speciation events. Assignment of orthology and paralogy is perhaps the best way of determining likely equivalences of function. Unfortunately, complete genomes are unavailable for most organisms. Some rough rules of thumb can be used: function is often conserved down to 40 per cent protein sequence identity, with the broad functional class being conserved to 25 per cent identity (Wilson *et al.*, 2000).

When considering a mutation, it is important to consider how conserved the position is within other homologous proteins. Conservation across all homologues (paralogues and orthologues) should be considered carefully. These amino acids are likely to play key structural roles, or a role in a common functional theme (i.e., catalytic mechanism). Other amino acids may play key roles only in the particular orthologous group (i.e., they may confer specificity on a substrate), thus meaning they vary when considering all homologues.

13.2.4 Protein function

Protein function is key to understanding the consequences of amino-acid substitution. Enzymes such as trypsin (Figure 13.1) tend to have highly conserved active sites involving a handful of polar residues. In contrast, proteins that function primarily only to interact with other proteins, such as fibroblast growth factors (Figure 13.2), interact over a large surface, virtually any amino acid being important in mediating the interaction (Plotnikov *et al.*, 1999). In other cases, multiple functions make the situation even more confusing; for example, a protein kinase (Hanks *et al.*, 1988) can both catalyse a phosphorylation event and bind specifically to another protein, such as cyclin (Jeffrey *et al.*, 1995).

It is not possible to discuss all of the possible functional themes here, but we emphasize that functional information, if known, should be considered whenever studying the effects of substitution.

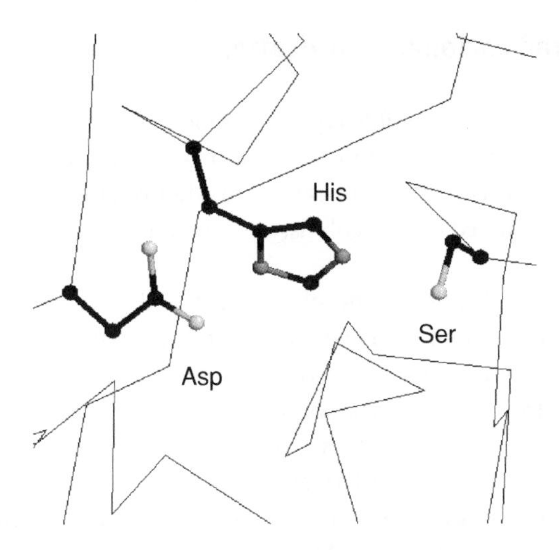

Figure 13.1 RasMol (Sayle and Milner-White, 1995) figure showing the catalytic Asp/His/Ser triad in trypsin (PDB code 1mct (Berman *et al.*, 2000)). Figure generated by the authors using data from Sayle and Milner-White, 1995. Permission not required

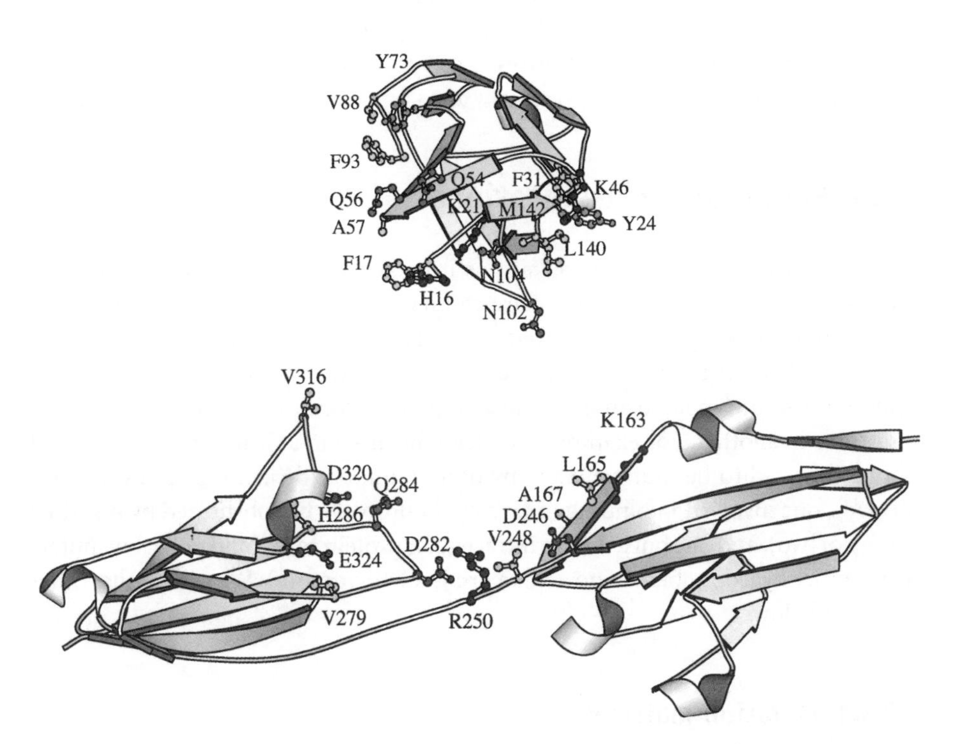

Figure 13.2 Molscript (Kraulis, 1991) figure showing fibroblast growth factor interaction with its receptor (code 1cvs (Plotnikov *et al.*, 1999)). Residues at the interface are labelled. The two molecules have been pulled apart for clarity. Figure generated by the authors using data from Kraulis, 1991, and Plotnikov *et al.*, 1999. Permission not required

13.2.5 Post-translational modification

Although there are only 20 possible types of amino acid that can be incorporated in a protein sequence upon translation of DNA, there are many more variations that can occur through subsequent modification. In addition, the gene-specified protein sequence can be shortened by proteolysis, or lengthened by addition of amino acids at either terminus.

Two common modifications, phosphorylation and glycosylation, are discussed in the context of the amino acids where they most often occur (tyrosine, serine, threonine and asparagine; see below). We direct the reader to the review by Krishna and Wold (1993) for more information on many other known types and specific examples. The main conclusion is that modifications are highly specific, with specificity provided by primary, secondary and tertiary protein structure, although the detailed mechanisms are obscure. The biological function of the modified proteins is also summarized, from the reversible phosphorylation of serine, threonine, and tyrosine residues that occurs in signalling, to the formation of disulphide bridges and other cross-links that stabilize tertiary structure, and on to the covalent attachment of lipids that allows anchorage to cell membranes. More detail on biological effects is given by Parekh and Rohlff (1997), especially where it concerns possible therapeutic applications. Many diseases arise by abnormalities in post-translational modification, and these are not necessarily apparent from genetic information alone.

13.3 Amino-acid classifications

We have a natural tendency to classify, as it makes the world around us easier to understand. As amino acids often share common properties, several classifications have been proposed. This is useful, but a little dangerous if over-interpreted. Always remember that, for the reasons discussed above, it is very difficult to put all amino acids of the same type into an invariant group. A substitution in one context can be disastrous in another. For example, a cysteine involved in a disulphide bond would not be expected to be mutatable to any other amino acid (i.e., it is in a group on its own), one involved in binding to zinc could probably be substituted by histidine (group of two), and one buried in an intracellular protein core could probably mutate to any other hydrophobic amino acid (a group of 10 or more). We will discuss other examples below.

13.3.1 Mutation matrices

One means of classifiying amino acids is a mutation matrix (or substitution or exchange matrix). This is a set of numbers that describe the propensities of exchanging one amino acid for another (for a thorough review and explanation, see Durbin *et al.*, 1998). These are derived from large sets of aligned sequences by counting the number of times that a particular substitution occurs, and comparing this to what would be expected by chance. High values indicate that a substitution is seen often in nature and so is favourable, and vice versa. The values in the matrix are usually calculated on some model of evolutionary time, to account for the fact that different pairs of sequences are at different evolutionary distances. Probably the best-known matrices are the point accepted mutation (PAM) matrices of Dayhoff *et al.* (1978) and the BLOSUM matrices (Henikoff and Henikoff, 1992).

Mutation matrices are very useful as rough guides to how good or bad a particular change will be. Another useful feature is that they can be calculated for different data sets to account for some of the protein features that affect amino-acid properties, such as cellular locations (Jones *et al.*, 1994) or different evolutionary distances (e.g., orthologues or paralogues (Henikoff and Henikoff, 1992)). Several mutation matrices are reproduced in the appendix to this chapter.

13.3.2 Classification by physical, chemical and structural properties

Although mutation matrices are very useful for protein sequence alignments, especially in the absence of known three-dimensional structures, they do not precisely describe the likelihood and effects of particular substitutions at particular sites in the sequence. Position-specific substitution matrices can be generated for the family of interest, such as the profile-HMM models generated by HMMER (Eddy, 1998) and provided by Pfam (Bateman *et al.*, 2000), and those generated by PSI-BLAST (Altschul *et al.*, 1997). However, these are automatic methods suited to database searching and identification of new members of a family, and do not really give any qualitative information on the chemistry involved at particular sites.

Taylor presented a classification that explains mutation data through correlation with the physical, chemical and structural properties of amino acids (Taylor, 1986). The major factor is the size of the side chain, closely followed by its hydrophobicity. The effects of different amino acids on protein structure can account for mutation data when these physicochemical properties do not. For example, hydrophobicity and size differ widely between glycine, proline, aspartic acid and glutamic acid. However, they are still closely related in mutation matrices because they prefer sharply turning regions on the surface of the protein; the phi and psi bonds of glycine are unconstrained by any side chain, Proline forces a sharp turn because its side chain is bonded to the backbone nitrogen as well as to carbon, and aspartate and glutamate prefer to expose their charged side chains to solvent.

The Taylor classification is normally displayed as a Venn diagram (Figure 13.3). The amino acids are positioned on this by multidimensional scaling of Dayhoff's mutation matrix, and then grouped by common physicochemical properties. Size is subcategorized into small and tiny (with large included by implication). Affinity

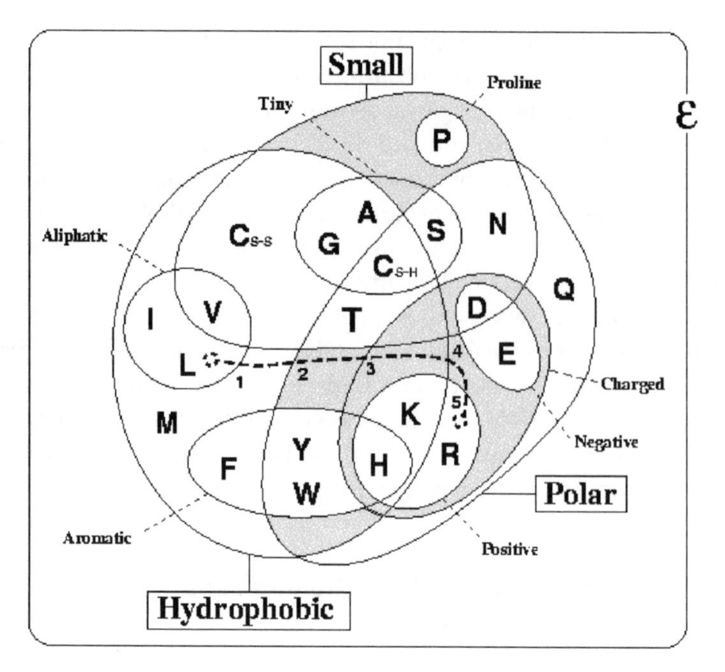

Figure 13.3 Venn diagram illustrating the properties of amino acids

for water is described by several sets: polar and hydrophobic, which overlap, and charged, which is divided into positive and negative. Sets of aromatic and aliphatic amino acids are also marked. These properties were enough to distinguish between most amino acids. However, properties such as hydrogen-bonding ability and the previously mentioned propensity for sharply turning regions are not described well. Although these factors are less important, on average, and would confuse the effects of more important properties if included on the diagram, the dangers of relying on simple classifications are apparent. This can be overcome somewhat by listing all amino acids that belong to each subset (defined as an intersection or union of the sets) in the diagram – for example, 'small and non-polar' – and including extra subsets to describe important additional properties. These subsets can be used to give qualitative descriptions of each position in a multiple alignment, by associating the positions with the smallest subset that includes all the amino acids found at that position. This may suggest alternative amino acids that could be engineered into the protein at each position.

13.4 Properties of the amino acids

The sections that follow will first consider several major properties that are often used to group amino acids together. Note that amino acids can be in more than one group, and that sometimes properties as different as 'hydrophobic' and 'hydrophilic' can be applied to the same amino acids. These properties are summarized interactively at http://www.russell.embl-heidelberg.de/aas/.

13.4.1 Hydrophobic amino acids

Probably the most common broad division of amino acids is into those that prefer to be in an aqueous environment (hydrophilic) and those that do not (hydrophobic). The latter can be divided according to whether they have *aliphatic* or *aromatic* side chains.

Aliphatic side chains

Strictly speaking, aliphatic means that the side chain contains only hydrogen and carbon atoms. By this strict definition, the aliphatic side chains are alanine, isoleucine, leucine, proline and valine. The extreme shortness of alainine's side chain means that it is not particularly hydrophobic, and proline has an unusual geometry that gives it special roles in proteins, as we shall discuss below. Although it also contains a sulphur atom, we often conveniently consider methionine in the same category as isoleucine, leucine and valine. The unifying theme is that they contain largely non-reactive and flexible side chains that are ideally suited for packing in the protein interior.

Aliphatic side chains are very non-reactive, and are thus rarely involved directly in protein function, though they can play a role in substrate recognition. In particular, hydrophobic amino acids can be involved in binding/recognition of hydrophobic ligands such as lipids.

Several other amino acids also contain aliphatic regions. For example, arginine, lysine, glutamate and glutamine are *amphipathic*, meaning that they contain hydrophobic and polar parts. All contain two or more aliphatic carbons that connect the protein backbone to the non-aliphatic portion of the side chain. In some instances, it is possible for such amino acids to play a dual role, with part of the side chain being buried in the protein, and another being exposed to water.

Aromatic side chains

A side chain is aromatic when it contains an aromatic ring system. The strict definition has to do with the number of electrons contained within the ring. Generally, aromatic ring systems are planar, and electrons are shared over the whole ring structure. Phenylalanine and tryptophan are very hydrophobic aromatic side chains; tyrosine and histidine are less so. The latter two can often be found in positions somewhere between buried and exposed. The hydrophobic aromatic amino acids can

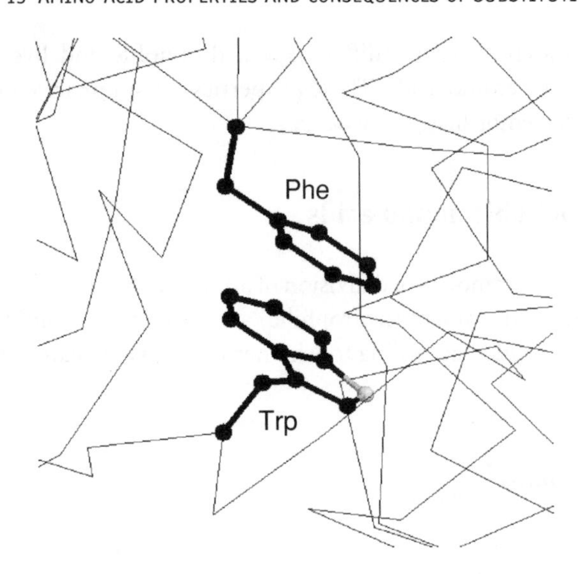

Figure 13.4 Example of aromatic stacking

sometimes substitute for aliphatic residues of a similar size; for example, phenylalanine to leucine, but not tryptophan to valine.

Aromatic residues have also been proposed to participate in 'stacking' interactions (Hunter *et al.*, 1991) (Figure 13.4). Here, numerous aromatic rings are thought to stack on top of each other such that their pI electron clouds are aligned. They can also play a role in binding to specific amino acids, such as proline. SH3 and WW domains, for example, use these residues to bind to their polyproline-containing interaction partners (Macias *et al.*, 2002). Owing to its unique chemical nature, histidine is frequently found in protein-active sites, as we shall see below.

13.4.2 Polar amino acids

Polar amino acids prefer to be surrounded by water. Those that are buried within the protein usually participate with other side chains, or the protein main chain, in hydrogen bonds that essentially replace the water. Some of these carry a charge at typical biological pH: aspartate and glutamate are negatively charged; lysine and arginine are positively charged. Other polar amino acids, histidine, asparagine, glutamine, serine, threonine and tyrosine, are neutral.

13.4.3 Small amino acids

The amino acids alanine, cysteine, glycine, proline, serine and threonine are often grouped together for the simple reason that they are all small in size. In some protein structural contexts, substitution of a small side chain for a large one can be disastrous.

13.5 Amino-acid quick reference

In the sections that follow, we discuss each amino acid in turn. For each, we will briefly discuss general preferences for substitutions, and important specific details of possible structure and functional roles. More information is found on the website that accompanies this chapter (http://www.russell.embl-heidelberg.de/aas). This website also features amino-acid substitution matrices for transmembrane, extracellular and intracellular proteins; these can be used to score numerically an amino-acid substitution, where unpreferred mutations are given negative scores, preferred substitutions are given positive scores, and neutral substitutions are given zero scores.

13.5.1 Alanine (Ala, A)

Substitutions

Alanine substitutes with other small amino acids.

Structure

Alanine is probably the dullest amino acid. It is not particularly hydrophobic and is non-polar. However, it contains a normal Cβ carbon, meaning that it is generally as hindered as other amino acids with respect to the conformations that the backbone can adopt. For this reason, it is not surprising to see alanine present in just about all non-critical protein contexts.

Function

The alanine side chain is very non-reactive, and is thus rarely directly involved in protein function, but it can play a role in substrate recognition or specificity, particularly in interactions with other non-reactive atoms such as carbon.

13.5.2 Isoleucine (Ile, I)

Substitutions

Isoleucine substitutes with other hydrophobic, particularly aliphatic amino acids.

Structure

Being hydrophobic, isoleucine prefers to be buried in protein hydrophobic cores. However, isoleucine has an additional property that is frequently overlooked. Like valine and threonine, it is Cβ branched. Whereas most amino acids contain only one non-hydrogen substituent attached to their Cβ carbon, these three amino acids contain two. This means that there is a lot more bulkiness near the protein backbone, and therefore that these amino acids are more restricted in the conformations the main chain can adopt. Perhaps the most pronounced effect of this is that it is more difficult for these amino acids to adopt an α-helical conformation, though it is easy and even preferred for them to lie within β-sheets.

Function

The isoleucine side chain is very non-reactive, and is thus rarely directly involved in protein functions such as catalysis, though it can play a role in substrate recognition. In particular, hydrophobic amino acids can be involved in binding/recognition of hydrophobic ligands such as lipids.

13.5.3 Leucine (Leu, L)

Substitutions (see isoleucine)

Structure

Being hydrophobic, leucine prefers to be buried in protein hydrophobic cores. It also shows a preference for being within alpha helices to being within beta strands.

Function (see isoleucine)

13.5.4 Valine (Val, V)

Substitutions (see isoleucine)

Structure

Being hydrophobic, valine prefers to be buried in protein hydrophobic cores. However, valine is also Cβ branched (see isoleucine)

Function (see isoleucine)

13.5.5 Methionine (Met, M)

Substitutions (see isoleucine)

Structure (see isoleucine)

Function

The methionine side chain is fairly non-reactive, and is thus rarely directly involved in protein function. Like other hydrophobic amino acids, it can play a role in binding/ recognition of hydrophobic ligands such as lipids. However, unlike the proper aliphatic amino acids, methionine contains a sulphur atom, which can be involved in binding to atoms such as metals. However, whereas the sulphur atom in cysteine is connected to a hydrogen atom, making it quite reactive, methionine's sulphur is connected to a methyl group. This means that the role that methionine can play in protein function is much more limited.

13.5.6 Phenylalanine (Phe, F)

Substitutions

Phenylalanine substitutes with other aromatic or hydrophobic amino acids. It particularly prefers to exchange with tyrosine, which differs only in that it contains a hydroxyl group in place of the *o*-hydrogen on the benzene ring.

Structure

Phenylalanine prefers to be buried in protein hydrophobic cores. The aromatic side chain can also mean that phenylalanine is involved in stacking (Figure 13.4) interactions with other aromatic side chains.

Function

The phenylalanine side chain is fairly non-reactive, and is thus rarely directly involved in protein function, though it can play a role in substrate recognition (see isoleucine). Aromatic residues can also be involved in interactions with non-protein ligands that themselves contain aromatic groups via stacking interactions (see above). They are also common in polyproline-binding sites, such as SH3 and WW domains (Macias *et al.*, 2002).

13.5.7 Tryptophan (Trp, W)

Substitutions

Tryptophan can be replaced by other aromatic residues, but it is unique in chemistry and size, meaning that often replacement by anything could be disastrous.

Structure (see phenylalanine)

Function

As it contains a non-carbon atom (nitrogen) in the aromatic ring system, tryptophan is more reactive than phenylalanine but less reactive than tyrosine. Tryptophan can play a role in binding to non-protein atoms, but such instances are rare. See also phenylalanine.

13.5.8 Tyrosine (Tyr, Y)

Substitutions (see phenylalanine)

Tyrosine substitutes with other aromatic amino acids.

Structure

Being partially hydrophobic, tyrosine prefers to be buried in protein hydrophobic cores. The aromatic side chain can also mean that tyrosine is involved in stacking interactions with other aromatic side chains.

Function

Unlike the very similar phenylalanine, tyrosine contains a reactive hydroxyl group, thus making it much more likely to be involved in interactions with non-carbon atoms. See also phenylalanine. A common role for tyrosines (and serines and threonines) within intracellular proteins is phosphorylation. Protein kinases frequently attach phosphates to these three residues as part of the signal transduction process. Note that, in this context, tyrosine rarely substitutes for serine or threonine since the enzymes that catalyse the reactions (the protein kinases) are highly specific (tyrosine kinases generally do not work on serines/threonines and vice versa (Hanks *et al.*, 1988)).

13.5.9 Histidine (His, H)

Substitutions

Histidine is a generally considered to be a polar amino acid; however, it is unique with regard to its chemical properties; therefore, it does not particularly substitute well with any other amino acid.

Structure

Histidine has a pKa near to that of physiological pH, meaning that it is relatively easy to move protons on and off the side chain (i.e., changing the side chain from neutral to positive charge). This flexibility has two effects. The first is ambiguity about whether it prefers to be buried in the protein core, or exposed to solvent. The second is that it is an ideal residue for protein functional centres (discussed below). It is false to presume that histidine is always protonated at typical pHs. The side chain has a pKa of approximately 6.5, which means that only about 10 per cent of molecules will be protonated. The precise pKa depends on local environment.

Function

Histidine is the most common amino acid in protein-active or -binding sites. It is very common in metal-binding sites (e.g., zinc), often acting together with cysteine or other amino acids (Figure 13.5) (Wolfe *et al.*, 2001). In this context, it is common to see histidine replaced by cysteine.

The ease with which protons can be transferred on and off histidine makes it ideal for charge relay systems such as those within catalytic triads, as found in many cysteine and serine proteases (Figure 13.1). In this context, it is rare to see histidine exchange for any amino acid at all.

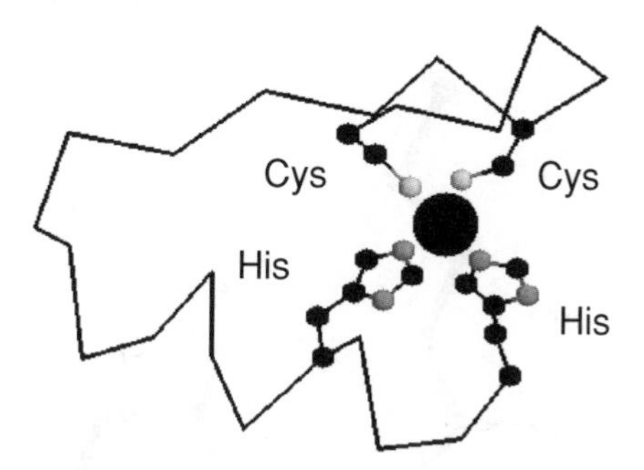

Figure 13.5 Example of a metal binding site coordinated by cysteine and histidine residues (code 1g2f (Wolfe *et al.*, 2001)). Figure generated by the authors using data from Wolfe *et al.*, 2001. Permission not required

13.5.10 Arginine (Arg, R)

Substitutions

Arginine is a positively charged, polar amino acid. Thus, it most prefers to substitute for the other positively charged amino acid, lysine, though in some circumstances it will also tolerate a change to other polar amino acids. Note that a change from arginine to lysine is not always neutral. In certain structural or functional contexts, such a mutation can be devastating to function (see below).

Structure

Arginine generally prefers to be on the surface of the protein, but its amphipathic nature can mean that part of the side chain is buried. Arginine is also frequently involved in salt bridges, where it pairs with a negatively charged aspartate or glutamate to create stabilizing hydrogen bonds that can be important for protein stability (Figure 13.6.

Function

Arginine is quite common in protein-active or -binding sites. The positive charge means that it can interact with negatively charged non-protein atoms (e.g., anions or carboxylate groups). It contains a complex guanidinium group on its side chain that has a geometry and charge distribution that is ideal for binding negatively charged groups on phosphates (it can form multiple hydrogen bonds). A good example can be found in the src homology 2 (SH2) domains (Figure 13.7) (Waksman *et al.*, 1992). The two arginines shown in Figure 13.7 make multiple hydrogen bonds with the phosphate. In this context, arginine is not easily replaced by lysine. Although lysine can interact with phosphates, it contains only a single amino group, meaning it is

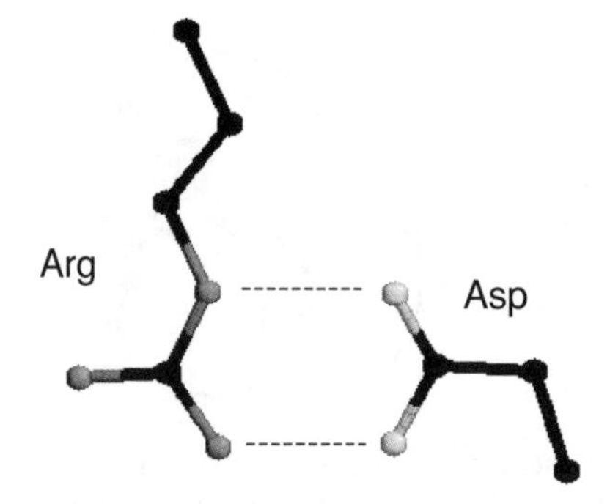

Figure 13.6 Example of a salt bridge (code 1xel)

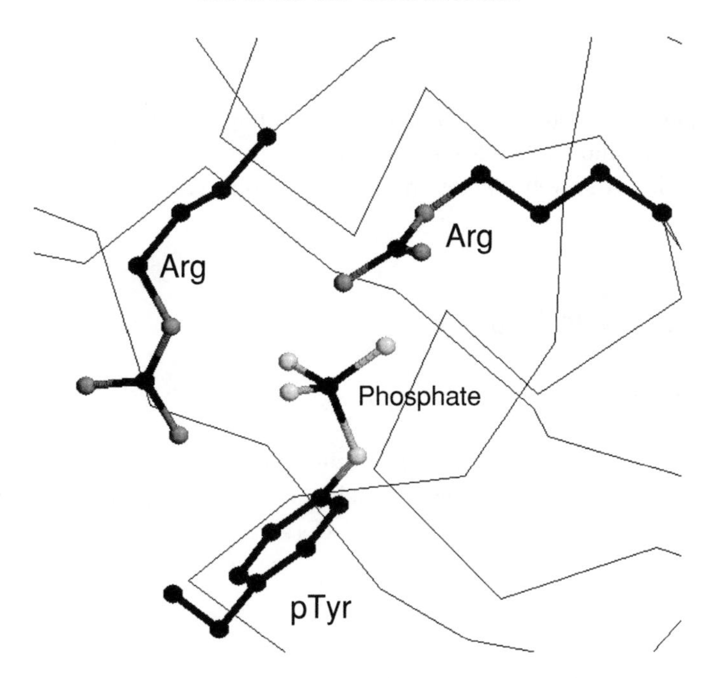

Figure 13.7 Interaction of arginine residues with phosphotyrosine in an SH2 domain (code 1sha (Waksman *et al.*, 1992)). Figure generated by the authors using data from Waksman *et al.*, 1992. Permission not required

more limited in the number of hydrogen bonds it can form. A change from arginine to lysine in some contexts can thus be disastrous (Copley and Barton, 1994).

13.5.11 Lysine (Lys, K)

Substitutions

Lysine substitutes with arginine or other polar amino acids.

Structure

Lysine frequently plays an important role in structure. First, it can be considered to be somewhat amphipathic as the part of the side chain nearest to the backbone is long, carbon containing and hydrophobic, whereas the end of the side chain is positively charged. For this reason, we can find lysine where part of the side-chain is buried, and only the charged portion is outside the protein. However, this is by no means always the case, and generally lysine prefers to be outside proteins. Lysine is also frequently involved in salt bridges (see arginine).

Function

Lysine is quite common in protein-active or -binding sites. Lysine contains a positively charged amino on its side chain that is sometimes involved in forming hydrogen bonds with negatively charged non-protein atoms (e.g., anions or carboxylate groups).

13.5.12 Aspartate (Asp, D)

Substitutions

Aspartate substitutes with glutamate or other polar amino acids, particularly asparagine, which differs only in that it contains an amino group in place of one of the oxygens found in aspartate (and thus also lacks a negative charge).

Structure

Being charged and polar, aspartate prefers generally to be on the surface of proteins, exposed to an aqueous environment. Aspartate (and glutamate) is frequently involved in salt bridges (see arginine).

Function

Aspartate is quite frequently involved in protein-active or -binding sites. The negative charge means that it can interact with positively charged non-protein atoms, such as cations like zinc. Aspartate has a shorter side chain than the very similar glutamate, meaning that it is slightly more rigid within protein structures. This gives it a slightly stronger preference to be involved in protein-active sites. Probably the most famous example of aspartate being involved in an active site is within serine proteases such as trypsin, within which it functions in the classical Asp-His-Ser catalytic triad (Figure 13.1). In this context, it is quite rare to see aspartate exchange for glutamate, though it is possible for glutamate to play a similar role.

13.5.13 Glutamate (Glu, E)

Substitutions

Glutamate substitutes with aspartate or other polar amino acids, in particular glutamine, which is to glutamate what asparagine is to aspartate (see above).

Structure (see aspartate)

Function

Glutamate, like aspartate, is quite frequently involved in protein-active or -binding sites. In certain cases, it can also perform a similar role to aspartate, in the catalytic site of proteins such as proteases or lipases.

13.5.14 Asparagine (Asn, N)

Substitutions

Asparagine substitutes with other polar amino acids, especially aspartate (see above).

Structure

Being polar, asparagine prefers generally to be on the surface of proteins, exposed to an aqueous environment.

Function

Asparagine is quite frequently involved in protein-active or -binding sites. The polar side chain favours interactions with other polar or charged atoms. Asparagine can play a similar role to aspartate in some proteins. Probably the best example is found in certain cysteine proteases, where it forms part of the Asn/His/Cys catalytic triad. In this context, it is quite rare to see asparagine exchange for glutamine. Asparagine, when occurring in a particular motif (Asn-X-Ser/Thr), can be *N*-glycosylated (Gavel and von Heijne, 1990). Thus, in this context, it is impossible to substitute it with any amino acid at all.

13.5.15 Glutamine (Gln, Q)

Substitutions

Glutamine substitutes with other polar amino acids, especially glutamate (see above).

Structure (see asparagine)

Function

Glutamine is quite frequently involved in protein-active or -binding sites. The polar side chain favours interactions with other polar or charged atoms.

13.5.16 Serine (Ser, S)

Substitutions

Serine substitutes with other polar or small amino acids, in particular threonine, which differs only in that it has a methyl group in place of the hydrogen group found in serine.

Structure

Being a fairly indifferent amino acid, serine can reside both within the interior of a protein and on its surface. Its small size means that it is relatively common within tight turns on the protein surface, where it is possible for the serine side chain hydroxyl oxygen to form a hydrogen bond with the protein backbone, effectively mimicking proline.

Function

Serine is quite common in protein functional centres. The hydroxyl group is fairly reactive, and can form hydrogen bonds with a variety of polar substrates. Perhaps the best-known role of serine in protein-active sites is in the classical Asp/His/Ser catalytic triad found in many hydrolases (e.g., proteases and lipases) (Figure 13.1). Here, a serine, aided by a histidine and an aspartate, acts as a nucleophile to hydrolyse (effectively cut) other molecules. This three-dimensional 'motif' is found in many non-homologous (i.e., unrelated) proteins, and is a classic example of molecular convergent evolution (Russell, 1998). In this context, it is rare for serine to exchange with threonine, but in some cases, the reactive serine can be replaced by cysteine, which can perform a similar role. Intracellular serine can also be phosphorylated (see tyrosine). Extracellular serine can also be *O*-glycosylated where a carbohydrate is attached to the side-chain hydroxyl group (Gupta *et al.*, 1999).

13.5.17 Threonine (Thr, T)

Substitutions

Threonine substitutes with other polar amino acids, particularly serine (see above).

Structure

As a fairly indifferent amino acid, threonine can reside both within the interior of a protein and on its surface. Threonine is also Cβ branched (see isoleucine).

Function

Threonine is quite common in protein functional centres. The hydroxyl group is fairly reactive, and can form hydrogen bonds with a variety of polar substrates. Intracellular threonine can also be phosphorylated (see tyrosine), and in the extracellular environment it can be *O*-glycosylated (see serine).

13.5.18 Cysteine (Cys, C)

Substitutions

Cysteine has no general preference for substituting with any other amino acid, though it can tolerate substitutions with other small amino acids. Its role is very dependent on cellular location, making substitution matrices dangerous to interpret (e.g., Barnes and Russell, 1999).

Structure

The role of cysteine in structure is very dependent on the cellular location of the protein in which it is contained. Within extracellular proteins, cysteine is frequently involved in disulphide bonds, where pairs of cysteines are oxidized to form a covalent bond. These bonds serve mostly to stabilize the protein structure, and the structure of many extracellular proteins is almost entirely determined by the topology of multiple disulphide bonds (Figure 13.8).

The reducing environment inside cells makes the formation of disulphide bonds very unlikely. Indeed, disulphide bonds in the intracellular environment are so rare that they almost always attract special attention. Disulphides are also rare within the membrane, though membrane proteins can contain disulphide bonds within extracellular domains. Disulphide bonds are such that cysteines must be *paired*. If one-half of a disulphide bond pair is lost, the protein may not fold properly.

In the intracellular environment, cysteine can still play a key structural role. Its sulphydryl side chain is excellent for binding to metals, such as zinc, meaning that cysteine (and other amino acids such as histidine) is very common in metal-binding motifs such as zinc fingers (Figure 13.5). Outside this context within the intracellular environment, and when it is not involved in molecular function, cysteine is a small, neutral amino acid, and prefers to substitute with other amino acids of the same type.

Function

Cysteine is also very common in protein-active and -binding sites. Binding to metals (see above) can also be important in enzymatic functions (e.g., metal proteases).

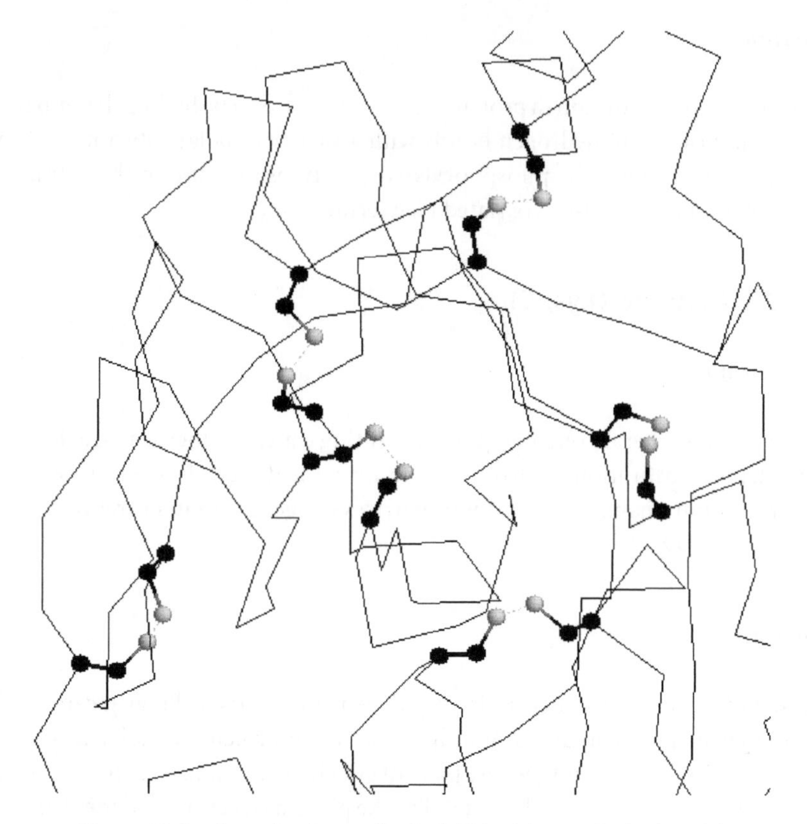

Figure 13.8 Example of a small, disulphide-rich protein (code 1tfx)

Cysteine can also function as a nucleophile (i.e., the reactive centre of an enzyme). Probably the best-known example of this occurs within the cysteine proteases, such as caspases, or papains, where cysteine is the key catalytic residue, being helped by histidine and asparagine.

13.5.19 Glycine (Gly, G)

Substitutions

Glycine substitutes with other small amino acids, but be warned that even apparently neutral mutations (e.g., to alanine) can be forbidden in certain contexts (see below).

Structure

Glycine is unique as it contains a hydrogen as its side chain (rather than a carbon as is the case for all other amino acids). This means that there is much more

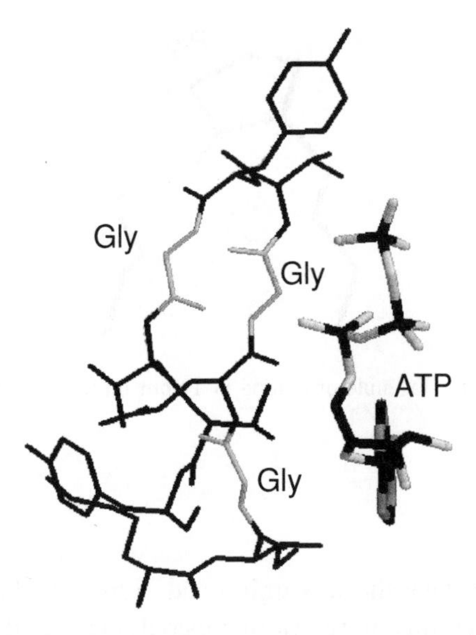

Figure 13.9 Glycine-rich phosphate-binding loop in a protein kinase (code 1hck (Schulze-Gahmen *et al.*, 1996)). Figure generated by the authors using data from Schulze-Gahmen *et al.*, 1996. Permission not required

conformational flexibility in glycine. Therefore, glycine can reside in parts of protein structures that are forbidden to all other amino acids (e.g., tight turns in structures).

Function

The uniqueness of glycine also means that it can play a distinct functional role, such as using its backbone (without a side chain) to bind to phosphates (Schulze-Gahmen *et al.*, 1996). This means that if one sees a conserved glycine changing to any other amino acid, the change could have a drastic impact on function. A good example is found in protein kinases. Figure 13.9 shows a region around the ATP-binding site in a protein kinase. The ATP is shown to the right of the figure, and part of the protein to the left. The glycines in this loop are part of the classic 'Gly-X-Gly-X-X-Gly' motif present in the kinases (Hanks *et al.*, 1988). These three glycines are hardly ever mutated to other residues; only glycines can function to bind to the phosphates of the ATP molecule with their main chains.

13.5.20 Proline (Pro, P)

Substitutions

Proline can sometimes substitute for other small amino acids, though its unique properties means that it does not often substitute well.

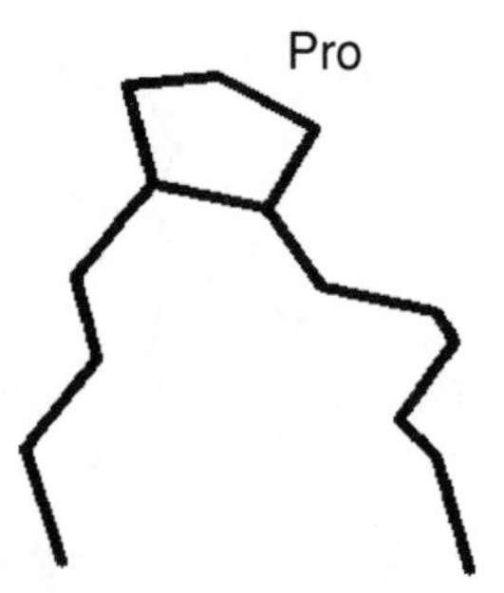

Figure 13.10 Example of proline in a tight protein turn (code 1ag6)

Structure

Proline is unique in that it is the only amino acid whose side chain is connected to the protein backbone twice, forming a five-membered ring. Strictly speaking, this makes proline an imino acid (since in its isolated form, it contains an NH_2^+ rather than an NH_3^+ group, but this is mostly just pedantic detail). This difference is very important as it means that proline is unable to occupy many of the main chain conformations easily adopted by all other amino acids. In this sense, it can be considered the opposite of glycine, which can adopt many more main-chain conformations. For this reason, proline is often found in very tight turns in protein structures (i.e., where the polypeptide chain must change direction) (Figure 13.10). It can also function to introduce kinks into α-thelices, since it is unable to adopt a normal helical conformation. Although it is aliphatic, the preference for turn structure means that proline is usually found on the protein surface.

Function

The proline side chain is very non-reactive. This, together with its difficulty in adopting many protein main-chain conformations, means that it is very rarely involved in protein-active or -binding sites.

13.6 Studies of how mutations affect function

Protein structures are plastic – they can tolerate many different substitutions in many different places – and some regions are more plastic than others. The most conserved region is generally the core (Lim and Sauer, 1989, PMID 2524006), although the main requirement is only that this should remain hydrophobic. Substitutions in specific solvent exposed residues can easily affect function but do not affect the ability of the mutant sequence to fold in the same way as the wild-type sequence (Hecht *et al.*, 1983, PMID 6221342). More radical substitutions are tolerated if less than wild-type activity is allowed. In other words, context is everything. This is not easily seen by relying on simple substitution matrices, which treat every position in the sequence in the same way as but independent from every other position. Several studies have examined mutations in the context of structure, function, disease, evolution, and other mutations. We review some of these below.

13.6.1 Site-specific substitution rates and correlated mutations

The two main assumptions behind most substitution tables are sites are independent of each other and that they are all equal. Exceptions to these assumptions are clearly demonstrated by the systematic mutagenesis of λ repressor by Hecht *et al.* and Lim and Sauer discussed above, and by other such studies. To address the first assumption one can allow for correlated mutations, where a substitution at one site affects the acceptability of a substitution at another site, although the results of such studies are unclear at best. To address the second assumption one can use site-specific substitution matrices.

An increase in residue size at one position, for example, could be compensated for by a decrease at site close in three dimensions. However, correlated mutations such as this will be less significant in the plastic regions of the structure, and the initial substitution would only be tolerated in more conserved regions if it's effect on stability or function was small. Several other factors may also make them difficult to detect: long-range correlation, correlation with multiple positions with the particular positions used differing in different lineages, and correlation by chance between independent sites. Of the many methods developed (Pollock and Taylor, 1997), the best ones use phylogeny to reduce the effect of chance correlations (Pollock *et al.*, 1999; Fukami-Kobayashi *et al.*, 2002), and can detect correlations in simulated data. Meaningful correlations, mainly between sites brought in to contact by the periodicity of alpha helices, have also been detected (Pollock *et al.*, 1999). Others studies have not directly modelled correlations but have used trees and structures in methods that cut away chance and uninteresting correlations, and used the rest to predict functional sites (Lichtarge *et al.*, 1996; Aloy *et al.*, 2001).

Different rates of substitutions at different sites can be allowed for by pre-classifying sites based on their position in the protein structure, and then using substitution matrices specifically constructed for the classes identified (Overington *et al.*, 1992). These methods obviously require some knowledge other than the sequence, such as a known or predicted structure. Other methods are able to divide sites in to different classes at the same time as constructing the substitution matrices for these classes, and have been used to predict functional sites (Soyer and Goldstein, 2004). An easier to use classification is simply to define your protein as intracellular, extracellular or membrane, and then to use the appropriate matrix from the appendix.

13.6.2 Single nucleotide polymorphisms (SNPs)

A SNP is a point mutation that is present at a measurable frequency in human populations. It can occur either in coding or non-coding DNA. Non-coding SNPs may have effects on important mechanisms such as transcription, translation, and splicing. However, the effects of coding SNPs are easier to study, and are potentially more damaging, and so they have received considerably more attention. They are also more relevant to this chapter. Coding SNPs can be divided into two main categories, synonymous (where there is no change in the amino acid coded for), and non-synonymous. Non-synonymous SNPs tend to occur at lower frequencies than synonymous SNPs. Minor allele frequencies also tend to be lower in non-synonymous SNPs. This is a strong indication that these replacement polymorphisms are deleterious (Cargill *et al.*, 1999).

To examine the phenotypic effects of coding SNPs, Sunyaev *et al.* (2000) studied the relationships between non-synonymous SNPs and protein structure and function. Three sets of SNP data were compared: disease-causing substitutions, substitutions between orthologues, and substitution represented by human alleles. Disease-causing mutations were more common in structurally and functionally important sites than were variations between orthologues, as might be expected. Allelic variations were also more common in these regions than were those between orthologues. Minor allele frequency and the level of occurrence in these regions were correlated, another indication of evolutionary selection of phenotype. The most damaging allelic variants affect protein stability, rather than binding, catalysis, allosteric response or post-translational modification (Sunyaev *et al.*, 2001). The expected increase in the number of known protein structures will allow other analyses and refinement of the details of the phenotypic effects of SNPs.

Wang and Moult (2001) developed a description of the possible effects of missense SNPs on protein structure, and used it to compare disease-causing missense SNPs with a set from the general population. Five general classes of effect were considered: protein stability, ligand binding, catalysis, allosteric regulation, and post-translational modification. The disease and population sets of SNPs contain those that can be mapped on to known protein structures, either directly or through homologues of known structure. Of the disease-causing SNPs, 90 per cent were explained by the description, the majority (83 per cent) being attributed to effects on protein stability, as seen by Sunyaev *et al.* (2001). The 10 per cent that are not explained by the description may cause disease by effects not easily identified by structure alone. Of the SNPs from the general population, 70 per cent were predicted to have no effect. The remaining 30 per cent may represent disease-causing SNPs previously unidentified as such, or molecular effects that have no significant phenotypic effect.

13.6.3 Evaluating paired residues in protein structures

Studies of the structural impact of an amino-acid substitution should try to take into account the preferences for structural pairing between different amino acids as described earlier in this chapter. Fooks *et al.* (2006) examined statistical approaches to study amino-acid pairing preferences within parallel beta-sheets. The main-chain, hydrogen-bonding pattern in parallel beta-sheets means that, for each residue pair, only one of the residues is involved in main-chain hydrogen bonding with the strand containing the partner residue. For instance, they found that Asn-Thr and Arg-Thr were favoured pairs, where the residues adopted favoured rotamer positions that allowed side-chain interactions to occur. In contrast, Thr-Asn and Thr-Arg were not significantly favoured, and could form side-chain interactions only if the residues involved adopted less favourable conformations. Cysteine–cysteine pairs were also significantly favoured, although these did not form intrasheet disulphide bridges. This research provides rules that could technically be applied to protein structure prediction, and comparative modelling of amino substitutions. The methods used to analyse the pairing preferences are automated and detailed; results are available from the following URL: http://www.rubic.rdg.ac.uk/betapairprefsparallel/.

13.6.4 Site-directed mutagenesis

Site-directed mutagenesis is a powerful tool for discovering the importance of an amino acid to the function of the protein. Gross changes in amino-acid type can reveal sites that are important in maintaining the structure of the protein. Conversely, in investigating functionally interesting sites, it is important to choose replacement residues that are unlikely to affect structure dramatically; for example, by choosing ones of a similar size to the original. Peracchi (2001) reviews the use of site-directed mutagenesis to investigate mechanisms of enzyme catalysis, in particular those studies involving mutagenesis of general acids (proton donors), general bases (proton acceptors) and catalytic nucleophiles in active sites. These types of amino acid could be considered to be the most important to enzyme function, as they directly participate in the formation or cleavage of covalent bonds. However, studies indicate that they are often important but not essential – rates are still higher than the uncatalysed reaction even when these residues are removed, because the protein is able to use an alternative mechanism of catalysis. Moreover, direct involvement in the formation and cleavage of bonds is only one of a combination of methods that an enzyme can use to catalyse a reaction. Transition states can be stabilized by the complementary shape and electrostatics of the binding site of the enzyme, and substrates can be precisely positioned, lowering the entropy of activation. These factors can also be studied by site-directed mutagenesis, consideration of the physical and chemical properties of the amino acids again guiding the choice of replacements, along with knowledge of the structure of the protein.

13.6.5 Key mutations in evolution

Golding and Dean (1998) reviewed six studies that demonstrate the insight into molecular adaptation that is provided by combining knowledge of phylogenies, site-directed mutagenesis, and protein structure (Golding and Dean, 1998).

Many changes can occur over many generations, only a few being responsible for changes in function. For example, the sequences of lactate dehydrogenase (LDH) and malate dehydrogenase (MDH) from *Bacillus stearothermophilus* are only about 25 per cent identical, but their tertiary structures are highly similar. Only one mutation, that of uncharged glutamine 102 to positive arginine in the active site, is required to convert LDH in to a highly specific MDH. Arginine is thought to interact with the carboxylate group that is the only difference between the substrate/products of the two enzymes (Figure 13.11) (Wilks *et al.*, 1988).

Thus, amino-acid changes that appear to be radical or conservative from their scores in mutation matrices or amino-acid properties may be the opposite when their effect on protein function is considered. Glutamine to arginine has a score of 0 in the PAM250 matrix, meaning that it is neutral. The importance of the mutation at position 102 in LDH and MDH could not be predicted by this information alone.

Reconstruction of an ancestral ribonuclease showed that the mutation that causes most of the fivefold loss in activity toward double-stranded RNA is of Gly38 to Asp, more than 5 Å from the active site (Golding and Dean, 1998).

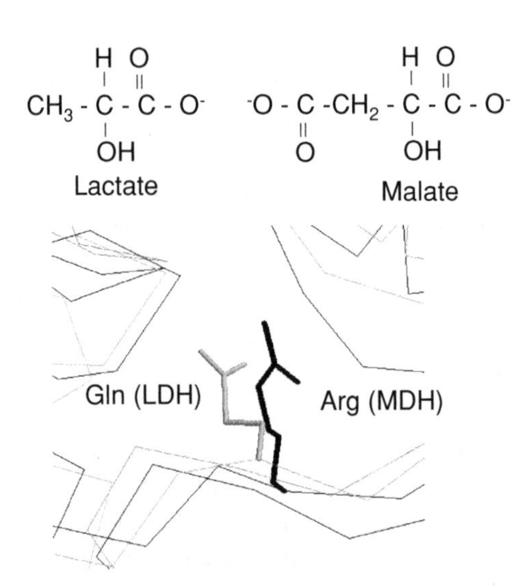

Figure 13.11 Lactate and malate dehydrogenase specificity (codes 9ltd and 2cmd (Wilks *et al.*, 1988)), Figure generated by the authors using data from Wilks *et al.*, 1988. Permission not required

Two different mutations in different locations in the haemoglobin genes of bar-headed goose and of Andean goose give both breeds a high affinity for oxygen. Structural studies showed that both changes remove an important van der Waals contact between subunits, shifting the equilibrium of the haemoglobin tetramer toward the high-affinity state. These studies emphasise the importance of both protein structure and phylogeny when considering the effects of amino acid mutations (Golding and Dean, 1998).

13.7 A summary of the thought process

We hope that this chapter has given you some guidelines for interpreting how a particular mutation might affect the structure and function of a protein. We suggest that you ask the following questions.

First about the protein

1. What is the cellular environment?
2. What does it do? Is anything known about the amino acids involved in its function?
3. Is there a structure known or one for a homologue?
4. What protein family does it belong to?
5. Are any post-translational modifications expected?

Then about a particular amino acid

1. Is the position conserved across orthologues? Across paralogues?
2. If a structure is known: is the amino acid on the surface? Buried in the core of the protein?
3. Is it directly involved in function, or near (in sequence or space) other amino acids that are?
4. Is it an amino acid that is likely to be critical for function? For structure?

Once you have answers to these questions, you should be more able to make a rational guess or interpretation of the effects seen with an amino-acid substitution, and select logical amino acids for mutagenesis experiments.

Appendix: tools

Protein sequences

- http://www.expasy.ch/
- http://www.ncbi.nlm.nih.gov/

Amino-acid properties

- http://russell.embl-heidelberg.de/aas/

Domain assignment/sequence search tools

- http://www.ebi.ac.uk/interpro/
- http://www.sanger.ac.uk/Software/Pfam/
- http://smart.embl-heidelberg.de/
- http://www.ncbi.nlm.nih.gov/BLAST/
- http://www.ncbi.nlm.nih.gov/COG/
- http://www.cbs.dtu.dk/TargetP/

Protein structure

- databases of 3-D structures of proteins: http://www.rcsb.org/pdb/
- structural classification of proteins: http://scop.mrc-lmb.cam.ac.uk/scop/

Protein function

- http://www.geneontology.org/

References

Altschul, S. F., Madden, T. L., Schaffer, A. A. *et al.* (1997). Gapped BLAST and PSI-BLAST: a new generation of protein database search programs. *Nucleic Acids Res* **25**(17), 3389–3402.

Ashburner, M., Ball, C. A., Blake, J. A. *et al.* (2000). Gene ontology: tool for the unification of biology. The Gene Ontology Consortium. *Nat Genet* **25**(1), 25–29.

Barnes, M. R. and Russell, R. B. (1999). A lipid-binding domain in Wnt: a case of mistaken identity? *Curr Biol* **9**(19), R717–719.

Bateman, A., Birney, E., Durbin, R. *et al.* (2000). The Pfam protein families database. *Nucleic Acids Res* **28**(1), 263–266.

Berman, H. M., Westbrook, J., Feng, Z. *et al.* (2000). The Protein Data Bank. *Nucleic Acids Res* **28**(1), 235–242.

Cargill, M., Altshuler, D., Ireland, J. *et al.* (1999). Characterization of single-nucleotide polymorphisms in coding regions of human genes. *Nat Genet* **22**(3), 231–238.

Cavallo, A. and Martin, A. C. (2005). Mapping SNPs to protein sequence and structure data. *Bioinformatics* **21**(8), 1443–1450.

Copley, R. R. and Barton, G. J. (1994). A structural analysis of phosphate and sulphate binding sites in proteins. Estimation of propensities for binding and conservation of phosphate binding sites. *J Mol Biol* **242**(4), 321–329.

Dayhoff, M. O., Schwartz, R. M. and Orcutt, B. C. (1978). A model of evolutionary change in proteins. In *Atlas of Protein Sequence and Structure*, vol. 5, pp. 345–352, M. O. Dayhoff (ed.). Washington, DC: National Biomedical Research Foundation.

Durbin, R., Eddy, S., Krogh, A. *et al.* (1998). *Biological Sequence Analysis. Probabilistic Models of Proteins and Nucleic Acids.* Cambridge: Cambridge University Press.

Eddy, S. R. (1998). Profile hidden Markov models. *Bioinformatics* **14**(9), 755–763.

Fooks, H. M., Martin, A. C., Woolfson, D. N. *et al.* (2006). Amino acid pairing preferences in parallel beta-sheets in proteins. *J Mol Biol* **356**(1), 32–44.

Gavel, Y. and von Heijne, G. (1990). Sequence differences between glycosylated and non-glycosylated Asn-X-Thr/Ser acceptor sites: implications for protein engineering. *Protein Eng* **3**(5), 433–442.

Gibson, T. J. and Spring, J. (1998). Genetic redundancy in vertebrates: polyploidy and persistence of genes encoding multidomain proteins. *Trends Genet* **14**(2), 46–49; discussion 49–50.

Golding, G. B. and Dean, A. M. (1998). The structural basis of molecular adaptation. *Mol Biol Evol* **15**(4), 355–369.

Gupta, R., Birch, H., Rapacki, K. *et al.* (1999). O-GLYCBASE Version 4.0: a revised database of *O*-glycosylated proteins. *Nucleic Acids Res* **27**(1), 370–372.

Hanks, S. K., Quinn, A. M. and Hunter, T. (1988). The protein kinase family: conserved features and deduced phylogeny of the catalytic domains. *Science* **241**(4861), 42–52.

Henikoff, S. and Henikoff, J. G. (1992). Amino acid substitution matrices from protein blocks. *Proc Natl Acad Sci U S A* **89**(22), 10915–10919.

Hunter, C. A., Singh, J. and Thornton, J. M. (1991). Pi–pi interactions: the geometry and energetics of phenylalanine–phenylalanine interactions in proteins. *J Mol Biol* **218**(4), 837–846.

Jeffrey, P. D., Russo, A. A., Polyak, K. *et al.* (1995). Mechanism of CDK activation revealed by the structure of a cyclinA-CDK2 complex. *Nature* **376**(6538), 313–320.

Jones, D. T., Taylor, W. R. and Thornton, J. M. (1994). A mutation data matrix for transmembrane proteins. *FEBS Lett* **339**(3), 269–275.

Karchin, R., Diekhans, M., Kelly, L. *et al.* (2005) LS-SNP: large-scale annotation of coding non-synonymous SNPs based on multiple information sources. *Bioinformatics* **21**(12), 2814–2820

Kraulis, P. J. (1991). MOLSCRIPT: a program to produce both detailed and schematic plots of protein structures. *J Appl Cryst* **24**, 946–950.

Krishna, R. G. and Wold, F. (1993). Post-translational modification of proteins. *Adv Enzymol Relat Areas Mol Biol* **67**, 265–298.

Macias, M. J., Wiesner, S. and Sudol, M. (2002). WW and SH3 domains, two different scaffolds to recognize proline-rich ligands. *FEBS Lett* **513**(1), 30–37.

Pal, C., Papp, B. and Lercher, M. J. (2006). An integrated view of protein evolution. *Nat Rev Genet* **7**(5), 337–348.

Parekh, R. B. and Rohlff, C. (1997). Post-translational modification of proteins and the discovery of new medicine. *Curr Opin Biotechnol* **8**(6), 718–723.

Peracchi, A. (2001). Enzyme catalysis: removing chemically 'essential' residues by site-directed mutagenesis. *Trends Biochem Sci* **26**(8), 497–503.

Plotnikov, A. N., Schlessinger, J., Hubbard, S. R. *et al.* (1999). Structural basis for FGF receptor dimerization and activation. *Cell* **98**(5), 641–650.

Russell, R. B. (1998). Detection of protein three-dimensional side-chain patterns: new examples of convergent evolution. *J Mol Biol* **279**(5), 1211–1227.

Sayle, R. A. and Milner-White, E. J. (1995). RASMOL: biomolecular graphics for all. *Trends Biochem Sci* **20**(9), 374.

Schoorlemmer, J. and Goldfarb, M. (2001). Fibroblast growth factor homologous factors are intracellular signaling proteins. *Curr Biol* **11**(10), 793–797.

Schulze-Gahmen, U., De Bondt, H. L. and Kim, S. H. (1996). High-resolution crystal structures of human cyclin-dependent kinase 2 with and without ATP: bound waters and natural ligand as guides for inhibitor design. *J Med Chem* **39**(23), 4540–4546.

Sunyaev, S., Lathe, W., 3rd and Bork, P. (2001). Integration of genome data and protein structures: prediction of protein folds, protein interactions and 'molecular phenotypes' of single nucleotide polymorphisms. *Curr Opin Struct Biol* **11**(1), 125–130.

Sunyaev, S., Ramensky, V. and Bork, P. (2000). Towards a structural basis of human non-synonymous single nucleotide polymorphisms. *Trends Genet* **16**(5), 198–200.

Taylor, W. R. (1986). The classification of amino acid conservation. *J Theor Biol* **119**(2), 205–218.

Waksman, G., Kominos, D., Robertson, S. C. *et al.* (1992). Crystal structure of the phosphotyrosine recognition domain SH2 of v-src complexed with tyrosine-phosphorylated peptides. *Nature* **358**(6388), 646–653.

Wang, Z. and Moult, J. (2001). SNPs, protein structure, and disease. *Hum Mutat* **17**(4), 263–270.

Wilks, H. M., Hart, K. W., Feeney, R. *et al.* (1988). A specific, highly active malate dehydrogenase by redesign of a lactate dehydrogenase framework. *Science* **242**(4885), 1541–1544.

Wilson, C. A., Kreychman, J. and Gerstein, M. (2000). Assessing annotation transfer for genomics: quantifying the relations between protein sequence, structure and function through traditional and probabilistic scores. *J Mol Biol* **297**(1), 233–249.

Wolfe, S. A., Grant, R. A., Elrod-Erickson, M. *et al.* (2001). Beyond the 'recognition code': structures of two Cys2His2 zinc finger/TATA box complexes. *Structure* **9**(8), 717–723.

14

Non-coding RNA Bioinformatics

James R. Brown[1], Steve Deharo[2], Barry Dancis[1], Michael R. Barnes[2] and Philippe Sanseau[3]

Discovery and Analysis Bioinformatics, GlaxoSmithKline Pharmaceuticals, [1] Upper Providence, PA, USA, [2] Harlow, Essex, UK, [3] Stevenage, Hertfordshire, UK

14.1 Introduction

The study of human disease genetics has largely focused on the protein-coding gene as the primary functional unit in disease phenotypes. Despite the fundamental tenet of genes coding for protein via mRNA, it is becoming increasingly apparent that a significant proportion of the transcriptional output of metazoan genomes is actually devoted to the production of functional RNA molecules that do not encode proteins (Kampa *et al.*, 2004). Many of these non-coding RNAs (ncRNAs) are tightly regulated and appear to have a protein-coding gene regulatory function. Our understanding of ncRNA (and its role in disease) is currently in a state of extreme flux. The concept of antisense regulation of RNA has been around for several decades; however, small regulatory endogenously produced RNAs were not described in detail until 1993, when the 22nt micro-RNA (miRNA) *lin-4* was found to regulate *lin-14* mRNA in *C. elegans* via an antisense interaction (Lee *et al.*, 1993). Despite the interest that this paper generated, it took at least 7 years before research into this area really took off, mainly on the back of the analysis of the newly completed human and mouse genomes. Now, an increasingly complex picture of ncRNAs is emerging from complex bioinformatics analysis, suggesting that up to 98 per cent of the transcriptional output of the human genome may be non-protein-coding RNA (Mattick, 2003). This appears to be borne out by extensive gene duplication and divergence into distinct functional families. To date, most focus has been on miRNAs; as many as 800 miRNAs have

Bioinformatics for Geneticists, Second Edition. Edited by Michael R. Barnes
 ISBN 978-0-470-02619-9 (HB) ISBN 978-0-470-02620-5 (PB)

been predicted to exist in man (Bentwich *et al.*, 2005), although fewer than 400 have been deposited in the databases (Griffiths-Jones *et al.*, 2006). Interest in ncRNAs may in part be due to the way that they help to explain the complexity of mammals (and metazoans in general), which is not accounted for by the modest number of protein- and large RNA-coding genes. A picture is emerging of miRNA and other ncRNAs as an additional layer of complexity in the spatial and temporal regulation of gene expression that explains the breath-taking complexity of metazoan life. Naturally, genetics is likely to play a huge role in this complexity, as the driving forces of evolution cause these systems to evolve and mutate at the DNA level, with disease being the flip side of the evolutionary coin.

In this chapter we will review methods for analysis of mammalian ncRNAs, with a focus on miRNAs; however, we will also describe wider ncRNA database resources. The chapter will also review the nascent field of RNomics and the role of ncRNAs in mammalian gene regulation, disease and development. ncRNAs are already causing some considerable excitement, as reflected by the burgeoning literature in this area. But, surprisingly, the role of ncRNA variation in human disease is still very much unexplored (perhaps with the notable exception of oncology), so in this chapter we will try to illustrate the ways in which this role could be investigated.

Bioinformatics and genetic methods can ultimately identify candidate ncRNA for testing – there are a number of biochemical and genomic methods that can be employed to characterize further the role of these ncRNAs in disease. We refer the reader to Huttenhofer and Vogel (2006) for a more detailed review of these methods.

14.2 The non-coding (nc) RNA universe

14.2.1 ncRNA – what we know so far

Our current understanding of ncRNAs in eukaryotes is revealing a rapidly expanding universe of key components in a wide range of cellular mechanisms. These range from critical elements in core cellular systems, such as the ribosome and the spliceosome, to key regulators of these systems that guide essential modifications of ribosomal and spliceosomal RNAs, culminating in the discovery of a huge new class of mRNA-regulating miRNAs with roles as diverse as the mRNA transcripts that they regulate (Bartel, 2004; Carthew, 2006; Kim, 2006).

The rapid emergence of miRNAs and the hints of further complexity in ncRNA that we have seen in the ENCODE regions (see below) have shown that our understanding of ncRNA is still quite incomplete. Classification of ncRNA, can be a problem in itself. Non-coding RNAs are united by what they do not do (i.e., code for protein) but by little else. Nomenclature is somewhat confusing. ncRNA groups may be named by their cellular localizations, such as small nucleolar RNAs (snoRNAs); by their function, as in transfer RNAs (tRNA); or by their size, such as miRNA. In fact, ncRNAs come in many flavours (Eddy, 2001; Costa, 2005). It would be somewhat futile to attempt

Table 14.1 Tools for analysis of ncRNA in a genetic context

Tool	URL
miRNA gene analysis	
MirScan	http://genes.mit.edu/mirscan/
miRBASE	http://microrna.sanger.ac.uk/sequences/
BayesMiRNAfind	http://wotan.wistar.upenn.edu/miRNA
miRNAMap	http://mirnamap.mbc.nctu.edu.tw/
Argonaute	http://www.ma.uni-heidelberg.de/apps/zmf/argonaute/interface/
miRNA target analysis	
TargetScan	http://genes.mit.edu/targetscan/index.html
PicTar	http://pictar.bio.nyu.edu/
MiRanda: Targets	http://microrna.sanger.ac.uk/targets/v2/
Software	http://www.microrna.org/miranda_new.html
RNAHybrid	http://bibiserv.techfak.uni-bielefeld.de/rnahybrid/
Other ncRNA analysis	
NONCODE	http://noncode.bioinfo.org.cn/
Rfam	http://www.sanger.ac.uk/Software/Rfam/
RNAdb	http://research.imb.uq.edu.au/rnadb/
siRNAdb	http://sirna.cgb.ki.se/help.html
UTRsite	http://www2.ba.itb.cnr.it/UTRSite/
NcRNAdb	http://biobases.ibch.poznan.pl/ncRNA/
MFOLD	http://www.bioinfo.rpi.edu/applications/mfold/
RNAFold	http://rna.tbi.univie.ac.at/cgi-bin/RNAfold.cgi

to catalogue all ncRNAs in this chapter; instead, we direct the reader to some of the databases that attempt to catalogue and classify ncRNA superfamilies (Table 14.1), the most comprehensive of which is probably Rfam (Griffiths-Jones *et al.*, 2005).

Many databases attempt to classify ncRNAs, but most are hindered by the lack of sequence similarity between RNA genes. The Rfam database takes a computationally intensive approach to this problem by examining the largely sequence independent, conserved secondary structure that exists within ncRNA families. This is entirely analogous to the conservation that is seen between the fold patterns of proteins that share functional roles, but very little sequence similarity, as used in the Structural Classification of Proteins database (SCOP) (see Chapter 13; Andreeva *et al.*, 2004). Rfam captures the combined secondary structure and primary sequence profile of a multiple sequence alignment of ncRNAs, using a statistical model, known as profile stochastic context-free grammar (SCFG) – which is analogous to the hidden Markov models (HMMs) used for protein alignments. The user interface to Rfam is fairly self-explanatory; the user can browse by ncRNA type (Figure 14.1) or family. Following a link to a ncRNA family returns a family view (Figure 14.2), which offers a full alignment of family members. This function is valuable for assessing the function of variants in miRNA, as the degree of conservation throughout a ncRNA family is a good indicator of functional constraint (see Section 14.3.2 below).

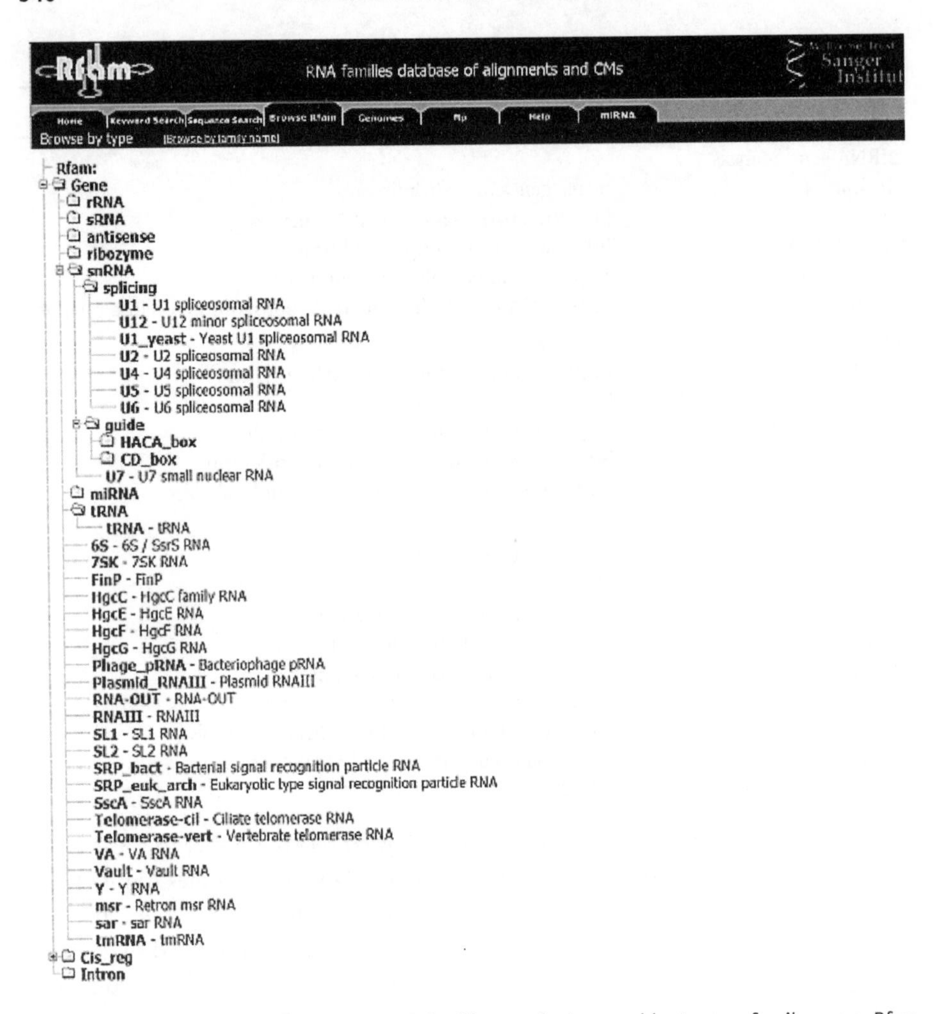

Figure 14.1 The Rfam database – ncRNA families can be browsed by type or family name. Rfam Database, http://www.sanger.ac.uk/Software/Rfam/. Described in Griffiths-Jones, S., Bateman, A., Marshall, M., Khanna, A. and Eddy, S. R. Rfam: an RNA family database. *Nucleic Acids Res* (2003) 33(1), 439–441.

Functionally, ncRNAs have been implicated in a wide range of normal biological processes such as X chromosome activation and inactivation (Rastan, 1994; Lee *et al.*, 1999), genomic imprinting (Wylie *et al.*, 2000) and bone differentiation (Takeda *et al.*, 1998). More recently, it appears that miRNAs, a large growing family of ncRNAs, are implicated in regulation of multiple genes (see Section 14.3.1 below). Excluding miRNAs, ncRNAs have also been implicated in a range of disorders such as a variety of cancers (Srikantan *et al.*, 2000; Tam *et al.*, 2002; Manoharan *et al.*, 2003) but also neurological diseases (Millar *et al.*, 2000; Runte *et al.*, 2004). Taken together, the functional data on ncRNAs clearly indicate that they are major components of biology across many species (Costa, 2005).

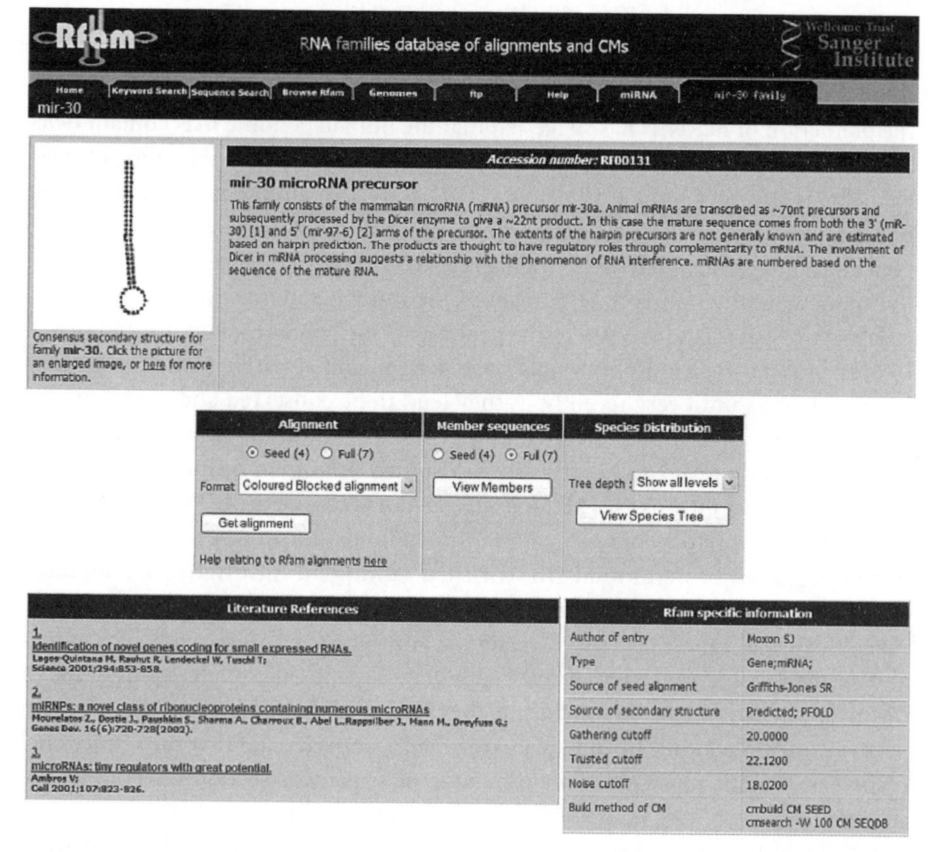

Figure 14.2 The Rfam database – miR30 family view. Rfam Database, http://www.sanger.ac.uk/Software/Rfam/. Described in Griffiths-Jones, S., Bateman, A., Marshall, M., Khanna, A. and Eddy, S. R. Rfam: an RNA family database. *Nucleic Acids Res* (2003) 33(1), 439–441.

14.2.2 Getting a grip on unknown ncRNA – the Dark Continent of the genome?

Databases like Rfam are already helping to curate and catalogue our knowledge of ncRNA, but it is likely that this knowledge will remain at best partial for a long while to come. Quite naturally, most of our understanding of the genome and disease has tended to focus on what we know best. Genes code for proteins, and protein malfunction causes disease. This is something we are comfortable with. So, quite understandably, analysis of disease genetics has focused most heavily on the ~30 000 protein-coding genes in the genome and to a lesser extent their regulatory regions. The collective ignorance of miRNA, which has prevailed until just a few years past, underscores the importance of considering the 'unknown unknown' elements of the human genome in such analyses (Lai, 2003). We know that less than 1.5 per cent of the genome is protein coding, and the rest, frankly, is little more than a jungle of unknown function. Recent advances in the understanding of miRNA should teach us to respect all of the genome as a potentially functional element. Recent research in the prediction of ncRNAs has suggested that the human genome may contain twice as many ncRNAs as it does protein-coding mRNA, but our understanding of these elements is still very limited (Bejerano *et al.*, 2004). To emphasize the importance of those elements we do not yet know in the genome, Sironi *et al.* (2005) completed an interesting analysis of multispecies conserved sequence (MCS) elements that punctuate the non-coding portion of the human genome. MCS elements are reliable surrogates for function in the genome (See Chapter 6), and it is important to bear in mind that miRNAs would have been included within these anonymous MCS sequences until very recently, as they tend to be conserved and are frequently located in the introns of protein-coding genes. The study by Sironi *et al.* (2005) demonstrated that MCSs are unevenly distributed in human introns, the majority of relatively short introns (<9 kb long) displaying no or a few MCSs, and that MCS density reached up to 10 per cent of total size in longer introns. After correction for intron length, MCSs were found to be enriched within genes involved in development and transcription, and depleted in immune-response loci. They also found that human disease and cancer genes showed significantly enriched MCSs. This enrichment also seemed to correlate significantly with gene functional complexity in terms of distinct protein domains. In conclusion, they suggested that evolution acts on human genes as integrated units of coding and regulatory capacity and that such functional complexity might represent a major source of negative selection on non-coding sequences.

Although the observed conservation of MCS elements in certain classes of genes suggests that ncRNAs may in some instances be expressed in a coordinate manner with the genes that they regulate, their action is unlikely to be limited to these genes. Again in the case of miRNA, several independent computational predictions (MiRANDA, TargetScan, see below) have shown that a single miRNA may bind and regulate as many as 200 target genes. These predictions of miRNA binding have estimated that over a third of human protein-coding genes are regulated by miRNA. Again, this just represents our current knowledge of this area; only time will tell what role other as yet uncharacterized ncRNAs play in this grand schema. Thus, it appears that ncRNAs (some of which are still unknown to us) are likely routinely to modulate levels of protein expression in the cell by dampening the translation of thousands of mRNAs. Evidence all points to ncRNA-mediated gene regulation as a fundamental layer of the genetic program that operates at the post-transcriptional level as a balancing mechanism to the complex mechanisms that operate at the transcriptional level. Therefore, it is clear that ncRNAs have a huge potential for involvement in the pathology of human diseases, and the probable magnitude of this, considering the relatively short ncRNA research timeline, is just beginning to emerge.

14.3 Computational analysis of ncRNA

14.3.1 Principles of miRNA function

The mature miRNAs of 21-22 nucleotides (nt) in length are originally derived from larger precursors of ~60–70 nt, the pre-miRNAs (Bartel, 2004). These precursors can fold into imperfect stem-loop structures, and in animals they are themselves the cleavage products of a primary miRNA (pri-miRNA) transcript by a multicomplex protein, the microprocessor (Denli *et al.*, 2004; Gregory *et al.*, 2004; Gregory and Shiekhattar, 2005). For more details on microRNA biogenesis, the reader should refer to the recent review by Kim (2006).

The first miRNA discovered in *C. elegans*, *lin-4*, was shown to mediate repression of its target gene, *lin-14*, through partial antisense complementarity with the 3′ untranslated region (3′ UTR) of the *lin-14* mRNA. It was only several years later with the discovery of *let-7* in *C. elegans*, a 22-nt regulatory small RNA similar to *lin-4* and conserved in other species, that regulation of expression by homologous RNA–RNA interactions was identified as a more general mechanism of post-transcriptional repression. However important distinctions have to be made. In plants, miRNAs appear to bind their gene targets with near to perfect complementarity, resulting in target cleavage (Baulcombe, 2004). In animals, miRNAs typically make imperfect base pairings with their targets, usually in the 3′ UTRs. Initially, it was thought that miRNAs blocked protein synthesis through the elongation phase (Olsen and Ambros, 1999). However, several discoveries now suggest that other mechanisms of miRNA action could inhibit protein synthesis at the translational phase (Pillai *et al.*, 2005) or affect directly mRNA levels and stability (Bagga *et al.*, 2005; Jinq *et al.*, 2005). It has also been recently observed that miRNAs could reduce gene expression by sequestering their target mRNAs into P-/GW-bodies, where they are not accessible to the protein synthesis machinery (Liu *et al.* 2005; Sen and Blau, 2005).

Assessing the number of miRNA target genes is a major goal to understand the functional diversity of pathways possibly regulated by this class of ncRNAs. Only a very limited number of animal miRNAs have been functionally characterized. However, it appears that a wide range of functions could be affected by miRNA regulation in animals: developmental timing (Wightman *et al.*, 1993; Grosshans *et al.*, 2005), adipocyte differentiation (Esau *et al.*, 2004), proliferation (Calin *et al.*, 2004, O'Donnell *et al.*, 2005), cell death and fat metabolism (Xu *et al.*, 2003), insulin secretion (Poy *et al.*, 2004) or apoptosis (Leaman *et al.*, 2005).

Bioinformatics approaches have been very helpful in identifying both miRNA genes and their targets (Brown and Sanseau, 2005). Recent estimates are that at least 20–30 per cent of human genes could be regulated by miRNAs (Krek *et al.*, 2005; Lewis *et al.*, 2005).

14.3.2 Computational identification of miRNA genes

Approaches to miRNA gene prediction are generally based on machine-learning techniques (Lai *et al.*, 2003; Nam *et al.*, 2005). The algorithm is first run with known miRNA gene sequences as a training set, in order to create a profile or a matrix. The 'trained' program is used in a genome-wide scan, and can identify putative novel miRNA sequences. A large set of hairpin loops found randomly in the genome is first used as a negative training set. These hairpins are assumed not to be miRNA precursors. Known miRNA genes are then used as a positive training set. Although miRNA genes share little primary sequence similarity, they do share numerous properties that can be used to aid recognition in miRNA prediction algorithms. These include hairpin length, hairpin-loop length, thermodynamic stability, base-pairing, bulge size, bulge location, nucleotide content, sequence complexity and repeat elements. A miRNA prediction algorithm will give a score to a scanned sequence depending on its similarity to these distinctive properties. The algorithm is iteratively rerun, checked and improved by training on subsets of known miRNAs, as well as control groups of random hairpins, and checked on its scoring accuracy. This allows a fine-tuning of the weight of each distinctive property. These scores may be assessed for their sensitivity and specificity. Cross-species conservation filters may also be applied to the output set to filter out false positives, although this has immediate limitations, as it detects only conserved miRNAs and may exclude species-specific miRNA genes.

Several prominent miRNA gene detection algorithms have been developed. MiRseeker (Lai *et al.*, 2003) assesses the folding patterns of RNA sequences as predicted by MFold (Zuker 1989, 2003). MirScan (Lim, 2003) uses RNAFold (Hofacker *et al.*, 1994) to find hairpin structures in evolutionarily conserved sequences. Each conserved hairpin is further analysed to locate the miRNA within it. Yousef *et al.* (2006) developed BayesMiRNAfind, a method using a naive Bayes classifier, to generate a gene model from training data, based on the sequence and structure of known microRNAs from a variety of species. This method appeared to demonstrate a higher sensitivity and specificity than other methods. In analysis of known miRNAs in the mouse genome, BayesMiRNAfind demonstrated 97 per cent sensitivity and 91 per cent specificity. The method also predicted 244 putative miRNA genes in the mouse genome.

Although machine-learning approaches have dominated approaches to miRNA prediction, as discussed earlier, species conservation also has a part to play in the identification of new miRNA genes. Berezikov *et al.* (2005) employed a phylogenetic shadowing approach to detect possible novel miRNA genes. They first sequenced 122 miRNAs in 10 different species of primate. This revealed some valuable information about the conservation characteristics of miRNA genes. Strong conservation was observed in the stems of miRNA hairpins and increased variation was seen in the loop sequences. A striking drop in sequence conservation was observed in the regions immediately flanking the miRNA hairpins. This characteristic profile of conservation was used to predict novel miRNAs using cross-species comparisons. In this analysis they identified 976 vertebrate conserved candidate miRNAs (made available in supplementary materials with their publication) by scanning whole-genome human/mouse and human/rat alignments.

De novo prediction of miRNA genes is probably a tall order for the non-specialist, so it is probably worth focusing on what is already known. The miRBASE database at the Sanger Institute (Griffiths-Jones *et al.*, 2006) does a very good job of curating all known miRNAs, and also hosts the miRNA registry and miRNA target predictions. Each record has been manually curated and links to papers that support evidence for those miRNAs are available. miRNA genes sourced from miRBASE can also be viewed in the UCSC genome browser with the 'sno/miRNA' track. All validated miRNAs as well as orthologues of validated miRNAs in other species are available to view with this interface.

14.3.3 Computational identification of novel small ncRNAs

The UCSC human genome browser is a valuable resource for ncRNA analysis, as it contains genome-wide annotations of known and predicted ncRNA determined by a number of methods, each of which is well described in the track settings. These include catalogues of known ncRNA, such as 'miRNA genes' representing all the miRNAs from miRBase and 'RNA genes', a comprehensive mapping of known ncRNA genes, including tRNA, rRNA, scRNA, snRNA, snoRNA, miRNA and other known ncRNA (e.g., Xist). A number of novel prediction methods are also applied genome wide, including 'EvoFold', an RNA secondary structure prediction method that exploits the evolutionary signal of genomic multiple-sequence alignments for identifying conserved functional RNA structures, and 'RNAfold', a similar RNA secondary structure prediction method but without a requirement for conservation. Finally, the 'Rfam seed folds' track shows the secondary structure annotation of ncRNAs from the Rfam database.

The ENCyclopedia Of DNA Elements (ENCODE) Project has given a valuable insight into the potential abundance of ncRNA transcripts in the human genome (ENCODE Project Consortium, 2004). The pilot phase of this project has focused on 44 discrete regions covering 30 Mb (1 per cent) of the human genome, chosen to represent a range of genomic features. The ENCODE project is applying state-of-the-art bioinformatics and laboratory techniques to detect all sequence elements that confer biological function. The results of the pilot phase of ENCODE analysis are available to view at the UCSC human genome browser. The results make interesting viewing (Figure 14.3); alongside the mapping of known ncRNA and computational predictions, there are also tracks presenting results from the Affymetrix genome tiling arrays. These arrays capture transcription of polyadenylated and nonpolyadenylated RNAs for 10 human chromosomes, mapped at 5-bp resolution in eight cell lines. The 'Yale TAR' track shows the locations of transcriptionally active regions (TARs) for several samples.

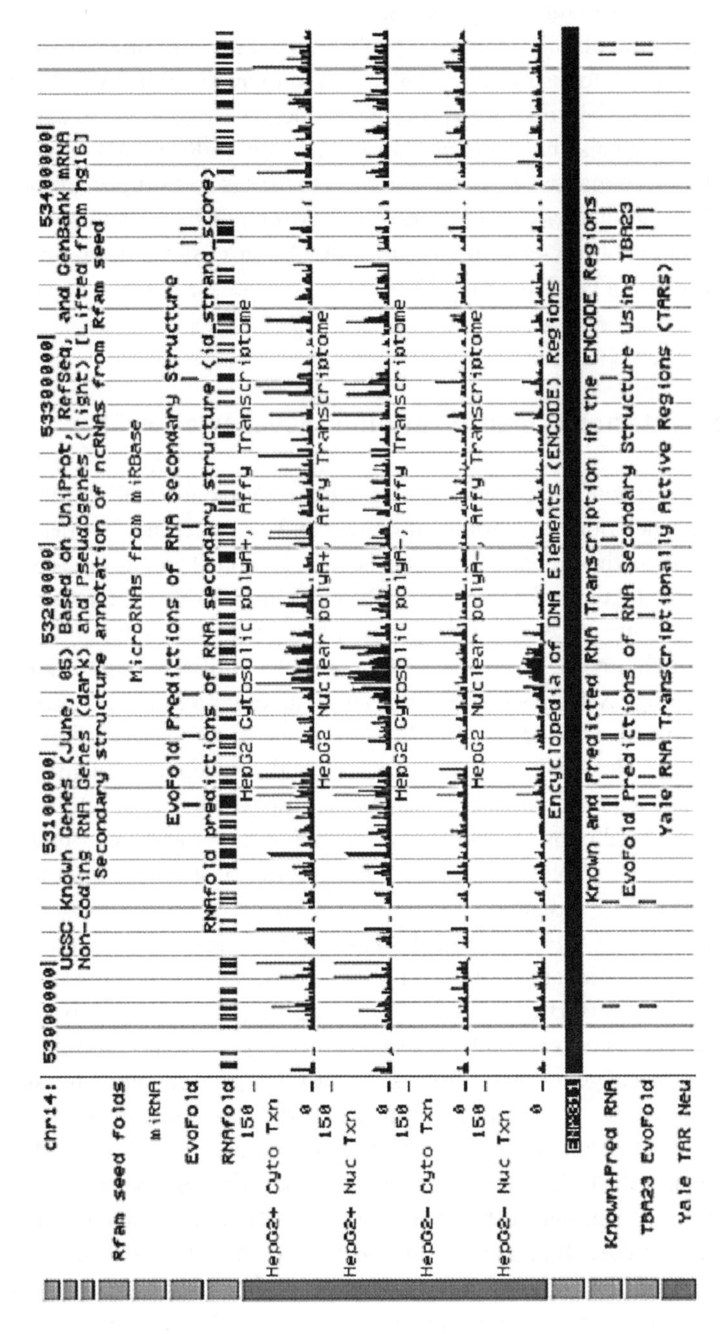

Figure 14.3 A view of ncRNA and expression data across the ENr311 ENCODE region in the UCSC genome browser

14.3.4 Computational identification of miRNA targets

Prediction of miRNA targets for translational regulation is challenging on many levels. Methods are evolving rapidly, reflecting the miRNA field as whole. Here are the problems. Firstly, the interaction with a target mRNA usually occurs via a non-strict base pairing. As a consequence, common alignment tools, such as Blast, are not appropriate for searching for miRNA targets. As the target site consists of non-strictly matching 8-23-bp sequences, there is also a very high potential for false-positive prediction. Moreover, a large set of laboratory-validated miRNA-binding sites that could be used as a training set for learning algorithms is not yet available.

Several miRNA target-prediction tools have been developed with the characteristics of binding of the miRNA to its target: The 5′ end of a miRNA called the 'seed', which is 6–8 nt long, has been shown to be critical and sometimes sufficient to repress target translation (Lewis *et al.*, 2003, 2005). There is also evidence that the 3′ end of miRNAs may compensate for imperfect 5′ binding. Moreover, repression may be increased by multiplicity (several binding sites in a transcript for a single miRNA) and cooperativity (several miRNAs bind to a single transcript). Further comparisons of repression by miRNAs bound to two, four and six different binding sites on a reporter construct indicate that translation decreases with each additional site (Doench *et al.*, 2003; Zeng *et al.*, 2003).

There are a number of published miRNA target-prediction tools, and all have their caveats. TargetScan (Lewis *et al.*, 2003) heavily weights its predictions toward the 5′ seed end. As a result, no prediction could be made if the complementarity in the 5′ end is imperfect. PicTar (Krek *et al.*, 2005) looks for a perfect seed (called the nucleus) and combines the output with cross-species conservation and multiplicity of binding sites within a single UTR. The predominant focus of these two algorithms on the 5′ seed for target prediction is a source of some concern. In evolutionary terms, if the 3′ end of a miRNA target was unimportant, miRNAs might have been expected to evolve into shorter molecules. Moreover, using conservation as an exclusion filter is not ideal, as Giraldez *et al.* (2006) showed, in a laboratory study of zebrafish miR-430, that a large fraction of laboratory-verified miR-430 targets showed little or no conservation. From this observation, it seems that conservation should only be used to add confidence to a target, but it should not be used to exclude targets. This is particularly relevant in man, where the complexities of gene regulation in some systems, such as the brain, are unlikely to be strongly conserved across species.

RNAHybrid (Rehmsmeier *et al.*, 2004) is another miRNA target-prediction tool, with an improved folding algorithm that searches for the best minimum free energy (MFE) of binding and combines advanced statistics (P values). The tool is freely available on the Web (see Table 14.1). Miranda (John *et al.*, 2004) is similar in principle to the Pictar algorithm; however, it also provides P values. It searches for the best MFEs regardless of the seed and is freely available as a package for one's own analyses (an option is available for strict base-paring binding in 5′ of the miRNA).

MiRBase itself is available as a MySQL database and is currently the only database with a pipeline behind it that will allow updates with current gene builds. It also has a very close link with the miRNA registry that deals with the official naming of miRNA genes.

Comparing available prediction methods on a given dataset shows worryingly little overlap. For instance, given a set of two genes, 1-500 targets can be found for each gene, depending on the prediction method used (David Bartel, personal communication). This just serves to highlight the point that miRNA prediction still has a very long way to go. It is difficult to see how these methods can be improved further until a substantially characterized set of laboratory verified target sites is identified. In the meantime, the best approach to finding miRNA targets that are likely to be 'real' is probably to use Miranda or RNAHybrid, with close attention to predicted *P* values.

As in the case of miRNA genes, a substantial number of predictions from some of the miRNA target prediction tools are available online. Predicted miRNA targets from TargetScan and Pictar can be viewed in the UCSC browser, in the 'T-ScanS miRNA' and 'PicTar miRNA' tracks, respectively. Targets generated by MiRANDA can be viewed in Ensembl.

At the time of writing (June 2006), the conventional view is that miRNA generally bind to the 3′ UTR of mRNA transcripts; however, some literature is beginning to suggest miRNA binding within the coding sequence (CDS) (Chamary *et al.*, 2006). A study by Hurst (2006) suggests that miRNA binding to the CDS might explain the selection pressure that is observed to act on synonymous mutations. This is an issue that can be resolved only by further research. If miRNAs do target the coding sequences of genes, this increases the potential sites for miRNA binding by two- to threefold. This, in turn, would substantially increase the potential for miRNA regulation. However, it will make miRNA target prediction more difficult, as vertebrate conservation is widespread throughout coding sequences. Therefore, target prediction will need to focus on motif recognition alone, probably leading to a much high rate of false-positive identification.

14.3.5 Getting a handle on miRNA expression

The small size of mature miRNAs has presented considerable technical challenges in the detection of their expression patterns. Most miRNA-profiling studies include an initial size fractionation of total mRNA in order to enrich for small RNA species. Early methods used to determine the expression profile of miRNAs included northern blotting, quantitative PCR, dot blotting, and RNase protection assay. Large-scale cDNA analysis has also been used, but it is very laborious and highly dependent upon selectivity for small RNAs.

Recent miRNA expression studies have used microarray chips specifically designed to detect miRNAs. The laboratory of Carlo Croce published the first use of an miRNA microarray to study gene expression in mammalian embryonic and adult tissues (Liu *et al.*, 2004). However, the general drawbacks of early miRNA arrays are the use of 70-mer oligos to detect pre-miRNA products rather than the smaller mature miRNAs and biases possibly introduced from the methods used to make cDNA from the RNA pool (Esquela-Kerscher and Slack, 2004). Construction of microarrays with 22-mer antisense oligonucleotides corresponding to miRNAs, along with direct labelling of RNA, seems better to detect mature miRNAs (Miska *et al.*, 2004; Thomson *et al.*, 2004). A variant on this method is RNA-primed, array-based Klenow assay (RAKE), which selectively amplifies only mature miRNAs bound to an array with exact matching antisense oligonucleotides (Nelson *et al.*, 2004).

Cross-hybridization across oligonucleotide probes is also an issue, particularly for closely related miRNAs that differ by only one or a few nucleotides. In addition, miRNAs have wide-ranging GC content, making it difficult to optimize globally all hybridization conditions. Recently, locked nucleic acid or LNA probes have been developed that allow for the construction of overlapping short probes, spanning half of the mature miRNA sequence. Neely *et al.* (2006) successfully used spectrally distinguishable fluorescent LNA-DNA probe sets to determine the expression of 45 different miRNAs in the femtomolar range in 16 different tissues. Moreover, by having two different fluorescent-labelled probes to every miRNA, they were able to quantify closely related mature miRNAs, like let-7a and let-7c, which differ by only a single base pair. LNA-mediated probes have also been used for *in situ* detection of miRNAs in mouse and zebrafish embryos (Kloosterman *et al.*, 2006).

Of course, microarray or probe-based miRNA expression detection methods can only profile miRNA genes with known sequences. Other approaches have used expression profiling to characterize known miRNA genes as well as to discover novel miRNAs in purified RNA fractions. There are several recent examples. Cummins *et al.* (2006) characterized miRNAs present in the human colon by a modified SAGE (serial analysis of gene expression) technique, which they called miRAGE. Sequence analysis was performed on 273 966 small RNA tags from four human colorectal cancers and two matching samples of normal colonic muscosae. About 200 known miRNAs were identified as well as 133 novel miRNA candidates. Mineno *et al.* (2006) initially isolated all small RNAs in the 21–27-nt range from mouse embryo. Then, using massively parallel sequencing technology, they globally determined all nucleotide sequences in this pool. They found 390 expressed miRNAs, of which nearly half were novel mouse miRNAs.

Statistical analysis of miRNA profiles is similar to that of mRNA Taqman and microarray data, with some caveats. Until newer methods are validated, cross-hybridization of probes for similar miRNAs will be an issue. Thus, fine-scale detection of individual miRNAs from highly related families is somewhat provisional unless verified by more sensitive assays. Procedures for the normalization of miRNA expression data (to allow comparisons between experiments) are also unique. Housekeeping genes, such as GAPDH, are often used to normalize mRNA expression profiles among various samples. Small RNAs, such as 5S RNA and U6 RNA, are sometimes included in miRNA expression-profiling panels as putative small housekeeping RNAs. However, the expression profiles of these small RNAs are not fully understood and often show tissue-specific biases, limiting their reliability as housekeeping standards. Presently, most studies normalize miRNA gene expression data to median values, either per chip or across multiple panels.

Kits for miRNA detection are becoming commercially available from a number of molecular biology reagent vendors. Initially, these assays covered a limited number of known miRNAs. However, these products have been rapidly improved to cover the wider repertoire of newly discovered and validated miRNA genes. For example, the company Ambion has recently released both probe sets and preprinted miRNA arrays covering 384 miRNAs for man, mouse and rat (see http://www.ambion.com/). In addition, there have been advances in the isolation of miRNA expression profiles from clinically relevant sources, such as blood and formalin-fixed paraffin-wax-embedded (FFPE) samples. Although the technology is still nascent, miRNA expression profiling will no doubt become as common a molecular technique as mRNA microarray analysis.

14.4 ncRNA variation in disease

14.4.1 miRNA and ncRNA in cancer

Given the anticipated global cellular effect of ncRNAs on gene expression, it is not surprising they have been implicated in the pathogenesis of cancer – which is essentially a disease of dysregulated gene expression. Studies of cancer have helped to launch the entire field of analysis of ncRNAs, many of which have been rather fortuitously identified during the study of tumours, such as the imprinted transcript H19 (see Szymanski *et al.*, 2005, for a detailed review of this area). Small regulatory RNAs, including miRNAs, have also been shown to play an important role in cancer (Figure 14.4). Qualitative as well as quantitative differences in expression of miRNA have been observed in both cancer cell lines and tumours. Advances in understanding how miRNAs play a role in cancer, or act as so-called oncomirs, have been recently reviewed by Esquela-Kersher and Slack (2006). In the case of colorectal cancers, miR-24-2 showed a 50-fold difference between samples (Schmittgen *et al.*, 2004). The mechanisms underlying these observed differences are likely to be varied. To date, there is very little evidence of somatic point mutations in miRNA genes. Diederich and Haber (2006) analysed a panel of 91 cancer-derived cell lines for sequence variations in 15 miRNAs implicated in tumorigenesis. No mutations were detected within any of the short, mature miRNA sequences. One sequence variant was identified in a precursor miRNA, and 15 variants were isolated in primary miRNA (pri-miRNA) transcripts. Despite the possibility of an impact on the predicted secondary folding structure flanking putative cleavage sites in the pri-miRNA,

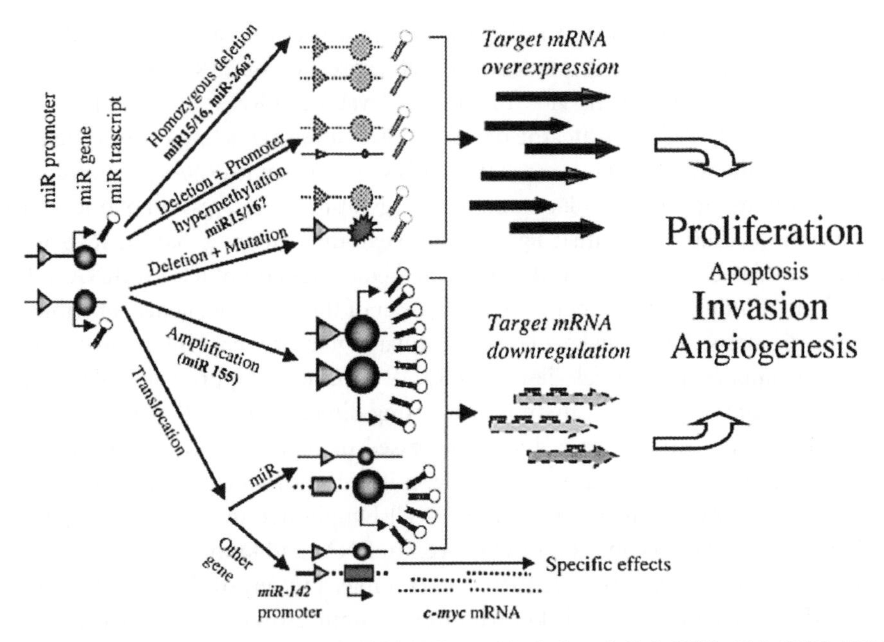

From Calin, George Adrian et al. (2004) Proc. Natl. Acad. Sci. USA 101, 2999-3004

Figure 14.4 The putative role of miRNA in cancer. From Calin *et al.* Human microRNA genes are frequently located at fragile sites and genomic regions involved in cancers. *Proc Natl Acad Sci U S A* 101(9): 3003. Copyright 2004 National Academy of Sciences, U.S.A.

processing and miRNA maturation were not affected *in vivo*. Thus, they concluded that genetic variants in miRNA precursors in cancer cells might not have a common physiological significance.

One of the commonest variants is probably alteration by chromosomal aberration, that is, deletion, amplification or translocation of chromosomal regions containing miRNA. Several miRNA genes are located in genomic regions that undergo frequent translocations or deletions in leukaemia. The miR-15a–miR-16-1 cluster is located on chromosome 13q14, a region that is frequently deleted in cases of chronic lymphocytic leukaemia (CLL), mantle-cell lymphoma, multiple myeloma and prostate cancer (Gregory and Shiekhattar, 2005). The miR-142 gene is another example, located at the break point of a t(8;17) translocation, which is believed to contribute to the progression of B-cell leukaemia (Calin *et al.*, 2004). There is compelling evidence to suggest a very widespread involvement of this mechanism in miRNA-mediated cancers in both a tumour-suppressive or oncogenic mode. In a detailed analysis of the chromosomal distribution of miRNAs, Calin *et al.* (2004) demonstrated that 98 out of 186 (52.5 per cent) miRNA genes were located within minimal tumour deletion, amplification or break-point regions. (These aberrant regions are catalogued in the Mitelman Database of chromosomal aberrations in cancer (see Chapter 17 for details).)

More subtle mechanisms of miRNA action in cancer also appear to exist. Most cancer cell types studied show reduced overall expression of miRNAs, and this generally appears to support the view that higher levels of miRNAs are associated with the differentiated cellular state, the observed lower levels reflecting the undifferentiated state of the tumour (Lu *et al.*, 2005). This suggests that miRNAs predominantly play a tumour-suppressive role, probably by regulating the expression of oncogenes. Johnson *et al.* (2005) identified a good example of this, showing that the *RAS* oncogene is regulated by the let-7 miRNA. Decreased expression of let-7 miRNA in human lung tumours caused increased expression of the *RAS* oncogene, contributing to tumorigenesis. Additional evidence of similar regulatory relationships is accumulating to support the general view of miRNAs as tumour suppressors (Morris and McManus, 2005). However, there is also strong direct evidence that miRNAs may act as oncogenes, in this case due to increased or inappropriate expression regulating tumour suppressors. For example, mir-155 shows increased expression in some B-cell lymphomas, such as large B-cell lymphoma and Burkitt lymphoma (Hammond, 2006). Feedback loops between miRNAs and known tumour suppressors or oncogenes might be extensive. For example, a recent study suggests that the transcription factor p53, a well-known tumour suppressor, regulates the expression of a large number of miRNA genes (Xi *et al.*, 2006).

Knowledge of miRNA expression in tumours is also having some unexpected benefits. Expression profiles of different miRNAs appear to discriminate between tumours that originate from different tissues as well as between normal and malignant cells (Lu *et al.*, 2005). These expression profiles can be highly effective in classifying human cancers and predicting the developmental origins of cancers, and this could be clinically important in tracing the origin of a metastatic tumour. Lu *et al.* (2005) demonstrated that miRNA microarrays are more effective in tumour classification than mRNA microarrays containing more than 16 000 protein-coding genes. Similarly, clinically derived solid tumours samples also have distinctive miRNA signatures (Volinia *et al.*, 2006).

While the *in vivo* and *in vitro* evidence supporting the role of miRNA in cancer mounts, the *in silico* evidence is even stronger. If bioinformatics predictions of target binding are taken into account, a quick review of a selection of oncogenes and tumour suppressors in miRBase Targets (http://microrna.sanger.ac.uk/targets/) reports that miRNA target-binding sites are widespread in both oncogenes and tumour-suppressor genes. If these predictions are accurate, the role of miRNA (and other ncRNAs) in cancer is likely to be ubiquitous. Clearly, these miRNA target-binding sites are likely to be strong candidates for mutation in tumours; however, it is very difficult to get an idea of how prevalent this is likely to be. Somewhat ironically, in perhaps the final twist of the dogma supporting coding sequence in disease, there are very few somatic mutation data available from the UTR sequences of genes. All major tumour mutation discovery projects (e.g., the Sanger Institute's Cancer Genome Project (see Chapter 17)) have focused on the coding regions of genes.

14.4.2 ncRNA in complex disease

Beyond the gross amplifications and deletions seen in miRNA in cancer, there are very few reports of variations in small ncRNAs in complex diseases. Broadening the view to all ncRNAs is a slightly different matter, as a number of larger ncRNAs have been implicated in a range of neurobehavioural and developmental syndromes and diseases (see Szymanski *et al.*, 2005, for a review). However, keeping our focus on small ncRNA, the clinical evidence is still largely absent, although the rationale is not lacking. This is probably due to the recent nature of our understanding of small RNA, so variations in these pathways have probably gone undetected. The potential for polymorphism and mutation in ncRNA genes, such as miRNA, may be limited due the strong selective constraint that exists at many of these loci, particularly in the conserved mature miRNA sequences. However, when potential polymorphisms in the mRNA target sequences are taken into account, the odds begin to look more favourable.

Genetic variants in miRNA genes as potential disease alleles

A deleterious mutation in a single miRNA gene might affect all the genes regulated by the given miRNA, possibly hundreds of genes. However, evolution has led to a fail-safe mechanism in many miRNA that may ameliorate such events by introducing functional redundancy into miRNA gene families. This redundancy can be absolute; 42/313 mature human miRNAs are identically encoded by more than one pre-miRNA gene. This redundancy can be up to fourfold (although usually it is two- to threefold); for example, the 22-nt, mature miR-30c miRNA is identically encoded by two pre-miRNA genes, on chromosomes 1 and 6 respectively. This redundancy is conveyed by the naming convention; distinct pre-miRNA loci that give rise to identical mature miRNAs have numbered suffixes, such as hsa-miR-30c-1 and hsa-miR-30c-2 (Figure 14.5A). To further complicate matters, opinion on the degree of conservation of mature sequence required for functional redundancy varies (Griffiths-Jones *et al.*, 2006). Some recent studies suggest that only the 5′ seed region (nucleotides 2–7) of the sequence forms a tight duplex with the target mRNA (Lewis *et al.*, 2005). If so, one might expect compensation for mutations by paralogous miRNA sequences whose mature miRNAs differ at only one or two positions. These paralogues are given lettered suffixes – for example, hsa-miR-30a or hsa-miR-30b, c, d and e (Figure 14.5B). If we take this into account, 163/313 mature miRNAs appear to be redundantly coded; this figure may increase as more pre-miRNA genes are identified. It is not clear how much redundancy of coding would compensate for loss of function at individual miRNA loci. It is possible that the redundant pre-miRNA genes are regulated differently and expressed in different tissues, in which case compensation for null alleles might not be possible.

A – Redundant genes encode miR-30c

Homo sapiens miR-30c-1 stem-loop (MI0000736)

```
a     cu    ugu u        u   aca           ---g  a
 ccaug  guag   g guaaaca ccu    cucucagcu      ug g
 |||||  ||||   | ||||||| |||    |||||||||      ||
 gguac  cguc   c cauuugu ggg    gagggucgg      ac c
a     --    uuc u        u   --a           ugga  u
```

Homo sapiens miR-30c-2 stem-loop (MI0000254)

```
         uacu        u   aca           guggaa
      aga    guaaaca ccu    cucucagcu       a
      |||    ||||||| |||    |||||||||
      ucu    cauuugu gga    gagggucga       g
         uucu        c   --a           aagaau
```

B – Multiple paralogues of miR-30

Homo sapiens miR-30a stem-loop (MI0000088)

```
   a            uc           -----    a
gcg cuguaaacaucc  gacuggaagcu     gug a
||| ||||||||||||  |||||||||||     |||
cgu gacguuuguagg  cugacuuucgg     cac g
   c            --           guaga    c
```

Homo sapiens miR-30b stem-loop (MI0000441)

```
a     uu      cau          u  -a        uaaua
 ccaag  ucaguu   guaaacaucc ac  cucagcug     c
 |||||  ||||||   |||||||||| ||  ||||||||
 gguuc  agucga   cauuuguagg ug  gggucggu     a
a     --      cuu          -  ga        uaggu
```

Homo sapiens miR-30d stem-loop (MI0000255)

```
guu  u         ccc          gua  ac
   gu guaaacauc   gacuggaagcu   ag  a
   || |||||||||   |||||||||||   ||
   cg cguuuguag   cugacuuucga   uc  c
cau  u         --a          --a  ga
```

Homo sapiens miR-30e stem-loop (MI0000749)

```
g        uu  ua            uu           g aag u
 ggcagucu  gc  cuguaaacaucc  gacuggaagcu u   g g
 ||||||||  ||  ||||||||||||  ||||||||||| |   |
 ccgucgga  cg  gacauuuguagg  cugacuuucga g   c u
a        --  gc            --           g aga u
```

C – Alignment of miR-30 paralogues

```
>hsa-miR-30a UGUAAACAUCCUCGACUGGAAG
>hsa-miR-30b UGUAAACAUCCUACACUCAGCU
>hsa-miR-30d UGUAAACAUCCCCGACUGGAAG
>hsa-miR-30e UGUAAACAUCCUUGACUGGAAG
>hsa-miR-30c UGUAAACAUCCUACACUCUCAGC
                      -
                      A*
```

*C>A SNP from Iwai & Naraba, (2005)

Figure 14.5 Redundancy, paralogy and the impact of variation in pre-miRNA genes. (A) Redundant pre-miRNA genes encode the miR-30c mature miRNA. Mature miRNA sequences are underlined. (B) Paralogous pre-miRNA genes encode four additional paralogues of miR-30c which may show functional redundancy. (C) Alignment of miR-30 paralogues identifies invariant nucleotide 1-11. C > A polymorphism is located at nucleotide 9. *Note:* Related hairpin precursor sequences may give rise to mature sequences with only marginal similarity (and vice versa). Therefore, miRNA gene names should not be relied upon to convey complex relationships between genes; instead, they should be used as a guide, backed up by detailed review of the sequences with miRBase (Table 14.1).

Despite the considerable redundancy seen between mature miRNA, genetic variants in miRNA genes still may affect many target genes with serious consequences. This would suggest that pre-miRNA genes are likely be under strong balancing selection. As if to confirm this, Iwai and Naraba (2005) sequenced 173 human pre-miRNA genome regions in 96 subjects and found only 10 polymorphisms; nine were located in the pre-miRNA hairpin regions and were predicted to have a limited effect on miRNA processing. However, one C to A polymorphism was identified in the miR-30c-2 mature miRNA sequence (Figure 14.5C). After further genotyping, they found the C > A miR-30c-2 SNP at a 0.0006 frequency in 3631 Japanese subjects. Considering the potential for redundant coding of the mature miR-30c transcript, they carried out Northern blot analysis, which indicated that miR-30c-2 was expressed in human tissues, but miR-30c-1 did not appear to be expressed in the tissues examined. Therefore, they predicted that it might not be possible for miR-30c-1 to compensate for loss of function at miR-30c-2, in which case the polymorphism might exert a significant biological effect.

The analysis by Iwai and Naraba is interesting, and it clearly illustrates some of the steps that need to be taken to assess the impact of variation on pre-miRNA genes. It may well be that the miR-30c-2 cannot be compensated for by miR-30c-1; in all tissues the two genes are not co-expressed, and, in view of its lack of expression, miR-30c-1 may be a pseudogene. However, the situation may be yet more complex. If conservation is only required at the 5′ seed region, as some have suggested, it may be possible for any of the five miR-30 paralogues to compensate for loss of function at the miR-30c locus, as all are completely conserved across the miRNA seed region (Figure 14.2). Another point to consider is that some miRNAs may be more vulnerable to mutation than others, especially when there are no other paralogues to compensate for expression. In this case, SNPs were also identified in the pre-miRNA hairpin regions of miR-149 and miR-217, neither of which has known paralogues. In such cases, even a subtle impact on function might have important consequences if no paralogues are available to compensate (we review the impact of these SNPs in Section 14.5.

Genetic variation in miRNA targets as potential disease alleles

By contrast to the scarcity of variation seen in miRNA genes, the potential for variation in miRNA target sites in the 3′ UTR of mRNA transcripts is very great indeed. Genetic association between complex diseases and miRNA target alleles may well be found in the near future, for a number of reasons. Firstly, the potential number of sites at which this might occur is very high; secondly, the impact of variation at these sites might be very subtle, fitting well the subtle dysregulation of genes that is sometimes seen in complex diseases. However, this has not been reflected in the literature so far. This is not due to a lack of association with UTR variants, of which there are many; the problem lies in proof of cause and effect. Explanations of functional impact in UTR have generally focused on RNA secondary structure and mRNA transcript stability, but this is difficult to demonstrate *in vivo*. In effect, genetic associations in the UTR of genes are usually the subject of a collective literary shrug; function remains undefined or a hidden causative variant in linkage disequilibrium with polymorphisms in the UTR will be suspected.

Now that some understanding of miRNA to mRNA targeting mechanisms exists, it may just be a matter of time before many good examples of miRNA target variation emerge in complex diseases. At the time of writing, however, there is only one known association between a variant in a miRNA target and a complex disease (other than cancer). Abelson *et al.* (2005) carried out this landmark study in patients suffering from Tourette's syndrome (TS). A G > A variant was identified in two unrelated patients in the 3′ UTR of the SLITRK1 gene in a highly conserved predicted binding site for the miRNA hsa-miR-189; this variant was absent from 3600 control chromosomes. The fact that this variant was in a conserved predicted binding site was not in itself enough to confirm the function of this mutation. However, Abelson *et al.* carried out a series of seminal experiments that convincingly argued that a variant in a miRNA target site may play a role in susceptibility to TS. Their experiments should form the basis of other follow-up studies. Firstly, they showed that SLITRK1 mRNA and hsa-miR-189 have an overlapping expression pattern in brain regions previously implicated in TS. Next they examined the impact of the G > A variant on binding of miR-189 to the UTR of SLITRK1, finding that the G > A variant replaced a G:U wobble base pair with an A:U Watson–Crick pairing at position 9 in the miRNA-binding domain. This G:U pairing was highly conserved in both SLITRK1 3′ UTR and miR-189. Other studies have also shown that G:U wobble base pairs inhibit miRNA-mediated protein repression to a greater degree than would be expected on the basis of their thermodynamic properties alone (Doench and Sharp, 2004). All of these factors strongly support the assumption that the G > A variant affects SLITRK1 expression. To confirm this hypothesis, Abelson *et al.* inserted the full-length SLITRK1 3′ UTR downstream of a luciferase reporter gene and transfected the construct into Neuro2a (N2a) cells. In the presence of miR-189, the expression of luciferase was significantly reduced, confirming the functional potential of the mRNA-miRNA duplex at this binding site. To evaluate the impact of the variant base, they constructed an identical construct differing only at the mutant base, and they recorded a modest but statistically significant and dose-dependent further repression of luciferase expression compared with that of the wild type. This confirms the subtle impact of miRNA target variants and their role in TS and serves as a template for future functional analysis of miRNA target binding.

14.5 Assessing the impact of variation in ncRNA

As the study by Iwai and Naraba (2005) demonstrated, in most cases, variation in ncRNAs is more likely to be localized in the less functionally constrained regions outside the mature RNA molecules. Getting an understanding of the impact of this variation is not easy; this is in part because we do not really have a good idea how small ncRNA genes are structured or regulated. Even for the well-studied miRNAs, we still have a somewhat limited understanding of their regulation; most do not have recognizable canonical promoters, although Cai *et al.* (2004) did demonstrate expression of some miRNAs from T1 promoter elements. We can assume that the regions outside the mature miRNA sequence, however, are still functionally significant by the simple fact that they are generally well conserved between species.

Zeng and Cullen (2003) gave some insight into the functional constraint of pre-miRNA genes. They demonstrated that the miRNAs, miR-30 and miR-21, could translationally inhibit an mRNA-bearing artificial target sites. They also demonstrated that mature miRNA production in both miRNA genes was highly dependent on the integrity of the precursor RNA stem, although the underlying sequence of the stem was not important as long as the integrity of the stem was retained. This suggests that variation within the stem loop would affect miRNA function only if the overall stem-loop structure was altered. The impact of such variants can be evaluated by RNA secondary structure prediction tools, such as RNAfold (Table 14.1), to fold the wild-type and mutant alleles of the miRNA stem-loop variants. Other parts of the miRNA stem loop may be less sensitive to mutation. Zeng and Cullen (2003) found that the changes in the sequence of the terminal loop affected miRNA production only moderately, suggesting that variants in this region are less likely to have a functional impact, unless major fold changes occur in the RNA. Again, RNAfold prediction may help in assessment of the impact of variants in the terminal loop.

14.6 Data resources to support small ncRNA analysis

A number of informatics resources are now available as databases for miRNAs and their targets and also as computational tools to identify miRNAs targets. The best-known miRNA resource is probably miRBase (http://microrna.sanger.ac.uk/) (Griffiths-Jones *et al.*, 2006), which is maintained at the Sanger Institute in the UK. miRBase provides a single comprehensive resource for miRNAs in the miRBase Sequences section and their predicted targets in the miRBase Targets section. At the time of writing, miRBase contains 3963 entries (Release 8.1 May 2006). miRBase provides also a service to register new miRNAs. Argonaute is another database for miRNAs (http://www.ma.uni-heidelberg.de/apps/zmf/argonaute/interface/) (Shahi *et al.*, 2006). Argonaute focuses on human, mouse and rat miRNAs and contains additional information on the origin of the miRNAs, such as their origin, tissue distribution and potential function. Other resources include miRNAMap (http://mirnamap.mbc.nctu.edu.tw/) (Hsu *et al.*, 2006), a comprehensive resource of experimentally verified miRNAs from four mammalian genomes (human, mouse, rat and dog). miRNAMap provides information on expression profiles of known miRNAs as well as links to other biological databases. In addition, various small companies, such as Actigenics (http://www.actigenics.com), Ambion (http://www.ambion.com) or RosettaGenomics (http://www.rosettagenomics.com), have private databases of miRNAs and their targets.

14.7 Conclusions

The landscape of genome regulation has been radically altered with the recent discovery of small ncRNAs which modulate transcription and translation. To date, new knowledge about ncRNAs has largely focused on miRNAs. Many studies suggest that miRNA expression profiles correspond to important stages in cellular development, growth, differentiation and apoptosis as well as wider embryogenesis and intercellular interactions. Moreover, there is growing evidence that certain miRNAs play at least some role in key diseases, particularly cancer.

However, there is much more to be learned about miRNA biology, especially from a genetics perspective. The mechanisms by which miRNAs exert their regulatory control, either on mRNA transcription or protein synthesis, are still unclear. This knowledge gap hinders the assessment of the functional effects of SNPs, mutations occurring in miRNA genes themselves, or the putative miRNA-binding sites in targeted coding genes. While many miRNA genes are evolutionarily conserved, extensive specialization of miRNA within species lineages is also evident. The entire complement of miRNA genes (miRome) in any genome has yet to be determined. In fact, several studies suggest that in man there might be more than double the current repertoire of validated miRNA genes. Finally, there is the issue of transcriptional regulation of miRNA genes. There needs to be more study of the transcription factors and the regulatory pathways which coordinate the expression and suppression of miRNA genes. Many miRNAs are encoded by redundant, duplicate copies on different chromosomes, suggesting more subtle levels of control.

Advances in technology to detect miRNA gene expression will lead to the wider use of miRNAs in genetic and genomic studies. The analysis and interpretation of ncRNAs on phenotype will be major challenge for geneticists in the years to come.

References

Abelson, J. F., Kwan, K. Y., O'Roak, B. J. *et al.* (2005). Sequence variants in SLITRK1 are associated with Tourette's syndrome. *Science* **310**(5746), 317–320.

Andreeva, A., Howorth, D., Brenner, S. E. *et al.* (2004). SCOP database in 2004: refinements integrate structure and sequence family data. *Nucleic Acids Res* **32**, D226–D229.

Bagga, S., Bracht, J., Hunter, S. *et al.* (2005). Regulation by let-7 and lin-4 miRNAs results in target mRNA degradation. *Cell* **122**, 553–563.

Bartel, D. P. (2004). MicroRNAs: genomics, biogenesis, mechanism, and function. *Cell* **116**, 281–97.

Baulcombe, D. (2004). RNA silencing in plants. *Nature* **431**, 356–363.

Bejerano, G., Haussler, D. and Blanchette, M. (2004). Into the heart of darkness: large-scale clustering of human non-coding DNA. *Bioinformatics* **20** Suppl 1, I40–I48.

Bentwich, I., Avniel, A., Karov, Y. *et al.* (2005). Identification of hundreds of conserved and nonconserved human microRNAs. *Nat Genet* **37**(7), 766–770.

Berezikov, E., Guryev, V., van de Belt, J. *et al.* (2005). Phylogenetic shadowing and computational identification of human microRNA genes. *Cell* **120**(1), 21–24.

Brown, J. R. and Sanseau, P. (2005). A computational view of microRNAs and their targets. *Drug Discov Today* **10**(8), 595–601.

Cai, X., Hagedorn, C. H. and Cullen, B. R. (2004). Human microRNAs are processed from capped, polyadenylated transcripts that can also function as mRNAs. *RNA* **10**(12), 1957–1966.

Calin, G. A., Dumitru, C. D., Shimizu, M. *et al.* (2002). Frequent deletions and down-regulation of micro-RNA genes miR15 and miR16 at 13q14 in chronic lymphocytic leukemia. *Proc Natl Acad Sci USA* **99**, 15524–15529.

Calin, G. A., Sevignani, C., Dumitru, C. D. *et al.* (2004). Human microRNA genes are frequently located at fragile sites and genomic regions involved in cancers. *Proc Natl Acad Sci U S A* **101**(9), 2999–3004.

Carthew, R. W. (2006). Gene regulation by microRNAs. *Curr Opin Genet Dev* **16**, 203–208.

Chalk, A. M., Warfinge, R. E., Georgii-Hemming, P. *et al.* (2005). siRNAdb: a database of siRNA sequences. *Nucleic Acids Res* **33**, D131–134.

Chamary, J.-V., Parmley, J. L. and Hurst, L. D. (2006). Hearing silence: non-neutral evolution at synonymous sites in mammals. *Nat Rev Genet* **7**, 98–108.

Costa, F. F. (2005). Non-coding RNAs: new players in eukaryotic biology. *Gene* **357**, 83–94.

Cummins, J. M., He, Y., Leary, R. J. *et al.* (2006). The colorectal microRNAome. *Proc Natl Acad Sci U S A* **103**, 3687–3692.

Denli, A. M., Tops, B. B., Plasterk, R. H. *et al.* (2004). Processing of primary microRNAs by the microprocessor complex. *Nature* **432**, 231–235.

Diederichs, S. and Haber, D. A. (2006). Sequence variations of microRNAs in human cancer: alterations in predicted secondary structure do not affect processing. *Cancer Res* **66**(12), 6097–6104.

Doench, J. G., Petersen, C. P. and Sharp, P. A. (2003). siRNAs can function as miRNAs. *Genes Dev* **17**, 438–442.

Doench, J. G. and Sharp, P. A. (2004). Specificity of microRNA target selection in translational repression. *Genes Dev* **18**(5), 504–511.

Eddy, S. R. (2001). Non-coding RNA genes and the modern RNA world. *Nat Rev Genet* **2**, 919–929.

ENCODE Project Consortium (2004). The ENCODE (ENCyclopedia Of DNA Elements) Project. *Science* **306**(5696), 636–640.

Esau, C., Kang, X., Peralta, E. *et al.* (2004). MicroRNA-143 regulates adipocyte differentiation. *J Biol Chem* **279**, 52361–52365.

Esquela-Kerscher, A. and Slack, F. J. (2004). The age of high-throughput microRNA profiling. *Nat Methods* **1**, 106–107.

Esquela-Kerscher, A. and Slack, F. J. (2006). Oncomirs – microRNAs with a role in cancer. *Nat Rev Cancer* **6**, 259–269.

Gregory, R. I., Yan, K. P., Amuthan, G. *et al.* (2004). The microprocessor complex mediates the genesis of microRNAs. *Nature* **432**, 235–240.

Gregory, R. I. and Shiekhattar, R. (2005). MicroRNA biogenesis and cancer. *Cancer Res* **65**(9), 3509–3512.

Griffiths-Jones, S., Grocock, R. J., van Dongen, S. *et al.* (2006). miRBase: microRNA sequences, targets and gene nomenclature. *Nucleic Acids Res* **34**, D140–144.

Griffiths-Jones, S., Moxon, S., Marshall, M. *et al.* (2005). Rfam: annotating non-coding RNAs in complete genomes. *Nucleic Acids Res* **33**, D121–124.

Grosshans, H., Johnson, T., Reinsert, K. L. *et al.* (2005). The temporal patterning microRNA let-7 regulates several transcription factors at the larval to adult transition in *C. elegans*. *Dev Cell* **8**, 321–330.

Hammond, S. M. (2006). MicroRNAs as oncogenes. *Curr Opin Genet Dev* **16**(1), 4–9.

Hayashita, Y., Osada, H., Tatematsu, Y. *et al.* (2005). A polycistronic microRNA cluster, miR-17-92, is overexpressed in human lung cancers and enhances cell proliferation. *Cancer Res* **65**(21), 9628–9632.

Hofacker, I. L., Fontana, W., Stadler, P. F. *et al.* (1994). Fast folding and comparison of RNA secondary structures. *Monatsh Chem* **125**, 167–188.

Hsu, P. W. C., Huang, H.-D., Hsu, S.-D. *et al.* (2006). miRNAMap: genomic maps of microRNA genes and theit target genes in mammalian genomes. *Nucleic Acids Res* **34**, D135–D139.

Hurst, L. D. (2006). Preliminary assessment of the impact of microRNA-mediated regulation on coding sequence evolution in mammals. *J Mol Evol* **63**(2), 174–182.

Huttenhofer, A. and Vogel, J. (2006). Experimental approaches to identify non-coding RNAs. *Nucleic Acids Res* **34**(2), 635–646.

Iwai, N. and Naraba, H. (2005). Polymorphisms in human pre-miRNAs. *Biochem Biophys Res Commun* **331**(4), 1439–1444.

Jing, Q., Huang, S., Guth, S. *et al.* (2005). Involvement of microRNA in AU-rich element-mediated mRNA instability. *Cell* **120**, 623–634.

John, B., Enright, A. J., Aravin, A. *et al.* (2004). Human MicroRNA targets. *PLoS Biol* **2**(11), e363.

Johnson, S. M., Grosshans, H., Shingara, J. *et al.* (2005). RAS is regulated by the let-7 microRNA family. *Cell* **120**(5), 635–647.

Kampa, D., Cheng, J., Kapranov, P. *et al.* (2004). Novel RNAs identified from an in-depth analysis of the transcriptome of human chromosomes 21 and 22. *Genome Res* **14**(3), 331–342.

Kern, W., Kohlmann, A., Wuchter, C. *et al.* (2003). Correlation of protein expression and gene expression in acute leukemia. *Cytometry* **55B**, 29–36.

Kim, V. N. (2006). MicroRNA biogenesis: coordinated cropping and dicing. *Nat Rev Mol Cell Biol* **6**, 376–385.

Kloosterman, W. P., Wienholds, E., de Bruijn, E. *et al.* (2006). *In situ* detection of miRNAs in animal embryos using LNA-modified oligonucleotide probes. *Nat Methods* **3**, 27–29.

Krek, A., Grun, D., Poy, M. N. *et al.* (2005). Combinatorial microRNA target predictions. *Nat Genet* **37**, 495–500.

Lai, E. C., Tomancak, P., Williams, R. W. *et al.* (2003). Computational identification of *Drosophila* microRNA genes. *Genome Biol* **4**, R42.

Lai, E. C. (2003). MicroRNAs: runts of the genome assert themselves. *Curr Biol* **13**(23), R925–936.

Leaman, D., Chen, P. Y., Fak, J. *et al.* (2005). Antisense-mediated depletion reveals essential and specific functions of microRNAs in *Drosophila* development. *Cell* **121**, 1097–1108.

Lee, J. T., Davidow, L. S. and Warshawsky, D. (1999). *Tsix*, a gene antisense to *Xist* at the X-incativation centre. *Nat Genet* **21**, 400–404.

Lee, R. C., Feinbaum, R. L. and Ambros, V. (1993). The *C. elegans* heterochronic gene lin-4 encodes small RNAs with antisense complementarity to lin-14. *Cell* **75**(5), 843–854.

Lewis, B. P., Burge, C. B. and Bartel, D. P. (2005). Conserved seed pairing, often flanked by adenosines, indicates that thousands of human genes are microRNA targets. *Cell* **120**, 15–20.

Lewis, B. P., Shih, I. H., Jones-Rhoades, M. W. *et al.* (2003). Prediction of mammalian microRNA targets. *Cell* **115**(7), 787–798.

Lim, L. P., Lau, N. C., Weinstein, E. G. *et al.* (2003). The microRNAs of *Caenorhabditis elegans*. *Genes Dev* **17**(8), 991–1008.

Liu, C., Bai, B., Skogerbo, G. *et al.* (2005). NONCODE: an integrated knowledge database of non-coding RNAs. *Nucleic Acids Res* **33**, D112–115.

Liu, C.-G., Calin, G. A., Meloon, B. *et al.* (2004). An oligonucleotide microchip for genome-wide microRNA profiling in human and mouse tissues. *Proc Natl Acad Sci USA* **101**, 9740–9744.

Liu, J., Valencia-Sanchez, M. A., Hannon, G. J. *et al.* (2005). MicroRNA-dependent localization of targeted mRNAs to mammalian P-bodies. *Nat Cell Biol* **7**, 719–723.

Lu, J., Getz, G., Miska, E. A. *et al.* (2005). MicroRNA expression profiles classify human cancers. *Nature* **435**(7043), 834–838.

Manoharan, H., Babcock, K., Willi, J. *et al.* (2003). Biallelic expression of the H19 gene during spontaneous hepatocarcinogenesis in the albumin SV40 T antigen transgenic rat. *Mol Carcinog* **38**, 40–47.

Mattick, J. S. (2003). Challenging the dogma: the hidden layer of non-protein-coding RNAs in complex organisms. *Bioessays* **25**(10), 930–939.

Mignone, F., Grillo, G., Licciulli, F. *et al.* (2005). UTRdb and UTRsite: a collection of sequences and regulatory motifs of the untranslated regions of eukaryotic mRNAs. *Nucleic Acids Res* **33**, D14114–14116.

Millar, J. K., Wilson-Annan, J. C., Anderson, S. *et al.* (2000). Disruption of two novel genes by a translocation co-segregating with schizophrenia. *Hum Mol Genet* **9**, 1415–1423.

Miller, V. M., Xia, H., Marrs, G. L. *et al.* (2005). Allele-specific silencing of dominant disease genes. *Proc Natl Acad Sci U S A* **100**(12), 7195–7200.

Mineno, J., Okamoto, S., Ando, T. *et al.* (2006). The expression profile of microRNAs in mouse embryos. *Nucleic Acids Res* **34**, 1765–1771.

Miska, E. A., Alvarez-Saavedra, E., Townsend, M. *et al.* (2004). Microarray analysis of microRNA expression in the developing mammalian brain. *Genome Biol* **5**, R68.

Morris, J. P. 4th and McManus, M. T. (2005). Slowing down the Ras lane: miRNAs as tumor suppressors? *Sci STKE* 16 Aug (297), pe41.

Nam, J. W., Shin, K. R., Han, J. *et al.* (2005). Human microRNA prediction through a probabilistic co-learning model of sequence and structure. *Nucleic Acids Res* **33**(11), 121–125.

Neely, L. A., Patel, S., Garver, J. *et al.* (2006). A single-molecule method for the quantitation of microRNA gene expression. *Nat Methods* **3**, 41–46.

Nelson, P. T., Baldwin, D. A., Scearce, L. M. *et al.* (2004). Microarray-based, high throughput gene expression profiling of microRNAs. *Nat Methods* **1**, 155–161.

O'Donnell, K. A., Wentzel, E. A., Zeller, K. I. *et al.* (2005). c-Myc-regulated microRNAs modulate E2F1 expression. *Nature* **435**, 839–843.

Olsen, P. H. and Ambros, V. (1999). The lin-4 regulatory RNA controls developmental timing in *Caenorhabditis elegans* by blocking LIN-14 protein synthesis after the initiation of translation. *Dev Biol* **216**, 671–680.

Pang, K. C., Stephen, S., Engstrom, P. G. *et al.* (2005). RNAdb – a comprehensive mammalian noncoding RNA database. *Nucleic Acids Res* **33**, D125–130.

Pillai, R. S., Bhattacharyya, S. N., Artus, C. G. *et al.* (2005). Inhibition of translational initiation by Let-7 microRNA in human cells. *Science* **309**, 1573–1576.

Poy, M. N., Eliasson, L., Krutzfeldt, J. *et al.* (2004). A pancreatic islet-specific microRNA regulates insulin secretion. *Nature* **432**, 226–230.

Rastan, S. (1994). X chromosome inactivation and the Xist gene. *Curr Opin Genet Dev* **4**(2), 292–297.

Rehmsmeier, M., Steffen, P., Hochsmann, M. *et al.* (2004). Fast and effective prediction of microRNA/target duplexes. *RNA* **10**(10), 1507–1517.

Runte, M., Kroisel, P. M., Gillessen-Kaesbach, G. *et al.* (2004). SNURF-SNRPN and UBE3A transcript levels in patients with Angelman syndrome. *Hum Genet* **114**, 553–561.

Schmittgen, T. D., Jiang, J., Liu, Q. *et al.* (2004). A high-throughput method to monitor the expression of microRNA precursors. *Nucleic Acids Res* **32**(4), e43.

Sen, G. L. and Blau, H. M. (2005). Argonaute 2/RISC resides in sites of mammalian mRNA decay known as cytoplasmic bodies. *Nat Cell Biol* **7**, 633–636.

Shahi, P., Loukianiouk, S., Bohne-Lang, A. *et al.* (2006). Argonaute – a database for gene regulation by mammalian micro RNAs. *Nucleic Acids Res* **34**, D155–118.

Sironi, M., Menozzi, G., Comi, G. P. *et al.* (2005). Analysis of intronic conserved elements indicates that functional complexity might represent a major source of negative selection on non-coding sequences. *Hum Mol Genet* **14**(17), 2533–2546.

Srikantan, V., Zou, Z., Petrovics, G. *et al.* (2000). PCGEM1, a prostate-specific gene, is overexpressed in prostate cancer. *Proc Natl Acad Sci U S A* **97**(22), 12216–12221.

Szymanski, M., Barciszewska, M. Z., Erdmann, V. A. *et al.* (2005). A new frontier for molecular medicine: noncoding RNAs. *Biochim Biophys Acta* **1756**(1), 65–75.

Takeda, K., Ichijo, H., Fujii, M. *et al.* (1998). Identification of a novel bone morphogenetic protein-responsive gene that may function as a noncoding RNA. *J Biol Chem* **273**, 17079–17085.

Tam, W., Hughes, S. H., Hayward, W. S. *et al.* (2002). Avian bic, a gene isolated from a common retroviral site in avian leukosis virus-induced lymphomas that encodes a noncoding RNA, cooperates with c-myc in lymphomagenesis and erythroleukemogenesis. *J Virol* **76**(9), 4275–4286.

Thomson, J. M., Parker, J., Perou, C. M. *et al.* (2004). A custom microarray platform for analysis of microRNA gene expression. *Nat Methods* **1**, 47–53.

Volinia, S., Calin, G. A., Liu, C.-G. *et al.* (2006). A microRNA expression signature of human solid tumors defines cancer gene targets. *Proc Natl Acad Sci U S A* **103**, 2257–2261.

Wightman, B., Ha, I. and Ruvkun, G. (1993). Postranscriptional regulation of the heterochronic gene *lin-14* by *lin-4* mediates temporal pattern formation in *C. elegans*. *Cell* **75**, 855–862.

Wylie, A. A., Murphy, S. K., Orton, T. C. *et al.* (2000). Novel imprinted DLK1/GTL2 domain on human chromosome 14 contains motifs that mimic those implicated in IGF2/H19 regulation. *Genome Res* **10**, 1711–1718.

Xi, Y., Shalgi, R., Fodstad, O. *et al.* (2006). Differentially regulated micro-RNAs and actively translated messenger RNA transcripts by tumor suppressor p53 in colon cancer. *Clin Cancer Res* **12**, 2014–2024.

Xu, P., Vernooy, S. Y., Guo, M. *et al.* (2003). The *Drosophila* microRNA Mir-14 suppresses cell death and is required for normal fat metabolism. *Curr Biol* **13**, 790–795.

Yousef, M., Nebozhyn, M., Shatkay, H. *et al.* (2006). Combining multi-species genomic data for microRNA identification using a Naive Bayes classifier machine learning for identification of microRNA genes. *Bioinformatics* **22**(11), 1325–1334.

Zeng, Y. and Cullen, B. R. (2003). Sequence requirements for micro-RNA processing and function in human cells. *RNA* **9**(1), 112–123.

Zeng, Y., Yi, R. and Cullen, B. R. (2003). miRNAs and small interfering RNAs can inhibit mRNA expression by similar mechanisms. *Proc Natl Acad Sci U S A* **100**, 9779–9784.

Zuker, M. (1989). On finding all suboptimal foldings of an RNA molecule. *Science* **244**, 48–52.

Zuker, M. (2003). Mfold web server for nucleic acid folding and hybridization prediction. *Nucleic Acids Res* **31**, 3406–3415.

Section V

Analysis at the Genetic and Genomic Data Interface

15

What Are Microarrays?

An Introduction to Microarray Methods for Measuring the Transcriptome

Catherine A. Ball[1] and Gavin Sherlock[2]

[1]*Department of Biochemistry and* [2]*Department of Genetics, Stanford University Medical School, Stanford, CA, USA*

15.1 Introduction

The field of genetics has traditionally aimed to detect genes whose mutations are responsible for particular allelic phenotypes or traits. The use of microarrays, however, provides geneticists with the ability to detect the entire complement of genes whose expression pattern is perturbed in an organism with a given phenotype or trait. Within a set of genes whose patterns of RNA abundance is changed when a mutant is compared to a wild type, we are likely to observe many more candidates than the mutant gene itself (in fact, the mutant gene many not have altered expression patterns at all). This technique provides us with a rapid, unbiased method of surveying how altered expression of many genes might contribute to an observed phenotype.

Although measuring transcript abundance is less direct than assaying protein activity in a biological sample, the predictable chemistry of nucleic acid populations provides the technical advantage of being convenient, simple to execute and relatively inexpensive. Not inconsequentially, the ability to amplify nucleic acids by the nucleic acid amplification allows mRNA abundance studies to be performed on very small samples. Some 30 years ago, biologists started measuring transcript abundance for a single sequence by Northern blot, an extension of the DNA-based Southern blot

Bioinformatics for Geneticists, Second Edition. Edited by Michael R. Barnes
 ISBN 978-0-470-02619-9 (HB) ISBN 978-0-470-02620-5 (PB)

developed by Southern (1975). Using filter-based blots, one could use the hybridization of nucleic acids to quantify a single species of nucleic acid from a large population. DNA microarrays, which have garnered a great deal of attention in the last few years, provide miniature, high-throughput platforms to perform hybridization-based assays of populations of nucleic acids. Instead of assaying a population of nucleic acids for those that hybridize to a single sequence, microarrays allow biologists to assay thousands of sequences simultaneously. An additional advantage is that the small surface areas associated with microarrays allow small hybridization volumes and therefore require much less nucleic acid than is commonly required for a Northern blot.

15.1.1 Microarray technologies

In the early 1990s, two groups pioneered microarray technology. Steve Fodor and co-workers at Affymetrix developed commercial microarrays, using photolithography and solid-phase chemical synthesis to build short oligonucleotides in high density on a solid surface (Fodor *et al.*, 1993; Pease *et al.*, 1994). Affymetrix still has the lion's share of the commercial microarray business, and, at the time of writing, was reported to have 70 per cent of the market. At the same time that Fodor and co-workers were developing their microarray, Patrick Brown and colleagues at Stanford University School of Medicine were developing a microarray that was manufactured by mechanically printing small spots of DNA solutions onto a glass microscope slide. Most of the early versions of these 'spotted' microarrays relied on PCR-amplified cDNAs or genomic DNA samples. Because Brown and co-workers openly disseminated directions for assembling microarray-printing machines (the MGuide, http://cmgm.stanford.edu/pbrown/mguide/), software for scanning the hybridized arrays (ScanAlyze, http://rana.lbl.gov/EisenSoftware.htm), and software for clustering and viewing microarray data (Cluster and TreeView; Eisen *et al.*, 1998), the spotted microarray technology was widely adopted. This was especially true in academic settings, where its relatively low cost and flexibility were important.

Now many companies market microarrays or other high-throughput methods to assay populations of nucleic acids by hybridization to thousands of reporter molecules. Some companies are even providing customized microarrays that allow researchers to determine the reporters used. Despite differences in technology or price, most of these products or 'home-made' arrays can be used for exactly the same experimental applications. Each platform has its advantages and disadvantages, but their recitation is beyond the scope of this chapter. Users should consider whether a research question can be adequately addressed by considering the following:

1) the types of molecules used for reporters (PCR products, short oligonucleotides or long oligonucleotides)

2) the number of channels measured (some platforms permit only one sample of labelled nucleic acid, while others require at least two differentially labelled samples)

3) the available resources for labelling, hybridization, scanning and data analysis.

15.2 Principles of the application of microarray technology

One of the major applications of microarray technology is to measure the abundance of mRNAs in an extract made from a sample of cells or tissue as a proxy assay of gene product activity. Other approaches, such as Serial Analysis of Gene Expression (SAGE) (Velculescu *et al.*, 1995), quantitative PCR or filter blots, also measure the abundance of RNA in an extract. Although it is an imperfect measure of gene expression or activity of gene products, RNA abundance is a useful proxy for both. The activity of a gene product is highly correlated to its cellular concentration – the more gene product there is, the higher the likelihood of observing its activity. Similarly, an increased level of translation of a gene product is generally correlated with an increased intracellular concentration. Finally, an increased concentration of transcript increases the translation rate. Of course, differential translational efficiency, post-translational modifications that affect activity, and protein degradation all confound this relationship, but that does not preclude the conclusion that, for most genes, an increase (or decrease) in mRNA abundance is going to result in an increase (or decrease) in gene product activity. Initial applications of microarray technology successfully examined relative transcript abundance in experiments designed to investigate many aspects of fundamental cellular processes (e.g., DeRisi *et al.*, 1997; Heller *et al.*, 1997; Spellman *et al.*, 1998; Iyer *et al.*, 1999; Ross *et al.*, 2000). At the time of writing, a PubMed search for the text term 'microarray gene expression' identified more than 9000 citations, showing how enthusiastically biomedical researchers have embraced this technology as a method to assay gene expression.

Because microarrays can simultaneously assay thousands of nucleic acids, this technology has also been applied to genomic-scale research questions beyond assaying RNA abundance (many of these alternative uses are reviewed in the next three chapters). Microarray-based comparative genome hybridization (array-CGH or aCGH) has been used to detect copy number changes in genomic DNA; for instance, to identify regions of recurrent deletion or amplification in various cancers (e.g., Pandita *et al.*, 1999; Pollack *et al.*, 1999, 2002; Forozan *et al.*, 2000; Linn *et al.*, 2003; see Chapter 17). Other groups have used microarrays to identify DNA-binding sites by assaying sequences that can be immunoprecipitated while cross-linked to

DNA-binding proteins. Such chromatin immunoprecipitation (chip-ChIP) experiments have been used to detect targets of transcription factors or other DNA-binding proteins on a genome-wide basis (e.g., Lee *et al.*, 2002; Harbison *et al.*, 2004). For the geneticist, microarray technology is now starting to have an enormous impact, as it has been adapted to sequence, genotype and identify SNPs. This area is covered in detail in Chapter 18. Other applications are sure to arise, as talented and motivated scientists attempt to answer research questions by novel, high-throughput methods.

15.2.1 The experimental process of microarray analysis

A typical microarray experiment follows several predictable steps. First, RNA is isolated from a biological sample of interest. Most often, the mRNA is separated from ribosomal RNA by a poly-A purification step. At this point, cDNA is usually synthesized, because DNA is more stable and easier to work with than RNA. Sometimes samples are amplified by linear amplification (Eberwine, 1992) or polymerase chain reaction (PCR) (Saiki *et al.*, 1985). Samples are then labelled with a fluorescent dye. In the case of two-channel platforms, a reference sample is also labelled with a different dye. The labelled DNA is then hybridized to the microarray surface, giving the reporters on the microarray opportunity to hybridize with their complements in the hybridization solution. After hybridization, the microarrays are washed to remove non-specific signal and then scanned with a confocal fluorescent microscope to obtain an image(s) at the wavelength of the label(s) used. In the case of two-channel platforms, scans are performed at two wavelengths and two images are obtained. These images show the level of fluorescent label hybridizing to each spot on a microarray. The images are then processed with one of a variety of data acquisition software packages that calculate important measurements for each spot on the array, such as total intensity, local background, or pixel-by-pixel intensity. These measurements are what are usually referred to as 'raw results' for microarray data. Raw results are used to calculate an indicator of mRNA levels in the original biological sample – as a function of either spot intensity (in the case of one-channel platforms) or of the ratio of intensity of the original biological sample to the intensity of a reference sample (as is common in the case of two-channel platforms). The other measurements (and there are typically several dozen) are useful for determining data quality or for performing data transformations (such as background subtractions).

Biomedical researchers using microarray technologies have specific needs to help them deal with the data they generate. Data need to be stored securely, viewed, transformed, edited, annotated, analysed, shared with collaborators and made available upon publication. Although this might seem a daunting task, there are many software packages (both commercial and freely available) that can handle these tasks admirably. A few of the most widely used, freely available software options are discussed at the end of this chapter.

15.2.2 Secure storage of microarray data

Most researchers want to store data in a way that prevents others from seeing them until they are published. Microarray data are not unique in this aspect, but the size and nature of the data necessitate storage on a computer, rather than on a page in a laboratory notebook. Since the scanned images and the resulting raw data are the end result of using rather expensive microarray technology, it is important that data be securely stored. Not only should the data be safe from corruption (for example, one ought to avoid storing data on a computer that receives email and is therefore prone to viruses), but also the data should be routinely saved to back-up media. In addition to ensuring that data are stable and can be recovered, most researchers need a method to control access to data. Chapter 2 of this book provides insight into how one might solve some of the problems encountered when handling large data sets.

15.2.3 Transforming microarray data

Raw microarray data formats are not universally useful. In order to compare data from one array to another in a meaningful way, researchers may need to implement one or more data transformation steps. Examples include background correction (there are several methods of varying sophistication and philosophy) and data normalization (to adjust the distribution of signal intensities so data from different arrays can be compared). It is often convenient for researchers to be able to access a suite of tools that perform data transformation steps.

15.2.4 Annotation of microarray data

Providing up-to-date and relevant biological annotations describing the reporter sequences on microarrays is crucial. Most microarray studies (indeed, most biomedical studies in general) occur over the course of months or even years. During such times, the accumulated knowledge about the genes of any organism is likely to change drastically. Without biological information, no one could obtain meaningful conclusions from any microarray-based experiment. Without timely and updated biological information, conclusions could be based on incomplete, stale or incorrect knowledge. However, obtaining and maintaining biological annotations for tens of thousands of genes is not trivial and should therefore be an important function for any software package used to analyse microarray data. Any number of websites provide biological annotation that can be downloaded (the list is too long to address comprehensively here). However, a software package that provides automated access to updated data will be far more convenient to researchers than one that requires them to retrieve and format annotation data repeatedly (a number of these tools are reviewed in Chapter 18 (Section 18.4.6)).

15.2.5 Filtering and selecting microarray data

Not all data obtained from a microarray are of high quality or relevant to a specific research problem. Therefore, software packages handling microarray data should allow researchers to choose which data to include based on various metrics, data measurements, or gene identity.

15.2.6 Analysis and visualization of microarray data

The number of algorithms and software packages for the analysis of microarray data increases nearly every day. Therefore, researchers should ensure that the software package they use to manage and analyse their microarray data permits needed analysis functions or at least exports data in a format that can be used. Similarly, researchers should select microarray data software packages that permit the visualization of analysed data as well as ways to examine the primary image of microarrays and microarray spots.

15.2.7 Sharing and publication of microarray data

Few researchers in the field of microarray analysis operate in a vacuum. Typically, microarray experiments are performed with a reasonably high number of collaborators and co-authors. In addition, most journals and funding agencies share the expectation that published microarray data be made freely available in one of the public microarray data repositories (ArrayExpress, GEO or CIBEX; Ikeo *et al.*, 2003; Barrett *et al.*, 2005; Parkinson *et al.*, 2005). Simply providing access to raw or processed data does not provide readers or future users of those data with the information required to replicate or reject a study. Indeed, most journals require that authors adhere to the Minimal Information About a Microarray Experiment (MIAME) (Brazma *et al.*, 2001), a simple list of the types of information that should be disclosed to explain fully the meaning and biological context of a study using microarray technology. In addition to the MIAME standard for annotation, there is a community-accepted format for expressing the experimental details about microarray studies (Microarray Gene Expression Markup Language (MAGE-ML) (Spellman *et al.*, 2002). All three public data repositories accept data submissions in MAGE-ML. Keeping in mind the need to share and publish microarray data, most users would take full advantage of microarray data software packages that help researchers share raw and/or processed data, the results of analyses, with collaborators. Most researchers would likewise benefit from a facile method of submitting MIAME-compliant and properly formatted MAGE-ML to one or more of the data repositories upon publication.

15.3 Complementary approaches to microarray analysis

In general, the relationship between the microarray signal for a nucleic acid species and its absolute abundance in a biological sample is not easily quantified. In addition, nucleic acids hybridized to microarrays are isolated from a potentially heterogeneous population of cells, providing little information about which subsets of cells are actually expressing any particular transcript. Accordingly, researchers are increasingly validating microarray data by different techniques, such as quantitative PCR (Arya *et al.*, 2005) to assay transcript abundance independently, or tissue microarrays (Braunschweig *et al.*, 2005) to examine the physical distribution of a message or antigen in normal or diseased tissue.

15.4 Differences between data repository and research database

As with DNA sequence or protein structure data, there are crucial differences between microarray research databases intended to help reach conclusions about current research and data repositories intended to be clearing houses for data that have already been published. Most software packages for the prepublication analysis of microarray data have some form of a database to help store such large and complex data sets. In addition, research databases should provide up-to-date biological annotations for features on microarrays, tools for prepublication exploration, analysis and visualization of microarray data, and methods to control access to sensitive prepublication data.

There are currently three public microarray data repositories accepting data. ArrayExpress at the European Bioinformatics Institute and the Gene Expression Omnibus (GEO) at the National Center for Biotechnology Information actively accept and disseminate published microarray data. The Center for Information Biology Gene Expression (CIBEX) is affiliated with the DNA Databank of Japan. The stated goal of each is to archive published (or soon-to-be-published) microarray data and provide public access to that data. A long-term (but as-yet pending) goal is to replicate the highly successful model of data exchange and interaction between the public DNA sequence data repositories of GenBank, EMBL and DDBJ, such that data submitted to one repository are soon mirrored at all three.

15.5 Descriptions of freely available research database packages

There are many possible options for installable research databases. Description of current commercial packages is beyond the scope of this discussion, but a (somewhat outdated) review by Gardiner-Garden and Littlejohn (2001) presents several options. In this section of the chapter, we present several freely available research database options. None offer all the desired features, and several require a considerable amount of expertise to install and operate, so it is important to keep in mind the scope of the microarray project when selecting one of these packages. For example, a single laboratory using microarrays to study one organism is more likely to need a database that can be installed on a desktop computer without extensive knowledge of database administration or programming. On the other hand, a research database that serves an entire institution will require the means to provide extensive biological annotation for several organisms, complex data access methods, a suite of analysis tools and the computational power to serve multiple users efficiently. In Table 15.1, we offer a direct head-to-head comparison of these tools.

Table 15.1 Key features of freely available microarray database packages

	BASE	Gecko	RAD	SMD	TM4
RDBMS	MySQL	Oracle	Oracle/ PostgreSQL	Oracle	MySQL
OS	Linux, Unix, MacOSX	Linux for server, Windows for client	Unix, Linux, MacOSX	Unix, MacOSX, Linux	Windows, Linux
Open source	Yes	Yes	Yes	Yes	Yes
Microarray platforms	Two-colour	Affymetrix	All	All	Two-colour
MIAME Compliance	Yes	No	Yes	Yes	Yes
MAGE-ML Export	Yes	No	Yes	Yes	Yes
MAGE-ML Import	No	No	No	Yes	No
Data-analysis tools	Yes	Many	Yes	Yes	Many
Biological annotation	Limited	Limited	Yes	Yes	Limited
LIMS	Yes	No	No	Yes	No
Public access		Yes	Yes	Yes	No
Total arrays stored (to date)	6000	30 000	2275	60 000	Unknown
Active development	Yes	Yes	Yes	Yes	Yes
QA tools	Yes	Yes	Plots	Yes	Yes
Microarray images	Unknown	Limited	No	Yes	Yes
aCGH	Unknown	Unknown	Yes	Yes	Unknown
Chip-chIP	Unknown	Unknown	Yes	Yes	Unknown
Other expression data (i.e., SAGE, RT–PCR)	Unknown	Unknown	Yes	No	Unknown

15.5.1 BASE

BASE (Saal *et al.*, 2002) is a MIAME-supportive system that provides an integrated framework for storing and analysing microarray data and related information.

Requirements and installation

BASE is written in PHP, with some additional C++ code for CPU-intensive tasks, and it uses the open-source MySQL for data storage. Typically, it is deployed on Linux, though there have also been successful reports of deployment on Sun's Solaris operating system, and MacOS X, and it can be modified for deployment on Windows with Cygwin. The BASE software itself is released under the GNU general public licence, and depends only on open-source software, such as Linux.

Features

BASE is capable of storing data from all aspects of the microarray process, from LIMS data (including tracking clones in microtitre plates, and recording details about microarray printing) to the loading and analysis of results data. Internally, the BASE data model closely resembles the MAGE Object Model (Spellman *et al.*, 2002) and allows users to specify relationships between various aspects of a microarray experiment. Although BASE does not support Affymetrix arrays, it supports homemade two-colour spotted arrays as well as commercial two-colour arrays.

BASE has a flexible configuration system that assists in the import of data from a variety of image-extraction packages, such as GenePix and QuantArray. BASE also has plug-in software architecture, so external developers can contribute additional software packages, and extracted data are typically produced in a BASEfile format that can be used by the plug-ins. Several plug-ins are already available, including ones that implement Lowess normalization, and multidimensional scaling. BASE also provides tools so users can allow others to view their data, so it can facilitate collaborative research.

Advantages

Since BASE does not require any purchased software, deployment is free of software licensing fees. BASE is MIAME-supportive and can export MAGE-ML files. BASE has a flexible data import tool and the ability to share data easily. BASE has been shown to store at least 6000 hybridizations, each with 50 000 features. BASE provides a flexible analysis and filtering pipeline that stores parameters for reuse. The plug-in architecture of BASE allows new analysis software to be easily added to BASE installations.

Disadvantages

The lack of support for Affymetrix data is a drawback. There is no easy way to transfer data between one BASE instance and another, and no MAGE-ML import abilities. The unproven scalability of MySQL might lead to performance issues for installations with a large volume of data.

15.5.2 *Gecko*

Gecko (http://sourceforge.net/projects/geckoe) (Theilhaber *et al.*, 2004) uses a client-server architecture, with a centralized repository that can store data from Affymetrix scans, and comes with an admirable suite of analysis tools.

Requirements and installation

Gecko requires a server running Linux or Solaris, and uses Oracle as its RDBMS. The Gecko client currently runs only on Windows, although work is underway to write a Java client, which theoretically could run on any platform.

Features

Gecko's most significant feature is the implementation of an admirable number of analysis tools. These include many two-class comparison tests (Student's t-test, SAM, comparison of variance and Mann–Whitney), as well as multiple-class and multiple-factor tests (one- and two-way ANOVA) and the ability to perform contrast calculations. Gecko provides a data representation called the 'Analysis Tree', which enables users to perform and save complex data-analysis work flows, which are stored as directed acyclic graphs (DAGs). While Gecko has largely been used on Affymetrix data, it would theoretically be able to store and process two-colour microarray data as well. While not fully MIAME compliant, Gecko implements some of the MIAME required annotation fields.

Advantages

Gecko has been demonstrated to be scalable to tens of thousands of arrays, has a comprehensive suite of analysis tools, and offers the Analysis Tree, which allows users to track past analyses and communicate results or work in progress across a large community. The Gecko server can be installed on a generic Linux platform.

Disadvantages

Gecko's software installation requires several steps (is not 'out of the box'), and the client software runs on Windows only. It is not fully MIAME compliant, and does not produce or read MAGE-ML files.

15.5.3 *RNA Abundance Database (RAD)*

RAD (Manduchi *et al.*, 2004) provides a MIAME-supportive infrastructure for gene-expression data management and makes extensive use of ontologies. RAD is part of the more general Genomics Unified Schema (GUS) (http://www.gusdb.org).

Requirements and installation

Because RAD relies on GUS, it is actually necessary to install GUS. GUS is supported by either Oracle 8i/9i/10g or PostgreSQL. Perl, PHP and Apache are required for installation, and there are some additional optional Java modules. GUS developers have recently produced a ready-to-use package that can be installed with relative ease.

Features

The RAD Study Annotator collects and records information about protocols, biological samples and study designs via Web-based annotation forms. The RAD Querier provides basic hierarchical clustering tools, plots for quality assessment of single arrays, and an in-house algorithm for detecting differentially regulated genes (PaGE), which is also available separately as a Perl program or a Java application (http://www.cbil.upenn.edu/PaGE/). All microarray platforms and image-analysis software are supported. In addition, RAD is being used for CGH, ChIP, and SAGE data. RAD can produce MAGE-ML files for export of data to other databases or software packages. RAD is part of a more general Genomics Unified Schema, which provides a platform to integrate gene and transcript data from a variety of organisms.

Advantages

RAD is a scalable, Web-accessible database that can accommodate data from several laboratories. The software is provided by an open-source method. The security features allow fine-tuned access for keeping unpublished data private, sharing data with collaborators, and making published data freely available. The methods to store and visualize information about protocols, biological samples and the design of experiments are excellent. RAD can produce MAGE-ML, easing submission of microarray data to public repositories. Since RAD and GUS are being actively developed (indeed, RAD was selected for the microarray data module of the Generic Model Organism Database project (GMOD) (http://www.gmod.org/), bugs are likely to be fixed, and new releases with additional features can be expected.

Disadvantages

Installation and maintenance of RAD can be rather work-intensive, but hardware, software and personnel requirements depend on the scale and scope of the project. RAD (as part of GUS) can be installed on a laptop and maintained by a single computer-savvy student, or it can be used to support cores and large bioinformatics resources. RAD has limited LIMS features and limited analysis tools available as part of the package.

15.5.4 Stanford Microarray Database (SMD)

The Stanford Microarray Database (SMD) (Ball *et al.*, 2005) provides a system for storing LIMS data, and storing and annotating microarray data from most platforms, and it provides a robust suite of tools for managing, sharing, selecting, processing, analysing and publishing microarray data. It can be accessed at http://genome-www.stanford.edu/microarray.

Requirements and installation

SMD installation requires Oracle Enterprise Edition server software, a Web server, Perl, and several Perl modules. Installation is currently not a simple task, though an installer script distributed with the software does take care of many of the details of getting the software running. Additional details, such as setting up the Oracle instance of the database, and creating all the tables and the relationships between them, does require a trained database administrator, though all the SQL scripts required to do this are distributed with the SMD package.

Features

SMD incorporates a LIMS tracking system, to track the 96- and 384-well plates in printing microarrays. It also allows loading of commercial array designs from Agilent,

Affymetrix, Combimatrix and Nimblegen (using MAGE-ML files). SMD allows entry of data derived from GenePix and ScanAlyze, as well as data extracted by Agilent's Feature Extraction software from Agilent arrays. In addition, SMD also provides native support for Affymetrix data, and users can upload CEL files and dChip files, and use a suite of tools especially for Affymetrix data. Microarray data for any of these packages can be uploaded in either single or batch mode. For each of the data-file types that SMD supports, all data are stored (in many cases, several dozen metric per spot). Data may be normalized by either global mean normalization or Lowess normalization, either globally or per print tip. Data can be retrieved for one or many experiments, with complex filtering, such that any spot metric, LIMS data or biological annotation may be used as filtering criteria, which may be combined in Boolean queries. In addition, SMD has many tools built on top of the database that may be used to assess array quality and experiment reproducibility, and visualize a representation of the original microarray, as well as tools for downstream analyses. These tools include hierarchical clustering, self-organizing maps, singular value decomposition (e.g., Alter *et al.*, 2000), and imputation of missing data (e.g., Troyanskaya *et al.*, 2001). SMD is MIAME-supportive, and can export data in MAGE-ML for import into the ArrayExpress (Parkinson *et al.*, 2005) or GEO (Barrett *et al.*, 2005) data repositories upon publication. SMD has well-developed data access methods, so that some data can be restricted to a few close collaborators, and other data can be made publicly available. SMD has been successfully used for gene expression, chromatin immunoprecipitation (ChIP), comparative genome hybridization (CGH) and protein microarrays, as well as other applications of microarray technology.

Advantages

SMD is a scalable solution for storing microarray data – the Stanford installation currently has data from over 60 000 microarrays, comprising data from ~1 800 000 000 spots – while a flexible security model allows fine-grained access control of both data and tools. Several tools are available with the database, and software for viewing proxy images of the microarray scans to evaluate visually the quality of the data is also available. Furthermore, SMD regularly updates the biological annotation of the human, mouse and yeast genes that are represented on the microarrays. One of the key advantages of SMD is that the software and database schemas are being actively developed so that new features and improvements are regularly available. Finally, SMD has support for both two-colour data and Affymetrix data.

Disadvantages

SMD is a 'heavy-weight' microarray database that is not easily implemented by a small operation. SMD may require expensive hardware and software, as well as trained staff (at least a database administrator and a programmer/curator), to keep it running, and it is not simple either to install or maintain. An offshoot of SMD, the Longhorn Array Database (LAD) (Killion *et al.*, 2003), was developed specifically because of these drawbacks of SMD. LAD can be deployed entirely with open-source software, its primary platform being PostGreSQL on Linux, and has been shown to be able to store data from several thousand microarrays. As of the time of writing (June 2006), LAD is based on a previous release of SMD, prior to the addition of Affymetrix, Agilent and MAGE-ML support.

15.5.5 TM4

TM4 (Saeed *et al.*, 2003)provides the basis for a suite of tools that can be used separately or together as a package that includes a MySQL database and a suite of data-analysis tools. TM4 and its associated tools are available from the Dana-Farber Cancer Institute through http://www.tm4.org.

Requirements and installation

The software associated with TM4 (TM4, MADAM, MeB, and MIDAS) runs on the Windows 2000/NT/XP systems, as well as Linux (theoretically, it should also run on MacOSX and Unix), and requires Java v1.4.1 or higher. It uses MySQL for its database.

Features

MADAM provides a Java-based user interface for entering microarray data and annotating it in a MIAME-supportive manner. While MADAM itself does not have any analysis tools, the TM4 package includes SpotFinder and MIDAS and MeV. SpotFinder is a microarray image-analysis package for examining two-colour arrays. MIDAS (Microarray Data Analysis System) and MeV (Multi-experiment Viewer) together provide the ability to execute many different types of normalization, quality-control steps, data transformation and data analysis. Because the tools are so tightly integrated via TM4, data from one component can easily be transferred into another component. TM4 provides a comprehensive set of open-source tools with responsive, attractive and easy-to-use interfaces.

Advantages

The user experience is one of the most attractive aspects of this suite of tools. The Java-based software runs on the user's desktop and provides quick responses. TM4 supports many different analysis methods and is completely open source. MADAM provides MIAME supportive annotation methods as well MAGE-ML export. TM4 and its partner software packages are undergoing active development in response to biomedical research, so valuable improvements can be expected on a regular basis.

Disadvantages

Because the TM4 suite is intended to run on users' desktops, it is not an ideal means for sharing data among collaborators. MADAM currently has support only for two-colour microarray experiments and can upload only the '.mev' file format. A utility called ExpressConverter can convert a wide range of formats, including GenePix, ImaGene, ScanArray, ArrayVersion, and Agilent, to '.mev' format for loading, but some information in these formats may be lost. Work is under way to allow more flexibility in uploading various data formats, and a second-generation release will also support single-colour arrays such as the Affymetrix GeneChip.

References

Alter, O., Brown, P. O. and Botstein, D. (2000). Singular value decomposition for genome-wide expression data processing and modeling. *Proc Natl Acad Sci U S A* **97**, 10101–10106.

Arya, M., Shergill, I. S., Williamson, M. *et al.* (2005). Basic principles of real-time quantitative PCR. *Expert Rev Mol Diagn* **5**(2), 209–219.

Ball, C. A., Awad, I. A., Demeter, J. *et al.* (2005). The Stanford Microarray Database accommodates additional microarray platforms and data formats. *Nucleic Acids Res* **33** (Database Issue), D580–582.

Barrett, T., Suzek, T. O., Troup, D. B. *et al.* (2005). NCBI GEO: mining millions of expression profiles – database and tools. *Nucleic Acids Res* **33** (Database Issue), D562–566.

Braunschweig, T., Chung, J. Y. and Hewitt, S. M. (2005). Tissue microarrays: bridging the gap between research and the clinic. *Expert Rev Proteomics* **2**(3), 325–336.

Brazma, A., Hingamp, P., Quackenbush, J. *et al.* (2001). Minimum information about a microarray experiment (MIAME)-toward standards for microarray data. *Nat Genet* **29**, 365–371.

DeRisi, J. L., Iyer, V. R. and Brown, P. O. (1997). Exploring the metabolic and genetic control of gene expression on a genomic scale. *Science* **278**, 680–686.

Eberwine, J., Spencer, C., Miyashiro, K. *et al.* (1992). Complementary DNA synthesis in situ: methods and applications. *Methods Enzymol* **216**, 80–100.

Eisen, M. B., Spellman, P. T., Brown, P. O. *et al.* (1998). Cluster analysis and display of genome-wide expression patterns. *Proc Natl Acad Sci U S A* **95**, 14863–14868.

Fodor, S. P., Rava, R. P., Huang, X. C. *et al.* (1993). Multiplexed biochemical assays with biological chips. *Nature* **364**, 555–556.

Forozan, F., Mahlamaki, E. H., Monni, O. *et al.* (2000). Comparative genomic hybridization analysis of 38 breast cancer cell lines: a basis for interpreting complementary DNA microarray data. *Cancer Res* **60**, 4519–4525.

Gardiner-Garden, M. and Littlejohn, T. G. (2001). A comparison of microarray databases. *Brief Bioinform* **2**, 143–158.

Harbison, C. T., Gordon, D. B., Lee, T. I. *et al.* (2004). Transcriptional regulatory code of a eukaryotic genome. *Nature* **431**, 99–104.

Heller, R. A., Schena, M., Chai, A. *et al.* (1997). Discovery and analysis of inflammatory disease-related genes using cDNA microarrays. *Proc Natl Acad Sci U S A* **94**, 2150–2155.

Ikeo, K., Ishi-i, J., Tamura, T. *et al.* (2003). CIBEX: center for information biology gene expression database. *C R Biol* **326**, 1079–1082.

Iyer, V. R., Eisen, M. B., Ross, D. T. *et al.* (1999). The transcriptional program in the response of human fibroblasts to serum. *Science* **283**, 83–87.

Killion, P. J., Sherlock, G. and Iyer, V. R. (2003). The Longhorn Array Database (LAD): an open-source, MIAME compliant implementation of the Stanford Microarray Database (SMD). *BMC Bioinformatics* **4**, 32.

Lee, T. I., Rinaldi, N. J., Robert, F. *et al.* (2002). Transcriptional regulatory networks in *Saccharomyces cerevisiae. Science* **298**, 799–804.

Linn, S. C., West, R. B., Pollack, J. R. *et al.* (2003). Gene expression patterns and gene copy number changes in dermatofibrosarcoma protuberans. *Am J Pathol* **163**, 2383–2395.

Manduchi, E., Grant, G. R., He, H. *et al.* (2004). RAD and the RAD Study-Annotator: an approach to collection, organization and exchange of all relevant information for high-throughput gene expression studies. *Bioinformatics* **20**, 452–459.

Pandita, A., Zielenska, M., Thorner, P. *et al.* (1999). Application of comparative genomic hybridization, spectral karyotyping, and microarray analysis in the identification of subtype-specific patterns of genomic changes in rhabdomyosarcoma. *Neoplasia* **1**, 262–275.

Parkinson, H., Sarkans, U., Shojatalab, M. *et al.* (2005). ArrayExpress – a public repository for microarray gene expression data at the EBI. *Nucleic Acids Res* **33** (Database Issue), D553–555.

Pease, A. C., Solas, D., Sullivan, E. J. *et al.* (1994). Light-generated oligonucleotide arrays for rapid DNA sequence analysis. *Proc Natl Acad Sci U S A* **91**, 5022–5026.

Pollack, J. R., Perou, C. M., Alizadeh, A. A. *et al.* (1999). Genome-wide analysis of DNA copy-number changes using cDNA microarrays. *Nat Genet* **23**, 41–46.

Pollack, J. R., Sorlie, T., Perou, C. M. *et al.* (2002). Microarray analysis reveals a major direct role of DNA copy number alteration in the transcriptional program of human breast tumors. *Proc Natl Acad Sci U S A* **99**, 12963–12968.

Ross, D. T., Scherf, U., Eisen, M. B. *et al.* (2000). Systematic variation in gene expression patterns in human cancer cell lines. *Nat Genet* **24**, 227–235.

Saal, L. H., Troein, C., Vallon-Christersson, J. *et al.* (2002). BioArray Software Environment (BASE): a platform for comprehensive management and analysis of microarray data. *Genome Biol* **3**, SOFTWARE0003.

Saeed, A. I., Sharov, V., White, J. *et al.* (2003). TM4: a free, open-source system for microarray data management and analysis. *Biotechniques* **34**, 374–378.

Saiki, R. K., Scharf, S., Faloona, F. *et al.* (1985). Enzymatic amplification of beta-globin genomic sequences and restriction site analysis for diagnosis of sickle cell anemia. *Science* **230**, 1350–1354.

Southern, E. M. (1975). Detection of specific sequences among DNA fragments separated by gel electrophoresis. *J Mol Biol* **98**, 503–517.

Spellman, P. T., Miller, M., Stewart, J. *et al.* (2002). Design and implementation of microarray gene expression markup language (MAGE-ML). *Genome Biol* **3**, RESEARCH0046.

Spellman, P. T., Sherlock, G., Zhang, M. Q. *et al.* (1998). Comprehensive identification of cell cycle-regulated genes of the yeast *Saccharomyces cerevisiae* by microarray hybridization. *Mol Biol Cell* **9**, 3273–3297.

Theilhaber, J., Ulyanov, A., Malanthara, A. *et al.* (2004). GECKO: a complete large-scale gene expression analysis platform. *BMC Bioinformatics* **5**, 195.

Troyanskaya, O., Cantor, M., Sherlock, G. *et al.* (2001). Missing value estimation methods for DNA microarrays. *Bioinformatics* **17**, 520–525.

Velculescu, V. E., Zhang, L., Vogelstein, B. *et al.* (1995). Serial analysis of gene expression. *Science* **270**, 484–487.

16

Combining Quantitative Trait and Gene-Expression Data

Elissa J. Chesler

Oak Ridge National Laboratory, Biosciences Division, Oak Ridge, TN, USA

16.1 Introduction: the genetic regulation of endophenotypes

The forward genetic, phenotype-driven analysis of gene function has led to the discovery of the heritable basis for many complex traits (Korstanje and Paigen, 2002), and the list of success stories is expanding rapidly thanks to the exciting new synergistic methods (DiPetrillo *et al.*, 2005; Flint *et al.*, 2005) covered elsewhere in this volume. The ability to discover new pathways for regulation of traits expressed by intact whole organisms with no prior molecular hypothesis has been exciting, challenging and increasingly rewarding. The advent of high-throughput molecular phenotypes, best exemplified by microarray measures of transcript abundance, present a new avenue for exploration of the basis of complex traits and the role of genetic polymorphisms in functional variability. Exploring the genetic regulation of these molecular traits gives us insight into the entire network of molecular-phenotypic variation that emanates from individual differences in DNA. These networks may be exploited to identify relationships among complex phenotypes, polymorphic and non-polymorphic therapeutic targets, and sources of genetic variability in drug response or disease. Understanding these networks will also allow us to understand how different individuals can use highly polymorphic networks to achieve very similar phenotypic states in many cases, and highly variable phenotypic states in others. Such analyses will necessarily require special adaptations of QTL analysis for gene expression, though, in

Bioinformatics for Geneticists, Second Edition. Edited by Michael R. Barnes
 ISBN 978-0-470-02619-9 (HB) ISBN 978-0-470-02620-5 (PB)

principle, gene expression is quite similar to other complex traits. For example, it is modulated by a variety of genetic and environmental factors with diverse modes of action (mis-sense, enhancer/promoter, distal modifier, nucleic acid structural, micro-RNA target and splice variant) and mode of inheritance (dominant/recessive or additive). The integration of genetic analysis of transcript abundance with other complex trait analysis provides another tool for nominating and electing candidate genes for the regulation of trait variability.

Perhaps the earliest genetic analysis of high-throughput molecular phenotypes was performed by Damerval *et al.* (1994). They applied the then emerging methods to detect quantitative trait loci to proteins separated by 2-D gel electrophoresis in a maize F2 population. This study was admirable for its sophisticated analysis of co-expression networks, epistasis and dominance effects, and gave us the first of many idiosyncratic terms for the QTLs regulating molecular phenotypes, the protein quantity locus (PQL). Some of the QTLs had effects that altered protein migration; these position shift loci (PSLs) presumably act through mis-sense polymorphism or variation in post-translational modification. Studies on the effects of polymorphisms on mRNA message reveal similarly diverse mechanisms of action.

Years after Damerval *et al.*'s groundbreaking study, Jansen and Nap (2001) gave this nascent field its name, 'genetical genomics'. This and other reviews at the time sparked tremendous excitement in the potential of genetic analysis of gene expression in diverse species. The experiments themselves rapidly followed, each of them involving microarray-based profiling of cell populations in F2 crosses, back-crosses, ascertained pedigrees and recombinant inbred lines, spanning eukaryotic species from yeast to man, starting with the first published study, performed in yeast (Brem *et al.*, 2002). Schadt *et al.* (2003) followed with a mammalian study of obesity and liver gene expression in F2 mice, CEPH pedigrees, and maize, giving us the term 'eQTLs' for expression QTLs. During the time of these early studies, several groups performed their genetic analysis of gene expression in recombinant inbred strains (Bystrykh *et al.*, 2005; Chesler *et al.*, 2005; Hubner *et al.*, 2005). Because these readily available lines are isogenic and phenotypic, genotypic and expression data can be integrated indefinitely. These data were made public in an Internet resource, WebQTL (Chesler *et al.*, 2004), now the mapping component of the more integrative http://www.GeneNetwork.org. The integrative network genetic analysis made possible by reference populations with well-characterized genomes, transcriptomes and phenomes has been coined 'systems genetics'. This is an exciting new field with the potential to integrate diverse biological data by evaluating the network of effects of genetic polymorphisms and their complex actions on genomes, transcriptomes, proteomes, and the higher-order structure and function of cells, tissues and organs.

16.2 Transcript abundance as a complex phenotype

The earliest use of gene expression microarrays was to identify the signature genes of tissues, cell populations and disease versus non-disease related samples. Genes were simply called 'present' or 'absent' to define the gene-expression signatures of a particular sample or tissue. This eventually gave rise to differential expression studies in which the quantitative expression differences between two classes of samples was compared. Several of these early studies took genetic variation into account, by carrying out expression profiling across mouse strains (Sandberg *et al.*, 2000; Pavlidis and Noble, 2001) and *Drosophila* genotypes (Jin *et al.*, 2001). Not surprisingly, these studies found strong genetic variation in the abundance of many transcripts on the array. This genetic variation is exploited by transcriptome QTL analysis to identify regulators of gene expression, and to construct genetic covariance networks from gene expression to complex phenotypes. Environmental variation, tissue variation, pharmacological treatment and many other sources of naturally occurring or experimental variance also influence gene expression. The steady reduction in the cost of microarrays over the past several years creates phenomenal potential to evaluate these multifactorial sources of variation in transcript abundance.

16.2.1 Sources of variation

Variation in gene expression is a highly complex phenotype caused by variation in specific regulatory pathways, and local environmental milieu, including tissue, organ and hormonal status; chromosomal sex; and epigenetic factors. Variation in mRNA is one of the functional manifestations of genomic activation; that is, it is one of the lowest levels through which environmental and physiological inputs result in moderate to long-term cellular change by activation of transcriptional and transcript regulatory processes. Thus, variation affecting any point of virtually any biological process coupled to gene expression, can alter transcript abundance. It is tempting to assume that performing QTL analysis of gene expression would identify many transcription factors as QTL candidate genes. However, in yeast, Kruglyak and colleagues have shown that most major expression-regulatory QTLs do not reside near the genomic location of transcription factors (Yvert *et al.*, 2003).

16.2.2 Heritability of gene expression

The proportion of trait variance that is accounted for by genetic factors, or heritability, is an important characteristic in determining which traits are amenable to genetic dissection. Heritability of expression variation is high for many genes, though median heritabilities in array experiments may be as low as 10 per cent. The high heritabilities of many transcript abundance phenotypes are due to a number of factors. Expression variation, though a complex trait, is a somewhat directly quantifiable phenotype. The relative number of copies of mRNA message is well estimated by microarray and quantitative PCR. For mRNA abundance under major genetic regulation by a large-effect locus, the expression distribution is practically bimodal. Another reason for the high heritability of expression variation is that this trait may be less constrained than other higher-order phenotypes. By allowing major variation at a molecular level, organismic structure and function can be constrained within a narrower range of variation by achieving a restricted range of phenotype, using radically different states of the underlying biological networks. What emerges from the genetic analysis of molecular endophenotypes is a picture of high biological complexity and the potential to understand the complex interplay of factors underlying phenotypic stability and phenotypic heterogeneity.

Estimating heritability in microarray analysis

The estimate of gene expression across a population has technical (between-array and between-sample), environmental (between individual) and heritable (between genotype) variation. Heritability is the proportion of this variation that is accounted for by genetic factors. Estimation of heritability in a microarray experiment is often not directly possible due to sample pooling. This methodological approach is quite useful in improving signal-to-noise ratios in an economical fashion by reducing environmental variance with multiple samples on a single array. It is often employed in the analysis of reference populations, such as the recombinant inbred strains, for which multiple individuals with identical genomes can be sampled. By pooling samples, environmental variability can be collapsed across the individuals, giving a more precise estimation of the genome-specific population mean. Because a single faulty sample within the pool will alter the trait values for that strain, an additional level of biological replication, the use of multiple pools, each on a separate array, allows the estimation of within-group variance in the mean. If we assume that environmental variation and technical variation in sample preparation are collapsed to near-zero by pooling, then the within-group variance actually reflects array processing-related technical variation. However, the pools are often so small that between-sample biological variation is not fully cancelled, resulting in a combined technical/residual environmental variance between arrays within a genotype class. The relative proportion of expression variation accounted for by genetic factors relative to the combined technical and residual environmental variation can be estimated. While the resulting value is not a true 'heritability', the genetic variation gives an idea of whether or not a trait is amenable to genetic dissection, and how reliable the results of such an analysis may be (Carlborg *et al.*, 2004). Under typical control of the genome-wise error rate at $P < 0.05$ per transcript, at least 1/20 mapping results are likely to be chance rejections of a null hypothesis of no linkage. Control of the false discovery rate (FDR) is a widely used method to deal with the multiplicity of tests across the microarray (Storey and Tibshirani, 2003a). However, it is impossible to determine biologically which of these results are true and which are false. One way to increase the proportion of true-positive results among the set of QTLs obtained in a microarray study is to exclude those phenotypes for which there is insufficient power to map a trait with the estimated genetic variance.

Data transformations and mapping quality

Microarray data may be transformed and normalized across two dimensions. Data compression and feature extraction can be achieved by a variety of algorithms, such as Affymetrix's MAS, RMA (Irizarry *et al.*, 2003), PDNN (Zhang *et al.*, 2003), dCHIP (Li and Hung Wong, 2001), and ANOVA (Kerr *et al.*, 2000, 2001). These approaches work to compose gene-expression means from sets of probes or spots on the microarray, and align the mean expression levels across all arrays. Some analysis methods perform both normalizations (or perhaps, more appropriately, standardizations) simultaneously. The methods work with varying degrees of success, and the best method may depend on the sample size of the study. For example, PDNN does not adjust across a set of arrays, but effectively reduces much array noise by incorporating hybridization kinetic information into the compression of microarray data. This method appears to work better in large-array data sets that can take advantage of additional arrays for obtaining precision in expression estimates. RMA reduces much technical variation between arrays, but can also reduce signal and decrease the apparent effect size. The increased precision obtained in this method results in a much higher rate of QTLs called for heritable traits.

Statistical transformation to normality should also be attempted across samples for each transcript abundance trait to be analysed. Most typical QTL mapping methods assume normality of the trait distribution within each genotype class. Because of the overwhelmingly large number of expression-related phenotypes on each microarray, these distributional assumptions are rarely checked for each individual phenotype. The impact of violation of this assumption is usually greatest on estimation of QTL effect size, but positional precision can also be affected. It is important to emphasize that this is a relative measure. In the case of microarrays, a fixed quantity of mRNA is hybridized to the array; thus, expression measures reflect the fraction of that total quantity that contains a particular mRNA species.

The aggressive transformation and normalization required by microarray measures, though necessary, can work against genetic analysis. Major QTLs have been observed which appear to regulate large numbers of transcripts. These QTLs have the capacity to alter the shape of the distribution of relative gene expression within each individual. This means that many measures of central tendency used to normalize array data sets may not correspond to one another. By aligning array means, the genetic signal responsible for trait variation is squashed and rendered more difficult to detect.

16.3 Scaling up genetic analysis and mapping models for microarrays

QTL mapping of microarray data has resulted in a need to scale up from single-trait analysis to tens of thousands of traits in a single study. This has necessarily resulted in the application of simple, fast mapping methods in many early studies. As computational power and QTL-mapping software improve, the careful attention paid to each single trait can be applied to the multitude of traits on the microarray. Exciting new approaches have taken into account the full set of information on the microarray, making use of many traits simultaneously in the mapping model, rather than treating traits in a list-based manner. Another unique challenge to QTL-microarray analysis is the need to control the statistical error rate in two dimensions, across the microarray, and across the genetic map.

16.3.1 Single-locus models

The first QTL-microarray studies employed simple mapping models, assuming a single regulatory locus (Brem *et al.*, 2002; Schadt *et al.*, 2003; Chesler *et al.*, 2005; Hubner *et al.*, 2005). Each trait was mapped separately, and attempts were made to quantify the number of QTLs on the genome by various methods to control the family-wise error rate (Li and Burmeister, 2005). For control of the genomewide error rate, some authors chose to use conventional significance thresholds based on theoretical distributions and assumptions of genome size. For the mouse, this significance threshold is LOD > 4.3 (Lander and Kruglyak, 1995). Other studies used distribution-free permutation analysis to control the genome-wide error rate (Churchill and Doerge, 1994). Neither of these approaches controls the number of QTLs on the microarray, for which some studies are now using control of the FDR (Storey and Tibshirani, 2003a, 2003b). All of these approaches assume the validity of the single-locus model, which, for complex traits presumably under polygenic control, is unlikely to hold.

Two types of regulatory loci were identified, *cis*-QTLs, which are due to polymorphisms at or near the transcript-coding region, and *trans*-QTLs, which are located distal to the coding region. *Cis*-QTLs may be due to polymorphisms in enhancer or promoter regions, thereby altering direct mechanisms of transcription control, such as activity of transcription activation, but they may also be present as mis-sense polymorphisms altering function of the gene product. These polymorphisms can activate indirect mechanisms of control, as, for example, by compensatory mechanisms that result in increased production of weakly functioning gene products. Other polymorphisms have been identified that do not regulate gene expression or protein product functionality, but they remind us that the trait being studied is actually not expression, but transcript abundance. These polymorphisms affect mRNA stability, but not the amino-acid sequence of the functional gene product. An example of such a polymorphism was found for the human dopamine receptor, *DRD2*, a finding that calls into question our ability to truly categorize 'silent' mutations (Duan *et al.*, 2003). Decay of mRNA is a well-regulated process that influences the steady-state abundance of many transcripts (Wang *et al.*, 2002). There is another, more trivial cause of *cis*-QTLs for transcript abundance that plagues most hybridization and antibody-based assays for high-throughput molecular phenotyping – polymorphisms in the probe targets themselves. Because the assays are typically developed to target sequences or structures derived from a single strain, the reaction often performs better for that strain. Different array platforms may be more robust to the effects of polymorphisms; for example, the Affymetrix system uses short 25-nucleotide probes, whereas many other systems rely on much longer probe sequences, and are therefore less affected by small differences in sequence. However, even these longer array probes are not a perfect solution. One particularly compelling case comes from the study by Schadt *et al.* (2003), in which a *cis*-QTL was found to alter a splice junction that was a polymorphic target of the probe sequence. In a later study, the impact of sequence variation in probe regions was systematically evaluated by various tests, and was found to be minimal in a bioinformatics analysis of mapping data, but an empirical '*cis*/*trans* test' by RT–PCR revealed that as many as 36 per cent of cis QTLs could not be confirmed (Doss *et al.*, 2005). In this test, mRNA from the two cross-progenitors and the F1 line are compared, to estimate the ratio of expression in an attempt to reproduce independently the QTL effect. It should be noted that the *cis*/*trans* test itself relies on a strong assumption of a single-locus model and of an additive mode of inheritance. Two parental lines and an F1 are not a model of a segregated mapping population.

Each of these early transcriptome-mapping studies reported the existence of major *trans*-regulatory QTLs, genome locations that are linked to many hundreds of gene expression levels. *Trans*-QTLs may be the site of transcription factors, but typically are not (Yvert *et al.*, 2003). The trans-acting loci could reflect the activity of single regulatory polymorphisms (gene pleiotropy) or may be due to the action of a large number of linked polymorphisms. Improved precision of the QTL study and techniques aimed at the reduction of QTL interval size will allow the dissection of large groups of linked regulatory loci. Gene density and SNP density analyses do not reveal significant associations with the number of QTLs mapping to each region. The presence of such loci reveals high expression covariance. This observation has many consequences for genetic analysis of transcriptome QTL mapping. First, efforts to control the FDR by analysing lists of significant QTL mapping results by transcript are destined to be far too conservative because of the presence of so many highly correlated statistical results. Second, substantial noise reduction and improved statistical power could be obtained by first decomposing the gene expression covariance matrix and then mapping QTLs for the reduced set of expression traits.

Table 16.1 Tools and databases related to expression QTL analysis

Tool	URL
Gene Network/Web QTL	http://www.genenetwork.org/
CTC-Oxford-Wellcome Trust SNP database	http://www.well.ox.ac.uk/mouse/INBREDS/
Celera Mouse SNP database	http://www.celera.org
Mouse Phenome Database SNPs	http://phenome.jax.org/pub-cgi/phenome/mpdcgi?rtn=snps/door

16.3.2 Multiple QTL models

In recent years, there has been significant attention to the development of QTL analysis tools (Table 16.1 that allow the capture of the true complexity of genetic regulation of phenotypic variation. Composite interval mapping is a technique which allows control of detected loci in the search for additional loci (Jiang and Zeng, 1995). Tools such as R/QTL (Broman *et al.*, 2003) and Pseudomarker (Sen and Churchill, 2001) use full genome-wise scans of all pairs of loci to identify pairs of loci that are responsible for trait variation. The set of single, joint and interacting loci can be incorporated into more complex, multiple-locus models that can better characterize the complexity of trait regulation (Broman and Speed, 2002). While the analysis required is slower and requires some supervision, it is clearly warranted. In a study of nearly 6000 yeast transcripts, Brem and Kruglyak (2005) found that despite high heritability, the vast majority of traits are under the regulation of two or more loci. In our study of brain gene expression QTLs, we have found that numerous transcript traits are regulated by a combination of two or more of the major *trans*-QTLs (Chesler and Langston, 2005). When these additional genetic factors are not considered, estimation of the effect size and location of QTLs can be flawed, and single markers may act as surrogates for better fitting, multilocus models. With the growing number of transcriptome-QTL analyses, tools for rapid discovery of multiple-locus models are being developed.

16.3.3 Multi-trait mapping models and integrative approaches

Neither the single-locus nor multilocus models described above make use of information from multiple traits. Several methods have been developed for the analysis of related sets of traits, and the approach is beneficial in that it improves signal for correlated expression phenotypes by making use of more information. Many of these approaches utilize some type of data reduction, including principal component analysis (Lan *et al.*, 2003) or graph decomposition (Baldwin *et al.*, 2005; Chesler and Langston, 2005) prior to mapping traits. To obtain data about individual genes, we must return to factor loadings or subgraph composition. The mixture-over-markers model (MOM) (Kendziorski *et al.*, 2006) combines data from multiple transcripts and markers, but currently its implementation is limited to simple single-locus models. Expansion of this approach is promising.

16.3.4 Reducing the size of expression QTLs with SNP integration and haplotype analysis

Once expression regulatory loci are identified, the challenge is to refine the interval and identify the QTL candidate gene or genes. Many new tools and approaches can be applied to this task. For example, large-scale SNP-genotyping resources have dramatically facilitated QTL refinement. This has been both by follow-up experimental strategies, including recombinant inbred segregation tests and Yin–Yang crosses (Shifman and Darvasi, 2005), and by using SNP haplotypes in concert with standard inbred strain phenotypes to identify haplotype intervals that are best associated with the phenotypes (Pletcher *et al.*, 2004). Many of these methods remain controversial because they implicitly assume single-locus models and non-transgressive segregation, and that the QTL effects are due to segregating ancestral polymorphisms. Nonetheless, compelling successes have been reported, and while the methods may fail for some phenotypes and QTLs, they remain useful. This is in part because the proliferation of tools and appreciation of the complexity and diversity of QTL regulatory effects has allowed analysts systematically to seek and integrate converging evidence for each of the genes and other polymorphic genome features in a QTL confidence interval. In expression QTL analysis, QTL refinement and candidate gene information can make facile use of known gene product interactions, pathway co-memberships, and, most importantly, expression covariation analysis of traits being measured, and the candidate genes for such phenotypes.

16.4 Genetic correlation analysis

Transcriptome QTL data are often collected in a population that has been characterized in other phenotypes; for example, Schadt *et al.* (2003) included obesity-related measures that could be related back to the gene-expression regulatory QTLs in an F2 population. Genetic reference populations have been extensively characterized on a wide range of phenotypes. When gene-expression data are collected in these populations, entire gene-phenotype networks can be constructed (Chesler *et al.*, 2003). The genetic correlation matrix of gene expression is also inherently useful to identify co-expression networks (Baldwin *et al.*, 2005). Many biological networks are scale free, with a few highly connected vertices and many sparsely connected vertices (Jeong *et al.*, 2000). Indeed, a scale-free network of genetic co-expression is observed.

16.4.1 Impact of study design on the interpretation of expression correlations

The nature of the population and the study design determine the computation and interpretation of genetic correlations. Theoretically, the correlation between traits in a population can be partitioned into a genetic correlation and an environmental correlation. This type of partitioning can be achieved when large sets of traits are each measured in a sufficiently large group of related individuals, allowing an estimation of correlation within (environmental) and between (genetic) related groups. Microarray measurements are an extreme form of this multivariate phenotyping, and because the number of samples within each family is typically small, it is not possible to partition genetic correlations among gene-expression traits. However, when traits are measured in independent groups of identical individuals (family means correlations), the trait correlation of gene expression and phenotype is considered purely genetic. In an F2 cross-population, the relationships among individuals are each equivalent, and therefore the correlations among traits cannot be interpreted as genetic or environmental in causality.

16.4.2 Genetic reference populations

A genetic reference population consists of a panel of isogenic lines typically bred from a small number of founders. A variety of such populations exist, including the standard inbred mouse strains, consomic lines, which contain a chromosome from one genetic background introgressed on another genetic background, and recombinant inbred strains, which derive from sib mating or selfing of the progeny of a segregating cross. The value of a reference population is that it serves as a tractable model of genetic diversity in a population. Mice from these populations can be bred indefinitely and repeatedly phenotyped. Precision can be obtained by performing replicate sampling within each line, giving enhanced efficiency for costly QTL mapping studies (Belknap, 1998). However, the value of these populations goes far beyond their utility for QTL mapping. Data from these populations can be aggregated multiplicatively. Any attribute that is studied in these populations can be related to all other attributes, making them a highly integrative and intrinsically collaborative resource (Chesler *et al.*, 2003) (Figure 16.1).

Currently, the largest widely used genetic reference population is the BXD recombinant inbred strains. These were originally bred by B. A. Taylor in the late 1970s, and were later expanded in 1999 (Williams *et al.*, 2001) and again in recent years (Peirce *et al.*, 2004), bringing the full set to a size of 80 lines. An equally large set, the LXS panel, was recently derived from an eight-way heterogeneous stock population selected for response to alcohol (Williams *et al.*, 2004). A major drawback to the use of these populations is their finite size, which ultimately limits mapping precision and the ability to discriminate between pleiotropy and linkage as sources of genetic

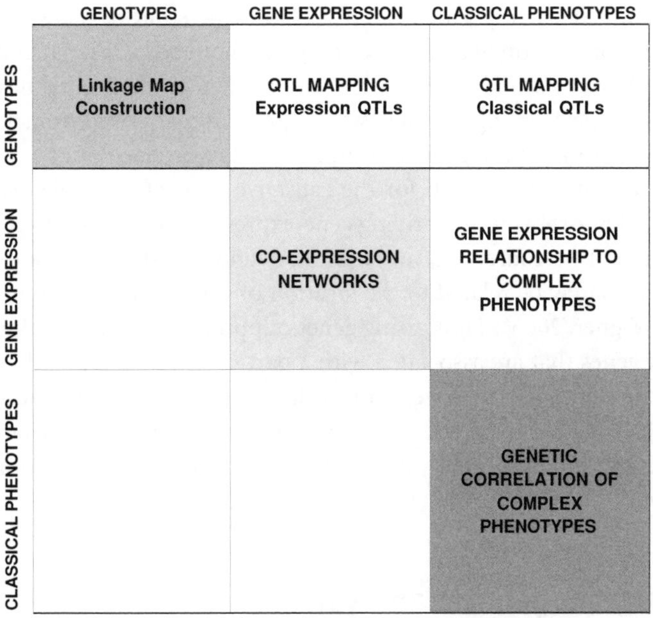

	GENOTYPES	GENE EXPRESSION	CLASSICAL PHENOTYPES
GENOTYPES	Linkage Map Construction	QTL MAPPING Expression QTLs	QTL MAPPING Classical QTLs
GENE EXPRESSION		CO-EXPRESSION NETWORKS	GENE EXPRESSION RELATIONSHIP TO COMPLEX PHENOTYPES
CLASSICAL PHENOTYPES			GENETIC CORRELATION OF COMPLEX PHENOTYPES

Figure 16.1 Simplification of data integration in a reference population. Genotypes, gene expression and complex phenotypes are all attributes of individuals or lines in the reference population. Linkage analysis, co-expression networks, genetic correlations and QTL analysis can each be conceived of as specialized correlations over the population, each computed on specific data submatrices

correlation. An exciting new effort by the Complex Trait Consortium involves the creation of 1000 recombinant inbred mouse lines from eight genetically diverse inbred strains (Churchill *et al.*, 2004). This set will have high power, precision for mapping, and genetic correlation-based approaches. Other large recombinant inbred panels of plant species also exist.

While the standard inbred mouse strains are a useful reference population, they pose significant challenges as a genetic reference population. The finite population size and idiosyncratic breeding history render them non-equidistant genetically. As such, their use as a mapping population is controversial (Chesler *et al.*, 2001; Darvasi *et al.*, 2001; Mhyre *et al.*, 2005). Nevertheless, QTL success stories have been reported in these populations, when coupled with additional mapping crosses (Pletcher *et al.*, 2004). The non-random mating history has led to large blocks of long-range linkage disequilibrium (Petkov *et al.*, 2005), resulting in many false-positive linkage results.

Relations of gene expression to complex phenotypes

Several recent studies have exploited a reference population strategy for the integration of transcript abundance and higher order phenotypes (Carter *et al.*, 2001;

Chesler *et al.*, 2003; Kempermann *et al.*, 2006). The strategy was successfully deployed in the nomination of glyoxylase and glutathione reductase as candidate genes for anxiety (Hovatta *et al.*, 2005). This study made use of a limited number of inbred strains. By performing the analysis in a population derived from a segregating cross, determination of the genetic source of trait covariance can be made. This is especially relevant in the search for the causative basis of heritable disease. While the precise polymorphisms underlying gene-expression correlation to complex phenotypes may not be conserved, mutation sites and mutated genes are likely to be conserved. This may be related to the location of genes in biological networks and pathways (Wagner, 2005). Thus, using genetic approaches to narrow the list of large numbers of genes that are associated with a particular phenotype is likely to result in the detection of candidate genes for translational research on the cause of disease. It is debatable whether these genes are necessarily strong therapeutic targets, as one may wish to develop drugs that will be effective in all members of a population. For this, a network-based approach will prove quite fruitful.

16.5 Systems genetic analysis

Transcriptome QTL analysis in genetic reference populations render possible an exciting new approach to biology referred to as 'systems genetics'. This approach derives from systems biology in which the network of interactions among genes, gene products and systems-level phenotypes is discovered by systematic network perturbation. In systems genetic analysis, the naturally occurring polymorphisms are the network perturbations that are being explored. These network analyses take on two forms, either to construct networks from polymorphism to higher-order phenotype or to use gene expression in concert with genetic information to refine the set of QTL candidate genes rapidly. The tools of systems genetic analysis are rapidly expanding. Several approaches have been developed to integrate multiple data types across the biological scale, particularly gene expression and genotype information. Both Bayesian (Li *et al.*, 2005, 2006) and combinatorial network-analysis methods (Baldwin *et al.*, 2005; Chesler *et al.*, 2005; Chesler and Langston, 2005) are being applied to genetical genomic data sets. The Bayesian approaches to microarray data may allow a causal interpretation (Friedman *et al.*, 2000; Pe'er *et al.*, 2001). This is particularly true when additional information, such as SNP distributions, is incorporated directly in the modeling (Li *et al.*, 2005, 2006). By integrating genotype and gene-expression data in a single model, Kulp and Jagular (2006) have drastically reduced QTL candidate genes, and perhaps inferred the structure of networks from expression-regulatory polymorphisms to variation. Combinatorial algorithms are applied to networks that are drawn by thresholding the gene-expression correlation matrix by high-pass filtering criteria. Edges in the graph represent high genetic correlation, and the vertices or nodes represent traits. Maximal cliques, the largest possible sets of completely connected (perfectly intercorrelated) traits, and other dense subgraphs are extracted and annotated from these gene sets (Baldwin *et al.*, 2005). Although the conversion of a correlation matrix into a discrete graph may seem like a major disadvantage to genetic analysis, extracting genetic correlates by combinatorial algorithms is advantageous for several reasons. Analysis of genome-scale data sets is computationally feasible by graphical methods; the associations that are detected are highly pure, unlike those obtained from K-means or K-nearest neighbours clustering; and there are no constraints on the size or number of cliques required. The resulting dense subgraphs represent highly co-regulated gene sets. These sets can be combined into larger structures, reflecting broader expression co-regulation. QTL mapping can be performed on these reduced multitrait sets, providing more precise and robust mapping, and identification of joint modifiers of expression. Annotation of these gene sets by Gene Ontology tools, transcription factor-binding site information, pathway analysis, and other approaches to understanding co-expression is a powerful method to identify regulators, and determine which biological processes are connected to higher-order phenotypes.

16.5.1 Pulling it all together in GeneNetwork.org

GeneNetwork.org is a public Internet resource for the analysis of gene expression and other complex phenotypes, with special emphasis on genetic reference populations. This resource, a broader system of tools around the original genetical genomics site, WebQTL, began service in 2001 (Wang *et al.*, 2003; Chesler *et al.*, 2004). This tool allows users to search for gene expression or traits of interest, find QTLs, evaluate QTL regions for candidate genes, and perform multivariate and network analyses. At the time of writing, expression data are available for the BXD RI forebrain (Chesler *et al.*, 2005), hippocampus, cerebellum, striatum, liver, eye and haematopoietic stem cells (Bystrykh *et al.*, 2005); AKXD mammary tumours; and HXB/BXH liver and intraperitoneal fat mRNA (Hubner *et al.*, 2005). Legacy phenotypes, hand curated from over 20 years of literature (Chesler *et al.*, 2003), and newly collected phenotypes for the BXD, AXB/BXA, LXS, CXB, and BXH recombinant inbred mice; HXB rats; and several panels of *Arabidopsis* are also incorporated in the website.

Users can search for traits of interest, including a wealth of alcohol and immune response data in the BXD RI lines, and diabetes- and obesity-related phenotypes in the HXB/BXH panel. Full-text searching of the complete MEDLINE abstracts, titles and authors associated with each trait is included for all published phenotypes. Gene-expression records can also be searched by gene symbol, probe set ID and GO term. Advanced search features also allow the retrieval of significant gene-expression QTLs, or QTLs within a particular region. Each trait has its own detail page, with an annotation, analysis, and data section. From this page, users can evaluate probe information and hybridization data, determine basic statistics on trait distributions, generate trait-to-trait correlations, perform QTL mapping with single locus and multilocus models, transform trait data and remove outliers. From the search page and many analytic tools that generate trait sets, entire groups of traits can be selected and added to a trait collections page. From this page, a host of multitrait tools can be executed, including computation of the genetic correlation matrix, principal component analysis, and simultaneous clustering and mapping of sets of phenotypes with the QTL Cluster Map. Network Graphs can be drawn with an integrated Graph Viz module, and connections of sets of traits in one data set. For example, brain gene expression can be correlated with sets of traits in another data set, such as behavioural phenotypes. Built in to this resource are tools such as QTL Analyst to identify candidate genes and polymorphisms; literature correlation analysis by the latent semantic indexing tool, Semantic Gene Organizer (Homayouni *et al.*, 2005); and integrated analysis with many large SNP sets, Gene Ontology overrepresentation and pathway matching (Zhang *et al.*, 2004), and annotation resources. By coupling flexible analytic tools with both molecular endophenotypes and higher-order phenotypic data, users can pursue a wealth of integrative systems genetics queries. Through the search features and tools of a resource like GeneNetwork.org, powerful systems genetics techniques are made readily available to public users, who can collaborate and share data in a purely analytical fashion.

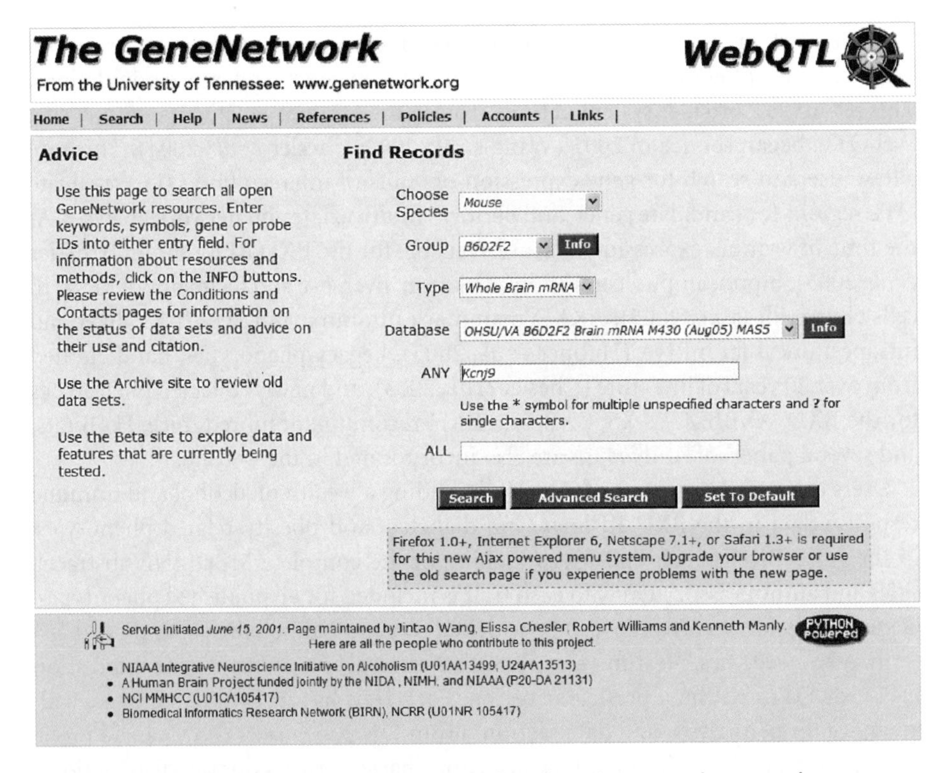

Figure 16.2 The GeneNetwork.org WebQTL query interface. Users can select a species, an appropriate data set and an analysis method; in this case, *Kcnj9* was used as a query gene against the B6D2F2 mouse brain mRNA data set. These data are from an F2 cross of C57BL/6J and DBA/2J. Web QTL, www.genenetwork.org

16.6 Using expression QTLs to identify candidate genes for the regulation of complex phenotypes

Few readers will themselves perform transcriptome QTL-mapping studies. Though these studies are rapidly expanding in number and are being performed in many tissues in several mapping populations, they are still quite large in scale and scope. Several hundred samples and at least 100 microarrays are required for even minimal

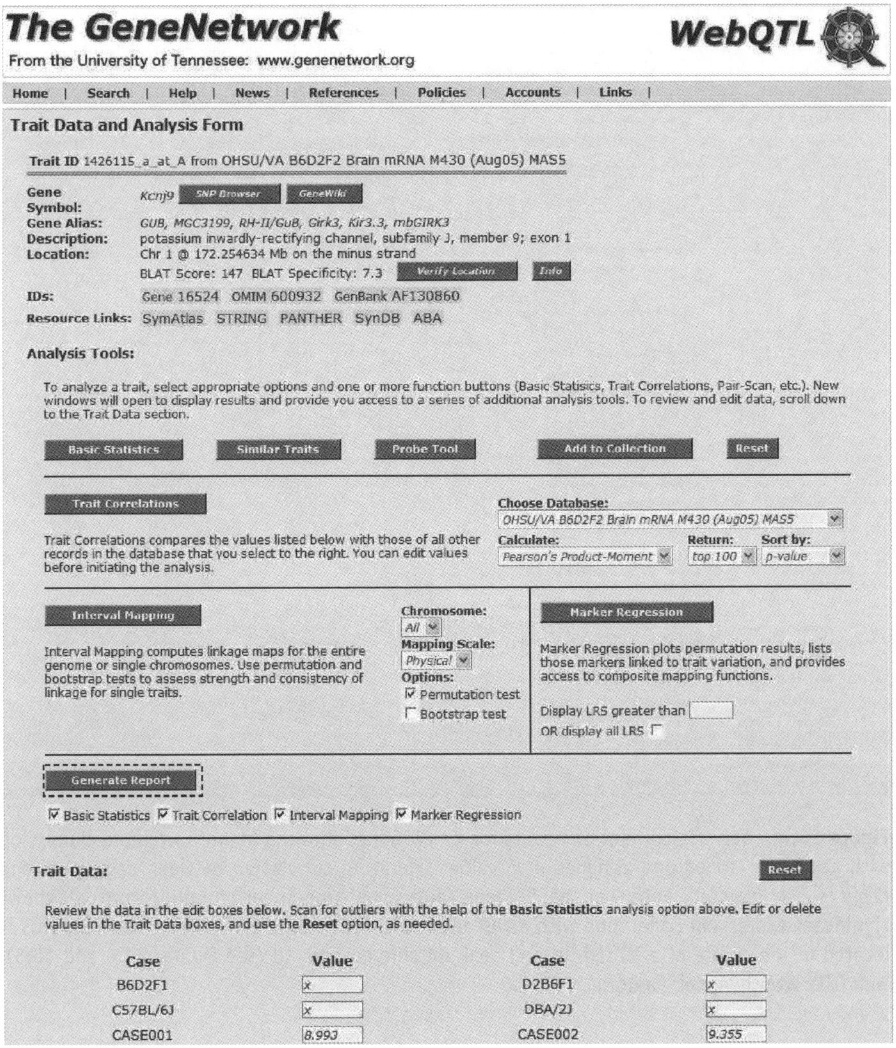

Figure 16.3 WebQTL trait data page for *Kcnj9* expression as a quantitative trait. Each trait data page contains informative links to annotation and probe data, analytical tools, and condensed probe level data for each strain. Detailed results of the QTL analysis can be viewed by following the 'generate report button'. Web QTL, www.genenetwork.org

Correlation Table

Values of Record 1426115_a_at_A in the OHSU/VA B6D2F2 Brain mRNA M430 (Aug05) MAS5 database were compared to all 45137 records in the OHSU/VA B6D2F2 Brain mRNA M430 (Aug05) MAS5 database. The top 100 correlations ranked by the Pearson's product-moment correlation are displayed. You can resort this list using the small arrowheads in the top row.
Click the correlation values to generate scatter plots. Select the Record ID to open the Trait Data and Analysis form. Select the symbol to open NCBI Entrez.

WebGestalt | Gene Ontology | QTL Cluster Map | Download Table | Add Correlation with OHSU/VA B6D2F2 Brain mRNA M430 (Aug05) PDNN

Select All | Clear | Invert | Add to Collection | Select Traits with r > -1.0 AND r < 1.0

	Record ID	Symbol	Description	Chr	Megabase	Mean Expr	Correlation	N Cases	p Value	Lit Corr
1	1426115_a_at_A	*Kcnj9*	potassium inwardly-rectifying channel, subfamily J, member 9	1	172.254634	8.770	1.0000	56	0.00e+00	1.0000
2	1427465_at_A	*Atp1a2*	ATPase, Na+/K+ transporting, alpha 2 polypeptide	1	172.203103	11.826	-0.7313	56	1.21e-11	0.2004
3	1460336_at_A	*Ppargc1a*	peroxisome proliferative activated receptor, gamma, coactivator 1 alpha	5	50.268705	9.445	0.7199	56	4.00e-11	0.1409
4	1429040_at_A	*Cdh11*	cadherin 11	8	101.956669	10.454	0.7060	56	1.59e-10	0.1995
5	1455693_x_at_A	*Rps6*	ribosomal protein S6	4	85.840881	12.539	-0.6977	56	3.43e-10	0.1636
6	1424414_at_A	*Ogfrl1*	opioid growth factor receptor-like 1	1	23.614552	9.155	0.6958	56	4.06e-10	N/A
7	1419168_at_A	*Mapk6*	mitogen-activated protein kinase 6	9	75.525656	9.159	0.6953	56	4.26e-10	0.2091
8	1458615_at_B	*Depdc5*	DEP domain containing 5	5	31.401584	8.323	0.6812	56	1.46e-09	N/A
9	1430980_a_at_A	*Eif4a1*	eukaryotic translation initiation factor 4A1	11	69.394172	11.568	0.6809	56	1.50e-09	0.1846
10	1432430_a_at_A	*1700081L11Rik*	RIKEN cDNA 1700081L11 gene	11	104.156084	9.209	0.6795	56	1.68e-09	N/A
11	1438099_at_B	*Trio*	triple functional domain (PTPRF interacting)	15	27.821685	7.383	0.6793	56	1.72e-09	0.1540
12	1435859_x_at_A	*Psmc2*	proteasome (prosome, macropain) 26S subunit, ATPase 2	5	20.256045	11.321	-0.6779	56	1.94e-09	0.2608
13	1450253_a_at_A	*Map3k4*	mitogen activated protein kinase kinase kinase 4	17	10.871939	9.418	0.6770	56	2.07e-09	0.0907
14	1430123_a_at_A	*Akr1a4*	aldo-keto reductase family 1, member A4 (aldehyde reductase)	4	115.595498	12.599	-0.6767	56	2.13e-09	0.2104
15	1420971_at_A	*Ubr1*	ubiquitin protein ligase E3 component n-recognin 1	2	120.375525	9.551	0.6761	56	2.25e-09	0.1647
16	1422249_s_at_A	*Zfa*	zinc finger protein, autosomal	10	52.780662	6.637	0.6747	56	2.50e-09	0.1745
17	1460305_at_A	*Itga3*	integrin alpha 3	11	94.86757	7.404	0.6741	56	2.64e-09	0.1416
18	1415900_a_at_A	*Kit*	kit oncogene	5	74.489095	8.639	0.6733	56	2.82e-09	0.1104
19	1419291_x_at_A	*Gas5*	growth arrest specific 5	1	160.939892	11.181	-0.6713	56	3.31e-09	0.1943
20	1452218_at_A	*BC018601*	cDNA sequence BC018601	11	5.424989	6.771	0.6711	56	3.37e-09	N/A
21	1421480_a_at_A	*Adarb1*	adenosine deaminase, RNA-specific, B1	10	77.405815	9.261	0.6702	56	3.61e-09	0.3073
22	1427967_at_A	*Cdk5rap2*	CDK5 regulatory subunit associated protein 2	1	131.19874	7.184	0.6676	56	4.45e-09	0.1684
23	1452512_a_at_A	*Ank1*	ankyrin 1, erythroid	8	21.875164	7.326	0.6613	56	7.26e-09	0.2074
24	1451897_a_at_A	*Nbr1*	neighbor of Brca1 gene 1	11	101.397392	10.036	0.6603	56	7.87e-09	0.1945
25	1417998_at_A	*Tebp*	TEA domain family member 4	10	127.813388	10.976	-0.6581	56	9.28e-09	0.2506
26	1453426_a_at_A	*Wdfy1*	WD repeat and FYVE domain containing 1	1	80.056909	7.023	0.6581	56	9.32e-09	0.1129
27	1425677_a_at_A	*Ank1*	ankyrin 1, erythroid	8	21.902895	9.758	0.6579	56	9.45e-09	0.2074
28	1449629_s_at_A	*Snrpd3*	small nuclear ribonucleoprotein D3	10	75.638756	10.393	-0.6577	56	9.59e-09	0.1550
29	1417747_at_A	*Cplx1*	complexin 1 (syntaxin-binding SNARE complex involved in vesicle fusion, synaphin 20)	5	107.595819	12.147	0.6547	56	1.20e-08	0.1302
30	1452413_at_A	*C230081A13Rik*	RIKEN cDNA C230081A13 gene	9	56.314258	8.209	0.6546	56	1.21e-08	N/A
31	1460221_at_A	*Tebp*	TEA domain family member 4	10	127.81303	11.231	-0.6527	56	1.39e-08	0.2506
35	1437257_at_B	*Wdr47*	WD repeat domain 47	3	108.414572	9.731	0.6471	56	2.04e-08	N/A
36	1431530_a_at_A	*Tspan5*	tetraspanin 5	3	137.786041	9.370	0.6465	56	2.14e-08	0.2108
37	1425748_at_A	*Diras1*	DIRAS family, GTP-binding RAS-like 1	10	81.155667	8.883	0.6464	56	2.16e-08	0.1357
38	1432026_a_at_B	*Herc5*	hect domain and RLD 5	6	57.770046	6.157	0.6423	56	2.89e-08	0.1127
39	1431062_a_at_A	*Sec8l1*	SEC8 like 1 (exocytosis complex subunit)	6	33.382382	9.040	0.6421	56	2.93e-08	0.1956
40	1419190_at_A	*Vti1a*	vesicle transport through interaction with t-SNAREs homolog 1A (yeast)	19	54.961183	8.595	0.6409	56	3.19e-08	0.0948
48	1449228_at_A	*Sh3gl2*	SH3-domain GRB2-like 2 (hippocampal CA3 expression signature)	4	84.365516	11.296	0.6365	56	4.32e-08	0.1345
49	1445359_at_B	*Adcy1*	adenylate cyclase 1	11	7.068581	10.229	0.6365	56	4.33e-08	N/A
50	1418621_at_A	*Rab2*	RAB2, member RAS oncogene family	4	8.505617	11.674	0.6355	56	4.64e-08	0.1149
51	1425339_at_A	*Plcb4*	phospholipase C, beta 4	2	135.485611	9.165	0.6349	56	4.83e-08	0.2884
52	1439443_x_at_A	*Tkt*	transketolase	14	28.708017	11.672	-0.6347	56	4.91e-08	0.1888
53	1417029_a_at_A	*Trim2*	tripartite motif protein 2	3	83.930477	10.146	0.6335	56	5.34e-08	0.1515
54	1421673_s_at_A	*Stx1b2*	syntaxin 1B2	7	121.860466	6.661	0.6332	56	5.44e-08	0.0910

Figure 16.4 Web QTL correlation results for *Kcnj9*. Genes showing highly correlated expression with *Kcnj9* are ranked and assigned a *P* value. Literature correlation between each gene and *Kcnj9* is also assessed. Note that the Zfa gene (autosomal zinc-finger protein; record 10) shows significant expression correlation with *Kcnj9* and is located at 52 Mb on chromosome 10. This is located in the centre of a strong linkage peak on chromosome 10 (see Figures 16.4 and 16.5). Web QTL, www.genenetwork.org

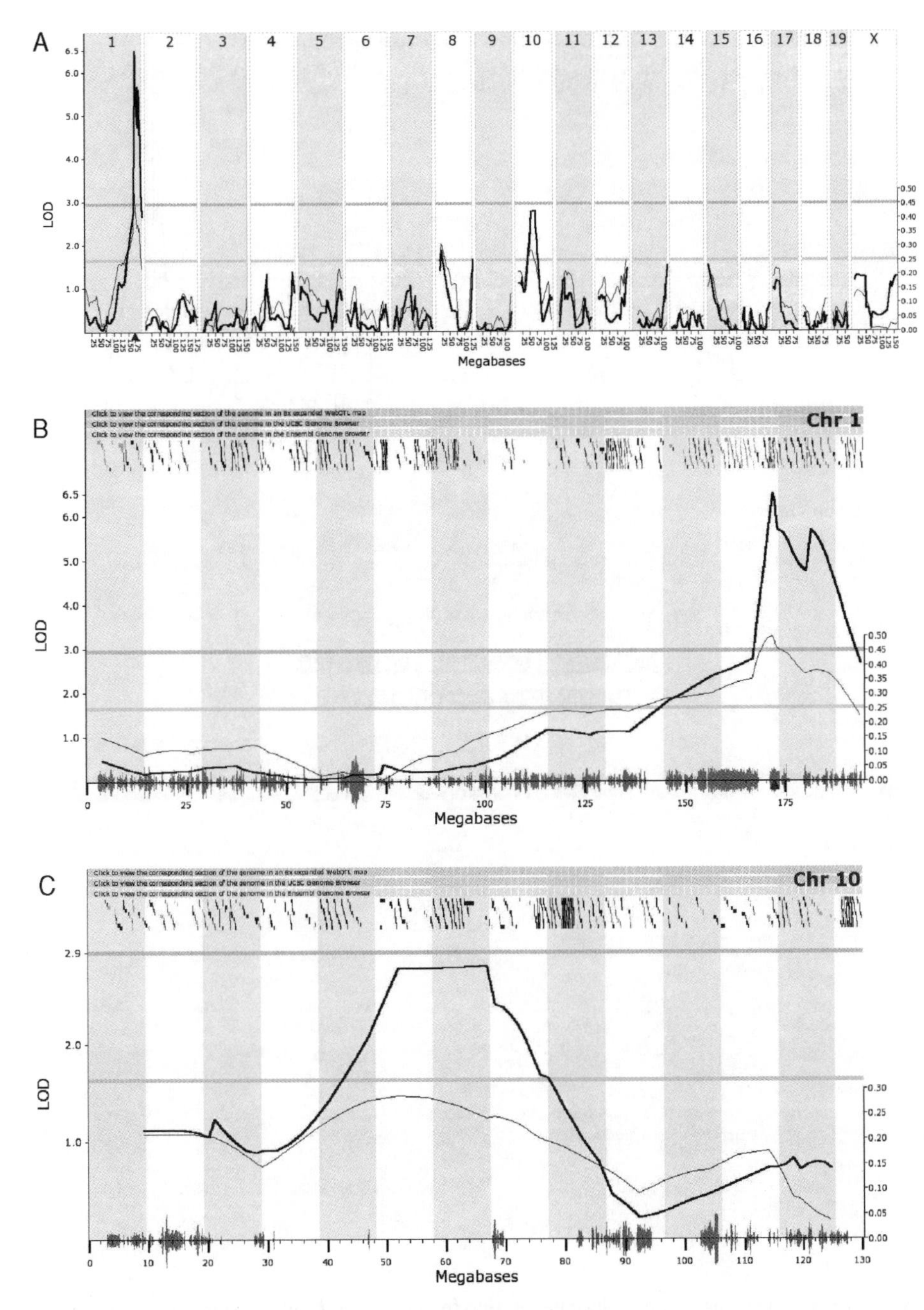

Figure 16.5 Web QTL interval mapping of *Kcnj9* expression analysed by linkage as a QTL. (A) Linkage data are displayed across all chromosomes. (B) The strongest signal is a *cis*-QTL seen on chromosome 1 centred on the *KcnJ9* gene. (C) Another strong signal is a *trans*-QTL seen on chromosome 10 centred at 52 Mb over the *Zfa* gene, suggesting possible *trans*-regulation of *Kcnj9* by the DNA-binding protein *Zfa*. This possible relationship may be worthy of further investigation. Web QTL, www.genenetwork.org

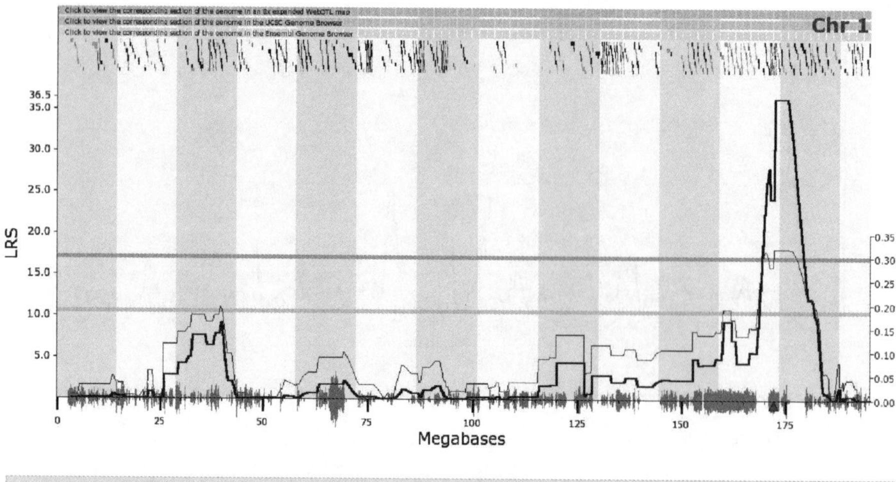

Correlation Table

Values of Record 1426115_a_at_A in the INIA Brain mRNA M430 (Jan06) RMA database were compared to all 856 records in the BXD Published Phenotypes database. The top 100 correlations ranked by the Pearson's product-moment correlation are displayed. You can resort this list using the small arrowheads in the top row.

Clicking on the record ID will open the published phenotype data for that publication. Click on the correlation to see a scatter plot of the trait data.

Multiple Mapping | QTL Cluster Map | Download Table

Select All | Invert | Clear | Add to Collection

☑ Display strain names in correlation plot
☑ Display fit line in correlation plot

	Record ID	Phenotype	Authors	Year	Correlation	N Cases	p Value
1 ☐	10043	Ectromelia virus-induced mortality males [mortality number]	*Brownstein DG, Bhatt PN, Gras L, Jacoby RO*	1991	0.9371	8	0.00013
2 ☐	10430	Tumor growth 2 weeks post-implantation [mm3]	*Mountz JD, Van Zant GE, Zhang H-G, Grizzle WE, Ahmed R, Williams RW, Hsu H-C*	2001	-0.9364	8	0.00013
3 ☐	10237	Proliferation of JTL-G12.8 (Tcell clone) without 50 ug/ml GAT (Glu60, Ala30, Tyr10) [cpm]	*Jenkins MK, Melvold RW, Miller SD*	1984	0.6877	22	0.00024
4 ☐	10577	Plasma corticosterone levels 7 hr post 4 g/kg ethanol [ug/dl]	*Roberts AJ, Phillips TJ, Belknap JK, Finn DA, Keith LD*	1995	0.7076	20	0.00027
5 ☐	10235	Proliferation of JTL-G12 (Tcell clone) without 50 ug/ml GAT (Glu60, Ala30, Tyr10) [cpm]	*Jenkins MK, Melvold RW, Miller SD*	1984	0.6468	22	0.00079
6 ☐	10414	Morris water maze-log latency 2	*Milhaud JM, Halley H, Lassalle JM*	2002	-0.6693	20	0.00084
7 ☐	10044	Ectromelia virus-induced mortality females [mortality number]	*Brownstein DG, Bhatt PN, Gras L, Jacoby RO*	1991	0.9008	8	0.00096
8 ☐	10429	Endogenous serum amyloid P-component (SAP) levels [ug/ml]	*Mortensen RF, Le PT, Taylor BA*	1985	0.5876	22	0.00331
9 ☐	10395	High frequency hearing loss and cochlear pathology; acoustic startle response to white-noise bursts at 110 dB SPL [g]	*McCaughran J, Bell J, Hitzemann R*	1999	0.6099	20	0.00347

Figure 16.6 The value of a reference population for analysis of *Kcnj9*. Analysis of *Kcnj9* expression in the BXD recombinant inbred lines. A dense aggregation of genotype and phenotype data is feasible, affording high-precision QTL analysis and rapid identification of a wealth of phenotypes that are genetically correlated to expression of *Kcnj9*. Web QTL, www.genenetwork.org

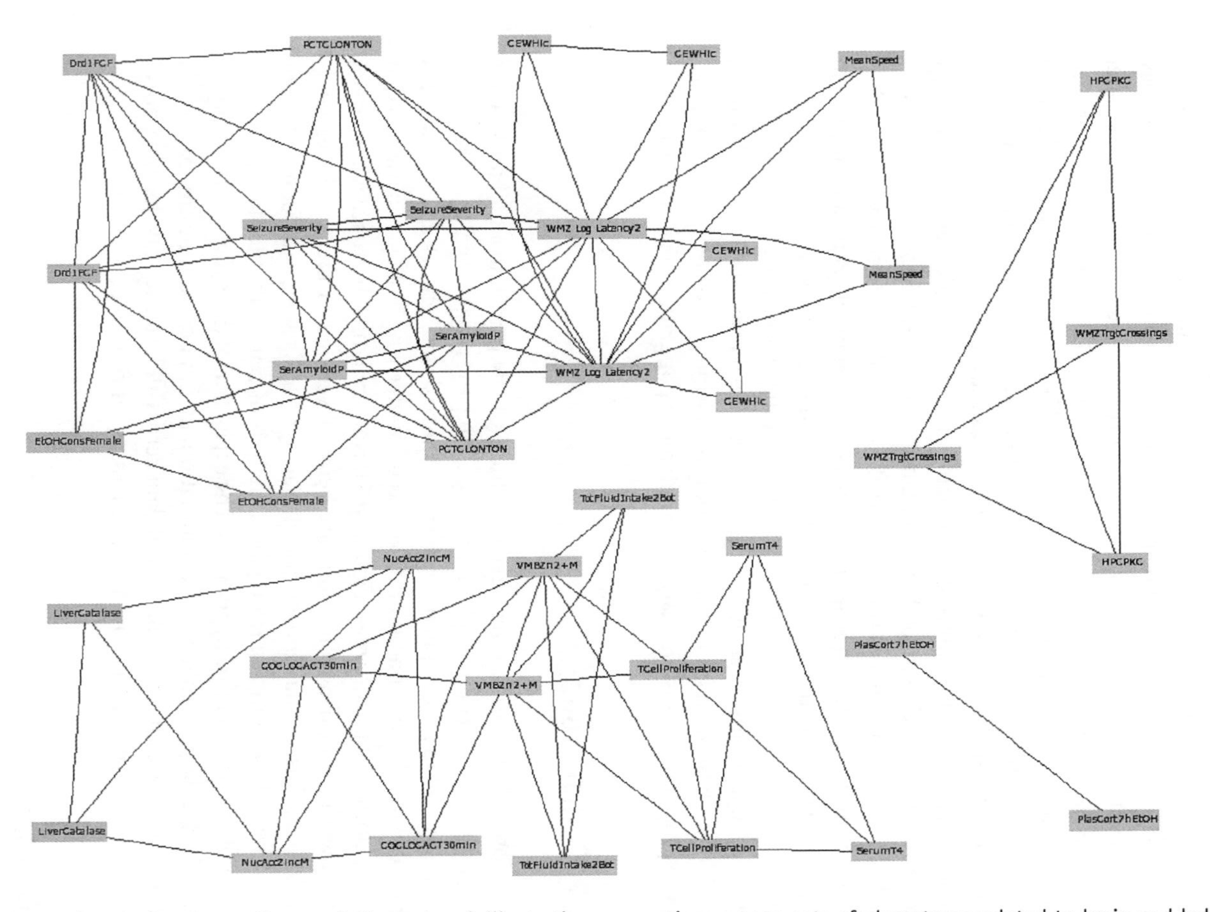

Figure 16.7 Network genetics. A genetic association network illustrating connections among sets of phenotypes related to brain and behaviour, all of which are correlated with brain expression of *Kcnj9*

designs. However, the results of many of these experiments are publicly available, and through these data repositories and their associated analytic tools, one can readily make use of GeneNetwork.org's public genetical genomic data sets in a variety of biological applications. Several researchers have begun to make use of integrated analysis of gene-expression QTLs and conventional QTL analysis of complex phenotypes to identify compelling candidate genes for higher-order complex phenotypes. The premise is that a polymorphism with phenotypic consequence should also alter the expression of trait-relevant genes. Therefore, *cis*-QTLs in the region of a phenotypic QTL are likely to be modifiers, particularly if they are also *trans*-regulators of trait-relevant genes. By coupling multiple cross-mapping, haplotype analysis, and expression QTL analysis, *Kcnj9* was implicated as the chromosome 1 QTL for basal locomotor activity in the mouse (Hitzemann *et al.*, 2003) (Figures 16.2–16.5). By gene-expression correlation analysis to ethanol preference, *Stxbp1* was identified as a likely candidate for ethanol preference (Fehr *et al.*, 2005), another widely studied complex trait. In another strategy, the results of a differential expression analysis in brain response to ethanol were filtered by expression correlation to alcohol-related phenotypes and then subjected to pathway matching analysis, leading to identification of *Sp1* as a putative regulator of the transcriptional response to ethanol (Rulten *et al.*, 2005). These analyses are made possible by the integration of deep genotyping, phenotyping and expression analysis in a reference population (Figure 16.6).

16.7 Conclusions

Expression QTL analysis has matured over the past few years into a tool for systems biological analysis of complex phenotypes spanning the range of biological structure and function. With genetic polymorphisms as the basis for network analysis, systems of traits can be related to one another and to the underlying biological networks that subserve clusters of phenotypes (Figure 16.7). New analytical approaches and larger, more powerful trait data collections are being developed in this rapidly expanding field. Web-based data repositories with analytic tools allow rapid integration of data, for hypotheses ranging from global analysis of genetical genomics, to specific gene- or phenotype-centred hypotheses. The expansion of these tools and techniques has resulted in larger explicit and computationally implicit community-based collaborative efforts to integrate biological data across all levels of scale for the understanding of susceptibility factors and mechanisms of complex disease.

References

Baldwin, N. E., Chesler, E. J., Kirov, S. *et al.* (2005). Computational, integrative, and comparative methods for the elucidation of genetic coexpression networks. *J Biomed Biotechnol* **2**, 172–180.

Belknap, J. K. (1998). Effect of within-strain sample size on QTL detection and mapping using recombinant inbred mouse strains. *Behav Genet* **28**, 29–38.

Brem, R. B. and Kruglyak, L. (2005). The landscape of genetic complexity across 5,700 gene expression traits in yeast. *Proc Natl Acad Sci U S A* **102**, 1572–1577.

Brem, R. B., Yvert, G., Clinton, R. *et al.* (2002). Genetic dissection of transcriptional regulation in budding yeast. *Science* **296**, 752–755.

Broman, K. W. and Speed, T. P. (2002). A model selection approach for the identification of quantitative trait loci in experimental crosses. *J R Stat Soc Ser B Stat Methodol* **64**, 641–656.

Broman K. W., Wu, H., Sen, S. *et al.* (2003). R/qtl: QTL mapping in experimental crosses. *Bioinformatics* **19**(7), 889–890.

Bystrykh, L., Weersing, E., Dontje, B. *et al.* (2005). Uncovering regulatory pathways that affect hematopoietic stem cell function using 'genetical genomics'. *Nat Genet* **37**, 225–232.

Carlborg, O., de Koning, D. J., Manly, K. F. *et al.* (2004). Methodological aspects of the genetic dissection of gene expression. *Bioinformatics* **21**(10), 2383–2393.

Carter, T. A., Del Rio, J. A., Greenhall, J. A. *et al.* (2001). Chipping away at complex behavior: transcriptome/phenotype correlations in the mouse brain. *Physiol Behav* **73**, 849–857.

Chesler, E. J., Rodriguez-Zas, S. L. and Mogil, J. S. (2001). *In silico* mapping of mouse quantitative trait loci. *Science* **294**, 2423.

Chesler, E. J., Wang, J., Lu, L. *et al.* (2003). Genetic correlates of gene expression in recombinant inbred strains: a relational model system to explore neurobehavioral phenotypes. *Neuroinformatics* **1**, 343–357.

Chesler, E. J., Lu, L., Wang, J. *et al.* (2004). WebQTL: rapid exploratory analysis of gene expression and genetic networks for brain and behavior. *Nat Neurosci* **7**, 485–486.

Chesler, E. J., Lu, L., Shou, S. *et al.* (2005). Complex trait analysis of gene expression uncovers polygenic and pleiotropic networks that modulate nervous system function. *Nat Genet* **37**, 233–242.

Chesler, E. J. and Langston, M. A. (2005). Combinatorial genetic regulatory network analysis tools for high throughput transcriptomic data. In *Proceedings, RECOMB Satellite Workshop on Systems Biology and Regulatory Genomics*, San Diego, California, December, 2005.

Churchill, G. A., Airey, D. C., Allayee, H. *et al.* (2004). The Collaborative Cross, a community resource for the genetic analysis of complex traits. *Nat Genet* **36**, 1133–1137.

Churchill, G. A. and Doerge, R. W. (1994). Empirical threshold values for quantitative trait mapping. *Genetics* **138**, 963–971.

Damerval, C., Maurice, A., Josse, J. M. *et al.* (1994). Quantitative trait loci underlying gene product variation: a novel perspective for analyzing regulation of genome expression. *Genetics* **137**, 289–301.

Darvasi, A. (2001). *In silico* mapping of mouse quantitative trait loci. *Science* **294**, 2423.

DiPetrillo, K., Wang, X., Stylianou, I. M. *et al.* (2005). Bioinformatics toolbox for narrowing rodent quantitative trait loci. *Trends Genet* **21**, 683–692.

Doss, S., Schadt, E. E., Drake, T. A. *et al.* (2005). *Cis*-acting expression quantitative trait loci in mice. *Genome Res* **15**, 681–691.

Duan, J., Wainwright, M. S., Comeron, J. M. *et al.* (2003). Synonymous mutations in the human dopamine receptor D2 (DRD2) affect mRNA stability and synthesis of the receptor. *Hum Mol Genet* **12**, 205–216.

Fehr, C., Rademacher, B. L., Buck, K. J. *et al.* (2005). The syntaxin binding protein 1 gene (Stxbp1) is a candidate for an ethanol preference drinking locus on mouse chromosome 2. *Alcohol Clin Exp Res* **29**, 708–720.

Flint, J., Valdar, W., Shifman, S. *et al.* (2005). Strategies for mapping and cloning quantitative trait genes in rodents. *Nat Rev Genet* **6**, 271–286.

Friedman, N., Linial, M., Nachman, I. *et al.* (2000). Using Bayesian networks to analyze expression data. *J Comput Biol* **7**, 601–620.

Hitzemann, R., Malmanger, B., Reed, C. *et al.* (2003). A strategy for the integration of QTL, gene expression, and sequence analyses. *Mamm Genome* **14**, 733–747.

Homayouni, R., Heinrich, K., Wei, L. *et al.* (2005). Gene clustering by latent semantic indexing of MEDLINE abstracts. *Bioinformatics* **21**, 104–115.

Hovatta, I., Tennant, R. S., Helton, R. *et al.* (2005). Glyoxalase 1 and glutathione reductase 1 regulate anxiety in mice. *Nature* **438**, 662–666.

Hubner, N., Wallace, C. A., Zimdahl, H. *et al.* (2005). Integrated transcriptional profiling and linkage analysis for identification of genes underlying disease. *Nat Genet* **37**, 243–253.

Irizarry, R. A., Bolstad, B. M., Collin, F. *et al.* (2003). Summaries of Affymetrix GeneChip probe level data. *Nucleic Acids Res* **31**, e15.

Jansen, R. C. and Nap, J. P. (2001). Genetical genomics: the added value from segregation. *Trends Genet* **17**, 388–391.

Jeong, H., Tombor, B., Albert, R. *et al.* (2000). The large-scale organization of metabolic networks. *Nature* **407**, 651–654.

Jiang, C. and Zeng, Z. B. (1995). Multiple trait analysis of genetic mapping for quantitative trait loci. *Genetics* **140**, 1111–1127.

Jin, W., Riley, R. M., Wolfinger, R. D.*et al.* (2001). The contributions of sex, genotype and age to transcriptional variance in *Drosophila melanogaster. Nat Genet* **29**, 389–395.

Kempermann, G., Chesler, E. J., Lu, L. *et al.* (2006). Natural variation and genetic covariance in adult hippocampal neurogenesis. *Proc Natl Acad Sci U S A* **103**, 780–785.

Kendziorski, C. M., Chen, M., Yuan, M. *et al.* (2006). Statistical methods for expression quantitative trait loci (eQTL) mapping. *Biometrics* **62**, 19–27.

Kerr, M. K., Martin, M. and Churchill, G. A. (2000). Analysis of variance for gene expression microarray data. *J Comput Biol* **7**, 819–837.

Kerr, M. K. and Churchill, G. A. (2001). Statistical design and the analysis of gene expression microarray data. *Genet Res* **77**, 123–128.

Korstanje, R. and Paigen, B. (2002). From QTL to gene: the harvest begins. *Nat Genet* **31**, 235–236.

Kulp, D. and Jagalur, M. (2006). Causal inference of regulator–target pairs by gene mapping of expression phenotypes. *BMC Genomics* **7**, 125.

Lan, H., Stoehr, J., Nadler S. T. *et al.* (2003). Dimension reduction for mapping mRNA abundance as quantitative traits. *Genetics* **164**, 1607–1614.

Lander, E. and Kruglyak, L. (1995). Genetic dissection of complex traits: guidelines for interpreting and reporting linkage results. *Nat Genet* **11**, 241–247.

Li, C. and Hung Wong, W. (2001). Model-based analysis of oligonucleotide arrays: model validation, design issues and standard error application. *Genome Biol* **2**, RESEARCH0032.

Li, H., Chen, L., Bao, K. F. *et al.* (2006). Integrative genetic analysis of transcription modules: towards filling the gap between genetic loci and inherited traits. *Hum Mol Genet* **15**, 481–492.

Li, H., Chen, L., Bao, K. F. *et al.* (2005). Inferring gene transcriptional modulatory relations: a genetical genomics approach. *Hum Mol Genet* **14**, 1119–1125.

Li, J. and Burmeister, M. (2005). Genetical genomics: combining genetics with gene expression analysis. *Hum Mol Genet* **14** (Spec No. 2), R163–169.

Mhyre, T. R., Chesler, E. J., Thiruchelvam, M. *et al.* (2005). Heritability, correlations and *in silico* mapping of locomotor behavior and neurochemistry in inbred strains of mice. *Genes Brain Behav* **4**, 209–228.

Pavlidis, P. and Noble, W. S. (2001). Analysis of strain and regional variation in gene expression in mouse brain. *Genome Biol* **2**, RESEARCH0042.

Pe'er, D., Regev, A., Elidan, G. *et al.* (2001). Inferring subnetworks from perturbed expression profiles. *Bioinformatics* **17** (Suppl 1), S215–224.

Peirce, J. L., Lu, L., Gu, J. *et al.* (2004). A new set of BXD recombinant inbred lines from advanced intercross populations in mice. *BMC Genet* **5**, 7.

Petkov, P. M., Graber, J. H., Churchill, G. A. *et al.* (2005). Evidence of a large-scale functional organization of mammalian chromosomes. *PLoS Genet* **1**, e33.

Pletcher, M. T., McClurg, P., Batalov, S. *et al.* (2004). Use of a dense single nucleotide polymorphism map for *in silico* mapping in the mouse. *PLoS Biol* **2**, e393.

Rulten, S. L., Ripley, T. L., Hunt, C. L. *et al.* (2006). Sp1 and NFkappaB pathways are regulated in brain in response to acute and chronic ethanol. *Genes Brain Behav* **5**(3), 257–273.

Sandberg, R., Yasuda, R., Pankratz, D. G. *et al.* (2000). Regional and strain-specific gene expression mapping in the adult mouse brain. *Proc Natl Acad Sci U S A* **97**, 11038–11043.

Schadt, E. E., Monks, S. A., Drake, T. A. *et al.* (2003). Genetics of gene expression surveyed in maize, mouse and man. *Nature* **422**, 297–302.

Sen, S. and Churchill, G. A. (2001). A statistical framework for quantitative trait mapping. *Genetics* **159**, 371–387.

Shifman, S. and Darvasi, A. (2005). Mouse inbred strain sequence information and yin–yang crosses for quantitative trait locus fine mapping. *Genetics* **169**, 849–854.

Storey, J. D. and Tibshirani, R. (2003a). Statistical significance for genomewide studies. *Proc Natl Acad Sci U S A* **100**, 9440–9445.

Storey, J. D. and Tibshirani, R. (2003b). Statistical methods for identifying differentially expressed genes in DNA microarrays. *Methods Mol Biol* **224**, 149–157.

Wagner, A. (2005). Distributed robustness versus redundancy as causes of mutational robustness. *Bioessays* **27**, 176–188.

Wang, J., Williams, R. W. and Manly, K. F. (2003). WebQTL: Web-based complex trait analysis. *Neuroinformatics* **1**, 299–308.

Wang, Y., Liu, C. L., Storey, J. D. *et al.* (2002). Precision and functional specificity in mRNA decay. *Proc Natl Acad Sci U S A* **99**, 5860–5865.

Williams, R. W., Gu, J., Qi, S. *et al.* (2001). The genetic structure of recombinant inbred mice: high-resolution consensus maps for complex trait analysis. *Genome Biol* **2**, RESEARCH0046.

Williams, R. W., Bennett, B., Lu, L. *et al.* (2004). Genetic structure of the LXS panel of recombinant inbred mouse strains: a powerful resource for complex trait analysis. *Mamm Genome* **15**, 637–647.

Yvert, G., Brem, R. B., Whittle, J. *et al.* (2003). *Trans*-acting regulatory variation in *Saccharomyces cerevisiae* and the role of transcription factors. *Nat Genet* **35**, 57–64.

Zhang, L., Miles, M. F. and Aldape, K. D. (2003). A model of molecular interactions on short oligonucleotide microarrays. *Nat Biotechnol* **21**, 818–821.

Zhang, B., Schmoyer, D., Kirov, S. *et al.* (2004). GOTree Machine (GOTM): a Web-based platform for interpreting sets of interesting genes using Gene Ontology hierarchies. *BMC Bioinformatics* **5**, 16.

17

Bioinformatics and Cancer Genetics

Joel Greshock

Translational Medicine, Clinical Pharmacology Division,
GlaxoSmithKline Pharmaceuticals, Upper Merion, PA, USA

17.1 Introduction

Cancer is widely recognized as the most common genetic disease. One in three people are afflicted with cancer during their lifetimes, and one in five will die of this condition. As cancer is a disease of the genome, much of oncology is aimed at discovering and describing the recurring molecular alterations and conditions that contribute to tumour genesis, proliferation and metastasis. This broadly includes identifying (i) which genes are altered in cancer, (ii) how many genes are altered in each cancer and (iii) which mechanisms drive these alterations (Weber, 2002). Effectively characterizing the cancer genome often requires multifaceted genetic data sets and sophisticated analyses.

Although the increasing role of bioinformatics in cancer research mirrors that of all genetics, the effects of the bioinformatics age on cancer research cannot be overstated. Bioinformatics has had such a profound influence on the field of cancer genetics that it has changed such fundamentals as cancer definition and diagnosis. Research in the field of cancer genetics has always generated an intricate and expansive body of data as a result of the dynamic and multifaceted genetic changes associated with cancer. The advent of the post-genome era and the improved capacity for high-throughput genomics assays has expanded this exponentially, necessitating the use of bioinformatics approaches to research the cancer genome. Bioinformatics has allowed the

Bioinformatics for Geneticists, Second Edition. Edited by Michael R. Barnes
 ISBN 978-0-470-02619-9 (HB) ISBN 978-0-470-02620-5 (PB)

molecular classification of cancers and the modelling of cellular circuits involved in tumour genesis and progression. It has been instrumental in guiding large-scale wet laboratory research (see Cancer Genome Atlas below) and has progressed to the juncture where novel bioinformatics-driven discovery research can be done with only data from public repositories. The ultimate result is more integrated research where primary data can be analysed from the perspective of previous knowledge. Bioinformatics has truly commenced a new era in cancer research.

This chapter will survey and highlight the role of bioinformatics in cancer genetics. It will focus on the diversity of Web databases that serve genetic assay data and genome annotations, as well as the algorithms and data-mining tools necessary for cancer analysis. Additionally, examples will be discussed that demonstrate the molecular characterization of tumours from gene-expression data and identification of tumour suppressor genes by genomic copy number alterations.

17.2 Cancer genomes

In considering the function of bioinformatics in cancer research, it is essential to understand the underlying principles of cancer genetics. A majority of research in the field of cancer genetics investigates the underlying processes that transform normal cells to malignant clones and enable their uncontrolled proliferation. The central focus is on the two conventional classes of cancer genes that are thought to prescribe these processes. Protective *tumour suppressor genes* typically restrict the growth of tumours, while *oncogenes* contribute to the creation of a cancer. These genes contribute to the biochemical changes, acquired abilities, and cellular traits shared by all cancers.

Tumorigenesis is an extremely complex, multistep process that is the product of genomes being altered in multiple and complementary ways, resulting in selective advantage for tumour cells. Although the ensuing 'cancer genomes' may retain many of the germ-line characteristics from which they have evolved, each represents the product of cells responding to selective pressures. The result is a relatively stable DNA 'fingerprint' that has a unique genetic make-up derived from the germ-line genome. There are many types of genetic alterations that can provide a selective advantage for tumour cells. Collectively, they contribute to the uniqueness of a cancer's genome. Examples include point mutations of tumour suppressor genes and chromosomal amplifications of oncogenes. In an attempt to codify the underlying principles of cancer, Hanahan and Weinberg (2000) put forth a popular model that breaks down cancer development and proliferation into six essential cellular capabilities, each of which requires genetic alterations. These acquired capabilities include:

1. evading apoptosis
2. self-sufficiency in growth signals

3. insensitivity to anti-growth signals
4. sustained angiogenesis
5. limitless reproductive power
6. tissue invasion and metastasis.

This commonly held notion suggests that the 100+ cancer types share a series of required phenotypes that are a result of multiple genetic alterations and conditions. Futreal *et al.* (2004) compiled a list of ~350 genes that are known to be implicated in one or more of the six phenotypes reviewed in Hanahan and Weinberg (2000). The genetic modifications that facilitate each of these capabilities may range from a single mutation of one gene, to a complex network of changes that contribute to a phenotype. Additionally, although a single genetic change can enable multiple cancer-related faculties, no single alteration is universally responsible for any one capacity. The result is a wide diversity of cancer genomes that require complex, data-rich approaches for effective characterization.

17.3 Approaches to studying cancer genetics

Investigation of somatic genetics of cancer is unlike the study of germ-line genetics. Cancer is a quickly evolving condition that is subject to unique selective pressures, such as the ability to evade apoptosis, that result in a series of genetic changes that distinguish it from a germ-line genome. Therefore, a core principle is that the tumour genome is distinct from the germ-line genome and is treated differently in cancer research. Studies of cancers often focus on the tumour genome's gradual divergence from the germ-line, where a study subject is normally a tumour rather than an individual. 'Somatic' alterations seen in cancers include various types of mutations and DNA copy number alterations. A clear distinction must be made between a somatic mutation and a germ-line polymorphism. The former could be designated a true 'mutation' in cancer, while the latter is an inherited polymorphism. Similarly, inherited copy number polymorphisms (CNPs) , some of which may be disease related, such as deletion in the *SNAI2* locus in Waardenburg syndrome (Sanchez-Martin *et al.*, 2002), must be distinguished from somatic copy number alterations unique to the diseased tissue (such as a complete loss of the *CDKN2A* gene described in several cancers). The pursuit of therapeutic targets concentrates on identifying these 'somatic' alterations, and drugs are designed in light of the phenotype resulting from them. Conversely, studies exclusively investigating germ-line cancer genetics often focus on predisposition and risk evaluation and can attempt to identify genuine germ-line mutations or high-risk genotypes. An example of a clinically relevant mutation for which high-risk individuals are screened is that of the *BRCA1* gene (Weber *et al.*, 1994).

17.3.1 Study design

Study design is an important aspect of any scientific investigation. The genetic analysis of cancers is not an exception. A thorough understanding of good study design in the field of cancer genomics is essential in planning wet laboratory studies for bioinformatics analysis, or doing *in silico* research with publicly available data resources. Study panels, typically composed of tumours as units, are interrogated for genetic alterations and variants. Proper studies will afford the best chance of making desired inferences about somatic alterations, such as their distinction from germ line, relative amounts of tumour diversity, and associations with a phenotype. Studies can range from truly exploratory, such as investigating gene-expression patterns between two phenotypes, to hypothesis-driven, as with the investigation of mutation frequencies of specific genes in a homogeneous population. In any case, the extent to which genetic alterations can be investigated is often constrained by the availability of tissues and the resources to process these tissues for analysis. Fresh tissues are often the focus, although tumour-derived cell lines are used extensively (see NCI-60 below). Confounding this general study design is the heterogeneous nature of cancer, which has a wide range of diagnoses, resulting in diversity of clinical phenotypes. Diversity is seen between sites on the body (such as brain versus colon cancers) as well as within diagnoses. For example, pathological variation described between two types of breast cancer (e.g., invasive versus lobular) may equal those seen between breast and ovarian cancers. Challenges arise with determining what constitutes reasonable groupings of cancers as test subjects. In other words, how can a homogeneous panel of cancers that can be expected to exhibit similar genetic mechanisms be identified? Moreover, a paradigm arises when the *objective* of basic research is the molecular characterization of a heterogeneous population of cancers and study panels. For example, in melanoma, ~70 percent of tissues carry mutations in the *BRAF* locus (Davies *et al.*, 2002). These are thought to constitute a genetically distinct group of cancers from *BRAF* wild types (Pavey *et al.*, 2004). Hereafter, if studying the somatic alterations contributing to the genesis and progression of melanomas, would it be proper to group *BRAF* mutants and wild types? This answer may seem obvious, but is typically less so where alterations are less well described (this is more common). Further, several cancer types have well described clinical subgroups. Breast cancers are typically grouped in terms of *ERBB2* expression, while neuroblastomas (a rare pediatric cancer) are grouped by a DNA amplification of the *N-myc* gene (Fong *et al.*, 1989). Specific research objectives may offer guidance in composing study panels; however, in some cases, there are no clear answers. Bioinformatics has, in part, assumed the responsibility of contributing to breaking this paradigm by providing complex molecular characterizations of cancers and designing computational methods for identifying unique subtypes.

17.3.2 Cancer cell lines

Panels of primary tissues serve as the basis for studying genetic changes in cancers. Tissues can be frozen at –80°C or embedded in paraffin wax for long-term storage. They are typically dissected, and DNA/RNA is extracted. Although these tissues are good for many studies of cancer genetics, they are less than ideal for others. Tumour-derived cancer cell lines offer many advantages over using fresh tumours. These are populations of cells drawn from a fresh tumour that are immortalized through an often difficult procedure that results in frozen cells that can be grown at will. Although the genetic make-up of these cell lines is thought to reflect that of the original tumours, several cell-line-specific traits have been described. Most notably, cell-line immortalization has been implicated as a source of cytogenetic changes, such as DNA copy number alterations. For example, recurring copy number gains of chromosome 20 and losses of chromosome 13 have been associated with the cell-line transformation process (Ratsch *et al.*, 2001; Jin *et al.*, 2004). Moreover, multiple growth passages (to which commercially available cell lines are routinely subjected) have been shown to be associated with random genomic instability (Meisner *et al.*, 1988). Finally, past studies have noted differences in gene-expression patterns between cell lines and their fresh/frozen tissue counterparts (Kees *et al.*, 1992; Mackay *et al.*, 1992; Lee and Maihle, 1998). For these reasons, a degree of vigilance must be practised when employing cell lines as tumour models.

Cell lines bear several advantages that make them attractive subjects for genomic analysis. Fresh tissue is often scarce and obtaining ample DNA is frequently difficult, while cell lines offer a replenishable source of DNA and RNA. Tumour heterogeneity is of chief importance when analysing tissues. The infusion of normal cells can dilute signal and reduce the possibility of detecting alterations (Garnis *et al.*, 2005). Alternatively, cell lines provide a more homogeneous cell population in which cell-to-cell variation in copy number is reduced. Most importantly, tumours cannot be analysed for genetic alterations *in vivo*. Using cell lines for genetic analysis makes possible time-course analysis (e.g. Singh *et al.*, 2000) and pharmacodynamic studies (e.g. Hattinger *et al.*, 2003, O'Toole *et al.*, 2005). Further, a cell line and its derivatives have a parent–child relationship whereby drug resistance can be engineered and studied (e.g. de Angelis *et al.*, 2004). The NCI-60 is a panel of publicly available tumour-derived cell lines. This is a well-described group that has been worked on extensively and provides a wealth of publicly available genetic data.

17.3.3 Recent technologies used to study cancer genetics

Cancer research has embraced recent technological advances that have expanded the number of data researchers have been able to collect. Specifically, the capacity of newer assays to measure three major somatically acquired alterations, (i) gene sequence mutations, (ii) DNA copy number aberrations, and (iii) gene-expression alteration, is particularly attractive to those studying cancer genetics. Mutation screening by high-throughput sequencing methods and the migration of DNA copy number and gene expression measurements to high-density microarray platforms (in most cases, oligonucleotide chip-based) have made this approach more data-rich, necessitating complex bioinformatics research. Much of this bioinformatics-related work has been devoted to serving up these data for public use. These efforts range from general repositories (e.g., Oncomine; Rhodes *et al.*, 2004) to those focusing on persistent mutations of a single gene (e.g., the IARC *TP53* mutation database; Olivier *et al.*, 2002)). Analytic and algorithmic efforts have focused on the molecular characterization and classification of tumours, cancer subtype discovery, modelling cancer progression, and drug response using heterogeneous data.

17.4 General resources for cancer genetics

The research landscape of the post-genome era has facilitated data-rich industrialized research projects aimed at identifying therapeutic targets through detailed cancer genome characterization by high-throughput technologies. Bioinformatics necessarily plays a central role in these initiatives and often their associated public Web resources, and published tools are an excellent starting point for collecting primary data and general cancer-related information. One such project is the Wellcome Trust Sanger Institute's Cancer Genome Project (CGP), a multidisciplinary effort to identify somatically acquired sequence variants and aberrations in human cancers. It has served as a model for coupling high-throughput screening of gene-mutation data with bioinformatics tools designed for access and analysis. The CGP's centralized Web resource (Table 17.1) offers a wealth of primary data, including COSMIC, a catalogue of gene mutations (see below), genome copy number data, and gene sequences from both tumours and cancer cell lines. Another project of this magnitude is the National Human Genome Research Institute (NHGRI) and National Cancer Institute's (NCI) Cancer Genome Atlas (TCGA). This program interactively couples clinical outcome data with experimental data characterizing genomic alterations in cancers and promises to be a rich source of data and tools.

The NCI Center for Bioinformatics (NCICB), which supports the broad, integrative research programmes put forth by the NCI, has a Web resource that hosts several cancer-related tools and data repositories. One of the highlights is the Web resource for the Cancer Genome Anatomy Project (CGAP). The CGAP is the collective effort

Table 17.1 Useful URLs for accessing cancer-related data, software tools and information

Resource	URL
General Resources	
National Cancer Institute Center for Bioinformatics	http://ncicb.nci.nih.gov
Cancer Genome Project	http://www.sanger.ac.uk/genetics/CGP
Cancer Genome Anatomy Project	http://www.cgap.nci.nih.gov
Cancer Genes/Mutations	
p53 Gene mutation database	http://www-p53.iarc.fr/index.html
Catalog of Somatic Mutations in Cancer	http://www.sanger.ac.uk/genetics/CGP/cosmic/
Cancer Gene Data Curation Project	http://ncicb.nci.nih.gov/NCICB/projects/cgdcp
Atlas of Genetics and Cytogenetics in Oncology	http://www.infobiogen.fr/services/chromcancer/
The Tumor Gene Database	http://www.tumor-gene.org
Human Gene Mutation Database	http://www.hgmd.cf.ac.uk/
Genetic Polymorphism	
NCI SNP Resource	http://gai.nci.nih.gov/html-snp/ts.html
SNP500Cancer	http://snp500cancer.nci.nih.gov
Chromsomal alterations/Cytogenetics	
Mitelman database	http://cgap.nci.nih.gov/Chromosomes/Mitelman
SKY/M-FISH and CGH Database	http://www..ncbi.nlm.nih.gov/projects/sky
Recurrent Chromosomal Aberrations in Cancer	http://cgap.nci.nih.gov/Chromosomes/RecurrentAberrations
Progenetix	http://www.progenetix.net
Gene Expression	
Oncomine	http://www.oncomine.org
NCI- maDb	http://nciarray.nci.nih.gov/
Theraputics	
Drug Adverse Reaction Database	http://www.fda.gov/cder/cancer/toxicityfram.htm
Chemotherapy	http://www.cancersourcemd.com/drugdb3
Clinical Trials	
PDQ	http://www.cancer.gov/clinicaltrials
Oncolink	http://www.oncolink.com
Centerwatch	http://www.centerwatch.com
Toxicology	
The Carcinogenic Potency Database	http://potency.berkely.edu/cpdb.html
Chemical Carcinogenesis Research Information	http://toxnet.nlm.nih.gov/cgi-bin/sis/htmlgen?CCRIS
Text Mining	
Bioie	http://bioie.ldc.upenn.edu

of a network of investigators aiming at the molecular characterization of cancers through transcriptional profiling. The Web resource contains both primary data for bioinformatics analyses and tools for scientists to design and conduct genetic studies of cancer. Researchers can use CGAP information to design assays such as gene-expression microarrays, analysis of copy number alterations, and RNAi. Other information served by CGAP includes annotations of cancer-related genes, and pathways and serial analysis of gene-expression (SAGE) data. A notable unique component of CGAP is the Web resource for SNP500Cancer, a project devoted to querying reference samples to find known or newly discovered single nucleotide polymorphisms (SNPs), which are of direct importance to molecular epidemiology studies in cancer (Packer *et al.*, 2004). Web resources such as the CGP and CGAP exemplify the role of bioinformatics in cancer research and provide rich repositories of data for *in silico* bioinformatics research.

17.5 Cancer genes and mutations

A diverse range of important somatic and germ-line sequence mutations has been described in cancers, including all types of substitutions, INDELS and amplifications. Ultimately, loss of function of one or more alleles of a gene with tumour suppressive qualities, or gain of function in a gene with oncogenic behaviour, can prove to be sufficient to promote a proliferative advantage in targettissues. Most studies of mutations in cancer aim to determine where they occur, the frequency at which they occur, and their phenotypic effects. It is important to be aware that somatic mutations can fall into two groups. Most notably, 'driver' mutations confer selective advantage and are implicated in the development of tumours. Alternatively, 'passenger' mutations are essentially a by-product of uncontrolled proliferation and failure in mismatch repair. These are not subject to selection and are not causally involved in oncogenesis or tumour progression (Davies *et al.*, 2005). Distinguishing the former from the later is far from trivial, and though bioinformatics analysis of the functional impact of wild-type and mutant alleles can help considerably in this task (this is covered in detail in Chapter 10), ultimately, laboratory study is the only way to establish the status of a tumour mutation as a driver in the process of cancer.

Mutations can not only enhance uncontrolled proliferation, but also influence clinical characteristics (e.g., constitutive target activation (Goemans *et al.*, 2005)) and drug response (Tokumo *et al.*, 2005). An example of this in cancer is the nuclear tumour protein *TP53*, which plays a vital role in cell-cycle regulation and transition from G_0 to G_1. Although this tumour suppressor gene is normally expressed at low levels in most cells, somatic and germ-line mutations (occurring at rates of over 50 per cent in some cancers) can cause inactivation, resulting in unregulated cell proliferation. Hence, mutation screening in this gene has been quite extensive.

Computational analyses of oncology data on a genome-wide basis (as in gene-expression microarrays) are often problematic without some prior assumptions of the potentially important pathways and genes to study. Identifying which genes are mutated in cancer is an important step toward this goal, which the major tumour sequencing programmes are beginning to address. Bioinformatics approaches can help to identify genes thought to be most involved with cancer. These are usually associated with cellular growth and apoptosis Thomas *et al.* (2003) found much higher intensities of purifying selection in oncogenes and tumour suppressor genes than other genes. This highlights some of the differences between genes involved in cancer and those implicated in other human diseases, there being a highly significant overlap between cancer genes and 'essential genes' that show lethal phenotypes in knockout experiments. Previously described causal relationships between genes and known associations of gene mutations with cancers can expand the cancer gene list further. Compiling a list of these genes is not a simple task, as it requires the extensive use of literature-mining tools such as PennBioIE (Table 17.1), and knowledge of the potential downstream effects of complex signalling cascades or loss of gene function. One of the principal goals of the CGP is to identify which sets of genes are altered in human cancer. This effort has used extensive literature review and mutation screening to identify genes whose alteration has been shown to recur in one or more cancers, and that might provide growth advantages to the afflicted cells or facilitate clonal expansion to surrounding tissues. A growing list of cancer-related genes (at present, this includes 347 genes; reviewed in Futreal *et al.*, 2004), along with their described alterations, has been compiled, and is served as part of the CGP. Other major efforts to identify a comprehensive list of cancer genes include the NCI's Cancer Gene Data Curation Project and the Atlas of Genetics and Cytogenetics in Cancer (Dorkeld *et al.*, 1999). Although these lists largely overlap, they vary due to the unique criteria used to define cancer involvement. They all make excellent references for laboratory screening and computational analysis of gene alterations.

17.5.1 Cancer mutation databases

As high-throughput mutation screening has become one of the most important exercises in cancer research, bioinformatics efforts have consolidated mutation data into several key databases and have developed useful tools for the analysis of this data. The Catalogue of Somatic Mutations in Cancer (COSMIC) is a bioinformatics tool that provides a Web-based query interface to the growing database of published somatic mutations of cancer genes (Bamford *et al.*, 2004). The central purpose of COSMIC is to provide somatic mutation frequencies of cancer genes. In doing so, this tool elegantly converges the CGP's mutation screening data with publicly available mutation data and genome annotations. The interface offers two points of entry where a user can either query a gene by name for its documented mutations in all cancers, or search for mutations of all genes seen in a specific cancer tissue (such as breast). A diverse set of cancerous tissue is represented in this database, including cell lines, benign neoplasms, *in situ* and invasive tumours, recurrences, and metastases. The results can be easily navigated as HTML or exported as text files for further analysis. COSMIC also provides direct access to supporting data, including cDNA and protein sequences, sequence alignments, links to OMIM databases and the Ensembl genome browser, as well as access to publications representing the data's origin. At

Table 17.2 Tyrosine kinase genes are included as part of the Cancer Genome Project's target list. Associated information includes the cancer histologies where alterations have been implicated and which types of alterations have been previously observed. AML = acute myelogenous leukaemia, ALL = acute lymphocytic leukaemia, AS = angiosarcoma, T-ALL = T-cell acute lymphoblastic leukaemia, JMML = juvenile myelomonocytic leukaemia, MDS = myeloproliferative disorder, EMC = extraskeletal myxoid chondrosarcoma, Mis = missense mutation, O = other, T = translocation, L = leukaemia/lymphoma, M = mesenchymal, E = epithelial

Symbol	Name	Chr band	Tumour types	Tissue type	Mutation type	Translocation partner
FLT3	fms-related tyrosine kinase 3	13q12	AML, ALL	L	Mis, O	
FLT4	fms-related tyrosine kinase 4	5q35.3	AS	M	Mis	
LCK	lymphocyte-specific protein tyrosine kinase	1p35-p34.3	T-ALL	L	T	TRB@
NTRK1	neurotrophic tyrosine kinase, receptor, type 1	1q21-q22	papillary thyroid	E	T	TPM3, TPR, TFG
NTRK3	neurotrophic tyrosine kinase, receptor, type 3	15q25	congenital fibrosarcoma, Secretory breast	E, M	T	ETV6
PTPN11	protein tyrosine phosphatase, non-receptor type 11	12q24.1	JMML, AML, MDS	L	Mis	
SYK	spleen tyrosine kinase	9q22	MDS	L	T	ETV6
TEC	tec protein tyrosine kinase	4p12	EMC	M	T	EWSR1, TAF15, TCF12
TTL	tubulin tyrosine ligase	2q13	ALL	L	T	ETV6

present, COSMIC catalogues over 20 000 mutations in 587 genes from approximately 125 000 unique tissues (May 2006, Release no. 18).

For example, tyrosine kinases form a prominent group of cancer-related oncogene transducer genes (reviewed by Stephens *et al.*, 2005; Di Nicolantonio and Bardelli, 2006) and compose a substantial portion of the CGP cancer-related gene list (Table 17.2). Interrogating kinase genes for molecular alterations is of obvious importance, as they control the continuous flow of chemical signals that instruct cell growth and proliferation. Harmful alterations to tyrosine kinases can lead to uncontrolled cell growth and ultimately to tumour formation. As expected, tyrosine

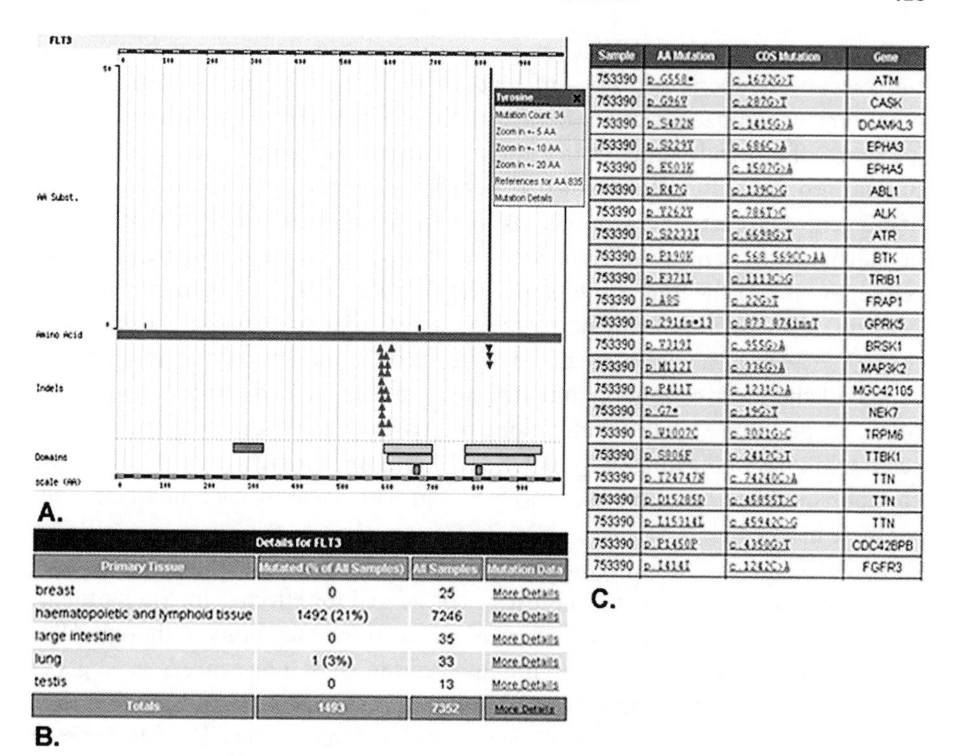

Details for FLT3			
Primary Tissue	Mutated (% of All Samples)	All Samples	Mutation Data
breast	0	25	More Details
haematopoietic and lymphoid tissue	1492 (21%)	7246	More Details
large intestine	0	35	More Details
lung	1 (3%)	33	More Details
testis	0	13	More Details
Totals	1493	7352	More Details

B.

Sample	AA Mutation	CDS Mutation	Gene
753390	p.G558*	c.1672G>T	ATM
753390	p.G96V	c.287G>T	CASK
753390	p.S472N	c.1415G>A	DCAMKL3
753390	p.S229Y	c.686C>A	EPHA3
753390	p.E503K	c.1507G>A	EPHA5
753390	p.R47G	c.139C>G	ABL1
753390	p.Y262Y	c.786T>C	ALK
753390	p.S2223I	c.6668G>T	ATR
753390	p.P190K	c.568_569CC>AA	BTK
753390	p.F371L	c.1113C>G	TRIB1
753390	p.A8S	c.22G>T	FRAP1
753390	p.291fs*13	c.873_874insT	GPRK5
753390	p.V319I	c.955G>A	BRSK1
753390	p.M112I	c.336G>A	MAP3K2
753390	p.P411T	c.1231C>A	MGC42105
753390	p.G7*	c.19G>T	NEK7
753390	p.W1007C	c.3021G>C	TRPM6
753390	p.S806F	c.2417C>T	TTBK1
753390	p.T24747N	c.74240C>A	TTN
753390	p.D15285D	c.45855T>C	TTN
753390	p.I15314L	c.45942C>G	TTN
753390	p.P1450P	c.4350G>T	CDC42BPB
753390	p.I414I	c.1242C>A	FGFR3

C.

Figure 17.1 (A) COSMIC query view of the tyrosine kinase *FLT3* indicates that mutations are relatively common in haematopoietic and lymphoid tissue (21 per cent). (B) However, they are absent in most other tissues. These mutations primarily consist of insertions at AA 598–599 (triangles) and substitutions of AA 835 (vertical bars). Additionally, repeated deletions of AA 836 were also noted (upside-down triangles). (C) A single lung tumor having a mutation in the *FLT3* gene shows mutations in 21 other genes. The mutation data were obtained from the Sanger Institute Catalogue of Somatic Mutations in Cancer website, http://www.sanger.ac.uk/cosmic. Bamford *et al.* (2004), The COSMIC (Catalogue of Somatic Mutations in Cancer) database and website. *Br J Cancer*, **91**, 355–358

kinase genes are well represented in COSMIC. With COSMIC's interface, it is simple to identify tissues cataloguing mutations of specific tyrosine kinases. For example, a query of the fms-related tyrosine kinase 3 (*FLT3*) demonstrates that sequence alterations of this gene are limited to haematopoietic and lymphoid tissue (1492/7246 tissues, 21 per cent; Figure 17.1). The lone exception is a single substitution of base 64 in a lung carcinoma noted by Davies *et al.* (2005). Viewing additional data about the individual samples that compose the results of this query can be done with the links provided by COSMIC. For example, it is simple to note that the lone lung tissue with a mutation in the *FLT3* gene also has mutations in 21 addition genes (COSMIC sample ID PD1362a). This rich hierarchy of data makes it possible to build complex models of the range of common pathway disruptions seen in different types of cancer, which might in turn be useful for interrogating other data, such as microarray experiments carried out on similar tumour types.

Given the pathological and phenotypic diversity seen in many cancers, histological subtypes are often scrutinized for differences in mutation frequency. Isolating groups of cancers for comparison can be done with COSMIC. In the case of all lung cancers, it may be of particular interest that the most common recorded gene mutation is tumour protein 53 (*TP53*), of which 48 per cent of all cases screened had mutations of this gene. Lung cancer diagnoses are diverse, and subtypes of lung carcinomas can be shown to exhibit many different gene-mutation patterns. For example, squamous cell carcinoma of the lung exhibits much lower rates of *TP53* mutations (6/19, 31 per cent) than small cell carcinoma of the lung (36/57, 63 per cent). Conversely, the *CDKN2A* gene, a negative regulator of the proliferation of normal cells, appears as one of the most commonly mutated genes in lung squamous cell carcinomas (4/12, 33 per cent), while no mutations of this gene have been detected in small cell carcinomas of the lung (0/57, 0 per cent). A more comprehensive profile of the common mutations seen in these cancers may provide a reasonable tumour classifier for these histologies.

While COSMIC is currently a standard for cataloguing and querying somatic mutations in cancers, its limitations must be considered for effective interpretation of its data. First, only a subset of genes were screened for mutations. Although *TP53* exhibits the overall highest mutation frequencies in all of lung cancers, most genes in the genome (e.g., those listed as RefSeq genes; http://www.ncbi.nih.gov/RefSeq/) were not screened. Secondly, not all genes considered were screened equally for each cancer type or histological group. For example, the Kirsten-ras oncogene homologue from the mammalian ras gene family, *KRAS*, has a well-documented mutation frequency in squamous cell carcinoma of the lung (67/1140, 5 per cent) due to extensive screening, while the apparently often mutated *CDKN2A* is based upon far less intensive screening in that histology (4/12, 33 per cent). Thirdly, mutation screening by the CGP has generally been limited to coding sequence; hence, there is little or no information on regulatory mutations in COSMIC, even though these are likely to play a significant role in cancer. Finally, not all genes are screened with a similar resolution for mutations. Large genes (e.g., *TP53*) are often screened for those regions that are most commonly mutated, referred to as 'mutation hotspots'. Designating a gene in a particular tumour as wild type (non-mutant) can be achieved only through a full gene sequence, and an adequate reference sequence. However, most mutation data in COSMIC are derived from disparate sources where no standards are enforced. Therefore, COSMIC is meant as a comprehensive consolidation of the current literature rather than a definitive source of mutation frequencies.

Various other oncology-specific databases and tools are devoted to serving gene-mutation data in cancers (Table 17.1). Like COSMIC, The Tumor-Gene Database serves mutation data from published studies. The Human Gene Mutation Database is a comprehensive database focusing on serving germ-line mutations, many of which have been associated with cancer predisposition and development (Stenson *et al.*, 2003). Locus- or disease-specific resources may offer expanded information about certain types of mutations or related clinical information when compared to global sources such as COSMIC. The IARC *TP53* Mutation Database is one example of a much more focused mutation database (Olivier *et al.*, 2002). This database, consisting of *TP53* mutations drawn from published literature, can be downloaded in its entirety or queried with a Web interface. Extending the previous example, in which COSMIC catalogues 19 *TP53* mutations in squamous cell carcinomas of the lung, this database contains 737 sequence variants. This resource can effectively query known mutations to specific codons and compare rates of occurrence between tissues and diagnoses. Other examples of gene-specific databases include *CDKN2A* (https://biodesktop.uvm.edu/perl/p16) and the Androgen Receptor Gene Mutation Database (http://www.androgendb.mcgill.ca). There are also resources serving disease-specific mutation data, including ones for breast cancer (http://condor.bcm.tmc.edu/ermb/bcgd/bcgd.html) and oral cancer (http://www.tumor-gene.org/Oral/index.html). These databases are all valuable for querying a gene's suspected involvement in one or more types of cancer.

17.6 Copy number alterations in cancer

DNA copy number in the genome is widely regarded as an important aspect of the aetiology of a range of human diseases (Weber, 2002). Complete and partial non-diploid genomes resulting from cytogenetic alterations have been implicated in the diagnosis of congenital disorders (Milunsky and Huang, 2003) as well as predictors of clinical outcomes of many cancer types (Look *et al.*, 1991). Aneuploidy (the occurrence of one or more extra or missing chromosomes) is common in tumour genomes, where one or both copies of a gene can be lost, or genes can exhibit DNA copy number gains exceeding 100 copies. Investigating somatic changes in DNA copy number is particularly useful in the detection of tumour suppressor loci. Known tumour suppressor genes such as *CDKN2A* (Kamb *et al.*, 1994) and *PTEN* (Li *et al.*, 1997) have been mapped to recurring homozygous deletions in several cancers. Hemizygous losses (loss of a single copy) may also harbour tumour suppressor genes where the one remaining copy has lost function by a mutation. Further, several types of oncogenes have been associated with gains in DNA copy number. For example, the common amplification of the proto-oncogene *N-myc* is a associated with poor prognosis in neuroblastomas (Mosse *et al.*, 2005) while DNA copy number amplifications of the *CYP24A1* locus have helped define it as a candidate oncogene (Albertson *et al.*, 2000).

17.6.1 Array comparative genomic hybridization (aCGH) technologies

Attaining accurate genome copy number measurements is very central to the molecular characterization of many cancers. Many tumour types show a very low rate of point mutation, but instead show extensive copy number changes; for example, Bignell *et al.* (2006) screened 13 testicular tumours, in 351 members of the protein kinase gene family and found only one somatic point mutation, whereas all tumours studied showed multiple copy number changes. Microarray-based comparative genomic hybridization (aCGH) has increased our ability to detect important copy number alterations (gains and losses) in the tumour genome. Where traditional metaphase CGH can detect alterations of approximately 10 Mb or greater, the printing of mapped sequences on a microarray chip has increased this by up to 100-fold, and this will increase further as oligonucleotide chip based methods advance. Copy number detection in the microarray format allows precise readings from individual mapped sequences. As a result, DNA copy number measurements in tumours are more accurate and data rich. This has necessitated the use of bioinformatics for the processing, storage, and analysis of DNA copy number data.

At present, there are three general types of microarray-based copy number assays.

Array-based comparative genomic hybridization (aCGH)

Genome-mapped sequences representing DNA extracted from artificially grown clones (typically either bacterial artificial chromosomes or cDNA clones) are printed on a glass slide. Many earlier developed platforms are composed of clones spaced at 1-Mb intervals across the entire human genome (e.g., Snijders *et al.*, 2001; Greshock *et al.*, 2004), while, more recently, assays with full genome coverage have been constructed (Ishkanian *et al.*, 2004). All are two-channel assays in which a tumour and normal diploid DNA are labelled with separate fluorescent dyes, typically Fluorolink Cy3 and Cy5 (GE Healthcare, Little Chalfont, UK), and subjected to a competitive hybridization in which the result is a tumour/normal hybridization ratio for each mapped probe.

SNP chips

The SNP chip, produced by Affymetrix (Sunnyvale, CA, USA) and Illumina (San Diego, CA, USA), is a microarray-based, high-throughput genotyping tool that can simultaneously measure up to 500 000 known SNPs (the use of these arrays in germline genetics is covered extensively in Chapter 19). Various computational methods have extended the utility of this single-channel assay to measuring genome copy number (Bignell *et al.*, 2004). Currently, an assay containing 25-mer oligonucleotide sequences (fabricated DNA sequence) representing ~120 000 known SNPs provides a copy number detection resolution of approximately 25 kb. While the mean probe spacing is vastly improved over the BAC clone aCGH, typically groups of SNPs are analysed together, thereby reducing resolution while improving specificity.

Oligonucleotide arrays

Oligonucleotide copy number detection assays are hybrids of the SNP chip and traditional aCGH assays. All are similar to traditional aCGH in that they employ competitive hybridization, while, as in the SNP chip, fabricated oligonucleotide sequences (typically under 100 bp) are printed on the microarray. These assays are produced both commercially (e.g., NimbleGen Systems, Madison, WI, USA; Agilent Technologies, Palo Alto, CA, USA) and privately (Lucito *et al.*, 2003). Resolution varies greatly, though the probe spacing tends to be intermediate to traditional aCGH and SNP chips.

There are several notable limitations of copy number measurement assays. First, they are unable to detect translocation and loci subject to loss of heterozygosity coupled with segmental duplication. Secondly, absolute copy number measurements can be problematic where complex, chromosomal duplications result in high variations in genome-wide aneuploidy (Davidson *et al.*, 2000).

Using bioinformatics approaches for copy number data analyses has successfully contributed to the fine-scale molecular characterization of cancers. This has included associating known cancer subtypes with specific copy number alterations (e.g., Hermsen *et al.*, 2005; Jones *et al.*, 2005) as well as discovering copy number profiles associated with clinical outcome and survival (e.g., Weiss *et al.*, 2004; Rubio-Moscardo *et al.*, 2005). More recently, pharmacogenomics approaches have shown biomarkers associated with drug response (Kokubo *et al.*, 2005; Xia *et al.*, 2005). All class differentiation approaches have employed computational techniques accounting for both the data scale and the vast number of measurements.

17.6.2 Public genome copy number data resources

The body of literature using aCGH has grown tremendously in recent years, as has the number of publicly available data. While presently no devoted resource exists to serve all published aCGH data, repositories such as NCBI's Gene Expression Omnibus (GEO) (Barrett *et al.*, 2005) and the Stanford Microarray Database (Gollub *et al.*, 2003) have a growing number of aCGH data sets. Several complementary Web resources are devoted to collating information about the DNA copy number alterations associated with cancer. The Mitelman Database (Table 17.1) is a resource manually culled from published literature that provides data about DNA copy number alterations from individual cases and primary associations (Mitelman *et al.*, 2005). Specifically, the Mitelman Database can relate chromosomal aberrations to tumour characteristics, structural changes in genomic sequence data, and clinical information. Mitelman maps are visible with NCBI's MapViewer. Another public resource is NCBI's SKY/M-FISH and CGH Database, which provides a public platform for investigators to submit and compare molecular cytogenetics data. Unlike the Mitelman database, this resource is not as carefully curated, though primary data sources are easily accessible. Both of these resources provide Web-based query tools capable of finding recurring aberrations across data submitted from multiple independent studies as well as associating these copy number alterations with a particular cancer diagnosis. While not geared to large-scale, bioinformatics-style data mining, this resource is invaluable for comparing raw data to existing knowledge and validating novel findings.

Another useful source of cancer genome copy number data is the Wellcome Trust Sanger Institute's Cancer Genome Project (CGP) (see above). The panel of cancer cell lines and primary tumours being analysed for somatic mutations by the CGP is also being interrogated for genome copy number alterations by the Affymetrix SNP chip (Bignell *et al.*, 2004). Raw data for over 350 tissues can be downloaded directly or queried via a Web resource capable of basic visualization and analysis.

Although the existence of recurring inherited germ-line CNPs has been well described (Sebat *et al.*, 2004), there is little evidence that familial cancers can be associated with inherited copy number alterations. For example, Kiemeney *et al.* (2006) demonstrated a lack of correlation between familial bladder cancers and germ-line CNPs. Alternatively, hereditary papillary renal carcinomas showed a tendency for similar patterns of somatic DNA copy number gains among related individuals (Prat *et al.*, 2006). Further, constitutional copy number changes in diseased individuals may occur at sites of more common somatic alterations that are thought to be of functional significance. Examples include a duplication of the *TOP3B and TAFA5* genes noted in glioblastoma patients (De Stahl *et al.*, 2005), and a loss of 11q14-23 seen in an infant with neuroblastoma (Mosse *et al.*, 2003). Each of these studies indicates that copy number alterations that commonly characterize a 'cancer genome' are preceded by a germ-line polymorphism in a small subset of patients. More studies will be required to determine exactly how germ-line copy number alterations influence cancer risk and phenotype.

17.6.3 Genome copy number data analysis tools

The expanded resolution of aCGH platforms and their widespread use has necessitated the development of analytical methods and accompanying software for visualization. Most packages employ all or part of a generic, commonly used analysis protocol. First, data are subject to some type of normalization step, followed by a processing procedure that determines where copy number break points occur in that tissue. Finally, these data can be visualized or queried as members of a panel of cancers. Free and commercial software packages have been developed to accomplish each of the steps. Most commonly, normalization and copy number break-point analysis are accomplished by one set of tools, while visualization and analysis of multiple samples is done by separate tools. Examples of both are noted below.

Estimating the genomic locations of copy number break points in a particular tumour is a very important step in aCGH analysis. Several methods exist to accomplish this. Most simply, this can be done by applying a simple ratio threshold for every clone (e.g., > 1.5 is a copy number gain; < 0.5 is a copy number loss). Alternatively, a more sophisticated algorithm that considers each probe's flanking clones may provide more accurate results. Several published examples that apply existing computational methods to this problem include the use of hidden Markov models (Willenbrock and Fridlyand, 2005), adaptive weights smoothing (Hupe *et al.*, 2004), and circular binary segmentation (Olshen *et al.*, 2004). Implementations of each of these are available through the R programming language's bioconductor initiative (Gentleman *et al.*, 2004).

Several freely available software packages are designed specifically for end-user analysis and visualization. These specialize in single-experiment visualization, such as SeeCGH (Chi *et al.*, 2004); data abstraction from a single platform, as with Caryoscope (Awad *et al.*, 2004); and multiple experiment analysis, including CGHPRO (Chen *et al.*, 2005) and CGHAnalyzer (Margolin *et al.*, 2005). The use of one over another depends entirely on the specific analytical objectives, as each was designed for a slightly different purpose. Most are able to load data in a generic spreadsheet-style format, where each row represents a single genome-mapped probe and its associated copy number estimate. As a result, they are largely compatible with most raw data posted in public repositories or output from the previously mentioned break-point detection packages.

Taking advantage of aCGH data in public repositories is generally simple. Like microarray-based, gene-expression data set postings in GEO, aCGH data sets are identified by unique accession numbers (normally from a single publication). Both sample data and platform information are also associated with accession IDs. In the case of aCGH data, it is particularly important to consider which platform the data were generated with, as this delineates the resolution of the data and dictates how accurately copy number break points can be defined. For example, the GEO data set from Curtin *et al.* (2005) (data accession GSE2631; platform accession GPL2024; described by Snijders *et al.*, 2001) represents aCGH data from 126 melanomas, each of which has data from 2462 uniquely mapped BAC clones. Prior to submission, these data were quality filtered by the degree of variation between replicate clones. Presubmission quality filtration is not implemented for every GEO data set and may be difficult to apply after submission. Each sample can be exported from GEO in a basic tab-delimited text format where every row has a unique probe that is annotated in the platform information.

These data are easily loaded into the CGHAnalyzer software, a freely available, open-source software suite that is designed specifically for analysis of multiple experiment copy number data (Margolin *et al.*, 2005). Loading is completed by formatting each tumour into separate files and placing all probe annotations into a single mapping file. CGHAnalyzer can execute the fundamental analyses normally performed on a panel of cancers, including displaying detailed copy number profiles for multiple experiments, querying large data sets for minimal common regions of aberration, and integrating other genomic features with copy number data (e.g., known/predicted genes). CGHAnalyzer can implement ratio threshold-based copy number break-point detection. In doing so, one must first select reasonable ratio thresholds for determining which probes demonstrate copy number gains and losses in each experiment. Common thresholds for two-channel (e.g., Cy3/Cy5) CGH arrays, where the probes are BAC clones, are 1.25 for copy number gains and 0.75 for copy number losses. This indicates that every probe with a tumour/normal hybridization ratio over 1.25 represents a copy number increase, and those under 0.75 denote copy number losses (these translate to 0.32 and –0.41 in the commonly used $\log_2$ scale, respectively). Probe genome sequence coverage on copy number assays is normally limited. For example, the array platform described by Snijders *et al.* (2001) (GEO accession GPL2024) has direct sequence coverage of approximately 15 per cent of the genome. It is often desirable to estimate the copy number status of genes that do not have direct sequence coverage on the array. As such, CGHAnalyzer estimates regions between clones by extending the sequence represented by aberrant probes to the borders of neighbouring probes of differing copy number. More sophisticated approaches are employed by most copy number break-point algorithms (e.g., HMMs; see above). The result of either approach is an estimated genome-wide copy number profile. Caution should be used in interpreting output from any method. For example, subtle differences in estimated copy number can be seen between the CGHAnalyzer method and the circular binary segmentation (CBS) algorithm (Olshen *et al.*, 2004) for chromosome 7 in a single melanoma (NCBI Sample Accession ID GSM50526) from Curtin *et al.* (2005) (Figure 17.2). Clearly, both detect

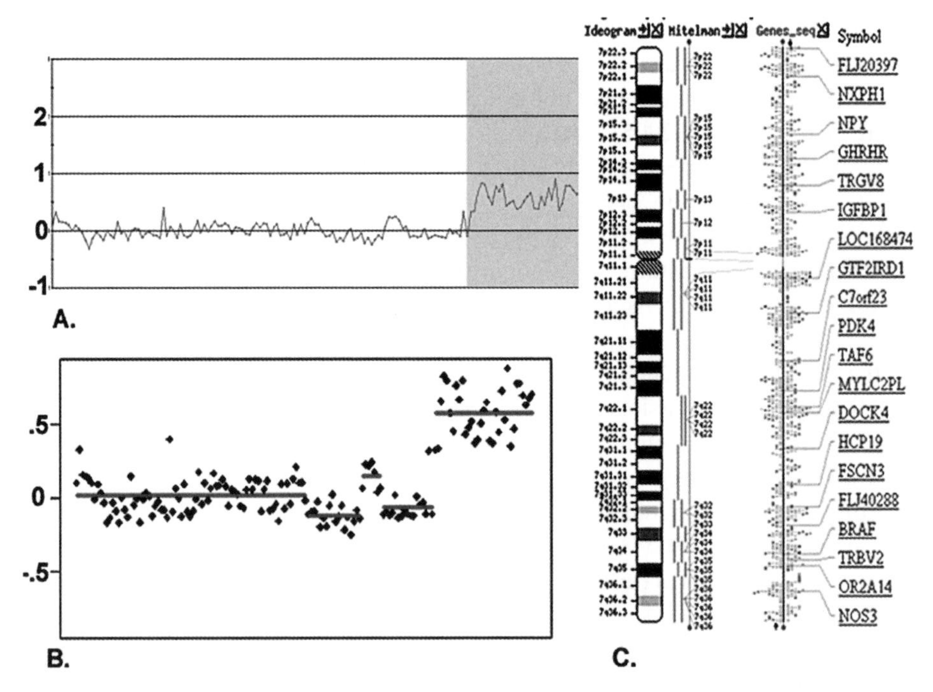

Figure 17.2 (A) A Scatterplot of data from a lung tumor highlights a q-arm deletion, using ration thresholds (highlighted region). (B) The same sample demonstrates a slightly different range for this amplification by the circular binary segmentation method. Transitions in the horizontal lines define the estimated copy number break points. (C) A NCBI MapViewer plot of the Mittleman database shows a known break point in this region

a copy number gain of the *BRAF* locus on 7q34, though the precise location of the copy number transition differs. The threshold method locates a gain extending from 121,395,000 bp-qter while the CBS estimates a smaller copy number gain, spanning 123,220,000-bp-qter (based on NCBI build 34). The contradictory results show that an approximately 2-Mb additional region of gain is predicted by the threshold method. The Mitelman Map describes a known break point that occurs closer to that seen in the CBS result, suggesting that the CBS may be more accurate in this case. Ultimately, the true status of this region in this melanoma can be most accurately determined by quantitative PCR of one or more of the 11 known genes in that region. Generally, the detection of small (< 5 Mb) changes is subject to the most variability between copy number break-point methods. Each mandates some trade-off between sensitivity to detect change and the specificity of those regions identified as aberrant.

The high-resolution analysis of somatic copy number changes in cancer has proven to be very fruitful and is now an essential component in the molecular profile of cancers. Publicly available data and accompanying software provide an expanding ability to incorporate copy number data into bioinformatics research.

17.7 Loss of heterozygosity in cancer

Loss of heterozygosity (LOH) is the most common molecular genetic alteration observed in human cancers. Although tumour suppressor gene loci are subject to LOH in many cancers, it is not clear exactly how early in tumorigenesis this takes place. The classic mechanism of tumour suppressor gene inactivation is the two-hit hypothesis, in which one allele is mutated and the other allele is lost, resulting in LOH at multiple loci. For example, mutational inactivation of tumour suppressor genes by LOH is a predominant mechanism in colorectal cancer (Fearon and Vogelstein, 1990). The mechanisms of LOH can be chromosome specific; some chromosomes display complete loss, and others show loss of only a part of the chromosome (Thiagalingam *et al.*, 2001). Regions shown to have heterozygous losses of DNA copy number necessarily exhibit LOH; however, the converse is not always true. Regional copy number losses can subsequently undergo segmental duplication, resulting in either diploidy or even allele-specific copy number gains (Herrick *et al.*, 2005). As a result, detection of LOH is widely used to identify genomic regions that harbour tumour suppressor genes and to characterize different tumour types, pathological stages and progression (Mei *et al.*, 2000; Hoque *et al.*, 2003).

17.7.1 Tools for analysing loss of heterozygosity data

Analysis of the data generated by the HapMap consortium has identified regions of extended linkage disequilibrium (LD) that manifest themselves as unusually long stretches of homozygosity in individual samples. Gibson *et al.* (2006) observed 1393 tracts of homozygosity exceeding 1 Mb in length among the 209 unrelated individuals used to construct the HapMap. The longest was an uninterrupted run of 3922 homozygous SNPs spanning 17.9 Mb in a Japanese individual. They found that these homozygous tracts were significantly more common in regions with high LD and low recombination, and the location of these tracts was similar across all populations. These data can help to distinguish between LOH that is potentially *de novo* and causal, and that which is simply commonly segregating in the population. LOH analysis methods that take into account HapMap patterns are currently under development (Mark Daly, personal communication).

As a growing number of LOH studies utilize microarray technologies such as the Affymetrix SNP chip, a host of software packages have been designed for their analysis. dChip can employ a reference file (e.g., matched non-cancerous tissue) or user-defined heterozygosity retention rates to calculate somatic LOH calls with a hidden Markov model (Lin *et al.*, 2004). This package also provides an interface for genome-wide LOH visualization. Alternatively, RLMM develops a robustly fitted linear model using a training sample set where genotypes are known (Rabbee and Speed, 2006). Others have developed protocols for identifying allele-specific amplification with SNP chip data. For example, LaFramboise *et al.* (2005) have developed PLASQ, a software package capable of detecting allele-specific copy number changes by first identifying regional LOH by an expectation-maximization approach and subsequently using these calls as input for a segmentation procedure that quantifies total copy number. Notably, both RLMM and PLASQ are freely available through the R-project's bioconductor initiative (Gentleman *et al.*, 2004).

17.8 Gene-expression data in cancer

Detailed quantifications of the cancer transcriptome by DNA microarrays have proven invaluable for the molecular characterization of cancers. This includes successfully associating patterns of gene expression with novel tumour subtypes, predicting drug response capacity, and characterizing numerous other tumour properties (reviewed in Chung *et al.*, 2002). Further, alterations in transcription can identify potential therapeutic targets in cancer. A range of public data resources, analysis tools and novel algorithms exemplifies the contribution that bioinformatics can make to transcriptome analysis. Many of these resources are reviewed in detail in Chapter 14; here we will review some of the specific issues that arise during cancer transcriptome analysis.

17.8.1 Public databases for gene-expression data in cancer

As most journals now require submission of gene-expression microarray data to a public repository, such as GEO (Edgar *et al.*, 2002) or ArrayExpress (Brazma *et al.*,

2003) prior to publication, the availability of cancer gene-expression data sets has improved considerably. However these databases provide only raw data and few analytical tools; therefore, several resources have been established for efficient collation and analysis cancer transcriptome data sets. These allow us to ask directed biological questions of gene-expression data sets. These can include complex problems, such as determining which gene families are most highly expressed in a given tissue, or which cancer types most commonly over-express a specific gene when compared to non-cancerous tissue. For ease of navigation, most resources employ well-described methods of normalization and standardized systems of ontology (Stoeckert *et al.*, 2002). Most provide data from microarray-based gene expression studies, though some serve serial analysis of gene expression (SAGE) data. Though not specifically designed for cancers, resources such as the Stanford Microarray Database and the Princeton University Microarray Database (described in Ball *et al.*, 2005) house a growing number of published cancer data sets and provide basic tools capable of recapitulating and extending their published analyses.

Several key features make Oncomine a leading public Web-based resource for querying and analysing published microarray-based gene-expression studies of cancer (Rhodes *et al.*, 2004). Powerful queries allow sample sets to be grouped by their origin of publication or diagnosis. Gene sets can be identified by ontologies, their association with published pathways, or relevance to cancer. Data queries can easily be subjected to a range of sophisticated analyses. For example, differentially expressed genes can be identified between tumour groups, or co-expressed gene groups can be queried within tumour sets. In addition, groups of over/under/co/expressed genes can be analysed for possible enrichment with published gene groups from such resources as Gene Ontology (Harris *et al.*, 2004) and KEGG (Kanehisa *et al.*, 2004). The volume of data in Oncomine is very high, presently containing data from 9900 microarray experiments representing 31 cancer types (Version 3.0, June 2006).

A sample query of Oncomine query yields 81 distinct studies of differential gene expression in breast cancers. The groups shown to have significant differentially expressed genes include those of known clinical parameters, such as oestrogen receptor (ER) status (Gruvberger *et al.*, 2001) and *BRCA1/BRCA2* gene mutation status (Hedenfalk *et al.*, 2001). For example, based on data from Gruvberger *et al.* (2001), a list of 114 genes (out of a total of 3369) appear upregulated in ER-positive breast tumours ($n = 28$) when compared to ER-negative tumours ($n = 30$). These analyses recapitulate the published analysis. The set of genes determined to be upregulated in ER-positive tumours by Oncomine is a near comprehensive superset of those presented as being upregulated in the analyses provided by Gruvberger *et al.* (2001). Cross-referencing these ER-positive upregulated genes with Oncomine's curated literature-derived gene sets shows that nine are cancer related genes and nine are documented therapeutic targets. Notably, the two genes, *CCND1*, a gene involved with cell-cycle progression, and *RET*, a gene that plays a crucial role in neural crest development, occur on both the cancer-related-gene and therapeutic target list. Individual genes can be queried across all data sets in Oncomine. A global Oncomine query of

Table 17.3 A reformatted sample output from an Oncomine query across all data sets for *CCND1* shows it to be upregulated in several cancers, including brain cancers, lymphomas and myelomas, represented as dark shaded cells. When compared to normal lung tissue, *CCND1*appears downregulated in lung cancers (represented by light shades). Cell numbers represent the number of primary data sources supporting the relative expression of *CCND1*

	Normal vs. normal	Cancer vs. normal		Cancer vs. cancer	Tumor stage	Molecular alteration		misc	
Endothelial	1								
Kidney	1								
Lung	1	2	4						
Prostate	1	2	2					2	
Vulvar	1								
Brain	2	1							
Lymphoma		1	1	4					
Myeloma		1							
Salivary gland		1							
Breast			1					2	1
Endocrine			1						
Renal			2	1					
Sarcoma			2	2	1			1	
Leukemia			1			2	1	1	
Normal									
Others									

the *CCND1* gene demonstrates that this is upregulated in other cancers, including brain cancers, lymphomas and myelomas (Table 17.3). These analyses demonstrate Oncomine's efficiency in regenerating published data analyses and extending these findings by putting them in the context of other studies of gene expression.

17.8.2 Gene-expression data analysis software

Tools

There is a wealth of open-source and commercially available stand-alone software tools for general microarray analyses. While each has certain strengths, several stand out as particularly amenable to cancer research. dChip is a freely available software package that can model higher-level gene expression and detect outliers (Li and Hung Wong, 2001). It is extremely flexible and can load raw data from a variety of formats, normalize expression levels, filter genes, and perform many applicable higher-level analyses. It offers robust methods for differentiating groups of samples by their gene-expression profiles. Its capability to map gene features to specific locations along chromosomes is often desirable in cancer analyses, as it allows easy comparison of gene-expression data with chromosomal alterations at the DNA level (e.g., genotyping, aCGH data). More recently, dChip has been extended to support the analysis of SNP chips (Affymetrix) (Lin *et al.*, 2004). ChARM (Myers *et al.*, 2005) is another freely available package that can analyse genes with respect to their chromosomal location. A noteworthy feature of ChARM is its implementation of an algorithm to detect positional biases in gene-expression levels. This feature is of particular interest in cancers, as relating recurring segmental aneuploidy or loss of heterozygosity to transcript levels is a basic means of identifying potential gene targets. Also of note is the Institute for Genomic Research's Multiple Experiment Viewer (Saeed *et al.*, 2003). This software contains implementations of most common algorithms used for microarray analysis, including clustering, class discrimination, and significance analysis. These packages offer several unique features that help answer common oncology-related questions of the transcriptome. However, there are many other freely available, proficient tools to analyse gene-expression data, and the selection of one over another depends entirely on the problems being investigated.

17.9 Multiplatform gene target identification

Computational approaches to cancer genomics are commonly directed at identifying and prioritizing a list of genes that are altered in a cancer. They may focus on the mechanisms of alteration and their resulting phenotype. While gene mutations, LOH, copy number alterations and gene-expression changes often serve as the primary information used to identify genes that participate in cancer genesis and progression, the effective identification of potential candidate genes requires the integration of two or more of these platforms as well as the incorporation of clinical data, pubic genome data, and annotations. One example of a bioinformatics-based, high-level, multiplatform investigation is determining which genes have somatic combinations of copy number losses and sequence mutations. Another example could be the identification of chromosomal locations (e.g., bands) that have an over-representation of highly expressed genes. A more involved example is determining where DNA copy number amplifications of genes for transcription factors influence the transcript abundance of their targets. Answering questions like these can be complicated. Although no single tool can solve these complex questions, a combination of published data and public software provides a reasonable approach to most problems. Which computational questions are addressed is dictated by the underlying biological hypotheses.

A panel of 126 primary melanomas presented by Curtin *et al.* (2005) provides a basic exercise in platform integration of publicly available bioinformatics resources. This set is composed of four distinct pathological subgroups, including sun-induced and non-sun-induced skin melanomas, mucosal melanomas and acral melanomas. Curtin *et al.* (2005) effectively demonstrate that the clinical heterogeneity seen between these groups is reflective of each having a distinct set of genetic alterations, including frequent occurrence of DNA copy number increases of the *CDK4* and

CCND1 loci in non-sun-induced melanomas. In other words, DNA copy number increases of both of these loci serve as biomarkers indicating melanomas not caused by solar damage. Both genes are downstream components of the *RAS-BRAF* pathway that is known to play a significant role in melanoma (Brose *et al.*, 2002). Notably, the *CDK4* gene, important for cell-cycle G_1 phase progression, is a member of the Ser/Thr protein kinase family that has been previously associated with tumorigenesis (Stephens *et al.*, 2005); it is on the CGP's cancer gene list and is represented in the COSMIC database.

By using aCGH data from Curtin *et al.* (2005) and the diagnostic tumour groupings provided as part of the supplemental information, these analyses could be further focused on identifying specific copy number alterations that may efficiently differentiate mucosal melanomas ($n = 20$) from all sun-induced melanomas ($n = 30$). As with other microarray data sets where multiple comparisons could lead to significant false discovery rates, true locus differentiators can be most effectively identified by a conservative step-down max-*T* adjustment (Dudoit *et al.*, 2003). This algorithm, as implemented in the CGHAnalyzer software (Margolin *et al.*, 2005, Saeed *et al.*, 2003), identified 10 genomic regions that were successful differentiators of mucosal and sun-induced melanomas ($P < 0.01$ based upon 1000 permutations). The results included regions encompassing 1q24.1–31.1, 8p12, and 10q24.32–25.1, all of which appear to be concordant with the broad analyses noted by Curtin *et al.* (2005). These regions harbour over 100 known or predicted genes. For the purposes of further refinement and gene target prioritizing, the genes occurring in these regions can be cross-referenced with those in the COSMIC database to determine whether any have been previously implicated in cancer. Three genes emerge as having copy number alterations that may be significant biomarkers capable of differentiating mucosal and sun-induced melanomas. First, the *FGFR1* gene, a member of the fibroblast growth factor receptor family, has more frequent copy number gains in mucosal melanomas than sun-induced melanomas. According to data in COSMIC, this gene has no recorded relevance to melanomas and only two noted missense substitutions (Bignell *et al.*, 2006). Conversely, *PIK3CA*, the catalytic subunit of *phosphatidylinositol* 3-kinase, has substantial mutation rates in several cancers and shows higher deletion rates in mucosal melanomas than in sun-induced melanomas. Most notably, this included somatic missense substitution rates of approximately 30 per cent in sporadic breast cancers (Bachman *et al.*, 2004; Lee *et al.*, 2005; Saal *et al.*, 2005) and hepatocellular carcinomas (Lee *et al.*, 2005). Other cancers, including ovarian cancers (Campbell *et al.*, 2004) and gastric cancers (Lee *et al.*, 2005), had mutation rates in this gene approaching 10 per cent. Despite this consistent deletion of the *PI3KCA* locus in mucosal melanomas, the analogous implications of its alterations seen in breast and hepatocellular carcinomas do not match. *PI3KCA* is a candidate oncogene, and alterations of this locus that would be most likely to affect a phenotype would be mutations and DNA amplifications. In fact, other aCGH screens of sporadic breast cancers demonstrated significant amplification rates at this locus, not deletions as seen in these melanomas (Pollack *et al.*, 2002; Naylor *et al.*, 2005).

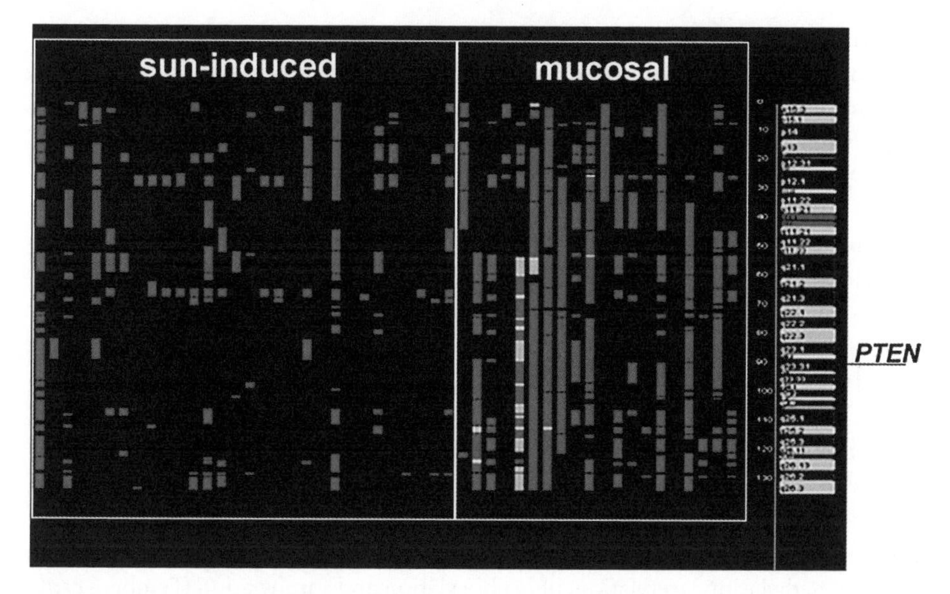

Figure 17.3 A CGHAnalyzer plot of the estimated deletions on chromosome 10 for 50 melanomas. This demonstrates that sun-induced melanomas appear to have lower deletion rates than mucosal melanomas. Deletion estimates for each sample appear as columns alongside the ideogram of chromosome 10

Therefore, the data supporting the deletion of *PI3KCA* as a potential biomarker for mucosal melanomas are not concordant with what is known about this locus.

The third cross-referenced gene, *PTEN*, is a known tumour suppressor and may provide the best candidate from these data. Queries of COSMIC yield a plethora of published *PTEN* sequence alterations in melanoma, including amino-acid deletions (Poetsch *et al.*, 2001) and missense substitutions (Guldberg *et al.*, 1997). Although many support the existence of alterations of *PTEN* in the melanoma genome, their occurrence appears variable, ranging from complete absence (Boni *et al.*, 1998) to sequence mutation rates of 20 per cent (Celebi *et al.*, 2000). The mucosal melanomas presented in Curtin *et al.* (2005) had estimated *PTEN* deletion rates of 55 per cent, including two possible homozygous deletions, while sun-induced melanomas had deletion rates of just 7 per cent (Figure 17.3). In the case of mucosal melanomas, loss of *PTEN* function through a homozygous deletion or hemizygous loss coupled with sequence-level alterations can promote tumour cell proliferation by regulating the *AKT/PKB* signalling pathway, which promotes cell survival. Curtin *et al.* (2005) support this assertion that *PTEN* alteration rates may vary between clinical subgroups of melanoma (ANOVA test, $P < 0.0001$). By focusing on the aCGH data, known mutation rates, and biological role, *PTEN* appears to be most plausible cancer-related candidate in differentiating mucosal and sun-induced melanomas. While Oncomine lacks data capable of distinguishing melanoma subtypes, it easily demonstrates *PTEN* as a tumour suppressor locus. *PTEN* downregulation is observed in prostate cancer and lymphoma, as well as in several cancer–cancer comparisons. Furthermore, studying patterns of co-expression of *PTEN* with other genes can help delineate those with similar tumour suppressor properties. For example, the significant downregulation of *PTEN* in prostate cancer and metastasis, when compared to normal prostate tissue (Dhanasekaran *et al.*, 2001), is mirrored by the expression patterns in 20 other genes. By integrating publicly available data from multiple platforms, we can make a convincing case that alterations of the tumour suppressor gene *PTEN* are important events associated with mucosal melanomas, but not sun-induced melanomas.

17.10 The epigenetics of cancer

Alterations in cancer cells are by nature genetic; however, one of the contributing causes of these alterations can also be epigenetic – heritable changes other than those in the DNA sequence. Epigenetics encompasses two major DNA or chromatin modifications: DNA methylation and post-translational modification of histones, including methylation, acetylation, phosphorylation and sumoylation (Feinberg *et al.*, 2006). Efforts to characterize the epigenetic alterations that may underlie cancer are progressing rapidly in the nascent field of epigenomics, which studies epigenetic alterations on a genome-wide scale (Rakyan *et al.*, 2004). Epigenetic changes in cancer can include global alterations, such as hypomethylation of DNA and hypoacetylation of chromatin, as well as gene-specific hypomethylation and hypermethylation. Global DNA hypomethylation can lead to chromosomal instability and increased tumour frequency, as in *in vitro* and *in vivo* mouse models (Eden *et al.*, 2003; Holm *et al.*, 2005), as well as gene-specific oncogene activation, such as *R-ras* in gastric cancer (Nishigaki *et al.*, 2005). In addition, the silencing of tumour suppressor genes is often associated with promoter DNA hypermethylation, as seen in the WRN gene in Werner syndrome (Agrelo *et al.*, 2006).

It is beyond the scope of this chapter to present a full review of the field of epigenomics; however, there are some tools that have direct relevance to the study of cancer genetics, most specifically the MVP viewer (http://www.epigenome.org), which presents the results of the Human Epigenome Project (HEP) (Rakyan *et al.*, 2004). For further coverage of epigenetics analysis methods, see Chapter 9.

17.11 Tumour modelling

Modelling tumour development and progression has become an increasingly common bioinformatics exercise. The general purpose of tumour models, which primarily work with known systems, is to determine whether applying experimental data to an empirical mathematical model can explain observed genotypic or clinical diversity. A true genetic model of cancer initiation and progression would be very complex, requiring almost universal knowledge of causal events and their genotypic and phenotypic effects. Although this is incredibly daunting, reasonable work has focused on modelling particular aspects of a tumour. This includes modelling the occurrence of genetic changes such as mutations (Natarajan *et al.*, 2003) and copy number alterations (Frigyesi *et al.*, 2003). Rudimentary tumour progression models have been developed by applying the principles of mutation rates and mutant phenotypes. One such model applies known mechanisms of tumorigenesis, cellular growth and apoptosis in colon cancer to their capacity to predict the emergence of copy number instability (Michor *et al.*, 2004). By principles of clonal evolution and known neuroblastoma cytogenetics, Bilke *et al.* (2005) infer a tumour progression model, using aCGH data to explain part of the heterogeneous clinical behaviour of neuroblastomas. Colon cancer and neuroblastoma provide ideal opportunities to fit tumour models, because genetically defined neoplasia, progression mechanisms and clinical subgroups exist for these cancers. Modelling tumours can be considerably more challenging for cancers where causal events and genetic subtypes are not well known. The growing number of studies describing tumours on a molecular level will facilitate more accurate models of cancer.

17.12 Conclusions

Bioinformatics now plays a central role in the study of cancer genetics. Databases, software tools, and statistical algorithms are now common components of any study of the cancer genome. Researchers can efficiently navigate large bodies of existing research and easily incorporate them into primary data. Public data resources facilitate the development of novel computational methods and make effective multiplatform studies of cancer genomics more feasible. Bioinformatics has been instrumental in effectively using both primary and secondary data in the accurate characterization of tumour genomes at many levels. As a result, its contributions to the field of cancer genetics have been outstanding, and its use has become convention. Bioinformatics research will continue to elucidate the genetic underpinnings of cancer development and progression. Ultimately, this will increase the accuracy of modelling the essential genetic alterations seen in cancers, leading to more effective, targeted therapies for this common genetic disease.

References

Agrelo, R., Cheng, W. H., Setien, F. *et al.* (2006). Epigenetic inactivation of the premature aging Werner syndrome gene in human cancer. *Proc Natl Acad Sci U S A* **103**(23), 8822–8827.

Albertson, D. G., Ylstra, B., Segraves, R. *et al.* (2000). Quantitative mapping of amplicon structure by array CGH identifies CYP24 as a candidate oncogene. *Nat Genet* **25**, 144–146.

Awad, I. A., Rees, C. A., Hernandez-Boussard, T. *et al.* (2004). Caryoscope: an open source Java application for viewing microarray data in a genomic context. *BMC Bioinformatics* **5**, 151.

Bachman, K. E., Argani, P., Samuels, Y. *et al.* (2004). The PIK3CA gene is mutated with high frequency in human breast cancers. *Cancer Biol Ther* **3**, 772–775.

Ball, C. A., Awad, I. A., Demeter, J. *et al.* (2005). The Stanford Microarray Database accommodates additional microarray platforms and data formats. *Nucleic Acids Res* **33**, D580–582.

Bamford, S., Dawson, E., Forbes, S. *et al.* (2004). The COSMIC (Catalogue of Somatic Mutations in Cancer) database and website. *Br J Cancer* **91**, 355–358.

Barrett, T., Suzek, T. O., Troup, D. B. *et al.* (2005). NCBI GEO: mining millions of expression profiles – database and tools. *Nucleic Acids Res* **33**, D562–566.

Bignell, G., Smith, R., Hunter, C. *et al.* (2006). Sequence analysis of the protein kinase gene family in human testicular germ-cell tumors of adolescents and adults. *Genes Chromosomes Cancer* **45**, 42–46.

Bignell, G. R., Huang, J., Greshock, J. *et al.* (2004). High-resolution analysis of DNA copy number using oligonucleotide microarrays. *Genome Res* **14**, 287–295.

Bilke, S., Chen, Q. R., Westerman, F. *et al.* (2005). Inferring a tumor progression model for neuroblastoma from genomic data. *J Clin Oncol* **23**, 7322–7331.

Boni, R., Vortmeter, A. O., Burg, G. *et al.* (1998). The PTEN tumour suppressor gene and malignant melanoma. *Melanoma Res* **8**, 300–302.

Brazma, A., Parkinson, H., Sarkans, U. *et al.* (2003). ArrayExpress – a public repository for microarray gene expression data at the EBI. *Nucleic Acids Res* **31**, 68–71.

Brose, M. S., Volpe, P., Feldman, M., *et al.* (2002). BRAF and RAS mutations in human lung cancer and melanoma. *Cancer Res* **62**, 6997–7000.

Campbell, I. G., Russell, S. E., Choong, D. Y. *et al.* (2004). Mutation of the PIK3CA gene in ovarian and breast cancer. *Cancer Res* **64**, 7678–7681.

Celebi, J. T., Shendrik, I., Silvers, D. N. *et al.* (2000). Identification of PTEN mutations in metastatic melanoma specimens. *J Med Genet* **37**, 653–657.

Chen, W., Erdogan, F., Ropers, H. H. *et al.* (2005). CGHPRO – a comprehensive data analysis tool for array CGH. *BMC Bioinformatics* **6**, 85.

Chi, B., Deleeuw, R. J., Coe, B. P. *et al.* (2004). SeeGH – a software tool for visualization of whole genome array comparative genomic hybridization data. *BMC Bioinformatics* **5**, 13.

Chung, C. H., Bernard, P. S. and Perou, C. M. (2002). Molecular portraits and the family tree of cancer. *Nat Genet* **32** (Suppl), 533–540.

Curtin, J. A., Fridlyand, J., Kageshita, T.*et al.* (2005). Distinct sets of genetic alterations in melanoma. *N Engl J Med* **353**, 2135–2147.

Davidson, J. M., Gorringe, K. L., Chin, S. F. *et al.* (2000). Molecular cytogenetic analysis of breast cancer cell lines. *Br J Cancer* **83**, 1309–1317.

Davies, H., Bignell, G. R., Cox, C. *et al.* (2002). Mutations of the BRAF gene in human cancer. *Nature* **417**, 949–954.

Davies, H., Hunter, C., Smith, R. *et al.* (2005). Somatic mutations of the protein kinase gene family in human lung cancer. *Cancer Res* **65**, 7591–7595.

De Angelis, P. M., Fjell, B., Kravik, K. L. *et al.* (2004). Molecular characterizations of derivatives of HCT116 colorectal cancer cells that are resistant to the chemotherapeutic agent 5-fluorouracil. *Int J Oncol* **24**, 1279–1288.

De Stahl, T. D., Hartmann, C., De Bustos, C. *et al.* (2005). Chromosome 22 tiling-path array-CGH analysis identifies germ-line- and tumor-specific aberrations in patients with glioblastoma multiforme. *Genes Chromosomes Cancer* **44**, 161–169.

Dhanasekaran, S. M., Barrette, T. R., Ghosh, D. *et al.* (2001). Delineation of prognostic biomarkers in prostate cancer. *Nature* **412**, 822–826.

Di Nicolantonio, F. and Bardelli, A. (2006). Kinase mutations in cancer: chinks in the enemy's armour? *Curr Opin Oncol* **18**, 69–76.

Dorkfeld, F., Bernheim, A., Dessen, P. *et al.* (1999). A database on cytogenetics in haematology and oncology. *Nucleic Acids Res* **27**, 353–354.

Dudoit, S., Shaffer, J. P. and Boldrick, J. C. (2003). Multiple hypothesis testing in microarray experiments. *Stat Sci* **18**, 71–103.

Eden, A., Gaudet, F., Waghmare, A. *et al.* (2003). Chromosomal instability and tumors promoted by DNA hypomethylation. *Science* **300**, 455.

Edgar, R., Domrachev, M. and Lash, A. E. (2002). Gene Expression Omnibus: NCBI gene expression and hybridization array data repository. *Nucleic Acids Res* **30**, 207–210.

Fearon, E. R. and Vogelstein, B. (1990). A genetic model for colorectal tumorigenesis. *Cell* **61**, 759–767.

Feinberg, A. P., Ohlsson, R. and Heniloff, S. (2006). The epigenetic progenitor origin of human cancer. *Nat Rev Genet* **7**(1), 21–33.

Fong, C. T., Dracopoli, N. C., White, P. S. *et al.* (1989). Loss of heterozygosity for the short arm of chromosome 1 in human neuroblastomas: correlation with N-myc amplification. *Proc Natl Acad Sci U S A* **86**, 3753–3757.

Frigyesi, A., Gisselsson, D., Mitelman, F. *et al.* (2003). Power law distribution of chromosome aberrations in cancer. *Cancer Res* **63**, 7094–7097.

Futreal, P. A., Coin, L., Marshall, M. *et al.* (2004). A census of human cancer genes. *Nat Rev Cancer* **4**, 177–183.

Garnis, C., Coe, B. P., Lam, S. L. *et al.* (2005). High-resolution array CGH increases heterogeneity tolerance in the analysis of clinical samples. *Genomics* **85**, 790–793.

Gentleman, R. C., Carey, V. J., Bates, D. M. *et al.* (2004). Bioconductor: open software development for computational biology and bioinformatics. *Genome Biol* **5**, R80.

Gibson, J., Morton, N. E. and Collins, A. (2006). Extended tracts of homozygosity in outbred human populations. *Hum Mol Genet* **15**, 789–795.

Goemans, B. F., Zwaan, C. M., Miller, M. *et al.* (2005). Mutations in KIT and RAS are frequent events in pediatric core-binding factor acute myeloid leukemia. *Leukemia* **19**, 1536–1542.

Gollub, J., Ball, C. A., Binkley, G. *et al.* (2003). The Stanford Microarray Database: data access and quality assessment tools. *Nucleic Acids Res* **31**, 94–96.

Greshock, J., Naylor, T. L., Margolin, A. *et al.* (2004). 1-Mb resolution array-based comparative genomic hybridization using a BAC clone set optimized for cancer gene analysis. *Genome Res* **14**, 179–187.

Gruvberger, S., Ringner, M., Chen, Y. *et al.* (2001). Estrogen receptor status in breast cancer is associated with remarkably distinct gene expression patterns. *Cancer Res* **61**, 5979–5984.

Guldberg, P., Thor Straten, P., Birck, A. *et al.* (1997). Disruption of the MMAC1/PTEN gene by deletion or mutation is a frequent event in malignant melanoma. *Cancer Res* **57**, 3660–3663.

Hanahan, D. and Weinberg, R. A. (2000). The hallmarks of cancer. *Cell* **100**, 57–70.

Harris, M. A., Clark, J., Ireland, A. *et al.* (2004). The Gene Ontology (GO) database and informatics resource. *Nucleic Acids Res* **32**, D258–261.

Hattinger, C. M., Reverter-Branchat, G., Remondini, D. *et al.* (2003). Genomic imbalances associated with methotrexate resistance in human osteosarcoma cell lines detected by comparative genomic hybridization-based techniques. *Eur J Cell Biol* **82**, 483–493.

Hedenfalkl, I., Duggan, D., Chen, Y. *et al.* (2001). Gene-expression profiles in hereditary breast cancer. *N Engl J Med* **344**, 539–548.

Hermsen, M., Snijders, A., Guervos, M. A. *et al.* (2005). Centromeric chromosomal translocations show tissue-specific differences between squamous cell carcinomas and adenocarcinomas. *Oncogene* **24**, 1571–1579.

Herrick, J., Conti, C., Teissier, S. *et al.* (2005). Genomic organization of amplified MYC genes suggests distinct mechanisms of amplification in tumorigenesis. *Cancer Res* **65**, 1174–1179.

Holm, T. M., Jackson-Grusby, L., Brambrink, T. *et al.* (2005). Global loss of imprinting leads to widespread tumorigenesis in adult mice. *Cancer Cell* **8**, 275–285.

Hoque, M. O., Lee, C. C., Cairns, P. *et al.* (2003). Genome-wide genetic characterization of bladder cancer: a comparison of high-density single-nucleotide polymorphism arrays and PCR-based microsatellite analysis. *Cancer Res* **63**, 2216–2222.

Hupe, P., Stransky, N., Thiery, J. P. *et al.* (2004). Analysis of array CGH data: from signal ratio to gain and loss of DNA regions. *Bioinformatics* **20**, 3413–3422.

Ishkanian, A. S., Malloff, C. A., Watson, S. K. *et al.* (2004). A tiling resolution DNA microarray with complete coverage of the human genome. *Nat Genet* **36**, 299–303.

Jin, Y., Zhang, H., Tsao, S. W. *et al.* (2004). Cytogenetic and molecular genetic characterization of immortalized human ovarian surface epithelial cell lines: consistent loss of chromosome 13 and amplification of chromosome 20. *Gynecol Oncol* **92**, 183–191.

Jones, A. M., Douglas, E. J., Halford, S. E. *et al.* (2005). Array-CGH analysis of microsatellite-stable, near-diploid bowel cancers and comparison with other types of colorectal carcinoma. *Oncogene* **24**, 118–129.

Kamb, A., Gruis, N. A., Weaver-Feldhaus, J. *et al.* (1994). A cell cycle regulator potentially involved in genesis of many tumor types. *Science* **264**, 436–440.

Kanehisa, M., Goto, S., Kawashima, S. *et al.* (2004). The KEGG resource for deciphering the genome. *Nucleic Acids Res* **32**, D277–280.

Kees, U. R., Rudduck, C., Ford, J. *et al.* (1992). Two malignant peripheral primitive neuroepithelial tumor cell lines established from consecutive samples of one patient: characterization and cytogenetic analysis. *Genes Chromosomes Cancer* **4**, 195–204.

Kiemeney, L. A., Kuiper, R. P., Pfundt, R. *et al.* (2006). No evidence for large-scale germline genomic aberrations in hereditary bladder cancer patients with high-resolution array-based comparative genomic hybridization. *Cancer Epidemiol Biomarkers Prev* **15**, 180–183.

Kokubo, Y., Gemma, A., Noro, R. *et al.* (2005). Reduction of PTEN protein and loss of epidermal growth factor receptor gene mutation in lung cancer with natural resistance to gefitinib (IRESSA). *Br J Cancer* **92**, 1711–1719.

Lafranboise, T., Weir, B. A., Zhao, X. *et al.* (2005). Allele-specific amplification in cancer revealed by SNP array analysis. *PLoS Comput Biol* **1**, e65.

Lee, H. and Maihle, N. J. (1998). Isolation and characterization of four alternate c-erbB3 transcripts expressed in ovarian carcinoma-derived cell lines and normal human tissues. *Oncogene* **16**, 3243–3252.

Lee, J. W., Soung, Y. H., Kim, S. Y. *et al.* (2005). PIK3CA gene is frequently mutated in breast carcinomas and hepatocellular carcinomas. *Oncogene* **24**, 1477–1480.

Li, C. and Hung Wong, W. (2001). Model-based analysis of oligonucleotide arrays: model validation, design issues and standard error application. *Genome Biol* **2**, RESEARCH0032.

Li, J., Yen, C., Liaw, D. *et al.* (1997). PTEN, a putative protein tyrosine phosphatase gene mutated in human brain, breast, and prostate cancer. *Science* **275**, 1943–1947.

Lin, M., Wei, L. J., Sellers, W. R. *et al.* (2004). dChipSNP: significance curve and clustering of SNP-array-based loss-of-heterozygosity data. *Bioinformatics* **20**, 1233–1240.

Look, A. T., Hayes, F. A., Shuster, J. J. *et al.* (1991). Clinical relevance of tumor cell ploidy and N-myc gene amplification in childhood neuroblastoma: a Pediatric Oncology Group study. *J Clin Oncol* **9**, 581–591.

Lucito, R., Healy, J., Alexander, J. *et al.* (2003). Representational oligonucleotide microarray analysis: a high-resolution method to detect genome copy number variation. *Genome Res* **13**, 2291–2305.

Mackay, A. R., Ballin, M., Pelina, M. D. *et al.* (1992). Effect of phorbol ester and cytokines on matrix metalloproteinase and tissue inhibitor of metalloproteinase expression in tumor and normal cell lines. *Invasion Metastasis* **12**, 168–184.

Margolin, A. A., Greshock, J., Naylor, T. L. *et al.* (2005). CGHAnalyzer: a stand-alone software package for cancer genome analysis using array-based DNA copy number data. *Bioinformatics* **21**, 3308–3311.

Mei, R., Galipeau, P. C., Prass, C. *et al.* (2000). Genome-wide detection of allelic imbalance using human SNPs and high-density DNA arrays. *Genome Res* **10**, 1126–1137.

Meisner, L. F., Wu, S. Q., Christien, B. J. *et al.* (1988). Cytogenetic instability with balanced chromosome changes in an SV40 transformed human uroepithelial cell line. *Cancer Res* **48**, 3215–3220.

Michor, F., Iwasa, Y., Rajagopalan, H. *et al.* (2004). Linear model of colon cancer initiation. *Cell Cycle* **3**, 358–362.

Milunsky, J. M. and Huang, X. L. (2003). Unmasking Kabuki syndrome: chromosome 8p22-8p23.1 duplication revealed by comparative genomic hybridization and BAC-FISH. *Clin Genet* **64**, 509–516.

Mitelman, F., Johansson, B. and Mertens, F. (2005). Mitelman Database of Chromosomal Aberrations in Cancer. http://cgap.nci.nih.gov/Chromosomes/Mitelman.

Mosse, Y., Greshock, J., King, A. *et al.* (2003). Identification and high-resolution mapping of a constitutional 11q deletion in an infant with multifocal neuroblastoma. *Lancet Oncol* **4**, 769–771.

Mosse, Y. P., Greshock, J., Margolin, A. *et al.* (2005). High-resolution detection and mapping of genomic DNA alterations in neuroblastoma. *Genes Chromosomes Cancer* **43**, 390–403.

Myers, C. L., Chen, X. and Troyanskaya, O. G. (2005). Visualization-based discovery and analysis of genomic aberrations in microarray data. *BMC Bioinformatics* **6**, 146.

Natarajan, L., Berry, C. C. and Gasche, C. (2003). Estimation of spontaneous mutation rates. *Biometrics* **59**, 555–561.

Naylor, T. L., Greshock, J., Wang, Y. *et al.* (2005). High resolution genomic analysis of sporadic breast cancer using array-based comparative genomic hybridization. *Breast Cancer Res* **7**(6), R1186–1198.

Nishigaki, M., Aoyagi, K., Danjoh, I. *et al.* (2005). Discovery of aberrant expression of R-RAS by cancer-linked DNA hypomethylation in gastric cancer using microarrays. *Cancer Res* **65**, 2115–2124.

O'Toole, S. A., Sheppard, B. L., Sheils, O. *et al.* (2005). Analysis of DNA in endometrial cancer cells treated with phyto-estrogenic compounds using comparative genomic hybridisation microarrays. *Planta Med* **71**, 435–439.

Olivier, M., Eeles, R., Hollstein, M. *et al.* (2002). The IARC TP53 database: new online mutation analysis and recommendations to users. *Hum Mutat* **19**, 607–614.

Olshen, A. B., Venkatraman, E. S., Lucito, R. *et al.* (2004). Circular binary segmentation for the analysis of array-based DNA copy number data. *Biostatistics* **5**, 557–572.

Packer, B. R., Yeager, M., Staats, B. *et al.* (2004). SNP500Cancer: a public resource for sequence validation and assay development for genetic variation in candidate genes. *Nucleic Acids Res* **32**, D528–532.

Pavey, S., Johansson, P., Packer, L. *et al.* (2004). Microarray expression profiling in melanoma reveals a BRAF mutation signature. *Oncogene* **23**, 4060–4067.

Poetsch, M., Dittberner, T. and Woenckhaus, C. (2001). PTEN/MMAC1 in malignant melanoma and its importance for tumor progression. *Cancer Genet Cytogenet* **125**, 21–26.

Pollack, J. R., Sorlie, T., Perou, C. M. *et al.* (2002). Microarray analysis reveals a major direct role of DNA copy number alteration in the transcriptional program of human breast tumors. *Proc Natl Acad Sci U S A* **99**, 12963–12968.

Prat, E., Bernues, M., Del Rey, J. *et al.* (2006). Common pattern of unusual chromosome abnormalities in hereditary papillary renal carcinoma. *Cancer Genet Cytogenet* **164**, 142–147.

Rabbee, N. and Speed, T. P. (2006). A genotype calling algorithm for Affymetrix SNP arrays. *Bioinformatics* **22**, 7–12.

Rakyan, V. K., Hildmann, T., Novik, K. L. *et al.* (2004). DNA methylation profiling of the human major histocompatibility complex: a pilot study for the Human Epigenome Project. *PLoS Biol* **2**(12), e405.

Ratsch, S. B., Gao, Q., Srinivasan, S. *et al.* (2001). Multiple genetic changes are required for efficient immortalization of different subtypes of normal human mammary epithelial cells. *Radiat Res* **155**, 143–150.

Rhodes, D. R., Yu, J., Shanker, K. *et al.* (2004). ONCOMINE: a cancer microarray database and integrated data-mining platform. *Neoplasia* **6**, 1–6.

Rubio-Moscardo, F., Climent, J., Siebert, R. *et al.* (2005). Mantle-cell lymphoma genotypes identified with CGH to BAC microarrays define a leukemic subgroup of disease and predict patient outcome. *Blood* **105**, 4445–4454.

Saal, L. H., Holm, K., Maurer, M. *et al.* (2005). PIK3CA mutations correlate with hormone receptors, node metastasis, and ERBB2, and are mutually exclusive with PTEN loss in human breast carcinoma. *Cancer Res* **65**, 2554–2559.

Saeed, A. I., Sharov, V., White, J. *et al.* (2003). TM4: a free, open-source system for microarray data management and analysis. *Biotechniques* **34**, 374–378.

Sanchez-Martin, M., Rodriguez-Garcia, A., Perez-Losada, J. *et al.* (2002). SLUG (SNAI2) deletions in patients with Waardenburg disease. *Hum Mol Genet* **11**, 3231–3236.

Sebat, J., Lakshmi, B., Troge, J. *et al.* (2004). Large-scale copy number polymorphism in the human genome. *Science* **305**, 525–528.

Singh, B., Kim, S. H., Carew, J. F. *et al.* (2000). Genome-wide screening for radiation response factors in head and neck cancer. *Laryngoscope* **110**, 1251–1256.

Snijders, A. M., Nowak, N., Segraves, R. *et al.* (2001). Assembly of microarrays for genome-wide measurement of DNA copy number. *Nat Genet* **29**, 263–264.

Stenson, P. D., Ball, E. V., Mort, M. *et al.* (2003). Human Gene Mutation Database (HGMD): 2003 update. *Hum Mutat* **21**, 577–581.

Stephens, P., Edkins, S., Davies, H. *et al.* (2005). A screen of the complete protein kinase gene family identifies diverse patterns of somatic mutations in human breast cancer. *Nat Genet* **37**, 590–592.

Stoeckert, C. J., Jr., Causton, H. C. and Ball, C. A. (2002). Microarray databases: standards and ontologies. *Nat Genet* **32** (Suppl), 469–473.

Thiagalingam, S., Laken, S., Willson, J. K. *et al.* (2001). Mechanisms underlying losses of heterozygosity in human colorectal cancers. *Proc Natl Acad Sci U S A* **98**, 2698–2702.

Thomas, M. A., Weston, B., Joseph, M. *et al.* (2003). Evolutionary dynamics of oncogenes and tumor suppressor genes: higher intensities of purifying selection than other genes. *Mol Biol Evol* **20**, 964–968.

Tokumo, M., Toyooka, S., Kiura, K. *et al.* (2005). The relationship between epidermal growth factor receptor mutations and clinicopathologic features in non-small cell lung cancers. *Clin Cancer Res* **11**, 1167–1173.

Weber, B. L. (2002). Cancer genomics. *Cancer Cell* **1**, 37–47.

Weber, B. L., Abel, K. J., Brody, L. C. *et al.* (1994). Familial breast cancer. Approaching the isolation of a susceptibility gene. *Cancer* **74**, 1013–1020.

Weiss, M. M., Kuipers, E. J., Postma, C. *et al.* (2004). Genomic alterations in primary gastric adenocarcinomas correlate with clinicopathological characteristics and survival. *Cell Oncol* **26**, 307–317.

Willenbrock, H. and Fridlyand, J. (2005). A comparison study: applying segmentation to array CGH data for downstream analyses. *Bioinformatics* **21**, 4084–4091.

Xia, W., Gerald, C. M., Liu, L. *et al.* (2005). Combining lapatinib (GW572016), a small molecule inhibitor of ErbB1 and ErbB2 tyrosine kinases, with therapeutic anti-ErbB2 antibodies enhances apoptosis of ErbB2-overexpressing breast cancer cells. *Oncogene* **24**, 6213–6221.

18

Needle in a Haystack? Dealing with 500 000 SNP Genome Scans

Michael R. Barnes and Paul S. Derwent

Bioinformatics, GlaxoSmithKline Pharmaceuticals, Harlow, Essex, UK

18.1 Introduction

As the recent initiation of a number of publicly funded genome-wide association projects testifies, the era of ultra-high-density genome scans for the genetic determinants of complex diseases appears to have arrived. A number of major studies are now under way and many more are in planning, involving the use of maps containing hundreds of thousands of SNPs to perform association scans using linkage disequilibrium (LD) to detect risk-associated variants in large, population-based sample collections (Thomas *et al.*, 2005; http://www.wtccc.org.uk/; http://www.ncbi.nlm.nih.gov/WGA/). These studies are stepping into virgin territory, because, no comprehensive, well-powered genome-wide association studies have yet been published. Considering the investments involved, the stakes are high – at the time of writing, we really do not know how successful this approach will be. Advocates of the genome-wide association approach argue that these studies will identify many variants that contribute to common disease, but others disagree. However, the sheer volume of the data outputs from these scans raise significant issues of analysis and interpretation, due to the large number of hypotheses tested, the comparatively small sample sizes that are generally employed, and the finite number of true gene effects that are likely to be detectable. This high level of risk coupled

Bioinformatics for Geneticists, Second Edition. Edited by Michael R. Barnes
 ISBN 978-0-470-02619-9 (HB) ISBN 978-0-470-02620-5 (PB)

with some over-optimistic projection about the potential success of the genome scan approach, have led some researchers to question the value and veracity of this admittedly expensive approach to genetic analysis (Weiss and Terwilliger, 2000).

What is certain is that designing, performing and successfully analysing the results of these studies presents some formidable challenges, including the identification of marker maps that capture a maximal amount of information with a minimal amount of redundancy; assaying these markers with a cost-effective, high-throughput genotyping technology; ascertaining a large number of well-phenotyped individuals; and, finally, developing computational and statistical methods to cope with these high volumes of data. Even if we manage to deal with these challenges, it is important to understand the fundamental limitations of the genome-wide association approach and some of the technical and analytical issues that might undermine the success that this approach may well offer.

18.1.1 Why use genome-wide association scanning to identify complex disease genes?

The primary objective of human complex disease genetics is to determine the molecular basis of common diseases with major clinical impact, such as diabetes, asthma, cardiovascular diseases and psychiatric diseases. These are likely to be based on a complex array of interactions among multiple genetic and environmental factors. Most studies to date suggest that individual genetic variants are likely to have a relatively modest effect on lifetime risk of susceptibility to complex diseases. This has been attributed to modifying factors affecting the penetrance and the relative risk of the allele. As relatively few complex disease-susceptibility genes have been identified to date, it is still unclear what the typical frequencies of variants underlying these traits are likely to be. There is some evidence that common variants (minor allele frequency (MAF) of >5 per cent) may play a major role in complex disease susceptibility (Reich and Lander, 2001). If so, then association analysis may be the method of choice for detection of these alleles, as linkage analysis is not well powered for detecting common alleles that have low penetrance. This was demonstrated clearly by the failure of multiple linkage studies to detect the peroxisome proliferator-activated receptor gamma (PPARγ) Pro12Ala variant, a moderate risk allele for type 2 diabetes, which was subsequently identified by association analysis (Altshuler *et al.*, 2000).

At the most basic level, association analysis compares the frequency of alleles of a variant between cases and controls to see whether a particular allele or haplotype is seen more often in cases than controls, suggesting an association between a particular allele and a disease phenotype. Most association studies have focused on candidate genes. Despite a somewhat chequered success rate, this approach has been modestly fruitful, producing most of the complex disease-susceptibility genes that have been identified. The primary drawback of candidate gene studies is that they rely on the appropriate selection of candidate genes by biological rationale. This approach has a few drawbacks. Firstly, the biology of some diseases, such as psychiatric diseases, is poorly understood, so selection of candidate genes on the basis of biology can be difficult. Secondly, the candidate approach hinders the identification of novel genes and pathways in disease. This is perhaps one of the most important roles of genetics in biology and may help to explain the enthusiasm that the genome-scan approach is generating.

18.2 Genome scan analysis issues

The key analytical considerations of genome-wide scans, such as power, type 1 error, sample size, population stratification and genotyping error, exert a vice-like grip on the design of these studies. Addressing all of these considerations is likely to be a tall order, but it is important to be aware of the issues so that resources are directed to maximal effect. Studies which consistently fail to address all of these issues are likely to meet with little else than failure. Addressing all of them in full would require either almost unlimited funds or the many minds and pooled resources of a well-organized research consortium. Whichever of these approaches is employed, the chances of success can only be improved further, or perhaps even accelerated, by bioinformatics analysis. Below, we will review the major issues of genome scan analysis, and where appropriate we will illustrate where bioinformatics approaches can help.

18.2.1 Common disease, common variants and the HapMap

The availability of the HapMap (reviewed in detail in Chapter 3) is the main reason for the recent acceleration in plans for genome-wide association studies. HapMap is an incredibly rich data source; however, there are some caveats which might hinder its success as the basis of genome-wide association studies. The HapMap approach is not free from controversy, and some researchers are highly sceptical about the fundamental assumptions underpinning the project (Terwilliger and Hiekkalinna, 2006). These arguments stem from a struggle between two hypotheses: the common disease, common variant (CD/CV) hypothesis (Reich and Lander, 2001) and the 'multiple rare variant' hypothesis (Pritchard, 2001). If the former holds true, markers selected by HapMap LD relationships will effectively tag disease alleles, while if the latter holds true, the LD relationships between disease variants and population variants may be very different, suggesting that markers in high LD with a causal variant might not actually show association with a disease as expected, even in infinite sample sizes (Wright and Hastie, 2001; Terwilliger and Hiekkalinna, 2006). These arguments, though complex, are quite reasonable. If we take a pragmatic view of both arguments, both are likely to be right some of the time, depending on study design and other factors (Pe'er *et al.*, 2006a). For the optimist, however, the HapMap approach is persuasive, and it is in use the world over (Kruglyak, 2005).

The HapMap SNP ascertainment strategy is another issue that has generated some debate. Phases I and II HapMap SNPs were prioritized for selection primarily on the basis of prior validation; failing this, they were also considered validated if they matched a variant in chimpanzee sequence data (Altshuler *et al.*, 2005). This means that the phase I, and to a lesser extent the phase II, data sets show some significant ascertainment bias toward ancestral (generally common) alleles (Clark *et al.*, 2005). The impact of this is complex and dependent on the specific analysis undertaken, but essentially it means that power to detect rare variants (MAF of <5 per cent) may be reduced, while power to detect common variants is increased (Pe'er *et al.*, 2006a).

18.2.2 On the likely nature of complex disease-susceptibility alleles

Aside from the CD/CV debate, there is also some debate about the probable molecular basis of complex disease-susceptibility alleles. If the wealth of mutations identified in Mendelian disorders is a guide, missense mutations may play a key role. Botstein and Risch (2003) proposed a concerted strategy to genotype all common missense variants. This led several companies to develop genotyping panels capturing a large proportion of known non-synonymous SNPs. Affymetrix (Santa Clara, CA, USA) attempted validation of ~60 000 missense variants from the dbSNP database (Ireland *et al.*, 2006). Only ~20K missense SNPs could be validated in the four HapMap populations; these were developed into a commercial oligonucleotide array (Human 20K cSNP kit). Although this approach of genotyping marker sets enriched for functional variants might well succeed in some instances, there is a strong argument that the mileage of this approach may be limited to simple disease mechanisms. The argument stems from the fact that monogenic disease alleles are highly penetrant, often severely affecting protein function; such mutations are likely to be subject to negative (purifying) selection in populations. By contrast, the alleles that underlie complex traits, by their very nature of low penetrance and modest effect, are likely to be subtle in impact. Intuitively, such polymorphisms are likely to include non-coding regulatory variants with a modest impact on expression and perhaps conservative amino-acid substitutions. This makes sense on a population scale, perhaps explaining the widespread nature of complex diseases, where alleles with modest impact or late onset are far less likely to be subject to strong negative selection, having little impact on reproductive fitness, in acute contrast to many Mendelian disorders.

18.2.3 On sample size and thresholds of significance

If variants that influence complex traits are likely to be modest in effect, the mandate to increase sample sizes is very strong. Sample sizes are important for minimizing both type 1 error (false-positive association) and type 2 error (false-negative association).

Although ultra-high-density genotyping potentially offers a level of genome-wide coverage that could detect most common alleles, sufficient sample sizes are required to detect disease alleles of modest effect or low frequency. Without adequate sample sizes, type 2 error would be an important problem. The imperative for large sample sizes is increased further in high-density association scans by the inherent increase in number of hypotheses tested and the type 1 error that this generates. Type 1 error is probably the most intractable problem that the high-density genome-wide association approach creates. Elements of study design are crucial in the correction of analysis results for type 1 error by methods such as Bonferroni correction (Balding *et al.*, 2003), false discovery rate (Benjamini and Hochberg, 1995) or permutation testing (Churchill and Doerge, 1994). Risch and Merikangas (1996) anticipated the problems of multiple testing at a very early stage, proposing that a P value of 5×10^{-8} is a conservative threshold for declaring a significant association in a genome-wide study when one million independent tests are undertaken.

The consequences of applying such stringent thresholds to association studies are stark. Wang *et al.* (2005) illustrated the influence of allele frequency and odds ratio of a disease allele on sample-size requirements in a quite sobering fashion (Figure 18.1).

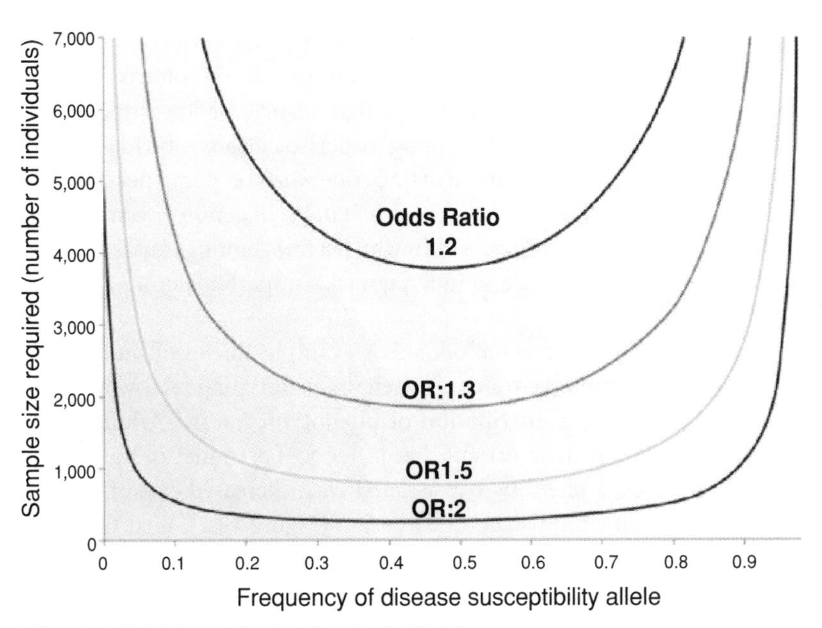

Figure 18.1 Effects of allele frequency on sample-size requirements. The numbers of cases and controls that are required in an association study to detect disease variants with allelic odds ratios of 1.2, 1.3, 1.5 and 2 are shown. Numbers shown are for a statistical power of 80 per cent at a significance level of $P < 10^{-6}$, assuming a multiplicative model for the effects of alleles and perfect correlative LD between alleles of test markers and disease variants. Reprinted by permission from Macmillan Publishers Ltd: *Nature Reviews Genetics* 6(2), 109–118, copyright 2005

In Figure 18.1, the lines represent the number of cases required to detect a specific allele with 80 per cent power and a value of $P < 10^{-6}$, assuming that the causal allele or a marker in perfect LD with the causal allele is tested. Many of the polymorphisms that have been consistently associated with complex diseases have relatively modest effects. For example, the PPARG (Pro12Ala) type 2 diabetes-associated allele, shows a Caucasian population frequency of 0.12 and an odds ratio of <1.3. Based on the projection in Figure 18.1, 4000–5000 cases would be needed to detect this association with $P < 10^{-6}$. This actually fits well with the empirical observation. Altshuler *et al.* (2000) evaluated a large number of conflicting published genetic associations between Pro12Ala and type 2 diabetes, and they confirmed the Pro12Ala association with $P = 0.002$ after analysing over 3000 individuals.

Collection and genotyping of thousands of samples can of course be prohibitively costly, but, arguably, reducing sample sizes to reduce cost is a false economy if the experiment fails or shows only limited success. Hirschhorn and Daly (2005) evaluated some alternatives, none of which is very encouraging. Setting a more liberal P value threshold of 0.05 would still require 1200 cases and 1200 controls to achieve 80 per cent power to detect a variant like the PPARG Pro12Ala allele. Lowering sample sizes below this threshold would run the risk of missing variants of moderate frequency and risk. In pragmatic terms, smaller sample sizes (e.g., 500 cases and 500 controls) are often employed in association studies, due to the practical and financial constraints of collecting large, well-characterized study cohorts. This inevitably forces investigators to relax thresholds of significance, or lower the stringency of Bonferroni correction to avoid discarding 'real associations' due to a lack of power. Unfortunately, at a P value threshold of 0.05, one would expect 5 per cent of all genotyped SNPs to be associated by chance. For a 500K SNP genome scan, this equates to 25 000 false-positive associations, within which a few genuine causal alleles are likely to exist – the proverbial needle(s) in a haystack, or, perhaps more appropriately, a needle in a needlestack.

There are ways to improve the odds. For example, alleles of modest effect may be detected by a quantitative trait approach, by genotyping relatively small sample sizes at the extremes of a distribution of phenotypic values. Arking *et al.* (2006) identified a common genetic variant that influences extremes of the electrocardiographic QT interval, a phenotype associated with increased risk of cardiovascular mortality. They used a multistage study design (Figure 18.2). First they carried out a 100K SNP chip genome-wide association study on 200 subjects at the extremes of a range of QT intervals seen in 3966 subjects. After identification of preliminary associations, they followed up markers in the remainder of the cohort and then independently replicated their findings in two independent samples of 2646 subjects from Germany and 1805 US subjects. The study identified a number of replicated loci, including the NOS1AP gene, a regulator of neuronal nitric oxide synthase. This approach could be applied to any other phenotype with a biomarker or other disease-relevant measure that shows a distributed range of values throughout the disease population.

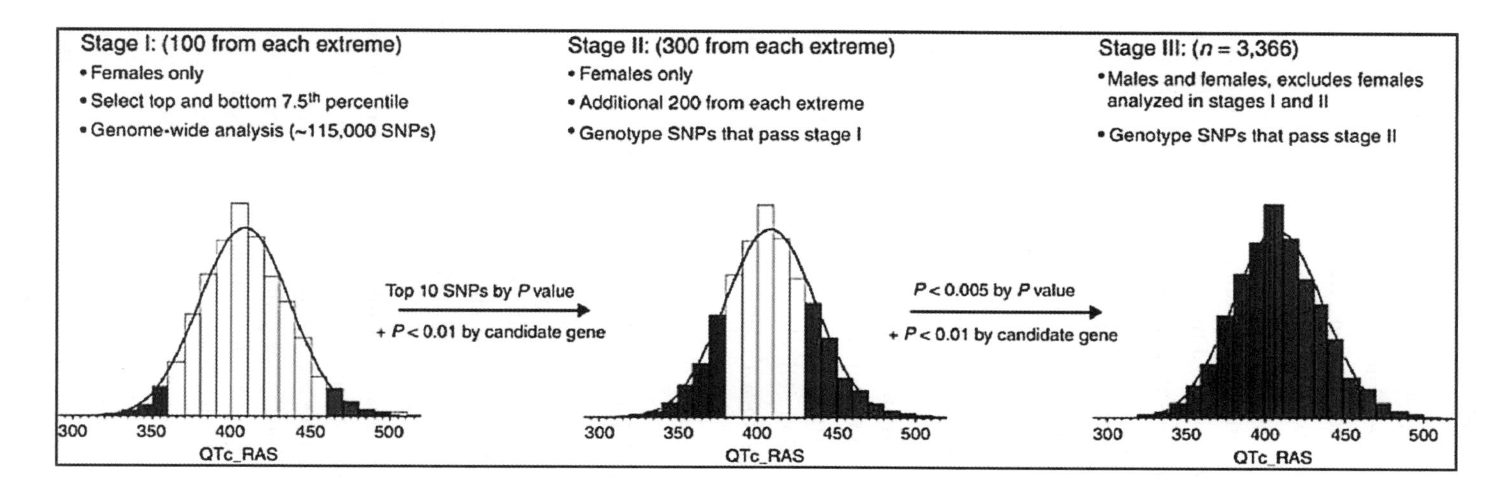

Figure 18.2 Using extremes of a QTL distribution for genome-wide association studies. A quantitative trait (QTL)-based genome-scan study design by Arking *et al.* (2006). In stage I, genome-wide genotyping was performed on 100 females from the extremes of the QTL distribution. In stage II, an additional 200 females from each extreme were genotyped, and the combined stages I and II samples were analysed. In stage III, all samples not typed in stages I and II were genotyped, including both males and females, and both combined and stratified analyses were performed. Reprinted by permission from Macmillan Publishers Ltd: *Nature Genetics* 38(6), 644–651, copyright 2006

18.2.4 On the causes of type 1 error

False-positive associations can arise by different mechanisms, which can be loosely divided into three main categories. The largest source of type 1 error probably reflects the statistical threshold used for inference. A significance level of 0.05 corresponds to a 5 per cent type 1 error rate, and this rate increases with multiple testing. Another contributor to type 1 error might be a systematic bias introduced by the study design, such as an underlying ethnic admixture. In this situation, strong false-positive associations may occur where a marker (unrelated to the phenotype) has different allele frequencies in different population groups. Many methods and analytical procedures have been developed to correct this problem (see Cardon and Palmer (2003) for a review). Technical artefacts are the final common source of type 1 errors, which may be caused by a range of subtle (and not so subtle) issues of genotyping and sample preparation. These can include issues such as preferential amplification of alleles or dropout of heterozygote calls. Some of these issues should be detected by testing for departure from Hardy–Weinberg equilibrium, a standard QC procedure; however, other issues, such as missing heterozygote calls during TDT analysis, can cause problems that are difficult to detect (Mitchell *et al.*, 2003). This latter category is essential to look out for as new technologies are employed for the first time; this is highly pertinent at the time of writing, as genotype-calling algorithms for the new-generation SNP chips and other support software are constantly being upgraded to keep up with rapid recent developments in genotyping technology.

18.2.5 Investigating positive findings

It is unlikely that it will ever be possible to resolve type 1 error completely from true positives in genetic experiments. However, there are some approaches that can help to uncover true associations. These approaches vary in their costliness in terms of resource requirements. The simplest, most inclusive approach is to accept initially a liberal P value threshold, say, a nominal $P < 0.05$, and then carry out follow-up studies to try to distinguish false positives from real associations. The nature of these studies could vary, the most conventional being independent replication. This again raises the issue of the cost of sample ascertainment, and, arguably, if the samples are available, they should be combined in the primary scan to increase power. However, a staged approach, in which a more modest threshold for association is used during the evaluation of the initial scan, has gained widespread acceptance as one practical solution of this problem (Lowe *et al.*, 2004; Hirschhorn and Daly, 2005; Wang *et al.*, 2006). In some instances, replication of association may not be an option, as in the case of rare population-based phenotypes such as adverse drug reactions, which might be observed in phase II or III clinical trials. Such events might lead to immediate termination of drug development, and hence there would be no further opportunity to collect cases (Camilleri *et al.*, 2002). Another possible solution is to seek cross-validation of associations with other biological data sources, including animal models (e.g., mouse knockout (KO) phenotypes similar to disease), gene-expression data (e.g., gene showing upregulation in disease), pathway information (e.g., gene in known disease pathway), or functional data (e.g., associated marker in LD with functional allele). Although this approach risks elevating genes that are actually false-positive associations, there have been some notable successes to support the use of prior knowledge to assist prioritization of associated genes (Mootha *et al.*, 2003). This is clearly a major area where bioinformatics analysis can make a real impact; therefore, we will review some of these approaches in detail below in Section 18.4.

18.2.6 Replication – the troublesome 'gold standard'

Efforts to identify complex disease-susceptibility alleles have been quite notable in the failure of most associations to replicate (Weiss and Terwilliger, 2000). While many of these failures can be expected to be explained by type 1 error and lack of power to replicate, others may not be so immediately obvious. Spector *et al.* (2006) reviewed some of the reasons that may explain these failures. They identified four key areas of concern; the first two are matters of study design, while the last two rely to a great extent on effective bioinformatics for resolution.

The first identified area was the overestimation of the genetic effect in the first report. Perhaps this is in the 'human error' category; pressures on investigators to publish statistically significant results can lead to a bias toward the most favourable odds ratio in an association study. This is sometimes called the 'winner's curse'. Therefore, a study to confirm a reported genetic association should have a sample size large enough to detect genetic effects that may well be smaller than the original association. Overestimation of effect was identified as a possible source of type 2 errors, as underpowered replication studies failed to detect a true association.

The second concern was the clinical definition of the phenotype between studies. Despite superficial similarities, subtly different phenotypes might be expected to have very different underlying molecular mechanisms and disease alleles. Spector *et al.* (2006) cited hip osteoarthritis and generalized osteoarthritis as a good example of related but distinct phenotypes with different underlying biology.

The third area of concern was inconsistent coverage of genetic variation in the gene. The concern here is that some replication studies genotype only the most significant SNPs from the original study, or, worse still, different SNPs from the original study. This approach risks a false-negative result, as it may fail to include, or indirectly tag, the relevant causal variant. The information that the HapMap provides may help to solve this problem, as full knowledge of haplotypes and LD across a locus should make it possible to relate associations among different SNPs and, better still, to select tag SNPs that capture most common variation.

The final concern was inherent genetic differences in the populations studied. Even if the same markers are analysed between a primary association and a replication, a true association may not be detected second time around. This may be due to subtle differences in LD between alleles or haplotypes, reflecting distinct population histories following a shared founder mutation, and it suggests that replication should be gene based, not allele based. One reasonable exception is where the same allele is estimated to have opposite effects between studies. Although there are theoretical arguments to suggest that different alleles of a marker may be in LD with the same causative allele in different populations, there have been no clear examples of this. Instead, association of different alleles of the same marker between studies should probably be accepted as a clear example of type 1 error.

When a replication at a gene, rather than allele, level is suspected, a review of the associated region by the HapMap or UCSC genome browsers may give clues of potential differences in LD between or within populations. Moreover, evidence of elevated recombination (viewable in both UCSC and HapMap viewers) may highlight regions where haplotypes and LD differ between populations. If two markers are separated by an area of elevated recombination, it is reasonable to expect to see heterogeneity between populations in haplotype structure across this area. Accepting that replication may be at a gene level does create problems; the most obvious one is that of multiple testing. If 100 markers are typed across a large gene, five markers would be expected to show association in both studies by chance alone; however, these results could not reasonably be proposed as evidence of replicated association. One solution of this problem is to apply a gene-based permutation test (Churchill and Doerge, 1994; see Dudbridge *et al.* (2006) for a review of these methods). This adjusts for multiple testing while preserving the correlation structure among markers that are physically close.

18.2.7 Detecting gene–gene interactions (epistasis)

An important consideration in the detection of complex disease-susceptibility alleles is the possible role of gene–gene interactions, otherwise known as epistasis. Analysis of how these interactions contribute to complex traits is very challenging (Cordell, 2002). The problem is one of dimensionality; even a genome-wide association analysis for a single-gene effect is over-burdened by hypotheses to test. Increasing the dimensionality of these hypotheses to account even for simple interactions between two genes is considered by some as a step too far, while others have suggested that these analyses may be possible within certain limits (Marchini *et al.*, 2005).

It may be reasonable to use a biological hypothesis-driven approach to identify epistatic interactions. In principle, modest individual effects detected during genome scan analysis could be queried for interactions, either among all the other positive markers or among specific biological candidates. For example, markers that show some evidence of interaction within the same pathway, or that share the same GO term (see Section 18.4.6 below) could be tested for epistasis. Hirschhorn and Daly (2005) pointed out that these types of approaches are likely to be important regardless of epistasis, as scans that are conditional on known positive results are likely to be more powerful for detecting novel effects once the variance explained by the major loci has been controlled. The hope is that this approach will help to unmask the signal from more minor contributing alleles.

18.2.8 Bayesian approaches to association analysis

An interesting recent response to the challenges of genome scan analysis is a move toward Bayesian analysis (de Bakker *et al.*, 2005; Morris, 2005; Pe'er *et al.*, 2006b). Bayesian and frequentist are two fundamentally different approaches to statistical analysis (see http://en.wikipedia.org/wiki/Bayesian_probability, for a lay review of both approaches). Essentially, Bayesian analysis uses prior information to weight probabilities. For example, an association with a SNP might be weighted (a 'prior weight' – often just termed a 'prior') according to the potential functionality of that SNP or other SNPs in LD with it; for example, a non-synonymous SNP might be weighted more highly than a synonymous SNP. Priors do not need to be limited to information about the SNP itself; they could also extend to the biological rationale of the gene or the wider locus.

The move toward Bayesian analysis in genome-wide association analysis is an obvious response to the problems of power and type 1 error. A key observation is the fact that SNP tests that capture many putative causal alleles have different statistical properties from tests capturing only a single site (de Bakker *et al.*, 2005). An immediate implication of this is that rather than treating all tests with the same significance threshold in a frequentist approach, we may increase power by applying Bayesian priors incorporating the number of sites captured by LD, together with each site's individual likelihood of being causal. The attraction of this approach is that it utilizes much more of the available information.

While the standard frequentist approach to genome-wide association applies a standard significance threshold across all markers tested based on their deviation from the expected null hypothesis, it takes no account of other information that may be relevant to this hypothesis. For example, a test that corresponds to a putative functional variant in a gene that shows strong rationale for the phenotype in question might be considered to have a stronger a priori rationale as a true positive than a test that corresponds to a test with no *known* function. There are caveats in such an assumption. Thus, all variation and LD relationships between variants have not been characterized, so it is not possible to state conclusively that a given marker and all other markers in LD with that marker have no function. *Known* is the key qualifier here; therefore, we must accept that Bayesian methods are generally limited by how much we know.

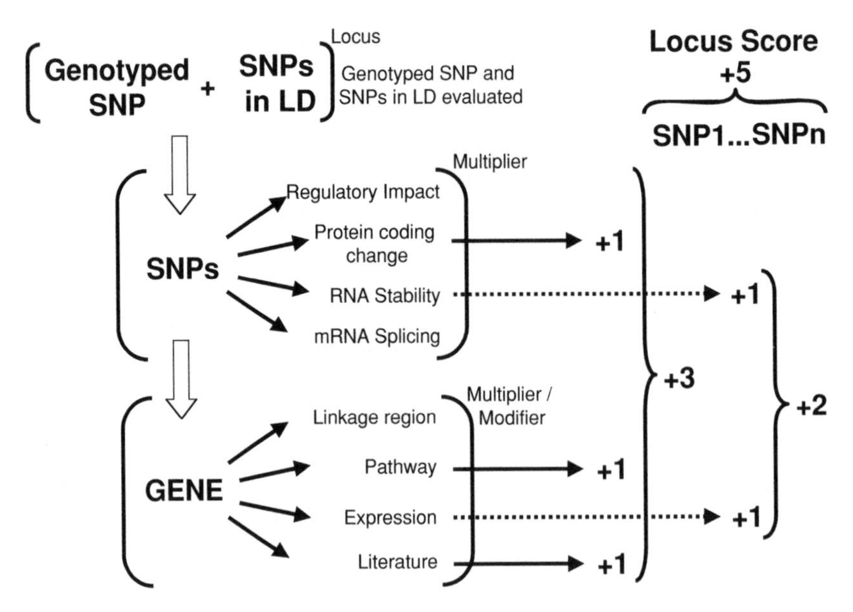

Figure 18.3 Using SNP- and gene-based biological function to set prior weights in Bayesian analysis

Clearly, this is an area where bioinformatics is critical to interpret biological data and determine potential functionality and other measures that could be used to apply weight to associations. In Figure 18.3, we describe a simple schema for determining scores that could form the basis of prior weights in Bayesian analysis. To be truly Bayesian in nature, these priors would need to be applied at the outset of statistical analysis; however, the same system could also be used for post hoc weighting of results obtained by frequentist analysis. The principle of the analysis is simple. To begin, all SNPs showing LD with the genotyped marker should be identified (e.g., by HapMap data). The genotyped marker and the SNPs in LD with it constitute the locus. Each marker in the locus should be evaluated for functional impact and scored (see Chapter 11 for an overview of this process). In each case, the functional score can be multiplied or modified according to the gene that is affected, by the rationale of the gene in the disease. For example, in a study of Parkinson's disease, expression in the brain might multiply the SNP functional weight, or lack of expression in the brain might reduce SNP functional weight. As shown in Figure 18.3, scores from all SNPs in the locus could be added together to give a combined locus score to be used in Bayesian analysis.

The reader may feel some intuitive resistance to some of these approaches, as they are clearly liable to investigator bias in setting up the priors. Care must be taken to ensure balanced incorporation of prior data; adjustments may need to be made in areas of high LD where many SNPs may be correlated or where contradictory information is available. However, there is some statistical support for the idea. Pe'er *et al.* (2006b) described an algorithm for such analyses and showed by simulated association studies that incorporation of prior probabilities (in this case assuming all SNPs to be equally likely to be causal) modestly but consistently improved power to detect association. At the moment, there is really no 'right way' to do these types of analysis. One of the most pragmatic approaches would probably involve testing a number of different weighting strategies to see which ones work best in practice according to analysis of a number of known genetic associations.

18.2.9 Genome scan analysis issues – conclusions

These are just a selection of the major issues that are confounding the effective identification and replication of disease alleles. At the most fundamental level, the greatest challenges concern the insufficient power that most studies have to detect modest genetic effects, and issues of multiple testing. Application of stringent P values or permutation testing (Dudbridge, 2006) may be a way to address multiple testing in genome scans, but it may not offer the best solution. The solution is intuitively simple, but not so easy to bring about. Geneticists need to collaborate and form larger and larger consortia to enable the collection of adequate sample sizes. Until this occurs, a combination of modest genetic effects and inadequate sample sizes will continue to confound identification of disease loci.

18.3 Ultra-high-density genome-scanning technologies

Despite early recognition of the potential of high-density association studies to detect disease-susceptibility alleles with weak to moderate effects (Risch and Merikangas, 1996), early attempts at genome-wide association analysis have been restricted by limited knowledge of common variation and LD and the lack of a robust physical mapping framework on which to place these variants. These clear technical limitations have been swept away by the chain of events that followed the sequencing of the human genome. The availability of a complete genome provided the robust, highly accurate physical map. The sequence information also provided information on several million common variants (SNP Consortium, 2001). The final critical piece to fall into place was the HapMap (Altshuler *et al.*, 2005; see Chapter 3), which characterized LD relationships among ~4 M SNPs in four population samples (Yorubans from Ibadan, Nigeria (YRI); CEPH trios of European ancestry from Utah, USA (CEU); Han Chinese from Beijing (CHB); and Japanese subjects from Tokyo (JPT)). With the availability of these resources, ultra-high-density genome scan maps have become a reality rather than a distant goal.

The genome-wide association approach relies on the assumption that most causal variants are likely to be in LD with nearby markers. To detect these causal variants, the markers selected for genome-wide association analysis must either be causal alleles or markers highly correlated (in LD) with the causal alleles (Kruglyak, 1999). Encouragingly, the analysis carried out during the human HapMap project suggests that most regions of the genome fall into segments of high LD, with limited haplotype diversity, within which variants are strongly correlated with each other. Specific analysis of the intensively studied HapMap-ENCODE regions showed that over 80 per cent of common variants (MAF > 5 per cent) are either represented directly in dbSNP or show tight correlation with other SNPs that are in dbSNP (Altshuler *et al.*, 2005). This analysis suggests that existing resources of SNPs could be used to design SNP genotyping panels to capture the vast majority of common variants and possibly a substantial proportion of rarer variants. Oligonucleotide-based SNP genotyping arrays have emerged as the effective technical solution to address this genotyping challenge; in this section, we review these new technologies and some of the bioinformatics issues that they raise.

18.3.1 Commercially available fixed-marker panels for genome scan analysis

Off-the-shelf SNP-based genotyping arrays are now being offered by companies such as Illumina (San Diego, CA, USA) and Affymetrix (Santa Clara, CA, USA). These technologies appear to be rapidly becoming standard genotyping tools, much as the U95 and U133 Affymetrix GeneChips have become standard gene-expression analysis tools. The content of these different array-based products varies quite widely. The Affymetrix 100K and 500K arrays were selected primarily on the basis of technical assay quality without the benefit of HapMap information, while the more recent Illumina HumanHap300 and 550 arrays were selected by a tag SNP approach based on HapMap LD information. Some of the key features of these different arrays are compared in Table 18.1.

18.3.2 Affymetrix SNP panel design

A detailed description of the Affymetrix 100K design process, which is similar to the 500K design process, is given by Matsuzaki *et al.* (2004); for the purposes of this chapter, we will give a brief overview of the process used for the design of both chips. The Mapping 100K and Mapping 500K GeneChip SNP arrays were both designed by a restriction enzyme-based method for SNP map construction. To be included on the array, a SNP must be within 2 kb of one of a given pair of restriction enzymes (XbaI and HindIII in the case of the 100K chip, and StyI and NspI in the case of the 500K chip). This is an important consideration of coverage. Although restriction enzyme sites generally exist at regular intervals throughout the genome (depending on the complexity of their cutting site), in some areas of reduced nucleotide complexity or low or high G/C ratio, the frequency of sites may vary, leaving severe constraints on

Table 18.1 Commercially available SNP genotyping microarrays

Array	SNP no.	Notes
Affymetrix Mapping 100K	116 945	Evenly distributed genome-wide mapping panel. SNP selection excluded markers showing complete LD ($r^2 = 1$) with other SNPs on panel. SNPs generally selected with Caucasian MAF > 5%
Illumina Sentrix Human-1	109 334	Exon centric content. >73 000 SNPs are within 10 kb of coding exons
Illumina Sentrix HumanHap300	317 502	Haplotype tag SNPs based on phase I HapMap. Tag SNPs biased toward genic and evolutionarily conserved regions. Over 7300 coding SNPs. Increased density across the MHC region. SNPs generally selected with MAF > 10%
Affymetrix Mapping 500K	500 566	Evenly distributed, genome-wide mapping panel. 500K SNPs selected with bias toward exons. Selection excluded markers showing complete LD ($r^2 = 1$) with other SNPs on panel. SNPs generally selected with MAF > 5%
Illumina Sentrix HumanHap550	561 299	Combines all content of the HumanHap300 chip with additional 240K tag SNPs derived from phase II HapMap. Also includes >4300 SNPs in recently reported copy number polymorphisms. SNPs generally selected with Caucasian MAF > 10%

SNP selection, and possibly leading to gaps in the mapping panel. Map construction for both the 100K and 500K was a three-step process (Matsuzaki *et al.*, 2004). Firstly, a larger number of candidate SNPs were genotyped in 54 ethnically diverse individuals; a subset of these was selected on the basis of MAF and call rate. Secondly, SNPs were genotyped on a total of 330 individuals, including various ethnicities. In the final stage, 116 945 SNPs that showed an acceptable call rate (>90 per cent) and a minor allele frequency of at least 1 per cent in at least one of the ethnic groups, were selected at evenly spaced intervals for inclusion on 100K production chip. A total of 500 566 SNPs with an acceptable call rate and a MAF over 1 per cent in one of the ethnic groups were selected for inclusion on the 500K production chip, with priority given to SNPs located in exons, while SNPs in complete LD ($r^2 = 1$) with a previously selected SNP were not selected. The genome-wide SNP spacing of the 100K map averages 1 SNP/24 kb, compared to the SNP spacing of the 500K map, which averages 1 SNP/5.4 kb. The genomic coverage of the 500K array is currently available to view in the 'Affy 500K' track at the UCSC genome browser; this enables viewing of the map coverage in the context of genes, genomic features and the HapMap (Figure 18.4). These data and data for the Illumina HumanHap arrays should be available on the UCSC production server, by the time this book is published. Coverage of other SNP panels can be viewed in a similar way by creating custom tracks based on the marker lists of each panel.

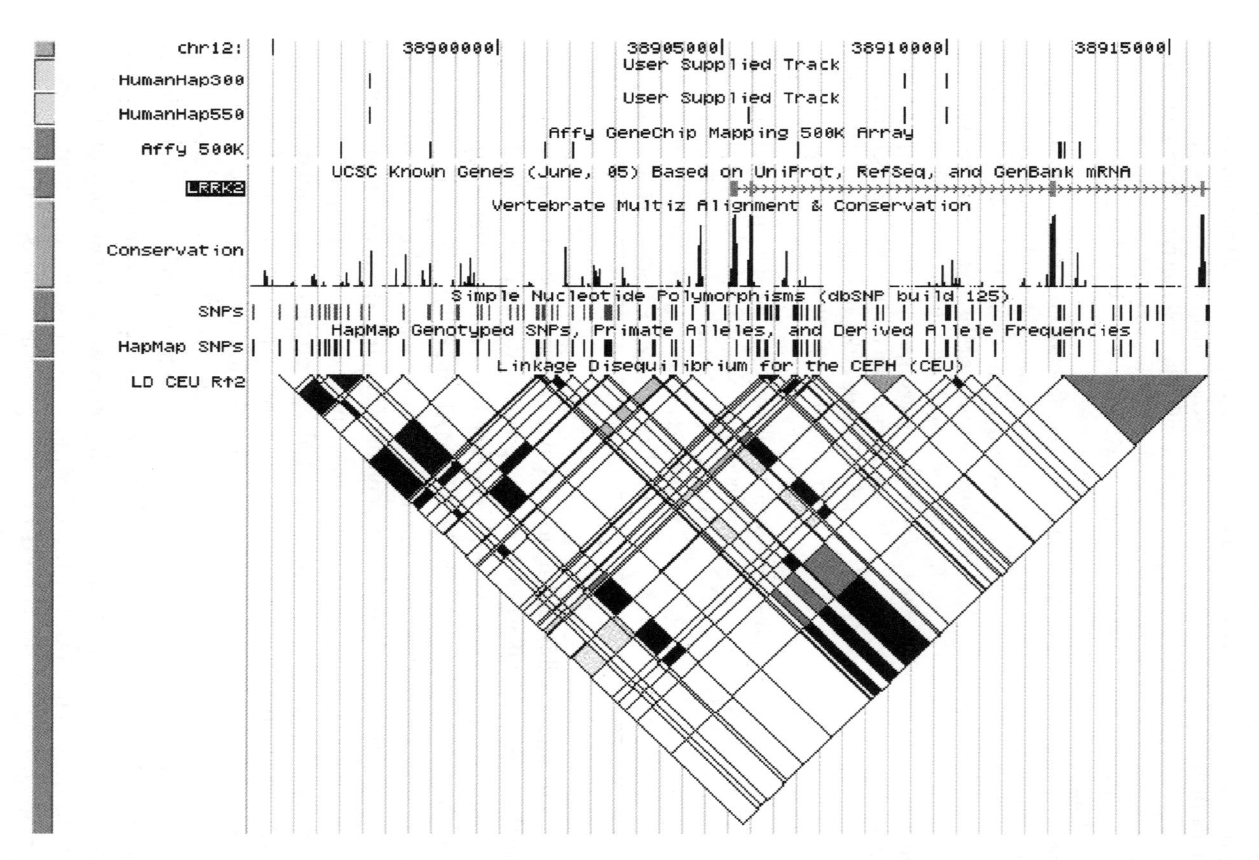

Figure 18.4 Viewing genomic coverage of the Affymetrix 500K GeneChip and Illumina BeadChips with the UCSC Genome Browser. Generated using the UCSC Genome Browser, http://genome.ucsc.edu

18.3.3 Illumina SNP panel design

Illumina currently markets three SNP panels for genome-wide association analysis. Details of these panels are available on the Illumina website (http://www.illumina.com); the genotyping technology is described in detail by Steemers *et al.* (2006). The Human-1 Genotyping BeadChip is an exon-centric panel with over 109 334 SNPs, more than 70 per cent of which are either in or within 10 kb of a transcript. The remaining SNPs were used to provide an even genome-wide coverage of 5 SNPs per 200 kb. SNPs were selected with a MAF of >0.03 after genotyping the 120 HapMap samples from four ethnic groups. Hot on the heels of the release of the phase I HapMap, the HumanHap300 Genotyping BeadChip was developed as a genome-wide SNP genotyping panel designed entirely from HapMap-derived data. This SNP panel contains over 317 000 haplotype tag SNPs selected by a pairwise correlation-based algorithm (Carlson *et al.*, 2004) applied to phase I CEU HapMap genotype data. The HumanHap300 chip is also slightly gene centric, with a higher density of tag SNPs within 10 kb of a gene or an evolutionarily conserved region (an important surrogate for an unknown gene or regulatory region). The panel has also been supplemented with 7300 non-synonymous SNPs (nsSNPs), and a panel of SNPs across the major histocompatibility complex (MHC) region on chromosome 6. The most recently released panel from Illumina, the HumanHap550 genotyping BeadChip, combines all the markers on the HumanHap300 panel with an additional 240 000 tag SNPs derived from the phases I and II HapMap data, and provides tag SNP coverage in regions of lower LD in the genome. In addition to the tagSNPs, there are also 180 mitochondrial SNPs, 11 Y-chromosome SNPs, and over 4300 SNPs in LD with genomic copy number polymorphisms (Newman *et al.*, 2006; see Chapter 9).

18.3.4 Evaluating genome-wide SNP genotyping panels

It is not immediately evident how these panels would compare in terms of overall capture of information on variation on a genome-wide scale. A number of empirical studies and theoretical simulations have provided estimates of the required density of markers to obtain effective genome coverage (see Judson *et al.* (2002) for a review). A recent empirical estimate, based on selection of haplotype-tagging SNPs from over 1.5 M SNPs, suggests that ~300 000 tag SNPs may be required to capture common variation in Caucasian populations (Hinds *et al.*, 2002). However, these figures cannot be considered accurate; the precise number of tag SNPs needed to capture common variation will depend on the methods used to select SNPs, the degree of long-range LD between blocks, and, perhaps most critically, the efficiency with which SNPs in regions of low LD can be tagged (Ke *et al.*, 2004).

In principle, evaluation of a genome-wide marker panel should seek to identify the extent to which each marker in the panel is correlated with all common alleles

(MAF > 5 per cent) in the genome – assuming that putative disease-causal alleles will be included among these alleles. Unfortunately, this ideal scenario is not yet possible on a genome-wide scale – although the data contained in the phase II HapMap are impressive in scale, it is still likely to represent a minority (perhaps 40–45 per cent) of all common SNPs in the four sampled populations (the phase II HapMap has characterized ~4 million SNPs of an estimated 11 million common SNPs).

However, there is a data set that approaches this ideal. The HapMap-ENCODE regions are ten 500-kb regions, selected to represent a genome-wide range of evolutionary conservation and gene density, that have been subjected to intensive SNP finding, to the point of near complete ascertainment of common variation (ENCODE Project Consortium, 2004). This depth of ascertainment makes the HapMap-ENCODE regions an ideal proving ground for genome-wide marker panels, offering an opportunity to evaluate directly the correlation between the marker panel and common variants across the regions (see de Bakker *et al.* (2005) for a review). The near complete ascertainment of the ENCODE regions also has an added benefit in the evaluation of maps that have not been selected on the basis of the HapMap data, as most marker should be represented across the ENCODE regions. For example, only ~75 per cent of the markers from the Affymetrix 500K array (which is not selected from HapMap) are represented in the genome-wide HapMap, whereas 93 per cent of markers are represented across the ENCODE regions.

The dense, but incompletely ascertained, genome-wide coverage of the phase II HapMap, and the almost completely ascertained, but restricted, coverage of the HapMap-ENCODE regions, give two options for evaluation of the power of genome-wide marker panels. Both approaches were used by Pe'er *et al.* (2006b) to evaluate the power of the Affymetrix 100K and 500K arrays, and the Illumina HumanHap300 array. For this evaluation, they calculated correlation between markers in each panel with markers in the HapMap ENCODE regions, and with common markers in the wider phase II HapMap. The results were quite encouraging, although they highlighted the Caucasian bias of these genotyping products. First, Pe'er *et al.* looked at the power of each array to capture information on common SNPs in phase II HapMap data across the HapMap population samples (Figure 18.5). In the case of the Affymetrix 100K, they found that the respective captures of common variation at $r^2 \geq 0.8$ across each population were 15 per cent (in YRI), 31 per cent (CEU) and 28 per cent (CHB plus JPT). In the case of the 500K chip, the captures of common variation were 44 per cent (YRI), 66 per cent (CEU) and 63 per cent (CHB plus JPT). Finally, in the case of the HumanHap300 chip, the captures of common variation were 31 per cent (YRI), 72 per cent (CEU) and 59 per cent (CHB plus JPT). As might be expected, the performance of all three panels was poorer in the non-Caucasian populations, as each panel had been selected on the basis of LD and frequency data from Caucasians. Population-specific differences were most acute in the case of the HumanHap300. Although the panel showed the highest performance in Caucasian populations, reflecting the tag-based nature of the panel, it also showed the biggest

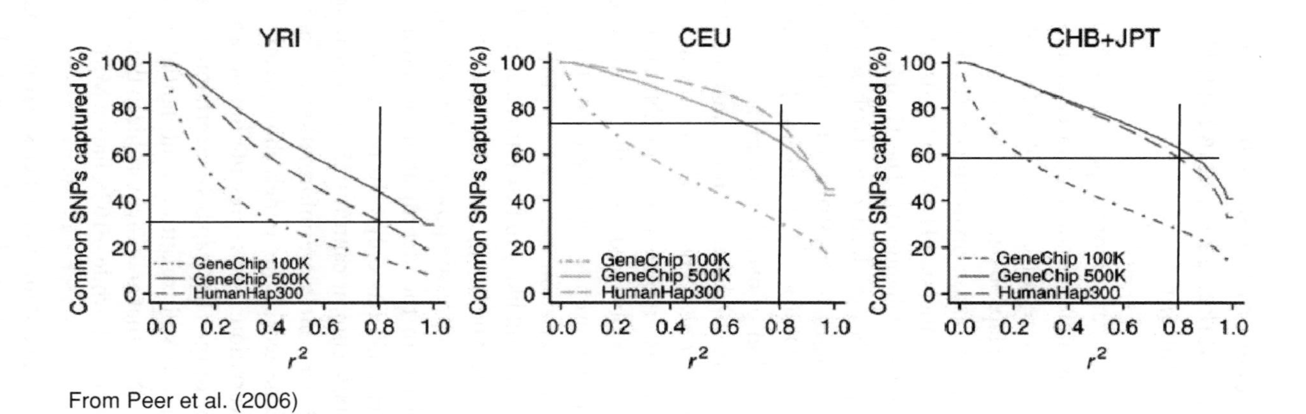

Figure 18.5 Fraction of common (MAF > 5 per cent) phase II HapMap SNPs (y-axis) captured by array SNPs as a function of the r^2 cut-off (x-axis). Data are presented for the GeneChip 100K, GeneChip 500K and HumanHap300 arrays, for each of the three HapMap analysis panels: Yoruba people ascertained in Ibadan, Nigeria (YRI); the CEPH-collected samples of European ancestry, ascertained in Utah, USA (CEU); and Han Chinese samples from Beijing with Japanese samples from Tokyo (CHB plus JPT). Reprinted by permission from Macmillan Publishers Ltd: *Nature Genetics* 38(6), 663–667, copyright 2006

drop in performance between populations, with a drop in performance from 72 to 31 per cent between Caucasian and Yoruban samples.

18.3.5 Case study – evaluation of genome-wide capture of variation by a fixed-marker panel using HapMap data in a database environment

Pe'er *et al.* (2006b) carried out a comparison of the performance of the Affymetrix 100K and 500K GeneChips across the common and rare variants in the phase II HapMap and the HapMap-ENCODE regions. To enable a comparison with the Illumina HumanHap300 and HumanHap550 panels, we carried out a similar analysis of both panels within an ORACLE database environment (see Box 18.1). The analysis was based on publicly available Caucasian HapMap LD data from the HapMap website (http://www.hapmap.org; see Chapter 3). The Hapmap LD data and the marker lists for both Illumina panels were loaded to database tables and queried with structured query language (SQL) to construct a report indicating all phase II HapMap markers that showed LD of $r^2 \geq 0.8$ with the Illumina markers. Further SQL queries enabled calculation of overall percentages of SNPs captured by each marker panel. The results were combined with the analysis of the Affymetrix 100K and 500K reported by Pe'er *et al.* (2006b) and are presented in Figure 18.5. Analysis of the close to completely ascertained HapMap-ENCODE regions appears to validate the genome-wide results obtained from the incompletely ascertained phase II HapMap. This suggests that even though the phase II HapMap may contain ~40–50 per cent of common variation (>5 per cent MAF), the unknown or ungenotyped common variants are still likely to be captured at modestly high rates by the three larger fixed genotyping panels. The performance of the HumanHap550 beadchip highlights the potentially impressive power of the tag-based approach for genome scan analysis, particularly for the identification of common variants. Although the tag-based HumanHap550 has only ~10 per cent more markers than the quasi-randomly selected 500K chip, it captures 25 per cent more information on common variation and 42 per cent more information on rare variation (<5 per cent MAF). As might be expected, all the panels show greatly reduced performance in the capture of rare variation. The representation of rare variants in the HapMap is limited by the scope of the HapMap SNP ascertainment strategy (Clark *et al.*, 2005). However, rare variants still constitute a substantial component of the phase II HapMap. Of the 2.5 M SNPs that are polymorphic in the Caucasian sample, ~416K have a MAF of $>0-5$ per cent, while ~107K have a MAF of $>0-1$ per cent (Figure 18.6).

Arguably, the reduced performance at low frequencies may not matter too much, as most modestly sized studies theoretically lack the power to detect associations with low-frequency variants (Figure 18.1). This assumption has had a great deal

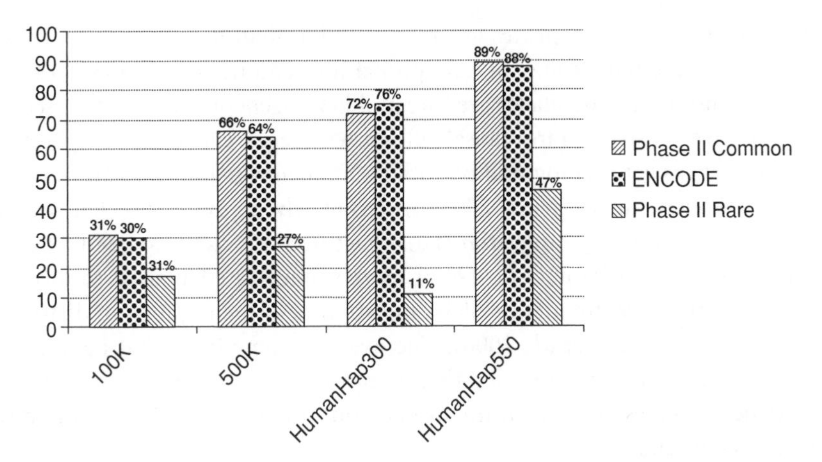

Figure 18.6 Fraction of Caucasian HapMap SNPs captured by SNPs on the Affymetrix 100K and 500K and Illumina HumanHap300 and HumanHap550 genome scan panels. Data are presented for common SNPs as observed in HapMap phase II and ENCODE, and for less common (MAF = 1–5 per cent) SNPs as observed in ENCODE

of influence on the design of genome-wide association panels, effectively focusing array content on common variants, using tag-SNPs with MAF of over 10 per cent. This approach effectively maximizes overall capture of variation at the expense of low-frequency variation, as exemplified by the performance of the HumanHap300 BeadChip (Figure 18.7), which captures a higher percentage of common variation

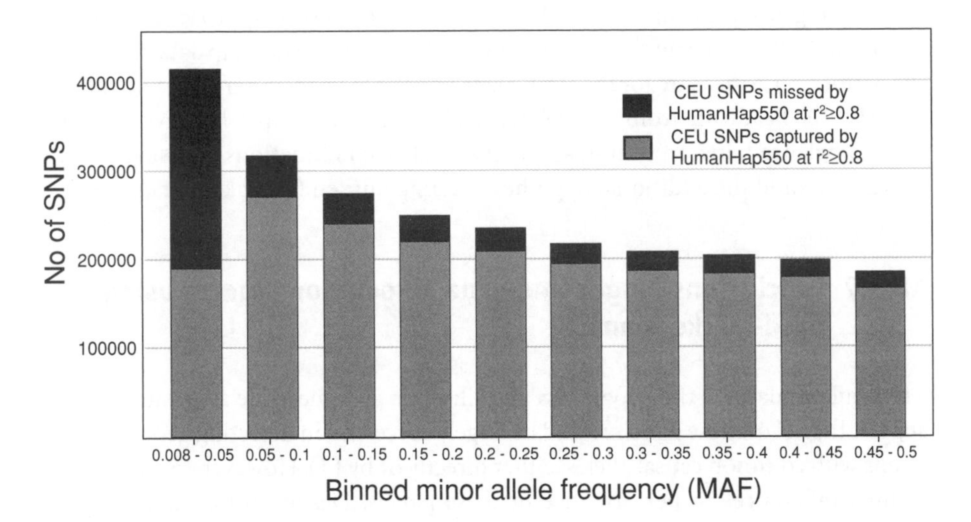

Figure 18.7 Minor allele frequency distribution of CEU polymorphic phase II HapMap SNPs. Grey and black shading indicates the fraction of SNPs captured or missed by the HumanHap550 genome-scan panel at $r^2 > 0.8$

than the 500K chip, but captures less rare variation than the 100K chip. It is clear that some of these panels may be suboptimal for genome scan analysis if the risk alleles of common disease alleles are rare and heterogeneous. Under that scenario, comprehensive scans of rare causal alleles will require a completely different approach, such as genotyping millions of low-frequency markers, or even complete genome resequencing (Service, 2006). Alternatively, a more directed approach might involve genotyping rare variants that have been predicted to be functional by *in silico* analysis (see Chapters 11–14) or more sophisticated analysis to investigate the enrichment of low-frequency alleles in extremes of the population distribution of certain traits (Cohen *et al.*, 2006). There is also some hope that haplotypes of common variants may capture slightly rare disease alleles (MAF <5 per cent) that are enriched in cases, as seen in the association of CARD15 with Crohn's disease (Vermeire *et al.*, 2002).

18.3.6 Alternative measures of marker panel performance

The pairwise measures of LD used by Pe'er *et al.* (2006b) to characterize the extent of LD between HapMap SNPs and given marker panels are widely used (Pritchard and Przeworski, 2001). However, other multilocus measures of LD have been developed recently, and these may further improve the power of analysis (de Bakker *et al.*, 2005; Nicolae *et al.*, 2006). They are somewhat analogous to pairwise measures, such as r^2, but may provide a more comprehensive measure of LD by taking into account the wider haplotype structure of a region. Pe'er *et al.* (2006b) evaluated multilocus predictors of LD against the r^2 measure and found that the multilocus approach captured an additional 9-25 per cent of SNPs in the ENCODE or phase II HapMap data, improving the performance measure for the 500K GeneChip from 66 to 78 per cent, and the HumanHap300 BeadChip from 72 to 86 per cent, in the CEU sample. However, we must strike a balance between testing all observed allele combinations and single SNPs, bearing in mind the additional hypothesis-testing burden (de Bakker *et al.*, 2005).

18.3.7 Conclusions on genome-wide association studies using fixed-marker panels

The available data on the Affymetrix and Illumina genome-wide association panels suggest that all of these panels would be largely successful in detecting genetic associations with common causal alleles, either directly or by LD. However, there are some distinct differences in performance of these panels. Despite some successes – for example, the identification of major susceptibility allele for macular degeneration (Klein *et al.* 2005) and a gene influencing QT interval (Arking *et al.*, 2006) – the Affymetrix 100K GeneChip is clearly underpowered for genome-wide association analysis of complex traits. However, the 100K is suitable for other applications that require lower-resolution genome coverage, such as chromosomal copy number analysis (Slater *et al.*, 2005; see Chapter 16).

The 500K GeneChip and the HumanHap300 BeadChip both offer broadly similar, effective genome-wide coverage, but each carries distinct advantages and disadvantages. The 500K chip shows good performance across all the HapMap population groups, outperforming the HumanHap300 chip in African and Asian populations due to the Caucasian bias in the HumanHap300 tag selection process. This feature is counter-balanced by the excellent performance of the tag-based HumanHap300 in Caucasian populations. Furthermore, in analytical terms, the HumanHap300 suffers less from multiple testing than the 500K. The statistical cost of performing this additional number of hypothesis tests is very high, leading to increased statistical significance thresholds to maintain constant type 1 error rates. One solution to this may be to run a marker-selection algorithm to identify tag SNPs that capture the majority of the variation in the entire 500K marker panel.

The performance of the HumanHap550 BeadChip (Figure 18.3) is clearly ahead of all the other panels, with an 89 per cent capture of common variation in Caucasians at an r^2 threshold of over 0.8. Clearly, this reflects use of the phase II HapMap in the design of the panel; it can be expected that the next-generation high-density arrays from Affymetrix will also be selected on the basis of empirical data from the HapMap.

In terms of cost, the economies of scale that these high-throughput genotyping technologies afford are bringing the cost of genotyping down dramatically. The 500K array breaks the 1 ¢/genotype barrier (the 500K array retails at ~\$1200, which translates to 0.24 ¢/genotype). The success of these arrays in genome-wide association studies is likely to depend, above all, on the prevailing nature of disease-susceptibility alleles. If the 'common disease, common variant' hypothesis (Risch and Merikangas, 1996) prevails, with disease alleles often showing over 5 per cent frequency, the level of performance seen with the larger arrays is likely to be adequate in most cases. If causal alleles are rarer, as others argue in the 'multiple rare variant' hypothesis (Pritchard, 2001), and we restrict our analysis to low-frequency SNPs (>0–5 per cent MAF), then the results from these panels may be at least erratic at the best of times.

18.4 Bioinformatics for genome scan analysis

Although the new ultra-high-density genotyping technologies, coupled with the HapMap, are enabling a comprehensive evaluation of association between common genetic variants and disease, these technologies are also creating huge new analytical challenges. Although we have reviewed many of these analytical challenges in Section 18.2, the best way to get a feel for the issues is to look at real data and examine the steps required to analyse them. In the following section, we will do exactly this; in a case study, we will progress from the preliminary results of a ~200K marker genome-wide association study to a final list of prioritized genes with supporting rationale in the disease under study. There are many steps along the way to this end point, many of which can be enhanced by bioinformatics analysis.

18.4.1 Case study – analysis of a 200K genome scan for Parkinson's disease

The example we will use for our case study is a somewhat pioneering one in terms of scale, carried out by Maraganore *et al.* (2005). This study involved a genome-wide association scan for susceptibility to Parkinson's disease (PD) (see Farrer (2006) for a review of the genetics and biology of PD), typing 248 535 SNPs in 443 sibling pairs ($n = 886$) discordant for PD. After removing SNPs that were monomorphic or out of Hardy–Weinberg equilibrium (based on a threshold of $P = 0.001$), the authors generated results on 198 345 SNPs. They followed up this initial scan by typing 1892 of the most strongly associated SNPs in 332 matched case-control unrelated pairs. Notably, all 198 345 single-SNP association results were made publicly available as supplementary information with the online publication of the study (including minor-allele frequencies, odds ratios (ORs), SNP to gene mapping and P values). The details of this study illuminate the probable direction of genome-wide association analysis on a number of levels – not all of which are encouraging.

Following the publication of the genome scan, four independent research teams sought to replicate the associations in the 11 most highly ranked SNPs. The results were mixed at best. Two groups (Clarimon *et al.*, 2006; Li *et al.*, 2006) reported statistically significant association between PD and one or more of these SNPs, but with different alleles, which does not represent a true replication. The other two groups found no statistically significant association between PD and any of the SNPs investigated (Farrer *et al.*, 2006; Goris *et al.*, 2006). In a meta-analysis of the data from three of the groups (Maraganore *et al.*, 2006), none of the 11 SNPs showed statistically significant association with PD. Taken together, these four studies do not appear to provide substantial evidence of replication for any of the SNPs originally identified as potential PD loci. The general conclusion drawn was that these 11 most convincing PD loci may all be false-positive associations (Myers, 2006).

The failure to replicate these associations is obviously a disappointment, so what are the weaknesses of this study and how might they be overcome? The first question is the completeness of the genomic coverage of the scan. This looks encouraging – Maraganore *et al.* evaluated the genomic coverage of their 248K panel, and its statistical power to detect unassayed, disease-associated variants. They did this by using the SNPs in their genome scan panel to genotype a set of 152 genes that had undergone intensive SNP discovery to a level at which all variants with over 10 per cent MAF had been identified (Hinds *et al.*, 2005). Using this set of known common variants, in a very similar manner to the way the HapMap-ENCODE regions have been applied to the evaluation of fixed marker panels (see Section 18.3.4), they were able to determine that the mean r^2 for unassayed SNPs was 0.57. This in itself seems reasonable coverage under the assumption that any undetected PD-susceptibility alleles have a MAF over 10 per cent and are in strong LD with alleles of over 10 per cent MAF; below this, the level of coverage is likely to decline relatively steeply.

The next question concerns power and population-attributable disease risk. Pathogenic PD mutations have so far been confirmed in seven genes, with familial mutations often turning up in sporadic population-based cases of PD (see OMIM http://www.ncbi.nlm.nih.gov/entrez/dispomim.cgi?id=168600). A number of further susceptibility loci have also been identified by linkage and association approaches (Farrer, 2006). The ORs for disease risk of PD genes identified to date are quite variable; in the case of parkin, a recessively inherited PD-susceptibility gene, West *et al.* (2002) reported an association with a core promoter SNP (–258G) in 296 Caucasian PD patients and 184 controls, with an OR of 1.52. Tan *et al.* (2005) reported a replicated association with the –258G promoter variant in 386 Chinese patients and 367 controls, with an OR of 1.83. Other genes may exert a stronger effect; for example, Skipper *et al.* (2005) found a haplotype, for the dominantly inherited LRRK2 susceptibility gene, which indicated a PD risk with an OR as high as 5.5. This information can be used to estimate the power of this study under several OR scenarios via the Genetic Power Calculator website (http://pngu.mgh.harvard.edu/~purcell/gpc/; Purcell *et al.*, 2003). The site includes a suite of power calculation tools, dealing with TDT and quantitative traits among others, but in this case, we can use the discrete trait case-control power calculator. If we follow the instructions on the website to enter study parameters and ORs, a picture rapidly emerges that bears a great deal of similarity to Figure 18.1 (Wang *et al.*, 2005). Assuming an OR of ~1.5, we would expect to see 80 per cent power with 598 cases, whereas an OR of 2 would give us similar power with 200 cases.

Myers (2006) commented that the lack of replication might also be due to the reduced penetrance seen in PD, a disease with a complex aetiology caused by interaction of genetic and environmental factors. He suggested that the randomly ascertained sporadic PD cases used in the replication study might have included a substantial proportion of cases with little or no genetic basis for disease.

If we account for all these factors and take sample size and the relatively thorough genome coverage of this study into consideration, it is reasonable to expect that true PD-risk alleles, with relatively high ORs, might be found among the associations generated by Maraganore and co-workers. The key question is how to identify them and effectively follow them up.

18.4.2 Dealing with genome-scale data sets

Before we delve into the complex possibilities for follow-up of this data, it is worth addressing some of the practicalities of analysis of 100K+ genome scans.

The Maraganore study was carried out on a genome-wide SNP panel just shy of 250K markers. In a foretaste of the sheer volumes of data that need to be processed in scans of this nature, it is worth noting that the scan generated 172 420 019 genotype calls, while the follow-up study generated 1 176 772 genotypes. Aside from the general processing, QC and analysis of these genotypes, this creates new data-handling and storage challenges at every stage. One of the most mundane problems relates to processing and viewing the association data by standard PC software. Microsoft Excel, the non-statistician's standard for handling and viewing data sets of this nature, has a row limit of 65 534 (although there are rumours that this limit will be increased to 1.1 M in Microsoft Office 12). This creates an immediate problem for data handling; in the case of the 198 345 single SNP associations, the file needs to be divided into four sections to allow viewing in Excel! Files of this nature can be processed by some MS-DOS text editors. FTE (http://fte.sourceforge.net/), a freeware text editor aimed at the development community, is also very useful for (relatively) painless editing of files containing millions of rows. A tool like FTE can divide large data files and load them in appropriately sized sections. Another, perhaps preferable alternative is to use commercial software intended for large-scale data analysis. Several software packages developed for gene-expression analysis generate data files of similar scale to genome scan analysis. Spotfire (http://www.spotfire.com/), an example of such a tool, can visualize and analyse millions of rows of data. This tool was used for some of the preliminary analyses and data visualization in this case study. Although Spotfire is useful for visualization and analysis of large data sets, it currently lacks the basic annotation and spreadsheet capabilities of Microsoft Excel. However, the two applications can be used in a complementary manner; for example, Spotfire can be used to identify and copy key subsets of data, such as all SNPs of $P < 0.05$. Once the data are pasted into Excel, they can be sorted and annotated. Incidentally, the functionality of Excel itself can be improved with commercially available macros that can also help greatly in these types of analyses. Spreadsheet Assistant (http://www.add-ins.com/assistnt.htm), a recommended example of such a tool, allows a wide range of formulas and functions, duplicate detection (e.g., finding duplicate SNPs or genes), and conditional selection queries.

The commercial software approach is the easiest avenue to genome-scale data analysis. However, the limitations of these tools are very substantial. Frankly, the approaches described in the previous 2-3 paragraphs should be avoided if possible (see Chapter 2 for reasons why). If you are interested in an approach with fewer limitations, or if Microsoft products and software are anathema to you, the only alternative is to deal with these data files with Perl or similar programming languages and a database environment (see Box 18.1). This approach is unquestionably the most powerful, integrative way of dealing with data of this nature, and it should be pursued whenever possible.

Box 18.1 Dealing with genome-scale data in a database environment

While a lot can be achieved in the analysis of genome-scale data by simply using data in flat files, the use of programming languages and a database environment adds power and flexibility to data manipulation and analysis tasks. Perl is a de facto programming standard in bioinformatics, being well suited to tasks such as genome scan analysis. Several fundamental bioinformatics functions are also publicly available written in Perl (see http://bioperl.org/wiki/How_Perl_saved_human_genome). Where statistical analysis is required, R is another commonly used language. More computationally intensive tasks may be best written or rewritten in other languages, C being one of the most common. Perl is an interpreted language, while C is compiled, and hence faster.

Perl, R and C are procedural languages; they work as a series of steps, some of which may be conditional, or repeated in a loop. Another, newer style of programming, object orientation, in which data-containing objects interact to achieve the required functionality, is also widely used. Perl can be written in an object-oriented way, and many Perl modules are written thus. C++ and Java are both object-oriented developments of C. Writing procedural programs is probably the best way to start programming; having gained experience, one moves on to object orientation.

Two relational database management systems (RDBMS) commonly used for bioinformatics are the open-source, MySQL (http://www.mysql.com) and the commercial ORACLE (http://www.oracle.com). MySQL is widely used, but tends to be limited to relatively small databases, while Oracle is used for very large databases. MySQL is used for the various EnsEMBL databases, but these databases are designed to be species, genome build and even data type specific, limiting their size. The UCSC Genome Browser also uses MySQL in a similar way.

At its simplest, a database is a grouping of tables, each of which contains rows and columns, with data of different types – broadly equivalent to flat files – and which are related to each other by key relationships. Good database design emphasizes the avoidance of data redundancy through normalization. Data in an RDBMS is accessed with Structured Query Language (SQL), which allows the user to create tables and indexes, insert data into tables, update/delete data in tables, and select data from one or more tables.

Perl programs can also access an RDBMS with SQL. This is achieved by using the Database Interface (DBI) Perl module, which allows a user to access a database of any type in a generic way through an application programming interface (API). When dealing with genome scale data sets, care should be taken when implementing SQL to ensure that indexes are utilized wherever possible, ideally by simple integer keys. This will have a major performance benefit. The minimum data possible should be selected, as moving large volumes of data across networks may well be slow. And consideration should be given to where the database you want to access is.

Finally, it is usually best to keep things – databases or programs – relatively simple, as this will make it much easier when you, or possibly someone else, need to maintain things in the future. Complex functionality can be built up by repeatedly using reusable modules. Start with something small and simple; you can always add complexity by extending it.

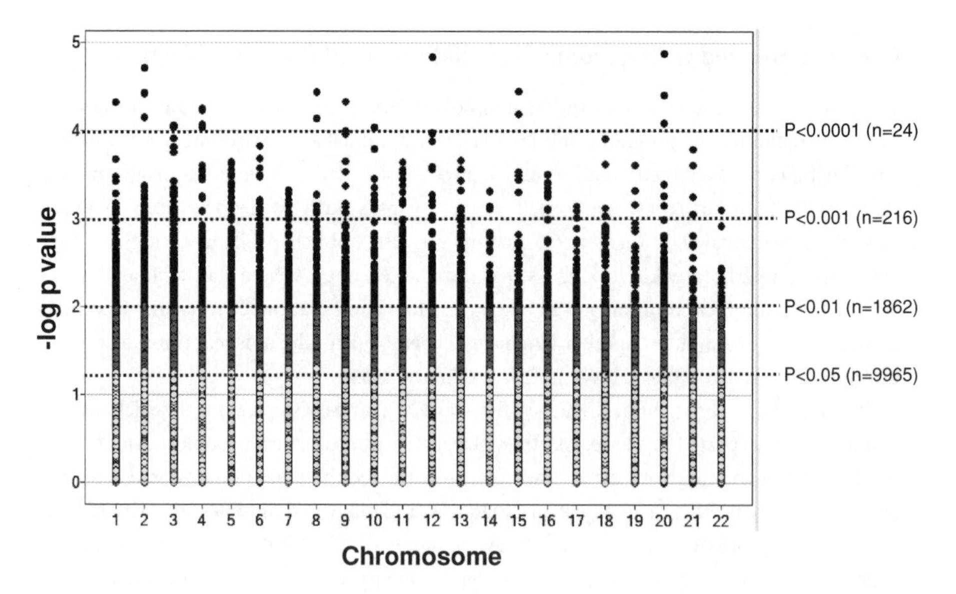

Figure 18.8 Genome scan for Parkinson's disease (PD) susceptibility: results across chromosomes 1–22. Results (log *P* values) of the first round (tier 1) genome-wide association analysis carried out by Maraganore *et al.* (2005) are plotted by chromosome. The number of SNPs meeting various *P* value thresholds is indicated in parentheses on the right hand side of the figure

18.4.3 Devising a robust follow-up strategy

Examination of the first-round association data set for the PD genome scan by Maraganore *et al.* (2005) quickly highlights the all-pervasive problem of the genome-wide association approach. Figure 18.8 shows the range of *P* values seen on a chromosome-by-chromosome basis. Of the 198 345 single-SNP associations, 9965 (5 per cent) show a nominally significant value ($p < 0.05$). Figure 18.8 shows the range of *P* values across chromosomes 1–22. The overwhelming majority of these associations are likely to be false. This highlights the need for a rigorous follow-up strategy. As discussed in Section 18.2, there are several possible approaches. Considering the limited power of this study, the correlation between the markers, and the benefit of hindsight in terms of the failure to replicate the most significantly associated loci (Myers, 2006), it seems reasonable to assert that a stringent *P* value threshold set by Bonferroni correction is inappropriate. Figure 18.8 gives an idea of the impact of altering the *P* value threshold for follow-up in the first round, at $P < 0.01$ (the threshold used by Maraganore *et al.*); 1862 associations are observed.

The original strategy employed by Maraganore *et al.* (2005) applied two different rationales to select SNPs for follow-up studies. The first group, which they called tier 2a, followed a purely statistical rationale. SNPs were selected on the basis of PD association with $P < 0.01$ (1862 SNPs). The second group, tier 2b, were all selected from nominally significant SNPs ($P < 0.05$), but on the basis of a priori biological or genetic hypotheses of PD susceptibility, and also by proximity to known genes. In this case, eight nominally significant SNPs were selected that were positioned within 10 kb of known PD-susceptibility genes, and 145 nominally significant SNPs were selected on the basis of their localization within PD loci previously identified by linkage studies, including PARK3 [OMIM 602404], PARK8 [OMIM 607060], PARK9 [OMIM 606693], PARK10 [OMIM 606852] and PARK11 [OMIM 607688]. A total of 589 SNPs (with $P < 0.05$) were selected on the basis of being within exons or within 10 kb of a known transcript. Both tier 2a and 2b, which comprised a total of 3148 SNPs, were tested for association with PD in the combined samples by the sib-TDT method (Schaid and Rowland, 1998). Robust data were obtained for 3034 SNPs, 174 of which showed nominal significance.

It is interesting to examine the P values obtained in the primary association screen and the subsequent replication screen. Nominally significant P values were seen in 9965/198345 (5 per cent) SNPs in the first-round association study, and 163/3034 (5.7 per cent) SNPs in the replication study. The numbers of nominally significant P values are reduced further to 77/3034 (0.8 per cent) if we require that SNPs show association with the same allele (direction of effect) between first round and replication. The 3034 SNPs analysed in the replication study are plotted in Figure 18.9 on a marker-by-marker basis, in the order of the significance of the

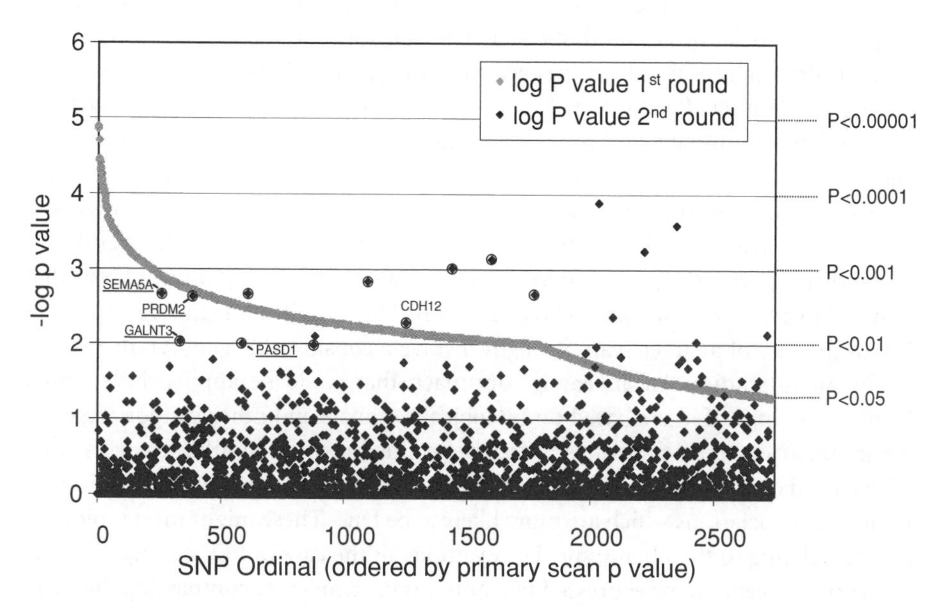

Figure 18.9 Comparison of P values from the primary association scan (tier 1) for PD susceptibility and the subsequent replication scan (tier 2) (Maraganore *et al.* 2005). SNP log P values are plotted on a marker-by-marker basis in order of the significance of the primary association. SNPs that showed association with different alleles in tiers 1 and 2a were removed. SNPs reported as replicated are highlighted with a circle and, where possible, genes mapping to these SNPs are indicated

primary association. SNPs that showed association with different alleles are removed from the second-round association data set. The most striking finding is that markers which showed an initially low P value in the primary screen appear to have no greater rate of replication than those with more modest P values in the primary screen. Taking the 200 lowest P values ($P < 0.001$) from tier 1, only three markers replicate in tier 2 and nominally so with no P value lower than 0.027. It is tempting to speculate that this may reflect the underlying nature of very low P values, possibly suggesting that they may have a tendency to contain artefacts of genotyping (confirmation genotyping by another technology might help to resolve this) or population stratification (see Cardon and Palmer, 2003). In the most pessimistic view, it is also possible that this study is simply insufficiently powered to detect alleles that truly influence the complex genetics of PD, and the results simply reflect a product of type 1 error.

18.4.4 Expanding follow-up into the entire 'SNP space' and 'gene space'

However the results presented in Figure 18.9 are viewed, they highlight the possible drawbacks of approaches that focus on following up only the lowest P values in genetic association data. Clearly, genome scans that are under-powered will fail to detect true associations of modest effect. This suggests that the true associations (of modest effect) may be nestling (like needles) somewhere at the bottom of the haystack among the nominal P values ($P = 0.01-0.05$). This raises a problem; considering the number of nominally significant associations that would be seen in a high density genome scan, such as 25 000 in a 500K scan, a balance needs to be struck between all-inclusive follow-up strategies and evidence-based follow-up strategies. Maraganore *et al.* (2005) tried to achieve this balance by including all nominally significant markers in known PD genes and loci as well as all nominally significant SNPs that mapped to an exon or within 10 kb of an exon. This is quite an inclusive approach, but in the case of a larger scan, it might involve a considerable genotyping burden in follow-up studies. An alternative approach that could be applied here would be to use a range of filters based on biological rationale and other supporting data. Figure 18.10 illustrates this concept. By intersecting statistical significance with other biological data, it may be possible to 'filter' the mass of false-positive associations to enrich for associations which are more likely to be true. These might meet biological criteria relating to the phenotype. For example, in the case of PD, we might expect susceptibility genes to be expressed in whole-brain samples, encompassing the main brain regions affected by PD (Farrer, 2006). Other criteria could also be applied simultaneously or consecutively. For example, genes could be evaluated on the basis of biological rationale, such as activity in pathways with a known role in PD or on the basis of positional rationale in known PD-linkage regions. Sometimes it can also help to review all PD genes identified to date in order to get some idea of what a

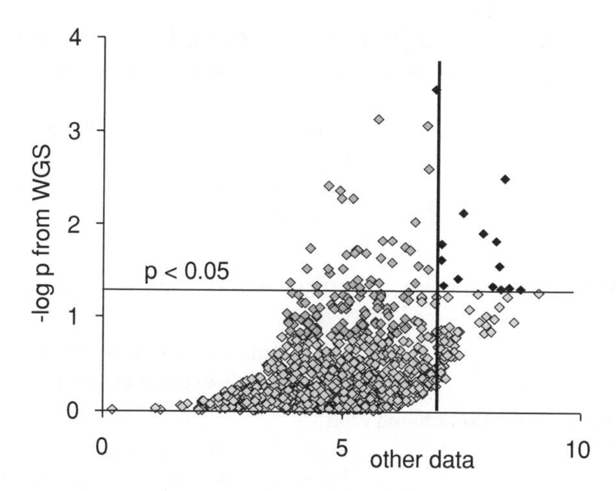

Figure 18.10 Four-box model approach to genome scan prioritization. A threshold is set on both dimensions to allow prioritization based on genetic and biological rationale (e.g., expression, pathways). A similar approach could be employed with *n* dimensions. SNPs denoted in black are candidates for further investigation (Figure courtesy of Dr Nick Galwey, personal communication)

PD-susceptibility gene looks like, in terms of expression profile, biological function and pathway involvement.

18.4.5 Using LD data to maximize inclusivity of analysis

Before applying filters to prioritize loci for follow-up analysis, it may be worthwhile to ensure that all genes in LD with associated markers are taken into account. The genome scan approach relies heavily on LD between markers and disease-susceptibility alleles; therefore, in biological terms, concentrating only on the genes that map to the marker panel may miss important data. The availability of the phase II HapMap LD reference data set makes it possible to make the link between associated markers and other genes in LD with these markers. In the example of the PD genome scan generated by Maraganore *et al.* (2005), there are nominally 9965 significantly associated markers. Approximately half of the associated SNPs map to intergenic regions and do not map to any known genes. It is possible to identify all HapMap SNPs that are in LD with the PD-associated SNPs by querying the HapMap data and thus to determine which genes might be implicated. For smaller numbers of SNPs, this can be achieved with tools such as HapMart. However, for thousands of SNPs, a database environment is required.

In this case, we queried an ORACLE database table (see Box 18.1) containing CEU LD data for all HapMap SNPs that showed LD ($r^2 > 0.8$) with the 9965 nominally significant SNPs. This identified an additional 54 252 SNPs in LD with the PD-associated SNPs. This expands our overall 'SNP space' of direct associations and indirect associations to 64 217 SNPs – all potential candidates for PD susceptibility.

Table 18.2 Useful bioinformatics tools and resources for analysis of genome scans

Data manipulation	
ID converter (GEPAS)	http://idconverter.bioinfo.cipf.es/
SNPPER	http://snpper.chip.org/
Spreadsheet Assistant	http://www.add-ins.com/assistnt.htm
Spotfire	http://www.spotfire.com
SNP/LD analysis	
HapMap/HapMart	http://www.hapmap.org/
Expression analysis	
Symatlas	http://symatlas.gnf.org/SymAtlas/
GEO	http://www.ncbi.nlm.nih.gov/geo/
Integrated genome-scale data annotation tools	
DAVID	http://david.abcc.ncifcrf.gov/
GEPAS	http://gepas.bioinfo.cipf.es/cgi-bin/anno
GFINDer	http://www.medinfopoli.polimi.it/GFINDer/
L2L	http://depts.washington.edu/l2l/
Specialist Gene Ontology (GO) analysis	
GO tools	http://www.geneontology.org/GO.tools.shtml
Gene Ontology Tree Machine	http://bioinfo.vanderbilt.edu/gotm/
FatiGO	http://fatigo.bioinfo.cipf.es/
Pathway analysis	
BIOCARTA	http://www.biocarta.com/genes/index.asp
KEGG	http://www.genome.ad.jp/kegg/
Cytoscape	http://www.cytoscape.org/
Disease biology and genetics	
OMIM	www.ncbi.nlm.nih.gov/entrez/query.fcgi?db=OMIM
Genetic Association db	http://geneticassociationdb.nih.gov/
Uniprot	http://www.uniprot.org
Whole-genome association resources	
NCBI Whole Genome Association resource	http://www.ncbi.nlm.nih.gov/WGA/
Wellcome Trust Case Control Consortium (WTCCC)	http://www.wtccc.org.uk/

As the general objective is to try to reduce the scale of the analysis, it is useful to move from the large 'SNP space' that we have defined to the total 'gene space' by filtering this information down to a non-redundant list of the genes covered by the SNPs. Mapping of SNPs to genes can be achieved by tools such as SNPPER (Table 18.2). To add some statistical ranking to these genes, we could apply permutation testing or false discovery rate testing; however, in this case, we will apply the simplest measure, by capturing the lowest P value recorded for each gene. The original set of 9965 associated SNPs mapped to 2878 genes. By including the genes that are covered by the extra SNPs in LD, an additional 1532 genes are added, expanding our total 'gene space' to 4410 genes with some evidence of association with PD susceptibility.

Since LD leads to a substantial increase in associated loci, this is obviously a concern, especially when the total numbers of loci are already at an untenably high level. But taking LD into account may help to clarify potentially important genetic associations from a mass of unpromising leads. For example, the SNP, RS16837037, which was identified in the first-round screen ($P = 0.03$), maps to the hypothetical gene FLJ20203. This SNP was not progressed to the second-round screen by Maraganore *et al.* (2005), because it did not show $P < 0.01$ and it did not map to a known gene. However, RS16837037 is in complete LD ($r^2 = 1$) with RS4971106, a marker located over 108 kb away, in the intron of synaptogamin XI (SYT11). Intriguingly, Huynh *et al.* (2003) showed that SYT11, an important synaptic vesicle-forming and docking protein, is bound and modified by the parkin protein, which is mutated in autosomal recessive juvenile PD. SYT11 is expressed in the core of the Lewy bodies in sporadic PD brain sections. The loss of parkin activity could affect multiple proteins controlling docking and release of synaptic vesicle pools, explaining the deficits in dopaminergic function seen in patients with parkin mutations.

This is just an example of the types of extra information that can be gleaned from LD data. Although the SNP in SYT11 is a tempting PD candidate, its P value is also well within the bounds of type 1 error, so it should not be over-interpreted. However, follow-up of both RS16837037 and RS4971106 in a second-round screen would probably be warranted. Unfortunately, phase II HapMap data were not available to Maraganore *et al.* (2005) when they formulated their follow-up study.

18.4.6 Filtering and annotating the output of genome scans

The large number of genes under consideration emphasizes the potential risks of an unfiltered approach to genome scan analysis. By our analysis, almost 15 per cent of human genes show some evidence of association with PD. Clearly, this is not a realistic reflection of PD pathology; instead, it primarily reflects the type 1 error of this experiment. The application of filters based on biological and other rationales to the output of genome scans can help to focus follow-up, and to annotate the output of a genome scan. Such annotation in itself can be useful for the formation of biological hypotheses, and also for identifying genes in shared pathways that could be subjected to further analysis to identify putative epistatic interactions.

Using expression data

Gene expression is one of the simplest filters that can be applied to the output of a genome scan. In the case of PD, risk-enhancing genes might be expected to be expressed in the brain in view of the neurodegenerative pathology of this disease (Farrer, 2006). Other diseases may not show such clear localization to one tissue, although primary tissue involvement is the norm in most of the commonest complex diseases and medically important traits, such as type 2 diabetes (pancreas, liver and muscle), asthma (lung), obesity (adipose and endocrine tissue), and adverse drug reactions (liver, gut and skin). There are many caveats in the application of such filters; for example, a gene might be expressed in the target tissue during a defined developmental stage, or under specific cellular conditions (e.g., cell stress). Therefore, expression data should probably not be applied as an absolute filter, but if we are faced with thousands of associations and resource constraints, there may be no choice but prioritization by this method.

In the case of the PD association scan, brain is an obvious tissue on which to filter by the pathology of PD (Farrer, 2006), so it may be valuable to identify all brain-expressed genes among the 4410 PD-associated genes. A good Web-based tool for this purpose is GNF SymAtlas (http://symatlas.gnf.org/SymAtlas/), which can be used to identify all genes that are expressed in a given tissue as follows. First load the list of gene symbols or accessions (obtained from the SNP annotation) to SymAtlas; the gene list should appear in the left-hand panel of the tool. Select the entire list of genes by the check box. Next follow the 'search expression' link and select a human expression data set from the drop-down box (e.g., Human GeneAtlas GNF1H, gcRMA). Then select 'intersect with previous' from the combined results drop-down menu, select 'whole brain' at the bottom of the list of tissues, and select a 'fold above median' value. A threshold of '>2-fold' above the median would generally find most genes that are significantly expressed in a tissue, although exact definition of what constitutes expression in a tissue is a matter of some debate. This debate largely concerns issues of normalization between genes to adjust for the innate variability in experimental conditions and probe efficiency, to name a few issues (see Royce *et al.* (2005) or Quackenbush (2001) for reviews). After you press the 'search' button, the uploaded gene list will be modified to show only those genes expressed in whole brain. The expression data and gene IDs for this list of genes can be downloaded by selecting the 'download data' pull-down menu. After you run this query on the 4410 genes, 633 genes show evidence of expression in whole-brain tissue at twofold over the median. This is quite a stringent threshold for expression, and these 632 genes are all likely to represent genes that are substantially expressed in the brain. Setting a lower threshold includes a larger number of genes. For example, >1.5-fold over the median includes 850 genes; >1-fold includes 1965 genes. These will include some genes that are genuinely expressed at low levels in the brain, which might still be significant in its biology, but at great risk of including false-positive expression calls.

If expression in multiple tissues needs to be evaluated, the entire unfiltered expression data set for your list of genes can be downloaded from SymAtlas by selecting the same pull-down menu, before applying the tissue-expression filter. Manual analysis of these data in a spreadsheet has some advantages, especially when large data sets are analysed; they can be slow to process over the Web. An identical median expression level analysis to that offered by the Web interface can be carried out by calculating a median value for all tissues on a gene-by-gene basis (by the Microsoft Excel MEDIAN function). Thresholds can then be applied to genes that are expressed at 2 × the median expression for a specific tissue. This also allows combining of data; so, for example, it is possible to identify genes that are expressed at 2 × the median in any one of a range of tissues. This allows a great deal of flexibility in the stringency of filtering that is applied to the data.

Tools such as SymAtlas can also be used to run more sophisticated queries of genome-scale data sets, which can help to identify genes sharing characteristics. For example, it is possible to identify genes that share a similar expression profile across all tissues by the 'find correlated' function in the SymAtlas expression profile view (Figure 18.11). Using this function, we can identify, on a genome-wide scale, genes that show similar expression characteristics to genes with known involvement in PD. For example, ubiquitin carboxyl-terminal esterase L1 (UCHL1) recycles polymeric ubiquitin to its monomeric form and is highly specific to neurons (OMIM: 191342). Variants of this gene are linked to increased susceptibility to PD, although they are not believed to be sufficient in themselves to cause PD (Liu *et al.*, 2002). UCHL1 may be a particularly interesting gene to evaluate in terms of expression, as the ubiquitination pathway seems to be key in PD (Liu *et al.*, 2002); two other gene products, α-synuclein (SNCA) and parkin (PARK2), which are linked to familial PD, are sensitive to or involved in this pathway (Liu *et al.*, 2002). Therefore, other genes showing nominal association with PD, and expressed in the same location at UCHL1, could be strong candidates for PD.

After running the 'find correlated' function for UCHL1 in SymAtlas on a genome-wide scale, 166 genes show a 0.8 correlation with UCHL1 expression. Cross-comparison of this list of genes with the 4410 genes with nominal association to PD identifies 57 genes that are correlated with UCHL1. This includes SYT11, the gene identified by LD in Section 18.4.5. Markers in these genes are plotted against the general background of associated SNPs from the first-round screen in Figure 18.12. Notably, 13 of these genes show $P < 0.01$ in the first-round screen, and again it is tempting to speculate about the potential rationale of these candidate genes in PD. These genes might be worthy of further analysis; for example, to evaluate epistasis.

Exploring pathways, Gene Ontology (GO) and other functional annotation in genome-scale data sets

Currently, many public databases focus on the functional annotation of genes, proteins and related, disease-specific data, Entrez Gene, UniProt and OMIM being notable tools that are leading the field in this area. However, most of these tools can be queried only on a gene-by-gene basis, making them unsuitable for analysis of genome-scale gene sets. The field of microarray analysis is one area of research with strong similarities to genome-wide association analysis – both deal with highly multidimensional data on a genome-wide scale, and both involve multiple testing, generating many thousands of results, with a large false-positive burden. Tools

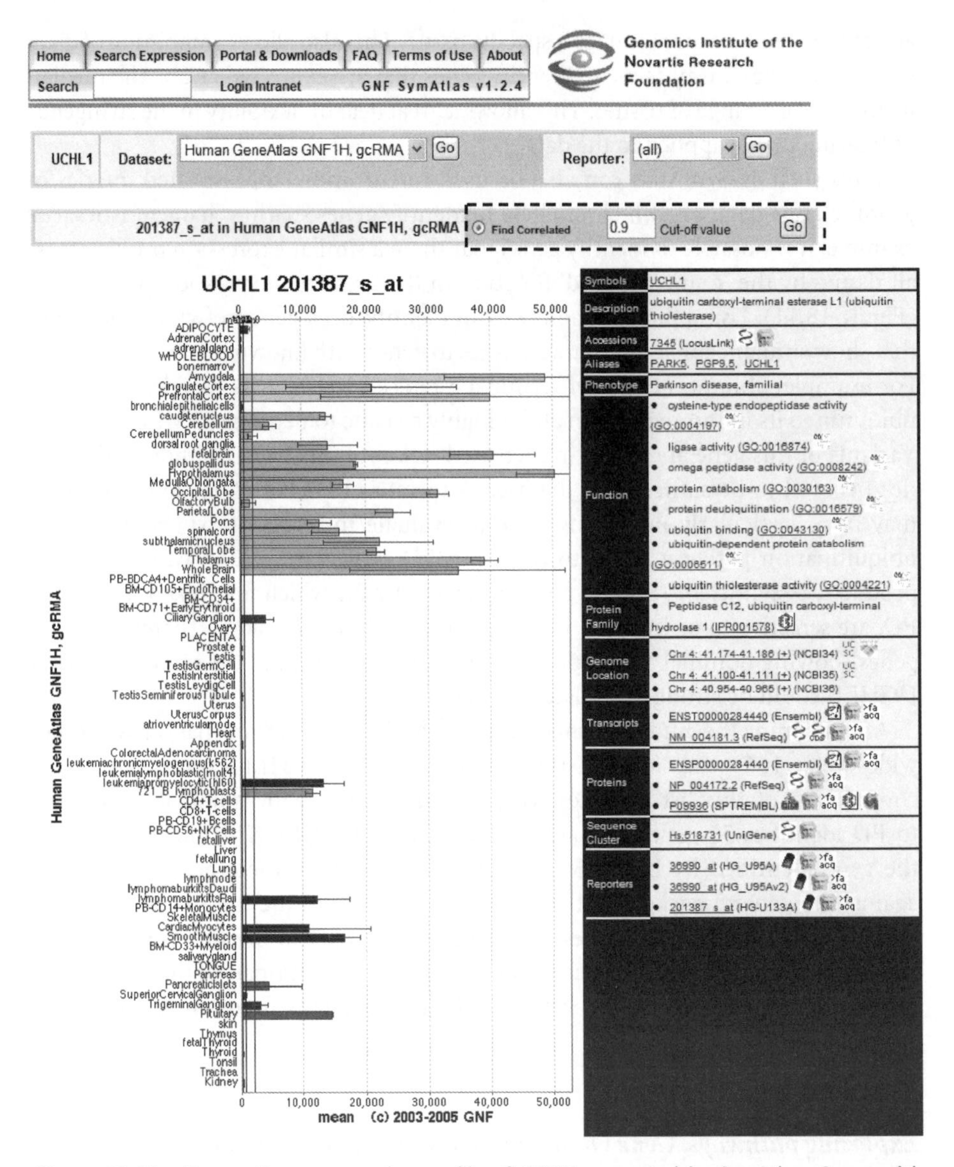

Figure 18.11 Human tissue expression profile of UCHL1 presented by SymAtlas. Genes with highly correlated expression profiles can be identified by the 'find correlated' tool. Human tissue expression profile of UCHL 1 presented by SymAtlas [SymAtlas © Genomics Institute of the Novartis Research Foundation (GNF), 2003. All rights reserved.]

specifically developed to deal with the output of genome scans are in their infancy: at the time of writing, there were no tools specifically developed for this purpose. Fortunately, tools focused on similar issues in the microarray domain are more mature, and several now available are worth exploring (Table 18.2; see Verducci *et al.* (2006) for a review).

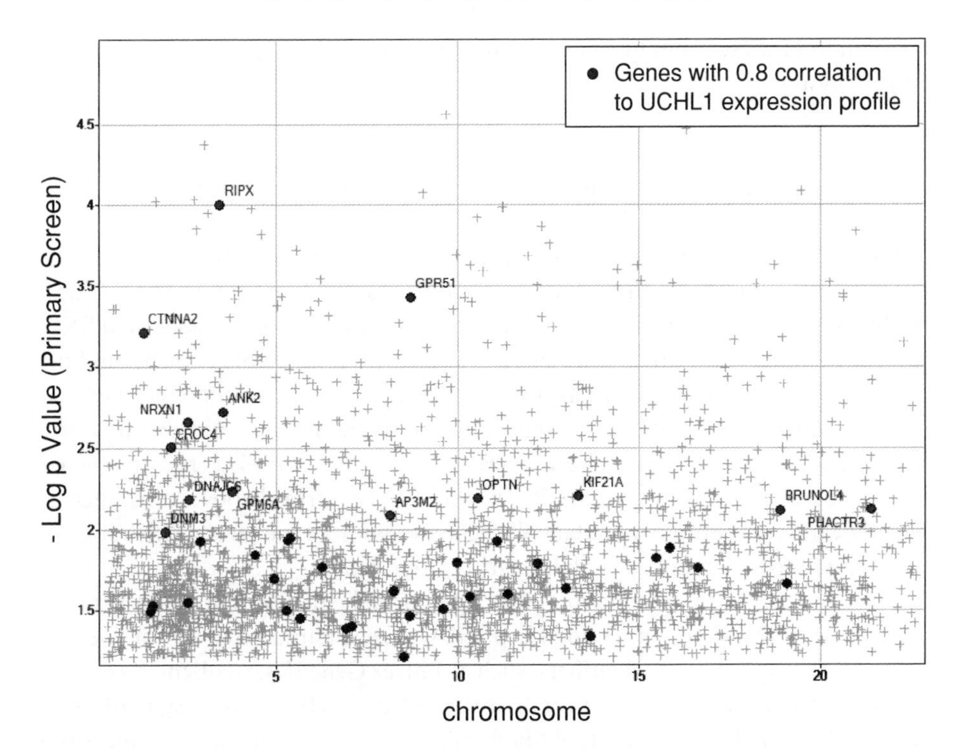

Figure 18.12 Visualization of markers in genes showing over 0.8 correlation in expression with UCHL1 plotted against the background of associated SNPs from the first-round PD association screen. Markers are plotted in chromosome order, with dithering to resolve data points for viewing

One of the most versatile tools for functional analysis of large gene sets is DAVID (Dennis *et al.*, 2003; http://david.abcc.ncifcrf.gov/), which provides a suite of data-mining tools that systematically combine functionally descriptive gene annotation based on GO (Ashburner *et al.*, 2000; see below), KEGG (Kanehisa *et al.*, 2004), Biocarta (http://www.biocarta.com) and other pathway tools with intuitive graphical displays. The tool provides exploratory visualizations of functional categories, pathways and GO terms that are enriched at statistically significant levels in the data set. Tools such as DAVID can be used in two distinct ways; simply to expedite the process of functional annotation and analysis of a list of genes for further analysis, or to identify genes that are significantly enriched in specific pathways or functional classes. The latter use is quite appropriate for gene-expression analysis, where whole pathways might be expected to show regulatory changes, and hence changes in gene expression in a specific disease state. However, in genome-wide association analysis, this is likely to be highly problematic due to the likelihood that only a limited number of genes may show genetic variation in a specific disease, whereas many may show differences in expression as a result of these variations. This is not to suggest that tools such as DAVID cannot be used to assess the significance of multiple genetic associations, but rather that their greatest strength lies in the effective annotation of these genes to enable further analysis, such as epistatic or functional SNP analysis.

Using DAVID to annotate and explore the results of genome scans

Returning to our PD genome scan case study, it would be valuable to explore the genes identified in the first round of the genome scan by Maraganore *et al.* (2005). At this stage, after initially identifying the genes on the basis of the associated SNP-to-gene mapping, very little information is available on the list of genes, beyond the HUGO identifiers. With SymAtlas, we have determined a subset of 632 genes that showed evidence of significant expression in the brain (see above). Deriving a real understanding of the biological processes and pathways of these genes on a gene-by-gene basis would be tremendously laborious. As described above, tools such as DAVID can systematically annotate all the genes with a biologically rich range of information from a number of bioinformatics resources.

The first stage in this annotation process requires the user to upload the genes of interest. The DAVID tool does not accept HUGO gene names, although it does accept a wide range of other identifiers, such as Entrez Gene IDs, GenBank accessions or Affymetrix probe IDs. These can be retrieved from HUGO gene symbols with tools such as the ID converter tool (Table 18.2). In our case, the data output from SymAtlas for the 632 brain-expressed genes includes Entrez Gene IDs, so these can be loaded directly in box A of the DAVID upload page. Once the list is submitted, a number of options are offered. In the first instance, we will use the 'functional annotation tool'. This returns an expandable list, annotating the input genes in order of significance of over-representation to GO terms, KEGG and BIOCARTA pathways (Figure 18.13). This annotation is immediately valuable, as it highlights key pathways that may be relevant to the phenotype under study. In our case, as the genes loaded are all expressed in the brain, neurological pathways are represented at highly significant levels (this is likely to be biased by the brain-expression filter rather than any rationale in PD!). These pathway classification data and functional information can be used to scan quickly for information relevant to the phenotype under analysis.

Using GO information

The controlled vocabulary of the GO Consortium provides a structured language that can be applied to the functions of genes and proteins in all organisms, even as knowledge of gene function continues to accumulate and evolve (Ashburner *et al.*, 2000; see Lomax (2005), for a review of the use of GO). The GO module in DAVID allows us to evaluate the distribution of submitted genes across three general types of classification: biological process (GOTERM_BP), cellular compo-

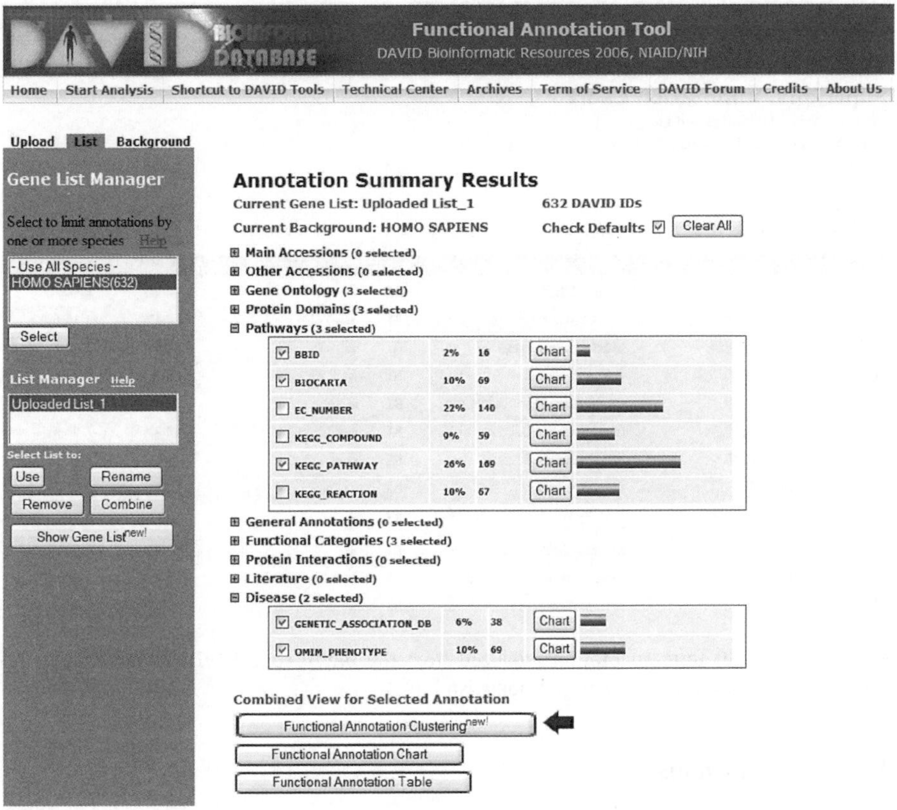

Functional Annotation Tool
DAVID Bioinformatic Resources 2006, NIAID/NIH

Home | Start Analysis | Shortcut to DAVID Tools | Technical Center | Archives | Term of Service | DAVID Forum | Credits | About Us

Upload | List | Background

Gene List Manager
Select to limit annotations by one or more species Help
- Use All Species -
HOMO SAPIENS(632)
Select

List Manager Help
Uploaded List_1
Select List to:
Use | Rename | Remove | Combine | Show Gene List new!

Annotation Summary Results
Current Gene List: Uploaded List_1 — 632 DAVID IDs
Current Background: HOMO SAPIENS — Check Defaults ☑ Clear All

⊞ Main Accessions (0 selected)
⊞ Other Accessions (0 selected)
⊞ Gene Ontology (3 selected)
⊞ Protein Domains (3 selected)
⊟ Pathways (3 selected)

☑ BBID	2%	16	Chart
☑ BIOCARTA	10%	69	Chart
☐ EC_NUMBER	22%	140	Chart
☐ KEGG_COMPOUND	9%	59	Chart
☑ KEGG_PATHWAY	26%	169	Chart
☐ KEGG_REACTION	10%	67	Chart

⊞ General Annotations (0 selected)
⊞ Functional Categories (3 selected)
⊞ Protein Interactions (0 selected)
⊞ Literature (0 selected)
⊟ Disease (2 selected)

☑ GENETIC_ASSOCIATION_DB	6%	38	Chart
☑ OMIM_PHENOTYPE	10%	69	Chart

Combined View for Selected Annotation
Functional Annotation Clustering new!
Functional Annotation Chart
Functional Annotation Table

Figure 18.13 Annotation summary from the DAVID tool. DAVID Bioinformatics Resources, http://david.abcc.ncifcrf.gov/home.jsp

nent (GOTERM_CC) and molecular function (GOTERM_MF). These are divided further into five levels of annotation of increasing specificity of term coverage. For example, at level 1 of the biological process classification, genes are just subdivided into the broad terms of 'cellular process' and 'development', covering ~65 per cent of submitted genes. However, at level 5, annotation is at its most specific and biologically meaningful, with disease-relevant terms such as 'neuron differentiation' or 'synaptic transmission', although this comes at a cost, as these terms cover only 2.5 per cent and 3.8 per cent of genes respectively. Intermediate-level annotation allows for broader terms, such as 'nervous system development', a term used at level 3 and covering 7.5 per cent of genes. These differing levels can be very useful for modifying the threshold of inclusion for selection of genes for follow-up based on biological rationale. Evaluation of the level 5 biological process term annotations quickly identifies a number of genes among the 632 brain-expressed genes that are involved in processes that are highly relevant to PD; these are summarized by a tabular visualization (Figure 18.14).

DAVID Bioinformatic Resources 2006
National Institute of Allergy and Infectious Disease (NIAID), NIH

Functional Annotation Chart
Current Gene List: Uploaded List_1
Current Background: HOMO SAPIENS
632 DAVID IDs

⊞ **Options**
Rerun Using Options | Create Sublist | Download File

Sublist	Category	Term	RT	Genes	Count	%	P-Value
☐	GOTERM_BP_5	neuron differentiation	RT		16	2.5%	3.4E-7
☐	GOTERM_BP_5	phosphate metabolism	RT		53	8.4%	1.1E-6
☐	GOTERM_BP_5	synaptic transmission	RT		24	3.8%	1.7E-6
☐	GOTERM_BP_5	cell-cell signaling	RT		20	3.2%	1.7E-6
☐	GOTERM_BP_5	transmission of nerve impulse	RT		19	3.0%	4.2E-6
☐	GOTERM_BP_5	neurophysiological process	RT		19	3.0%	1.5E-5
☐	GOTERM_BP_5	biopolymer modification	RT		71	11.2%	4.2E-5
☐	GOTERM_BP_5	neuron maturation	RT		7	1.1%	4.6E-5
☐	GOTERM_BP_5	cytoskeleton organization and biogenesis	RT		27	4.3%	1.0E-4
☐	GOTERM_BP_5	vesicle-mediated transport	RT		25	4.0%	3.8E-4
☐	GOTERM_BP_5	neurotransmitter secretion	RT		6	0.9%	4.1E-4
☐	GOTERM_BP_5	axon guidance	RT		7	1.1%	4.3E-4

Figure 18.14 GO term annotation produced from the DAVID tool. DAVID Bioinformatics Resources, http://david.abcc.ncifcrf.gov/home.jsp

Using pathway tools

DAVID also provides annotation on highly characterized pathways contained in KEGG (Kanehisa *et al.*, 2004), BioCarta (www.biocarta.com/) and a selection of other databases. While GO is based mainly on functional inference by homology, this information is based on experimental evidence and can be valuable for placing a gene in a validated pathway context. The volume of data in these databases is somewhat limited, but generally of high quality. Figure 18.15 shows some of the pathways identified among the 632 brain-expressed genes. Many of these pathways are also highly relevant to the PD phenotype (Farrer, 2006). For example, among the KEGG pathways, 12 genes are identified with a role in long-term potentiation (LTP), and nine genes are identified with a role in long-term depression (LTD), which are, respectively, a long-lasting increase, or an enduring decrease in synaptic strength and plasticity, known to play an important role in PD (Pisani *et al.*, 2005). Other KEGG pathways are also highlighted, with 22 genes identified in the axon-guidance pathway. In the case of the highly curated BioCarta pathways, five genes are highlighted with a role in the formation of synaptic junctions, which may also be important in PD. In each case, if the user follows the hyperlinked KEGG or BioCarta term, a detailed pathway model is returned, which can rapidly put a gene into full biological context.

It is worth re-emphasizing that strong conclusions should not be drawn from the results of this analysis of the first-round associations with PD susceptibility

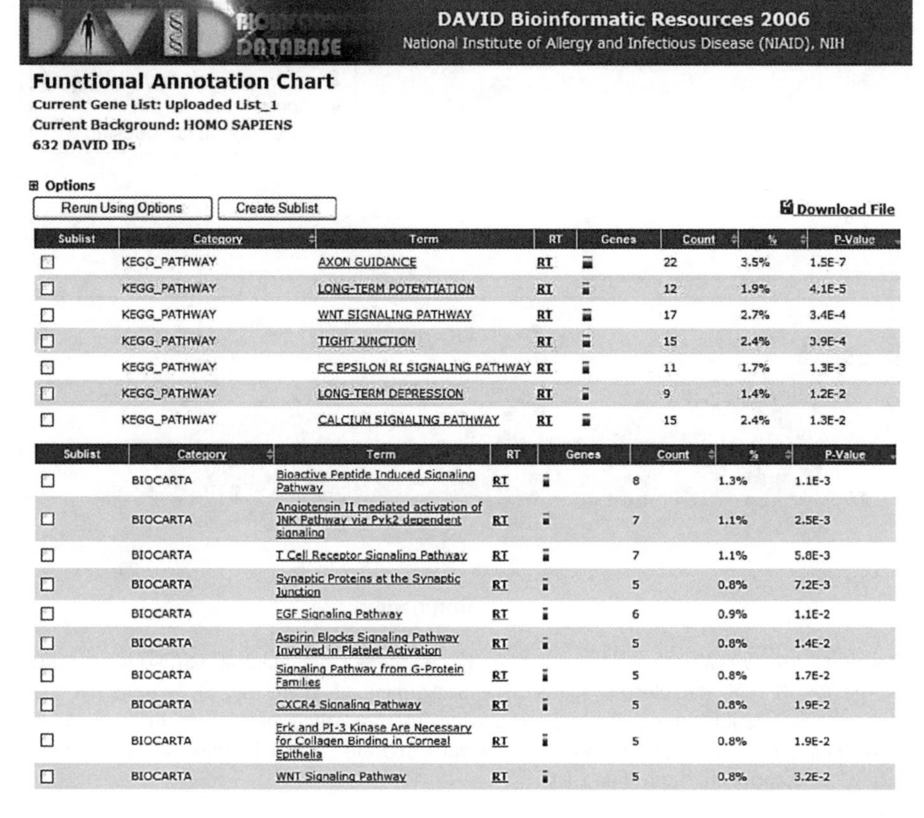
DAVID Bioinformatic Resources 2006
National Institute of Allergy and Infectious Disease (NIAID), NIH

Functional Annotation Chart
Current Gene List: Uploaded List_1
Current Background: HOMO SAPIENS
632 DAVID IDs

Options
Rerun Using Options | Create Sublist | Download File

Sublist	Category	Term	RT	Genes	Count	%	P-Value
☐	KEGG_PATHWAY	AXON GUIDANCE	RT		22	3.5%	1.5E-7
☐	KEGG_PATHWAY	LONG-TERM POTENTIATION	RT		12	1.9%	4.1E-5
☐	KEGG_PATHWAY	WNT SIGNALING PATHWAY	RT		17	2.7%	3.4E-4
☐	KEGG_PATHWAY	TIGHT JUNCTION	RT		15	2.4%	3.9E-4
☐	KEGG_PATHWAY	FC EPSILON RI SIGNALING PATHWAY	RT		11	1.7%	1.3E-3
☐	KEGG_PATHWAY	LONG-TERM DEPRESSION	RT		9	1.4%	1.2E-2
☐	KEGG_PATHWAY	CALCIUM SIGNALING PATHWAY	RT		15	2.4%	1.3E-2

Sublist	Category	Term	RT	Genes	Count	%	P-Value
☐	BIOCARTA	Bioactive Peptide Induced Signaling Pathway	RT		8	1.3%	1.1E-3
☐	BIOCARTA	Angiotensin II mediated activation of JNK Pathway via Pyk2 dependent signaling	RT		7	1.1%	2.5E-3
☐	BIOCARTA	T Cell Receptor Signaling Pathway	RT		7	1.1%	5.8E-3
☐	BIOCARTA	Synaptic Proteins at the Synaptic Junction	RT		5	0.8%	7.2E-3
☐	BIOCARTA	EGF Signaling Pathway	RT		6	0.9%	1.1E-2
☐	BIOCARTA	Aspirin Blocks Signaling Pathway Involved in Platelet Activation	RT		5	0.8%	1.4E-2
☐	BIOCARTA	Signaling Pathway from G-Protein Families	RT		5	0.8%	1.7E-2
☐	BIOCARTA	CXCR4 Signaling Pathway	RT		5	0.8%	1.9E-2
☐	BIOCARTA	Erk and PI-3 Kinase Are Necessary for Collagen Binding in Corneal Epithelia	RT		5	0.8%	1.9E-2
☐	BIOCARTA	WNT Signaling Pathway	RT		5	0.8%	3.2E-2

Figure 18.15 Pathway annotation produced from the DAVID tool. DAVID Bioinformatics Resources, http://david.abcc.ncifcrf.gov/home.jsp

obtained by Maraganore *et al.* (2005). There is no doubt that the vast majority of associations recorded here are likely to be false because of the influence of multiple testing alone. However, the rapid identification of genes in this first tranche of data that are expressed in the CNS or in pathways with a role in PD susceptibility is of great value for the design of follow-up studies, even if the preliminary associations are at the most nominal level. These tools allow the geneticist rapidly to dissect genes and pathways of interest from a very noisy genome-wide signal. Figure 18.16 shows the end result of this analysis: SNPs mapping to genes involved in PD-relevant pathways and biological processes highlighted against the genome-wide background of 9965 nominally associated SNPs. To simplify the whole process even further, Figure 18.17 shows the steps taken and demonstrates how at each stage the dimensionality of the analysis problem is reduced. In a world of unlimited resources, both financial and computational, these biological filtering steps might not be necessary. But in our world, filtering of results is necessary to improve the true-positive to false-positive ratio. In this case study, we move from a 'SNP space' of 64 217 nominally associated SNPs to a 'gene space' of 4410 genes, to a final list of 119 genes that make the cut as

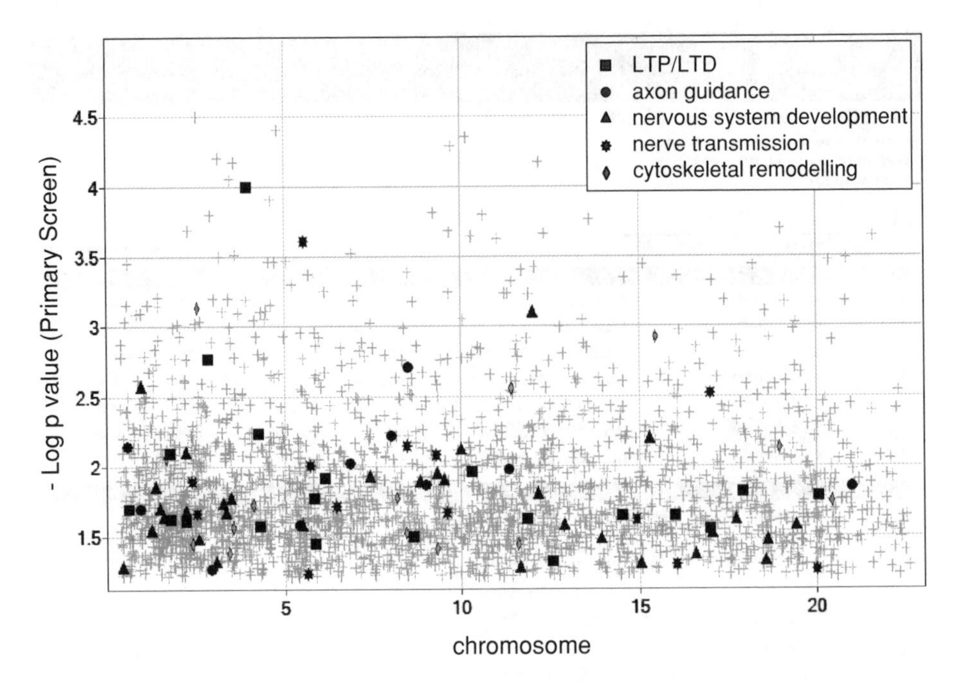

Figure 18.16 Visualization of markers in genes in key neurological pathways identified by the DAVID tool and plotted against the background of associated SNPs from the first-round PD association screen. Markers are plotted in chromosome order, with dithering to resolve data points for viewing. Data plotted with Spotfire. DAVID Bioinformatics Resources, http://david.abcc.ncifcrf.gov/home.jsp

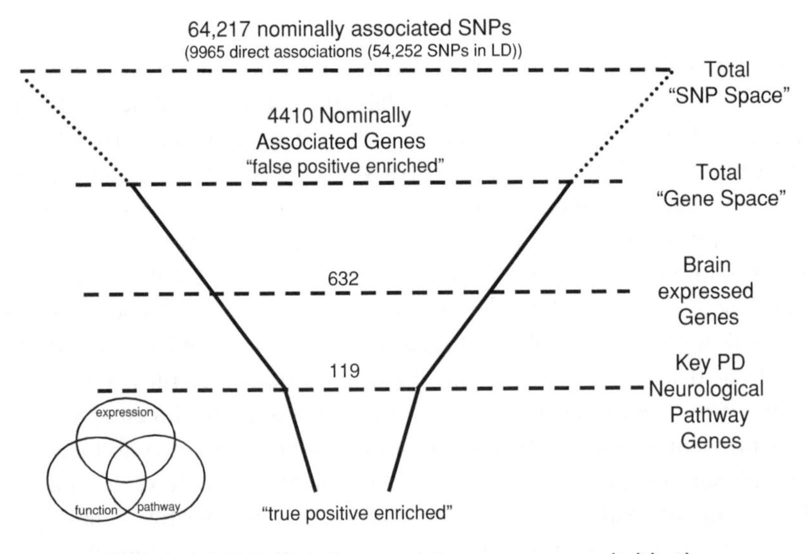

Figure 18.17 A filtered approach to genome scan prioritization

our most 'true-positive-enriched' set of genes. SNPs in only 42 of these genes were analysed in the tier 2 replication study by Maraganore *et al.* (2005).

Maraganore *et al.* (2005) identified 26 SNPs with P values of <0.01 in both tiers 1 and 2a of their study. Of these SNPs, 11 showed P values of <0.01 and the same associated alleles in tiers 1 and 2a. They considered these to be the only truly replicated loci in their study. Unfortunately, four independent research teams failed to replicate these associations (Myers, 2006). Given the problems in replicating associations with genes of modest effect (see Section 18.2.4), it is possible that these are in fact true associations, and with that in mind, we took a step back to re-examine the entire data set for other nominally replicated loci. Seventy-seven SNPs in 37 genes show nominally significant P values with the same alleles (replication) in both tiers 1 and 2a samples. Five of these genes are in our subset of 632 genes that are significantly expressed in the brain, and one gene, EphB6, is contained within the set of neurological pathway genes. This gene is modestly associated in both tiers, with $P = 0.017$ in tier 1 and $P = 0.037$ in tier 2a, but the rationale of the gene in PD is interesting. Very little is known about the biological role of EphB6, other than that it is predominantly expressed in the brain and pancreas, but, functionally, it is very interesting, as it has the primary structural features of an Eph family receptor tyrosine kinase, but it lacks several invariant residues that have been shown to be essential for tyrosine kinase activity (Matsuoka *et al.*, 1997). Although EphB6 has been shown to be catalytically inactive, it may not be biologically inactive. In a follow-up to their original study, Matsuoka *et al.* (2005) demonstrated that EphB6 exerted biphasic effects in response to different concentrations of its activating ligand ephrin-B2. At low ligand concentrations, EphB6 promoted cell adhesion and migration, but at higher ligand concentrations, it induced repulsion and inhibited migration. The authors suggested that this molecular switch for the functional transition of cells from an adhesive to a migratory state might play a key role in axon guidance. Considering this, it is particularly arresting that EphB6 is strongly expressed in the ganglionic eminence, a source of tangentially migrating neurons, which has been reported to improve symptoms when transplanted into PD patients (Lee *et al.*, 2003).

18.5 Conclusions

In the future, complete genome resequencing in a large population of patients and controls will be the most comprehensive approach to understanding complex disease susceptibility. This approach would cover the complete spectrum of coding and non-coding variants and, unlike the current use of LD-based genotyping panels, it would allow comprehensive testing of both rare and common variants in disease. Several technology companies are focused on making genome resequencing affordable (Service, 2006), suggesting that the '$1000 genome' may be more than a glint in Craig Venter's eye. If and when this breakthrough comes, genome resequencing may become the method of choice for genome scanning, although it is hard to overstate how challenging this is likely to be to interpret.

This chapter has reviewed the key steps in the design and implementation of genetic studies, using annotated genomic sequence as a template. As technologies advance, studies of the genome will become more and more precise, and much of the genetic analysis that we know today may become an increasingly *in silico* process. Ten years ago, it might have been difficult to believe that the genetic study process would have changed as dramatically to what we find today. Even as this chapter is being written, human genetics is clearly on the cusp of further great change. SNP chips have effectively removed traditional limitations on genotyping, although it is not clear how effectively they will detect rare disease alleles. Trends toward the formation of major research consortia such as the WTCCC (http://www.wtccc.org.uk/) promise to bring well-powered genetic association studies much closer to reality. As the genomic information wave continues to roll forward, we may be looking at much more intelligently designed genetic studies, with maps which account for local recombination, LD and a detailed knowledge of the impact of variation in genes and regulatory regions. With technology and samples in hand, we will just have to wait to see how close this brings us to the elucidation of the genetic basis of common human diseases.

References

Altshuler, D., Hirschhorn, J. N., Klannemark, M. *et al.* (2000). The common PPARG Pro12Ala polymorphism is associated with decreased risk of type 2 diabetes. *Nat Genet* **26**, 76–80.

Arking, D. E., Pfeufer, A., Post, W. *et al.* (2006). A common genetic variant in the NOS1 regulator NOS1AP modulates cardiac repolarization. *Nat Genet* **38**(6), 644–651.

Ashburner, M., Ball, C. A., Blake, J. A. *et al.* (2000). Gene ontology: tool for the unification of biology. The Gene Ontology Consortium. *Nat Genet* **25**(1), 25–29.

Balding, D. J., Bishop, M. and Cannings, C. (Eds) (2003). *Handbook of Statistical Genetics*, 2nd edn. Chichester: Wiley.

Benjamini, Y. and Hochberg, Y. (1995). Controlling the false discovery rate: a practical and powerful approach to multiple testing. *J R Stat Soc [Ser B]* **57**, 289–300.

Botstein, D. and Risch, N. (2003). Discovering genotypes underlying human phenotypes: past successes for mendelian disease, future approaches for complex disease. *Nat Genet* **33** (Suppl.) 228–237.

Camilleri, M., Atanasova, E., Carlson, P. J. *et al.* (2002). Serotonin-transporter polymorphism pharmacogenetics in diarrhea-predominant irritable bowel syndrome. *Gastroenterology* **123**(2), 425–432.

Cardon, L. R. and Palmer, L. J. (2003). Population stratification and spurious allelic association. *Lancet* **361**, 598–604.

Churchill, G. A. and Doerge, R. W. (1994). Empirical threshold values for quantitative trait mapping. *Genetics* **138**, 963–971.

Clark, A. G., Hubisz, M. J., Bustamante, C. D. *et al.* (2005). Ascertainment bias in studies of human genome-wide polymorphism. *Genome Res* **15**, 1496–1502.

Clarimon, J., Scholz, S., Fung, H.-C. *et al.* (2006). Conflicting results regarding the semaphorin gene (SEMA5A) and the risk for Parkinson disease. *Am J Hum Genet* **78**, 1082–1084.

Cohen, J. C., Pertsemlidis, A., Fahmi, S. *et al.* (2006). Multiple rare variants in NPC1L1 associated with reduced sterol absorption and plasma low-density lipoprotein levels. *Proc Natl Acad Sci U S A* **103**(6), 1810–1815.

Cordell, H. J. (2002). Epistasis: what it means, what it doesn't mean, and statistical methods to detect it in humans. *Hum Mol Genet* **11**, 2463–2468.

de Bakker, P. I., Yelensky, R., Pe'er, I. *et al.* (2005). Efficiency and power in genetic association studies. *Nat Genet* **37**(11), 1217–1223.

Dennis, G., Jr., Sherman, B. T., Hosack, D. A. *et al.* (2003). DAVID: Database for Annotation, Visualization, and Integrated Discovery. *Genome Biol* **4**(5), P3.

Dudbridge, F. (2006). A note on permutation tests in multistage association scans. *Am J Hum Genet* **78**(6), 1094–1095.

Dudbridge, F., Gusnanto, A., Koeleman, B. P. (2006). Detecting multiple associations in genome-wide studies. *Hum Genomics* **2**(5), 310–317.

ENCODE Project Consortium (2004). The ENCODE (ENCyclopedia Of DNA Elements) Project. *Science* **306**(5696), 636–640.

Farrer, M. J., Haugarvoll, K., Ross, O. A. *et al.* (2006). Genomewide association, Parkinson disease, and PARK10. *Am J Hum Genet* **78**, 1084–1088.

Farrer, M. J. (2006). Genetics of Parkinson disease: paradigm shifts and future prospects. *Nat Rev Genet* **7**(4), 306–318.

Goris, A., Williams-Gray, C. H., Foltynie, T. *et al.* (2006). No evidence for association with Parkinson disease for 13 single-nucleotide polymorphisms identified by whole-genome association screening. *Am J Hum Genet* **78**, 1088–1090.

Hinds, D. A., Stuve, L. L., Nilsen, G. B. *et al.* (2005). Whole-genome patterns of common DNA variation in three human populations. *Science* **307**, 1072–1079.

Huynh, D. P., Scoles, D. R., Nguyen, D. *et al.* (2003). The autosomal recessive juvenile Parkinson disease gene product, parkin, interacts with and ubiquitinates synaptotagmin XI. *Hum Mol Genet* **12**(20), 2587–2597.

Ireland, J., Carlton, V. E., Falkowski, M. *et al.* (2006). Large-scale characterization of public database SNPs causing non-synonymous changes in three ethnic groups. *Hum Genet* **119**(1–2), 75–83.

Judson, R., Salisbury, B., Schneider, J. *et al.* (2002). How many SNPs does a genome-wide haplotype map require? *Pharmacogenomics* **3**, 379–391.

Kanehisa, M., Goto, S., Kawashima, S. *et al.* (2004). The KEGG resource for deciphering the genome. *Nucleic Acids Res* 32(Database Issue), D277–280.

Ke, X., Hunt, S., Tapper, W. *et al.* (2004). The impact of SNP density on fine-scale patterns of linkage disequilibrium. *Hum Mol Genet* **13**(6), 577–588.

Klein, R. J., Zeiss, C., Chew, E. Y. *et al.* (2005). Complement factor H polymorphism in age-related macular degeneration. *Science* **308**(5720), 385–389.

Kruglyak, L. (1999). Prospects for whole-genome linkage disequilibrium mapping of common disease genes. *Nat Genet* **22**, 139–144.

Kruglyak, L. (2005). Power tools for human genetics. *Nat Genet* **37**, 1299–1300.

Lee, C. C., Lin, S. Z., Wang, Y. *et al.* (2003). First human ventral mesencephalon and striatum cografting in a Parkinson patient. *Acta Neurochir Suppl* **87**, 159–162.

Li, Y., Rowland, C., Schrodi, S. *et al.* (2006). A case-control association study of the 12 single-nucleotide polymorphisms implicated in Parkinson disease by a recent genome scan. *Am J Hum Genet* **78**, 1090–1092.

Liu, Y., Fallon, L., Lashuel, H. A. *et al.* (2002). The UCH-L1 gene encodes two opposing enzymatic activities that affect alpha-synuclein degradation and Parkinson's disease susceptibility. *Cell* **111**(2), 209–218.

Lomax, J. (2005). Get ready to GO! A biologist's guide to the Gene Ontology. *Brief Bioinform* **6**(3), 298–304.

Lowe, C. E., Cooper, J. D., Chapman, J. M. *et al.* (2004). Cost-effective analysis of candidate genes using htSNPs: a staged approach. *Genes Immun* **5**, 301–305.

Maraganore, D. M., Lesnick, T. G., Elbaz, A. *et al.* and UCHL1 Global Genetics Consortium (2004). UCHL1 is a Parkinson's disease susceptibility gene. *Ann Neurol* **55**, 512–521.

Maraganore, D. M., de Andrade, M., Lesnick, T. G. *et al.* (2005). High-resolution whole-genome association study of Parkinson disease. *Am J Hum Genet* **77**, 685–693.

Maraganore, D. M., de Andrade, M., Lesnick, T. G. *et al.* (2006). Response from Maraganore *et al. Am J Hum Genet* **78**, 1092–1094.

Marchini, J., Donnelly, P. and Cardon, L. R. (2005). Genome-wide strategies for detecting multiple loci that influence complex diseases. *Nat Genet* **37**(4), 413–417.

Matsuoka, H., Iwata, N., Ito, M. *et al.* (1997). Expression of a kinase-defective Eph-like receptor in the normal human brain. *Biochem Biophys Res Commun* **235**(3), 487–492.

Matsuzaki, H., Dong, S., Loi, H. *et al.* (2004). Genotyping over 100,000 SNPs on a pair of oligonucleotide arrays. *Nat Methods* **1**, 109–111.

Mitchell, A. A., Cutler, D. J. and Chakravarti, A. (2003). Undetected genotyping errors cause apparent overtransmission of common alleles in the transmission/disequilibrium test. *Am J Hum Genet* **72**, 598–610.

Mootha, V. K., Lepage, P., Miller, K. *et al.* (2003). Identification of a gene causing human cytochrome c oxidase deficiency by integrative genomics. *Proc Natl Acad Sci U S A* **100**(2), 605–610.

Morris, A. P. (2005). Direct analysis of unphased SNP genotype data in population-based association studies via Bayesian partition modelling of haplotypes. *Genet Epidemiol* **29**(2), 91–107.

Newman, T. L., Rieder, M. J., Morrison, V. A. *et al.* (2006). High-throughput genotyping of intermediate-size structural variation. *Hum Mol Genet* **15**(7), 1159–1167.

Nicolae, D. L., Wen, X., Voight, B. F. *et al.* (2006). Coverage and characteristics of the Affymetrix GeneChip Human Mapping 100K SNP Set. *PLoS Genet* **2**(5), e67.

Pe'er, I., Chretien, Y. R., de Bakker, P. I. *et al.* (2006a). Biases and reconciliation in estimates of linkage disequilibrium in the human genome. *Am J Hum Genet* **78**, 588–603.

Pe'er, I., de Bakker, P. I., Maller, J. *et al.* (2006b). Evaluating and improving power in whole-genome association studies using fixed marker sets. *Nat Genet* **38**(6), 663–667.

Pisani, A., Centonze, D., Bernardi, G. *et al.* (2005). Striatal synaptic plasticity: implications for motor learning and Parkinson's disease. *Mov Disord* **20**(4), 395–402.

Pritchard, J. K. (2001). Are rare variants responsible for susceptibility to complex diseases? *Am J Hum Genet* **69**, 124–137.

Pritchard, J. K. and Przeworski, M. (2001). Linkage disequilibrium in humans: models and data. *Am J Hum Genet* **69**(1), 1–14.

Purcell, S., Cherny, S. S. and Sham, P. C. (2003). Genetic power calculator: design of linkage and association genetic mapping studies of complex traits. *Bioinformatics* **19**(1), 149–150.

Quackenbush, J. (2001). Computational analysis of microarray data. *Nat Rev Genet* **2**(6), 418–427.

Reich, D. E. and Lander, E. S. (2001). On the allelic spectrum of human disease. *Trends Genet* **17**, 502–510.

Risch, N. and Merikangas, K. (1996). The future of genetic studies of complex human diseases. *Science* **273**, 1516–1517.

Royce, T. E., Rozowsky, J. S., Bertone, P. *et al.* (2005). Issues in the analysis of oligonucleotide tiling microarrays for transcript mapping. *Trends Genet* **21**(8), 466–475.

Schaid, D. J. and Rowland, C. (1998). Use of parents, sibs, and unrelated controls for detection of associations between genetic markers and disease. *Am J Hum Genet* **63**, 1492–1506.

Service, R. F. (2006). Gene sequencing. The race for the $1000 genome. *Science* **311**(5767), 1544–1546.

Skipper, L., Li, Y., Bonnard, C. *et al.* (2005). Comprehensive evaluation of common genetic variation within LRRK2 reveals evidence for association with sporadic Parkinson's disease. *Hum Mol Genet* **14**(23), 3549–3556.

Slater, H. R., Bailey, D. K., Ren, H. *et al.* (2005). High-resolution identification of chromosomal abnormalities using oligonucleotide arrays containing 116,204 SNPs. *Am J Hum Genet* **77**(5), 709–726.

SNP Consortium (2001). A map of human genome sequence variation containing 1.4 million SNPs. *Nature* **409**, 928–933.

Spector, T. D., Ahmadi, K. R. and Valdes, A. M. (2006). When is a replication not a replication? Or how to spot a good genetic association study. *Arthritis Rheum* **54**(4), 1051–1054.

Steemers, F. J., Chang, W., Lee, G. *et al.* (2006). Whole-genome genotyping with the single-base extension assay. *Nat Methods* **3**(1), 31–33.

Tan, E. K., Puong, K. Y., Chan, D. K. *et al.* (2005). Impaired transcriptional upregulation of parkin promoter variant under oxidative stress and proteasomal inhibition: clinical association. *Hum Genet* **118**(3–4), 484–488.

Terwilliger, J. D. and Hiekkalinna, T. (2006). An utter refutation of the 'fundamental theorem of the HapMap'. *Eur J Hum Genet* **14**, 426–437.

Thomas, D. C., Haile, R. W. and Duggan, D. (2005). Recent developments in genomewide association scans: a workshop summary and review. *Am J Hum Genet* **77**, 337–345.

Verducci, J. S., Melfi, V. F., Lin, S. *et al.* (2006). Microarray analysis of gene expression: considerations in data mining and statistical treatment. *Physiol Genomics* **25**(3), 355–363.

Vermeire, S., Wild, G., Kocher, K. *et al.* (2002). CARD15 genetic variation in a Quebec population: prevalence, genotype–phenotype relationship, and haplotype structure. *Am J Hum Genet* **71**(1), 74–83.

Wang, W. Y., Barratt, B. J., Clayton, D. G. *et al.* (2005). Genome-wide association studies: theoretical and practical concerns. *Nat Rev Genet* **6**(2), 109–118.

Wang, H., Thomas, D. C., Pe'er, I. *et al.* (2006). Optimal two-stage genotyping designs for genome-wide association scans. *Genet Epidemiol* **30**(4), 356–368.

Weiss, K. M. and Terwilliger, J. D. (2000). How many diseases does it take to map a gene with SNPs? *Nat Genet* **26**, 151–157.

West, A. B., Maraganore, D., Crook, J. *et al.* (2002). Functional association of the parkin gene promoter with idiopathic Parkinson's disease. *Hum Mol Genet* **11**(22), 2787–2792.

Wright, A. F. and Hastie, N. D. (2001). Complex genetic diseases: controversy over the Croesus code. *Genome Biol* **2**, 2007.

19

A Bioinformatics Perspective on Genetics in Drug Discovery and Development

Christopher Southan[1], Magnus Ulvsbäck[2] and Michael R. Barnes[3]

[1] *Global Compound Sciences, AstraZeneca R&D Mölndal, Sweden*

[2] *Molecular Pharmacology, AstraZeneca R&D Mölndal, Sweden*

[2] *Bioinformatics, GlaxoSmithKline Pharmaceuticals, Harlow, Essex, UK*

19.1 Introduction

Despite occasional lapses into hyperbole about the potential of genetics and genomics as a cure for all ills, there is no doubt that these rapidly maturing fields are beginning to edge beyond 'future promise' into a much more serious phase where tangible impact is becoming evident. Genetic and genomic approaches to drug discovery are starting to become an integral part of pharmaceutical research and development. Already there are a few drugs that have been developed against targets identified by 'genomics approaches'; for example, a small molecule inhibitor of cathepsin K was developed for treatment of osteoporosis after identification of this target by expression analysis in bone tissues, and subsequent mouse knockout analysis (Gowen *et al.*, 1999). Still, it is not easy to find a published example of a drug that has been identified by genetics, although, curiously, cathepsin K might also have been identified by this route, as it is also mutated in pycnodysostosis, a monogenic disease of bone (Gelb *et al.*, 1996). However, it may not be long before we are talking about concrete examples of genetics

Bioinformatics for Geneticists, Second Edition. Edited by Michael R. Barnes
 ISBN 978-0-470-02619-9 (HB) ISBN 978-0-470-02620-5 (PB)

playing a major role in target identification leading to novel drugs. Genetic association analysis of complex diseases is already yielding new disease-validated targets (Roses *et al.*, 2005); as these move through the drug discovery process, it will be only a matter of time before drugs developed against this new cohort of targets enter the clinic.

Genetics offers opportunities for the enhancement of drug discovery at almost every stage. As a source of target identification, it offers a unique opportunity to identify targets with no precedent rationale in disease. The hope is that such disease-validated targets will logically lead to more efficacious – and with effective development – safer new chemical entities (NCEs) that act upon these targets. This is quite an urgent hope, as, historically, more than 89 per cent of NCEs entering development have not reached the clinic (Kola and Landis, 2004). The failure of these compounds can be multifaceted. Firstly, they may show insufficient efficacy, often as a result of inadequate target validation. Secondly, they may have unacceptable toxicity profiles in animals or man. Finally, even after clinical trials involving hundreds of patients, they still retain the risk of unexpected side effects or toxicity in a subset of the population due to rare or population-specific adverse events. Genetics can be an underlying factor in all these issues of failure in the drug discovery process; pharmaceutical companies are now beginning to recognize this and invest their efforts accordingly in this area.

These multiple sources of attrition in drug discovery can be addressed, at least in part, by the integration of genetic analysis into the drug discovery process. Genetics is already yielding new disease-validated targets (Roses *et al.*, 2005). Further ahead in the development process, pharmacogenetics (PGx) is also having an impact by improving the understanding of the role of genetic variation in differential response and adverse reactions to drugs (Roses, 2002; Lindpaintner, 2003; Goldstein and Tate, 2004).

19.1.1 Genetics and bioinformatics in drug discovery – some unique challenges

Bioinformatics has traditionally led the search for new targets in the pharmaceutical industry by improving the understanding of disease biology and the criteria for druggability in target proteins. These bioinformatics-led concepts are also critical to the effective integration of genetics within the drug discovery and development process. A good understanding of disease biology and effective chemistry is not the only requirement for an efficacious drug; we also must understand how variation at the target affects drug action, and how variation in other genes affects the way drugs are absorbed, disseminated, metabolized and excreted. Genetic analysis in the drug development paradigm also faces some unique challenges; for example, the exquisite rarity of some adverse reactions makes collection of sufficient samples for well-powered genetic analysis almost impossible. Given some of these very special challenges, it is more important than ever that the right questions are asked and answered during genetic analysis. Bioinformatics is probably one of the most important means to ensure that this is done. This may include modelling the impact of variation on a drug target, identifying genes, which may be involved in the metabolism of a drug, or navigating a pathway to move from a disease-associated gene to a potential drug target. Many of the concepts underpinning these activities are covered in detail in other chapters in this book; this chapter focuses on some of the specific bioinformatics issues that arise in the application of genetics to drug discovery and development.

19.1.2 Genetics in the pharmaceutical research and development paradigm

Figure 19.1 illustrates how genetics can affect drug discovery and development. The figure shows some of the common genetics activities that are being incorporated into drug discovery and development pipelines. These activities include inputting targets identified by disease genetics at the start of the pipeline, screening patient populations for genetic variants that might alter drug efficacy or safety, and finally integrating a pharmacogenetic element into the clinical trial process during drug development. The last activity is looking increasingly like a possible future regulatory requirement as the US Food and Drug Administration (FDA) becomes more and more PGx focused. The FDA conducted a survey of recent investigational new drug (INDs) and new drug applications (NDAs) to identify the extent to which PGx is used in clinical studies (Lesko and Woodcock, 2002). The survey found over 15 applications in which PGx tests were reported, with all but one test related to pharmacogenetic variability in cytochrome P450 (CYP) enzymes. Despite the interest of the FDA, pharmaceutical companies are still moving forward quite cautiously in integrating PGx into their drug development programmes. In the case of the 15 applications mentioned above,

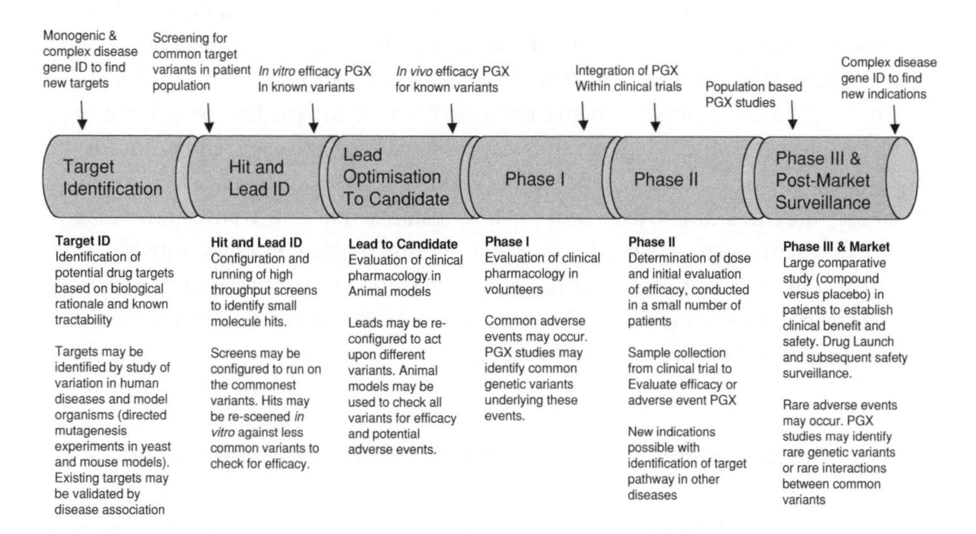

Figure 19.1 The impact of genetics on the drug discovery and development process

differences thought to be related to PGx subgroups were not used as a basis for any specific dosing recommendations on the product labels; however, it seems likely that this may well be a future direction for the industry.

19.2 Target genetics

19.2.1 Introduction to the use of genetics and bioinformatics for target discovery and progression

As genetic and genomic studies provide new insights into the molecular mechanisms of disease, it is possible that these insights will open up new opportunities for diagnosis, classification, prognosis, therapeutic intervention and outcome assessment of disease. While the intervention modalities of biopharmaceuticals, such as monoclonal antibody therapies or specific nucleic acid-based transcript suppression (e.g., anti-sense-based therapies), can be contemplated, the most common and desirable paradigm of therapy is still the development of a pharmaceutically tractable small molecule that modulates the activity of a protein, typically enzyme inhibitors or receptor antagonists. Such a protein is therefore, post hoc, classified as a drug target, and it is this subject that will be covered in the rest of this section.

Let us consider the context in which bioinformatics supports genetics for the exploitation of potential drug targets. It is certainly possible that a geneticist could be tasked with identifying drug target opportunities by simply walking up the keyboard and trawling external data rather than actually participating in a laboratory genetics project. In fact, the avalanche of public data referred to in many parts of this book is increasingly making such an *in silico* exercise plausible. But however compelling the accumulated evidence and inferences may be, such projects will eventually need to perform confirmatory experiments. It is therefore more practical to consider target bioinformatics in the context of supporting an experimental study where both external and internal data are being utilized. Because the median time for a drug discovery project is still 7–14 years, it is necessary to have a realistic end point for the target discovery aspects of a genetic study (Verkman, 2004). Even studies from which no target inferences can be drawn in the first instance may contribute knowledge of disease mechanisms from which a therapeutic intervention emerges only at a later date. The minimal end point within a project could be just a sentence on target inferences in a paper or grant application. The maximal could be a commitment to follow through to compound development.

19.2.2 The big issues in target identification and progression

The main task facing the bioinformaticist/geneticist is likely to be triaging genetic data down to plausible therapeutic intervention points. This immediately presents the following list of constitutive challenges that bioinformatics can play an important role in overcoming:

1) whether the study should be designed to pinpoint targets from the outset
2) dealing with type 1 error – risking progression of a false-positive association
3) localizing the result to a known coding exon
4) interpreting associations outside the proximity of exons or known functional elements
5) finding orthogonal data that supports a biochemically plausible mechanism for a causative contribution of the polymorphism to disease
6) assessing the feasibility of direct drug intervention where gain-of-function (GOF) is implicated
7) converting the more likely loss-of-function effects (LOF) to a GOF intervention point by a 'mechanistic walk'
8) selecting common, best-bet or multiple intervention points where results implicate multifactorial contributions from a number of genes
9) assessment of patent information associated with target candidates.

These points will be addressed in more detail below. Studies aimed at elucidating the genetic basis of any phenotype can be based on a specific hypothesis, or they can be hypothesis neutral, corresponding to the candidate gene approach and genome scans. To date, most genetic association studies have focused on the candidate gene approach, and this has been reflected in the pharmaceutical sector (Pettipher *et al.*, 2005; Roses *et al.*, 2005), where many studies have been designed from the outset for direct validation of specific drug target hypotheses. However, with the availability of HapMap data, efficient sets of tagging SNPs, and increased genotyping efficiency, genetic studies are now moving toward a more hypothesis-neutral approach. It seems likely that pharmaceutical sector activities in this area are likely to move in a similar direction. This is a potentially exciting prospect, as hypothesis-neutral studies to identify new drug targets have the potential to identify paradigm-breaking pathways and unprecedented new targets. These in turn could potentially be developed into more effective therapeutics for some of the most pressing unmet medical needs, such as the development of new drugs with better side-effect profiles for psychiatric diseases (Sundram *et al.*, 2003).

The lack of robustness of disease-association results (chiefly in the form of type 1 error) is covered elsewhere in this book (Chapter 18), but, clearly, PubMed is the central resource for information to compare positive and negative studies. The problem common to all literature mining is the choice between specificity (retrieving only relevant articles at the cost of missing some) and sensitivity (retrieving everything at the cost of including some that are irrelevant). For example, 38 610 PubMed articles are returned by the search term 'genetic association', whereas the term 'genetic disease association' retrieves 12 017 articles. Yet, 'genetic disease association drug target' returns only 17 articles (are pharmaceutical companies doing that badly?). However, a resource that already has much higher specificity is the Genetic Association Database (GAD), which so far contains 3613 PubMed-linked genetic study results (Becker *et al.*, 2004). While this may not have full sensitivity, the simple expedient of using the 'related articles' link in PubMed and 'Sort by – Pub Date' gives an instant update on both positive and negative associations. Although this database currently collapses to a small number of unique human genes, a major update is in progress (Kevin Becker, personal communication). The links to this database include Entrez Gene, under the 'Link out' button, so any gene entry can be checked for a GAD entry. Other portals are being developed in which association results with relevance to target identification are being pooled and thus will also facilitate comparisons. Other databases are being developed to capture highly curated information on genetic associations, most notably HGVBase (http://hgvbase.cgb.ki.se/). This database is currently being redeveloped as HGVBase-G2P, a genotype-to-phenotype association database, and by the time this book is published, it should be an active source of genetic association data. The Whole Genome Association database is another valuable site; this will contain clinical phenotype measures and associated whole-genome genotype data for several different studies, including the Framingham Heart Study (http://www.nhlbi.nih.gov/about/framingham/index.html). This is likely to be a more 'raw' source that the published studies collated in GAD. Another useful website, the Obesity Gene Map, includes detailed information on disease-related genes focused on one therapeutic area. Finally, the Wellcome Trust Case Control Consortium (WTCCC) is also generating high-density, genome-wide association data in more than 10 diseases (http://www.wtccc.org.uk/).

The majority of replicated genetic associations that have convincing mechanistic links to disease involve protein LOF effects. Unfortunately, this mechanism does not offer the best chance of chemical intervention, as antagonistic small molecules are much easier to generate than agonists. A series of PubMed queries makes this clear. The query *((inhibitor OR antagonist) AND ('Journal of medicinal chemistry'[Jour]))* gives 3358 hits. Substituting *((activator OR agonist) AND ('Journal of medicinal chemistry'[Jour]))* gives only 1256 hits. While these should not be considered fixed ratios, they clearly suggest that for receptor LOF the development of agonists is certainly feasible. However, the chances of therapeutically correcting an enzyme LOF by activator development seem much less likely.

Even if the more potentially tractable GOF were discovered, this would have to involve a druggable gene (a status assigned to ~10 per cent of genes—see below). If, as is stochastically more likely, the results point to the involvement of a non-druggable gene (~90 per cent of genes) in the disease, then a 'mechanistic walk' has to be considered whereby the biological system module in which the non-druggable gene participates has been analysed in the hope of revealing a druggable intervention point that compensates for the defect.

Because research to identify and exploit targets for human disease is highly competitive on a global scale, both the commercial sector and academic organizations have been filing patent claims at all stages of the process. While the consequences for freedom to operate are outside the scope of this chapter, what can be made clear is that patents are an increasingly important information source. Whereas biotechnology patent information has hitherto been mainly brokered by vendor databases, much of it is now publicly available. The Patent Abstracts DB is a set of over 1 million biology-related abstracts of patent applications derived from data products of the European Patent Office (EPO) together with US and world patent documents. An example of a specific query *(([patabs-Applicant:Decode*] & [patabs-Applicant:genetics*]) | [patabs-Applicant:Decode genetics*])* found 85 entries of patents filed by this company that are focused on the discovery of drug targets from genetic studies.

19.2.3 Target validation and tractability

Any consideration of drug discovery brings up the term 'target validation' sooner or later (Betz, 2005). Many commentators have made the obvious point that a target is only fully validated when compounds with a confirmed congruency between the *in vitro* and *in vivo* mechanism of action against that protein prove to be effective therapies. This could be extended to regulatory approval and marketing, but failures here can be compound-specific off-target effects or lack of advantages over other therapies, rather than target 'de-validation' per se (see Kola and Landis (2004) for a review of this complex area). In the earlier stages of target investigation, validation can be more pragmatically defined as the accumulated evidence package, from all sources, that pushes an initial target concept up the 'resourcing slope' of the various stages of drug discovery. Support for such decisions will not only include comprehensive bioinformatics analysis but also consider all available genetic data and inferences. While the great advantages of acquiring human genetic evidence have already been outlined at the beginning of this chapter and elsewhere in this book, it is also clear that many drug targets have been at least initiated without it, and few proceed where this remains the one and only source of validation data.

19.2.4 The targetome: the drugged and druggable target universe

The starting assumptions for this section are (a) that the genetic results and bioinformatics analysis have pinpointed a protein the expression or activity of which is causally related to a disease, and (b) that it should ideally be a GOF association.

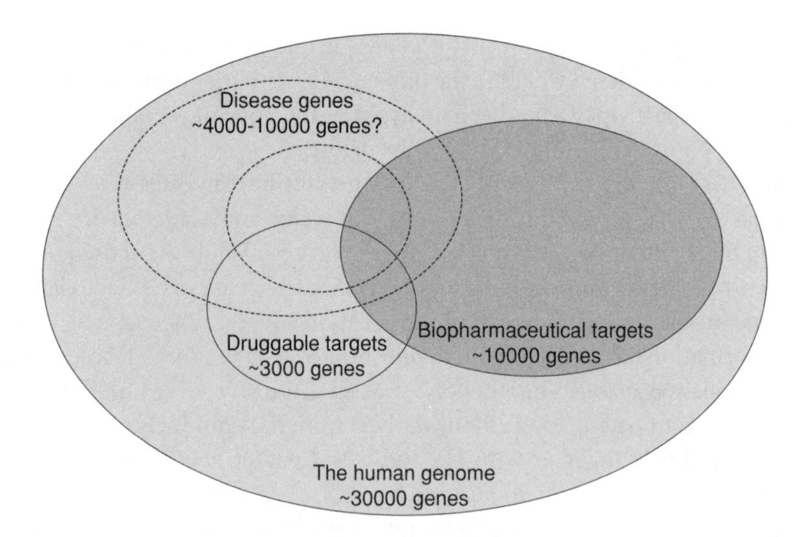

Figure 19.2 The 'target universe'

So how do we decide whether the development of a small molecule modulator as a therapy is feasible? As pointed out above, drug targets are generally defined retrospectively by previously successful translations of concept to effective therapies by medicinal chemistry. Generally, we can conceptualize in three ways the pool of genes, or rather their protein products, that may affect human disease (Figure 19.2):

1. There are proteins, specific variants of which may cause or modify disease.

2. There are proteins that can be targeted by biopharmaceutical approaches (non-small molecule approaches): monoclonal antibodies, antisense, siRNA, etc.).

3. Finally, there are proteins which can be targeted by small molecules, the 'druggable genome'.

These categories all overlap and their relative sizes are relatively fluid, but, generally, they illustrate the point that the therapeutic target space is rather finite.

The concept of a 'druggable genome' was introduced in 2002 as the group of proteins and their homologues that are known either to interact with, or be modulated by, drug-like chemical compounds (Hopkins and Groom, 2002). The initial total of known protein drug targets was 399, and they were distributed among 130 protein families, but only 120 of the non-redundant sequence set were the targets of marketed drugs. The wider set of 130 protein families was extrapolated on the basis of protein homology to 3051 potential druggable proteins, of which, unsurprisingly, G protein-coupled receptors (GPCRs), kinases and proteases constituted the largest part. Thus, if our GOF association is with one of these target homologues, it is likely to become a new target based on the retrospective success with old targets. However, as Hopkins and Groom (2002) pointed out, there are two key factors that suggest the currently achievable druggable genome will be much smaller. The first is that while genetics should eventually allow us to estimate the number of genes causatively involved in disease, these will show only partial overlap with the druggable families. The second is that the extrapolation from a very small number of targets to entire gene families is a broad-brush approach and in many cases is likely to include individual members that are sufficiently divergent from successful targets in their biochemical properties to render them unlikely intervention candidates. Examples of these would include protease paralogues that are intracellular rather than extracellular and those GPCRs classified as odour receptors.

So what do these numbers look like now and how can we find data sets to intersect with what our genetic studies have come up with? The most obvious of these would be information on known protein targets, not necessarily restricted to marketed drugs because these would effectively encompass 10 years of drug discovery, but also those that are 'declared' in the sense that they had already passed a substantial validation threshold, with small molecule modulator development well under way. Perhaps surprisingly, this information has hitherto been confined to vendors who have marketed compilations of targets together with the compounds that act on them. For marketed drugs, Inpharmatica (http://www.inpharmatica.com/biopendium.htm) has curated the targets down to 225 human sequences, below the Hopkins and Groom total of 399. To expand the druggable target domain a little more, Pharma Projects (http://www.pjbpubs.com/pharmaprojects5-1.htm) claims to have identified 1451 unique protein targets for drugs that have entered commercial pipelines (although this includes some microbial targets), and WOMBAT (http://sunsetmolecular.com/products/?id=4) includes 1320 target sequence entries.

Recently, two academic endeavours have produced the first comprehensive public drug target databases with explicit sequence links. The first of these, the Therapeutic Target Database (TTD), includes 1535 target protein sequence links, but the absence of a sequence search option limits its current utility. However, the authors of TTD have produced a comprehensive review of the distribution of targets by a range of different criteria (Zheng *et al.*, 2006). Their listing includes 997 distinct proteins with 268 targeted by at least one marketed drug; the remainder are targeted by investigational agents not yet approved for clinical use. The indications are that the current conversion rate of investigational to marketed (i.e., a 'declared' to 'successful' target transition) is low, and the overall family distribution of declared successful targets is roughly similar to the pattern of the 120 successful targets from 2002.

The second resource linking chemical data with comprehensive drug target sequence information is DrugBank(Wishart *et al.*, 2006). As well as linked molecular biological information about drug targets, this resource also offers extensive small-molecule drug information, including physical property data, compound structure, and pharmacological and physiological data about drug products. It has the major advantages of not only a local BLAST search that supports both single and multiple sequence queries but also links to PubChem. It also supports visualizing, querying and search options, including a structure similarity search tool and data extraction.

Usually when a disease association is traced to a protein, it is easy to recognize whether it is a GPCR, protein kinase, protease or nuclear hormone receptor (NHR). These protein classes are targeted by approximately 60 per cent of known drugs. However, it is still important to perform a BLAST search against DrugBank, not only to determine the level of homology to known targets and immediate links to chemistry but also to check the obvious, that is, whether the putative target has been targeted already! While it is also worth a search against the Druggable Genome website nucleotide sequences, the BLAT algorithm is not optimized to find anything more than close matches. However, this website does contain a very useful Excel download of 2935 sequences as a supplementary table with comprehensive sequence links and, crucially, a list of InterPro identifiers (Orth *et al.*, 2004). It is clear that druggable genome collection still needs be filtered beyond the simple homology arguments to bring it closer to the properties of the smaller subset of successful targets. While there is no bioinformatics equivalent of the Lipinski 'rule-of-five' (a widely used filter for drug-like properties) (Hopkins and Groom, 2002; Lipinski, 2004) by which we could set cut-offs, the following softer rules increase the likelihood of success:

1) If the investigation points unequivocally to an 'old' target, new genetic data could indicate a new therapeutic context for existing drugs; that is, an indication switch. A search against DrugBank is a good option here.

2) The sequence similarity with a known target should be solid rather than a 'twilight zone' match (same as above).

3) The protein should have an established biochemical and physiological function. OMIM includes very useful gene summaries. SwissProt entries have references, but they tend to focus on primary sequence submitters. The RIF entries in RefSeq are a good entry point, and can be expanded in PubMed by the 'related articles' (remember the green label gives full text access via PubMed Central).

4) If an enzyme is indicated, it should be secreted (this is usually clear from the SwissProt signal peptide annotation, but it is worth checking with SignalP http://www.cbs.dtu.dk/services/SignalP/, particularly in cases where it is difficult to distinguish between a signal peptide and an N-terminal transmembrane anchor.

5) Ideally, a target should have relatively few paralogues to avoid compound specificity issues. Both the 'Ensembl human family view' and the 'View other genes with this (InterPro) domain' are good options here.

6) Mammalian orthologues should be strictly 1:1 and show moderate sequence conservation (Ensembl 'Orthologue prediction' is useful here.

7) The target should be readily assayable; PubMed should come up with existing functional assays.

8) Ideally, a target should not be part of a major pathway interaction hub or complex.

9) Ideally, a target should show a restricted pattern of tissue expression.

10) A known structure or a homology model is a bonus allowing the possibility of rational drug design.

19.2.5 Target family databases: proteases as an example

Entry to a deeper level of collated information about druggability is facilitated by specific target family databases. Examples of these are given in Table 19.1, but it is instructive to examine just one of these in more detail. Proteases have a long history of being drug targets, and it is estimated that at least 10 per cent of this class of enzyme are under active investigation across a wide range of diseases (Southan, 2001; Puente *et al.*, 2003). However, the bioinformatician has to deal with both the advantages and disadvantages of such a large family of targets. The first problem is retrieval specificity, because they are united primarily under the mechanistic umbrella of peptide hydrolases rather than common ancestry. So, while it is trivial to identify all 48 known NHRs and fairly straightforward to retrieve GPCRs by three InterPro IDs, a very large number of InterPro IDs would be needed to retrieve all proteases. Fortunately, there are excellent secondary annotation sources to use (as for most target classes, some of which are represented in Table 19.1; the rest are just a short google away). The biggest and most comprehensive of these is MEROPS, which now lists 550 known and putative human peptidases. It is also unique among the target databases in linking not only to species orthologues, the available substrate assay, endogenous inhibitors, small molecule inhibitors and structural information, but also to the pharmaceutical relevance of many of the entries. However, because MEROPS is homology based and maximally inclusive, it also exemplifies the problem of broad categories of druggability, by including not only pseudogenes, retroviral components and hypothetical ORFs but also, as judged by the absence of critical active site residues or clear evolutionary shifts to non-proteolytic roles, over 400 entries that are not likely to be active proteases. But this depth of annotation and flagging of inactive homologues save a lot of bioinformatics analysis, and one should also be open to other potential catalytic or interaction functions. For example, some non-peptidase members of the S33 family with an

Table 19.1 Bioinformatics tools for target genetics

Tool	URL
Target genetics	
Genetic Association Database	http://geneticassociationdb.nih.gov/
Whole Genome Association	http://www.ncbi.nlm.nih.gov/WGA/
Obesity Gene Map	http://obesitygene.pbrc.edu/
Known targets with explicit medicinal chemistry links	
DrugBank	http://redpoll.pharmacy.ualberta.ca/drugbank/index.html
Therapeutic Target DB	http://xin.cz3.nus.edu.sg/group/cjttd/ttd.asp
PDSB Ki db	http://pdsp.cwru.edu/kidb.php
GLIDA GPCR-Ligand db	http://gdds.pharm.kyoto-u.ac.jp:8081/glida/ligand_classification.php
BindingDB	http://www.bindingdb.org/bind/index.jsp
Target families	
The Druggable Genome	http://function.gnf.org/druggable/index.html includes DNA search page and download of potential target listings
GPCRDB	http://www.gpcr.org/ also snake plots at http://www.gpcr.org/7tm/seq/snakes.html
List of SwissProt GPCR entries and servers	http://www.expasy.ch/cgi-bin/lists?7tmrlist.txt
Human, mouse and rat proteases	http://web.uniovi.es/degradome/index.htm
MEROPS (all proteases)	http://merops.sanger.ac.uk/
Peptidase entries in Swiss-Prot (from MEROPS)	http://www.expasy.ch/cgi-bin/lists?peptidas.txt
NucleaRDB (nuclear receptors)	http://www.receptors.org/NR/
Nuclear Receptor Signaling Atlas	http://www.nursa.org/
Compendium of Voltage-gated Ion Channels	http://www.iuphar-db.org/iuphar-ic/index.html (not a db but a useful journal listing)
Ligand-Gated Ion Channel DB	http://www.ebi.ac.uk/compneur-srv/LGICdb/LGICdb.php
Kinweb human protein kinases	http://bioinfo.itb.cnr.it/kinweb/index.htm (not updated recently)
Global analysis of kinases genes in genomes	http://kinase.com/
General resources	
siRNA results for some human genes	http://www.rnainterference.org/HumanSequences.html
InterPro protein families, domains and functional sites	http://www.ebi.ac.uk/interpro/index.html
PubMed	http://www.ncbi.nlm.nih.gov/entrez/query/static/overview.html
UCSC Genome Bioinformatics	http://genome.cse.ucsc.edu/
Human Proteomics Initiative (HPI)	http://www.expasy.ch/sprot/hpi/
International Protein Index (IPI)	http://www.ebi.ac.uk/IPI/IPIhelp.html
Patent Abstracts	http://srs.ebi.ac.uk/srsbin/cgi-bin/wgetz?-page+LibInfo+-id+1_uFX1SnYup+-lib+PATABS

alpha-beta hydrolase fold may have epoxide hydrolase or lipase activities that would still make them potentially tractable (http://merops.sanger.ac.uk/cgi-bin/make_frame_file?id=S33). MEROPS also shows a continual widening of the druggable envelope by updating new superfamily relationships that reveal hitherto undetected protease ancestry or novel catalytic mechanisms.

Surprisingly, perhaps the best characterized of all druggable enzyme families, namely, the S01 trypsin-like proteases, have proved particularly difficult to 'close' in terms of their final protein numbers. Thus, according to different annotation sources, these vary between 133 (MEROPS), 118 (Druggable Genome), 97 SRS (query *(((([swissprot-Species:homo sapiens*]) & ([swissprot-DbName:MEROPS*] > parent)) & [swissprot-DBxref_:S01*])* and only 81 (Ensembl IPR001254).

To establish whether an association result is in the vicinity of a protease, a BLASTX against MEROPS would be the first check. By including only the catalytic domain, this search has a higher specificity than a full-length database. The comprehensiveness of MEROPS curation also helps with two other key issues common to all target classes. These are orphan function (e.g., where the physiological role is unknown) and absence of any declared target precedents in certain subfamilies.

19.2.6 Disease case study: presenilins in Alzheimer's disease (AD)

Mutations in presenilin-1 (PS1; OMIM: 104311) and presenilin-2 (PS2; OMIM: 600759), which cause early-onset AD, both carry molecular defects that can be classified as GOF mutations. Mutations in both PS1 and PS2 result in the increased cleavage of the APP peptide (OMIM: 104760). In this case, the genetic studies were certainly the first to implicate PS1; prior to the genetic association of PS1, it was entirely unclear what role this polytopic membrane protein plays in AD pathology. PS1 was initially considered non-druggable, and it was some years later that the function of PS1 as part of an intramembrane aspartyl protease complex responsible for the gamma secretase cleavage was firmly established. Thus, genetic studies of PS1 mutations led to the identification of the gamma secretase protein complex (comprising PS1, PS2, NCSTN, APH1 and PEN2) as a drug target for AD, and the development of inhibitors is well advanced.

19.2.7 Outlook for genetics and bioinformatics in target discovery and progression

We can deduce a number of future trends in the contribution of genetics and bioinformatics to target discovery and progression. First, evidence presented elsewhere in this book shows that the combination of HapMap data and large academic consortia for genetic studies means that a substantial proportion of common variants causing common diseases may be discerned within a few years (if the common disease, common variant hypothesis holds up). It remains to be seen how effectively these results will flow into databases and metadata systems to facilitate bioinformatics interrogation and analysis to discern new details of molecular pathology and elucidate new therapeutic intervention points. It also seems likely that the commercial sector, by the application of target prioritization strategies based on diverse genomic data sources, will have triaged a substantial number of druggable targets against roughly the same set of diseases. While this may enhance the pharmaceutical companies' target portfolio, it is not clear to what extent these results will surface in the public domain and corroborate the academic output. This convergence is scientifically highly desirable and will go some way to counter the issues of independent replication. However, it also means that the chances of novel target discovery for major diseases will diminish over time. We can envisage a 'second wave' of genetic studies using increasingly complex techniques to detect epistasis, epigenetic effects and rare variation in less common diseases. We also expect to see more use of surrogate quantitative profiling, such as metabolite measurements, for clearer differentiation of disease states. Alongside advances in genetics, we also expect to see advances in functional genomics and systems biology 'filling in the gaps' and thus opening up more chances of postulating testable mechanistic connectivity from a genetic result. This should leave fewer genes that the bioinformatician cannot at least have a stab at fitting into the big jigsaw of genes and disease.

19.3 Pharmacogenetics (PGx)

It is well known that after exposure to a drug, almost any given cohort of patients show a wide variety of responses. In an ideal situation, patients show a beneficial response to the therapy, although they may also show no response or a weak response, and perhaps most worryingly, they may experience an adverse drug reaction (ADR), which in extreme situations could lead to serious illness or even death. ADR is an increasingly serious problem with a huge toll in lives and health-care costs every year. For example, in one year, 2 216 000 (6.7 per cent) hospitalized patients in the USA had serious ADR and 106 000 (0.32 per cent) had fatal ADR, making these reactions the fifth leading cause of death in the USA (Lazarou *et al.*, 1998). Providing patients with drugs that are most likely to be effective and least likely to cause harm is the primary objective of medicine development and perhaps the single most important area where genetics can make an impact.

A genetic contribution to variability in drug response has been recognized for many years. In 1902, Garrod reported a group of metabolic disorders, including alkaptonuria, that showed 'chemical individuality'; he accurately identified these disorders as 'inborn errors of metabolism'. Taking this concept a step further, he proposed that just as endogenous substrates undergo biotransformation by specific pathways, defects in such pathways could alter drug concentrations and effect. It took 94 more years to identify the defective gene responsible for alkaptonuria, homogentisate 1,2-dioxygenase (HGD), but we now know this to be a monogenic disease, and

Garrod's concept of chemical individuality is alive and well as the modern science of PGx.

19.3.1 An environmental or pharmacogenetic basis for drug efficacy and ADR?

Before getting into the complexities of PGx, it is important to recognize that many non-genetic factors also influence the efficacy of medications, including the patient's age, sex and general health, but also environmental factors, such as concomitant therapies, drug interactions and diet. To give a seemingly innocuous example, grapefruit juice is an inhibitor of intestinal cytochrome P-450 3A4, which is responsible for the first-pass metabolism of many medications. Inhibition of this enzyme by grapefruit juice leads to elevation of the serum concentrations of certain drugs, most notably antihistamine medications, which can lead to severe heart palpitations (Kane and Lipsky, 2000). Although these non-genetic factors may be very important, this is not within the scope of this chapter, so we direct the reader to more detailed reviews in this field (e.g., Sorenson, 2002).

19.3.2 Target genetics in the PGx paradigm

Variation in the drug target, preferably in the region where the drug binds, forms the basis of the classic paradigm of PGx in drug response (Figure 19.3). The simple concept here is that a therapeutic agent, which could be a small molecule, a peptide

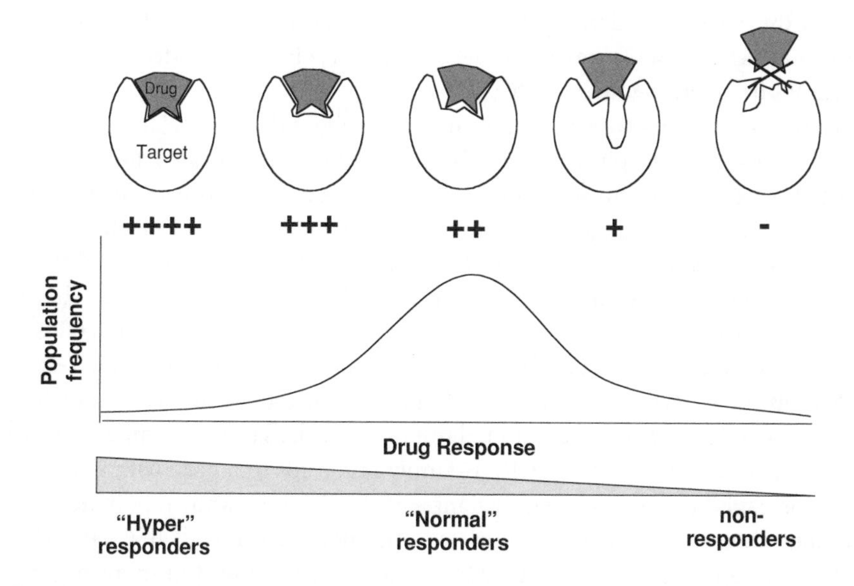

Figure 19.3 The classical PGx paradigm – target polymorphism and therapeutic response

or an antibody, binds to a specific region in the drug target (or in some cases several regions in target protein complexes). Assuming that variation exists in the drug-interacting target region, a range of responses is likely to be seen, depending on the impact of the variation on drug interaction. In most cases, this is a gross simplification of the mechanisms underlying a PGx phenotype, as differences in target response might also be seen in variants that destabilize the overall target protein in locations remote from the drug-interacting region(s). Variants might also affect downstream signalling of the target, or variants outside the target might alter the way that the drug is absorbed or metabolized (see below). Although in most instances efficacy or ADR issues are quite complex, there are known cases that fit the paradigm shown in Figure 19.3, a good example of which is the impact of polymorphism on β-agonist action in the β_2-adrenergic receptor (ADRB2).

Case study: analysis of ADRB2 receptor polymorphism

The ADRB2 plays an important role in vascular responses to physiological adrenergic stimulation, which, in turn, has a role in the pathogenesis of diseases such as hypertension, heart failure and asthma. Small-molecule ADRB2 agonists are one the most effective medications to treat acute asthma. A rare ADRB2 Thr164Ile polymorphism (found at frequencies ranging from 0.5 to 2.3 per cent in Caucasian populations) was found to have profound functional consequences in terms of isoproterenol action (a β_2-receptor agonist used for asthma). Dishy *et al.* (2004) carried out *in vivo* studies of the Thr164Ile polymorphism and found that it was associated with a five-fold reduction in sensitivity to β_2-receptor agonist-mediated vasodilatation, whereas vasoconstrictor sensitivity was increased. The overall effect of this polymorphism was to shift the balance of adrenergic vascular tone toward vasoconstriction. They proposed this as a mechanistic explanation for the clinically observed decreased survival of Thr164Ile heterozygotes with congestive heart failure.

These observations raise immediate questions about the molecular mechanism of the Thr164Ile polymorphism. Where is this polymorphism located and how might it affect receptor agonism so dramatically? These are all questions which bioinformatics can answer. To get an idea of the context of an amino-acid substitution, one of the best places to start is UniProt. A quick review of ADRB2 in the UniProt feature table (http://www.expasy.org/uniprot/P07550#features) shows that Thr164 is located in the fourth transmembrane domain of ADRB2. This protein is a GPCR, so UniProt provides a link to the GPCRDB, a valuable GPCR family-specific resource. GPCRDB allows us to review a range of information about this receptor, including sequences and a range of precomputed resources, such as protein family alignments and structural models. The GPCRDB-family link leads to a page with a link to an MSF-formatted multiple sequence alignment that includes known polymorphisms and mutations. A review of orthologues in the alignment reveals that Thr164 is in a strongly conserved region, with the Thr164 residue conserved in all organisms between man and Amphibia, although an Ile164 is seen in the fish genus, *Fugu*, suggesting that the Ile164 is potentially functional. In itself, this strong phylogenetic conservation testifies to strong selection for function at the Thr164 residue. In terms of amino-acid properties alone, a threonine to isoleucine substitution in a transmembrane environment is likely to be relatively neutral in effect (see Appendix II). However, threonine, is an uncharged polar residue, as would be expected in active sites and phosphorylation sites, whereas isoleucine is an aliphatic non-polar residue, which is much more hydrophobic and therefore much more likely to be buried in the hydrophobic core of a protein or, in this case, the membrane. Isoleucine residues are generally non-reactive, but they are known to play a role in the binding/recognition of hydrophobic substrates (see Chapter 13, or the amino-acids variation tool in SNPPER (http://snpper.chip.org/)).

It may be possible to clarify the role of Thr164Ile in agonist response by looking at this polymorphism at a structural level. A directly determined crystal structure for ADRB2 is not available; however, in the UniProt entry for ADRB2, the ModBase link leads to a database of comparative protein-structure models (Pieper *et al.*, 2004). Three-dimensional (3-D) structure models have been constructed by homology modelling, and here human ADRB2 has been aligned with the crystal structure of bovine rhodopsin. These two proteins share only 22 per cent sequence identity, so the accuracy of this model is likely to depend heavily on alignment quality. However, in our case, the alignment is likely to be reliable around Thr164, as it is located within the transmembrane regions of ADRB2 (these form reliable anchor regions for aligning GPCRs). We can retrieve the ADRB2 3-D structural model by selecting the coordinate file from the pull-down menu in the model information section. To view the structure, save the file as ADRB2.PDB and load it with DeepView, a protein-structure visualization tool (see Chapter 11 for details). After loading the ADRB2 structure to DeepView (Figure 19.4), we can highlight the Thr164 residue. To do this, select Window>Control Panel from the main menu. On the right-hand side, all amino acids in the ADRB2 model are displayed; click on Thr164 to highlight the residue in the structure and carry out a number of manipulations. First, we can identify all residues that might interact with the ADRB2 ligand-binding residues. To return to UniProt features for ADRB2, Asp113, Ser204 and Ser207 are all implicated in agonist binding. Select these residues in the DeepView control panel by holding down Control and clicking with the left mouse button. Once all agonist-binding residues are selected, go to Select>Neighbors of Selected AA on the main menu. You will now be asked to define the Ángstrom distance for neighbouring amino acids; the default is 6 Á (a probable maximum distance over which amino acids are likely to interact). Select this and press OK. This selects all residues in the ADRB2 model that are within 6 Á of the three agonist-binding residues in ADRB2. Perhaps unsurprisingly, Thr164 is one of the amino acids that are selected within 6 Á of the Ser207 agonist-binding residue. Ser207 is the most distant agonist-binding residue from Thr164 in linear terms, but in a 3-D environment, it is within 5.8 Á of Thr164. It can be useful to focus the model on the agonist-binding site residues only, by

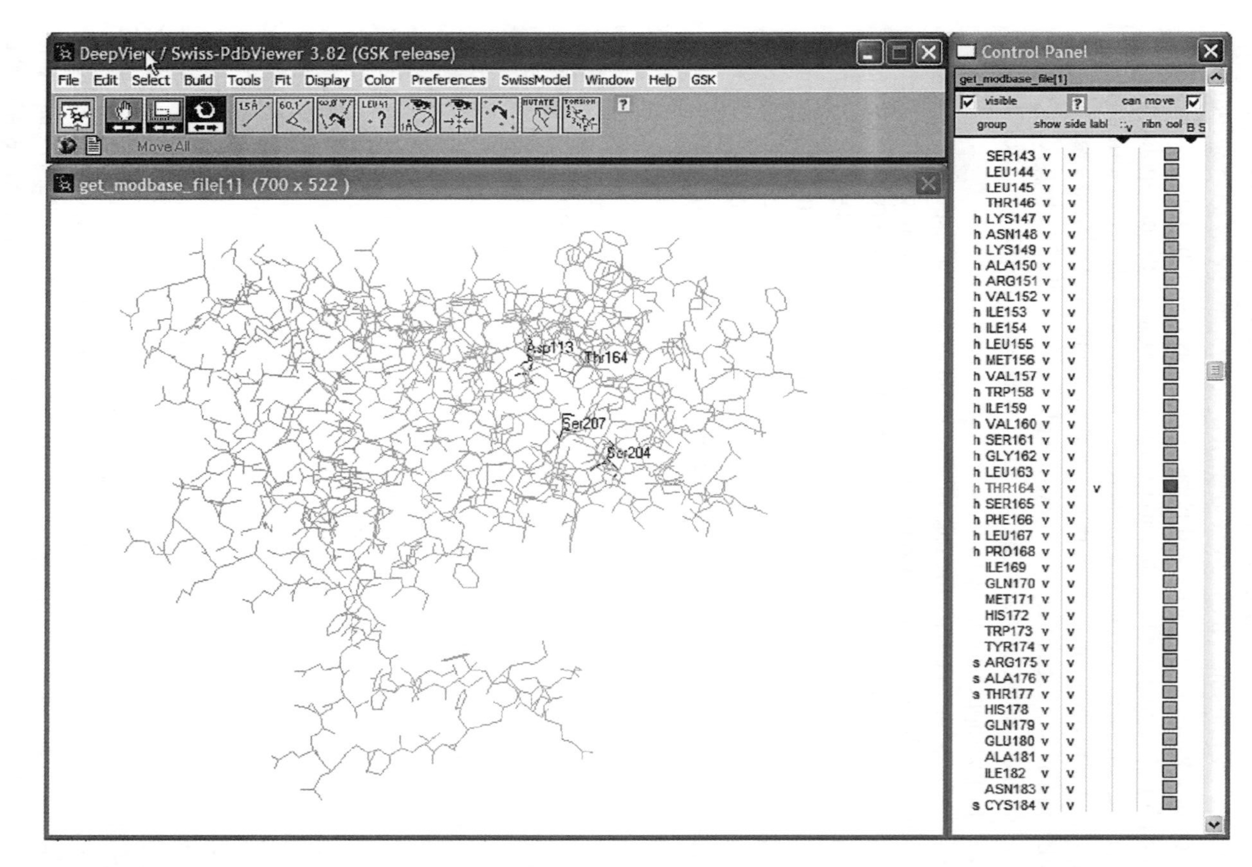

Figure 19.4 Viewing the Thr164Ile polymorphism in ADRB2. SwissPdb Viewer (http://www.expasy.org/spdbv/). Described in Guex, N. and Peitsch, M. C. (1997) SWISS-MODEL and the Swiss-PdbViewer: an environment for comparative protein modeling. *Electrophoresis* 18, 2714–2723

selecting File>Save>Save Selected Residues Only. This saves all residues within 6 Á of the agonist-binding residues as a separate 3-D structural model (Figure 19.5). The final step in this analytical process is to review the impact of the Ile164 allele on the agonist-binding site configuration. To do this, simply press the MUTATE icon and then left-click on the Thr164 residue in the structural view, revealing a menu of amino-acid changes – select Ile. The mutated residue is then displayed in the structural view. This clearly shows an alteration in the protein configuration that may alter the stearic configuration of the agonist-binding pocket, perhaps explaining the fivefold reduction in agonism seen with the Ile164 allele. As an *in silico* exercise using general Web-based bioinformatics resources, this matches quite closely the conclusions of Swaminath *et al.* (2005) in a detailed characterization of the ADRB2 agonist-binding pocket. They identified Tyr199 as a key residue involved in agonist binding, and inspection of this residue in the structural model shows that Tyr199 is adjacent to Thr164, with charge interactions within 3 Á, further supporting the probable direct role of Thr164 in agonist binding.

Exploring target PGx

The case study above is just one example of some of the types of analyses that can be carried out using purely *in silico* methods. Fortunately, there are some very rich sources of information relating to the major druggable protein families. The critical regions in drug targets that are involved in drug binding or protein structure and function are defined in several public resources. UniProt is always a good place to start looking for this kind of information; often this will annotate known functional residues in a protein. UniProt also links to other target specific resources, which often provide further detailed annotation; some of these are listed in Table 19.2. GPCRDB is a good example of a target class-specific resource, but others, such as MEROPS, focusing on human proteases; Kinweb, focusing on kinases; and NucleaRDB, focusing on NHRs, are also valuable. These resources provide detailed records that address many protein family specific issues, such as transmembrane structure, known mutations, active site configuration or phylogeny. For example, each MEROPS record contains a section defining the active site residues of each protease; interestingly, some, but not all, of these residues are annotated in UniProt, highlighting the need to consider these specialist resources when available.

19.3.3 Drug absorption, distribution, metabolism and excretion (ADME)

The mechanism by which a drug is absorbed, distributed, metabolized and excreted (ADME) from the cell ultimately determines its concentration both at the target and at off-target locations. Not surprisingly, this complex chain of events, alongside the

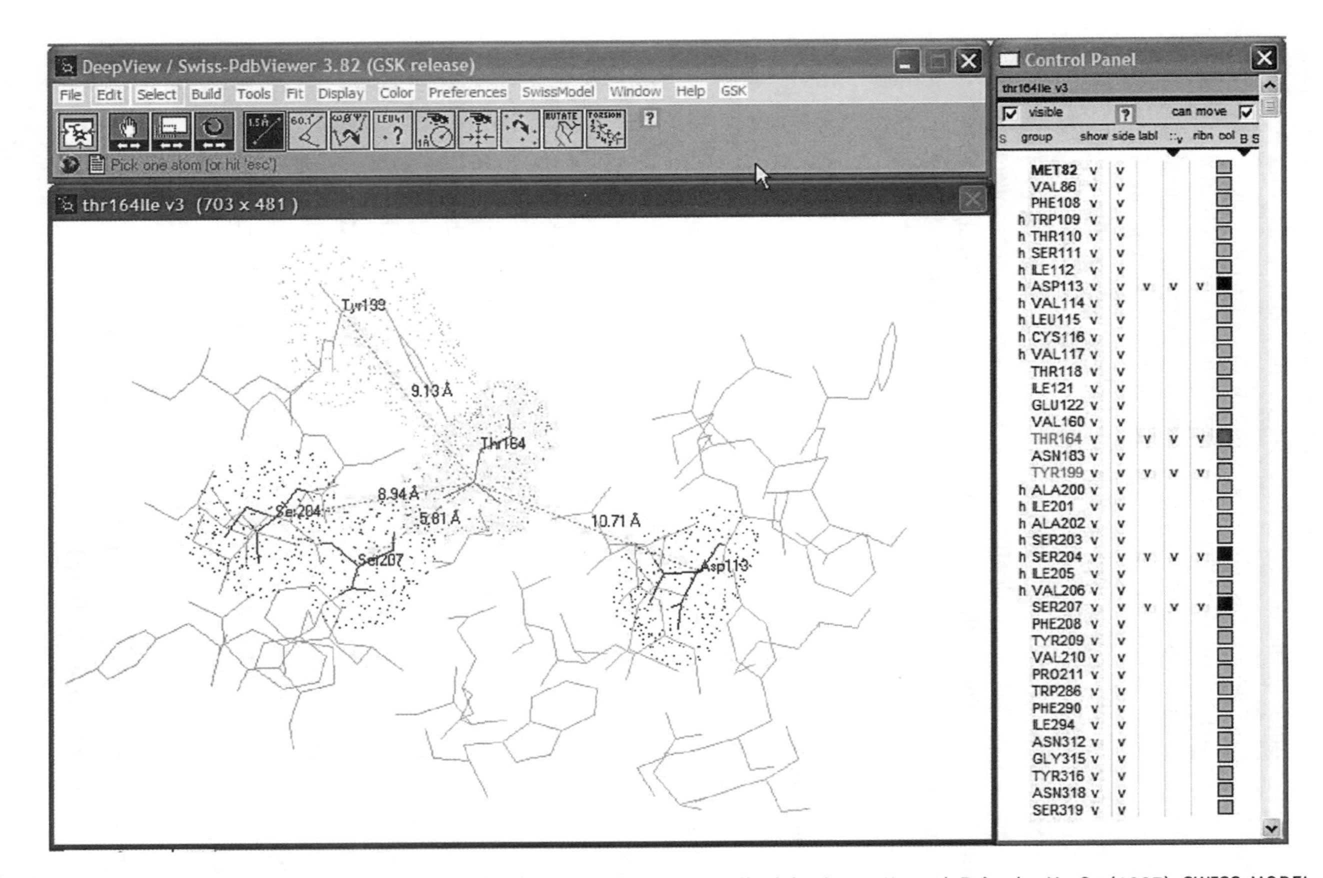

Figure 19.5 Viewing the ADRB2 agonist-binding residues with DeepView. Described in Guex, N. and Peitsch, M. C. (1997) SWISS-MODEL and the Swiss-PdbViewer: an environment for comparative protein modeling. *Electrophoresis* 18, 2714–2723

Table 19.2 Bioinformatics tools for genetic analysis in the context of drug discovery and development

Tool	URL
Pharmacogenetics (PGx)	
PharmGKB	http://www.pharmgkb.org/
PubChem	http://pubchem.ncbi.nlm.nih.gov/
DrugBank	http://redpoll.pharmacy.ualberta.ca/drugbank/
Pharmacogenomics	
Chemical Effects in Biological Systems (CEBS)	http://cebs.niehs.nih.gov/index.html
Edge[2] Environment, Drugs and Gene Expression	http://edge.oncology.wisc.edu/edge.php
GNF SymAtlas	http://symatlas.gnf.org/SymAtlas/
Target PGx	
GPCRDB	http://www.gpcr.org/
MEROPS (proteases)	http://merops.sanger.ac.uk/
Kinweb (Kinases)	http://bioinfo.itb.cnr.it/kinweb/index.htm
NucleaRDB (Nuc. receptors)	http://www.receptors.org/NR/
Ligand-gated ion channel DB	http://www.ebi.ac.uk/compneur-srv/LGICdb/
ADME gene families and pathways	
CypAlleles DB	http://www.imm.ki.se/CYPalleles/
Cytochrome P450 interaction table	http://medicine.iupui.edu/flockhart/table.htm
Directory of P450-containing systems	www.icgeb.org/p450/
Human membrane transporter database HMTD	http://lab.digibench.net/transporter/
Human ABC transporters database	http://nutrigene.4t.com/humanabc.htm
Resources for immune-mediated ADRs	
Immuno-polymorphism database	http://www.ebi.ac.uk/ipd/
MHC haplotype project	http://www.sanger.ac.uk/HGP/Chr6/MHC/
IMGT Immunoinformatics page	http://imgt.cines.fr/textes/Immunoinformatics.html

study of the target itself, forms the mainstay of most PGx research. The acronym *ADME* is becoming ubiquitous in representing these four key events in drug action. Sometimes the acronym *ADMET* is also used to capture toxicology within this paradigm, although toxicology can arguably be subsumed within ADME, as all drugs are toxic at sufficient dose, and of course dosage is usually related to one of these events. Essentially, the study of ADME genetics involves identifying the variants that influence these processes. Bioinformatics is a valuable tool in the design of ADME genetic studies, whether they are genome-wide scans or candidate gene focused. In fact, ADME is rather well suited to the candidate gene approach. A PGx phenotype can often give strong clues to the likely pathways that may be involved and also the number of genes involved in ADME processes is likely to be quite finite. Estimates vary, but it is possible that fewer than 500–600 genes are regularly involved in ADME processes. The ADME mechanisms of some classes of drugs are quite well known, and can be readily reviewed in resources such as PharmGKB or DrugBank (see below). Adsorption, distribution and excretion are often mediated by well-characterized drug-uptake or drug-efflux transporters, such as members of the OATP and p-glycoprotein families, while drug-metabolizing enzyme complexes, such as CYP, have been characterized for many drugs. This wealth of information creates opportunities for the creative use of bioinformatics to expand ADME candidate genes, using expression, pathway and literature information.

Existing knowledge of the routes of drug ADME in the patient also makes targeted queries of gene-expression data feasible. The route of a drug is well described by Goldstein *et al.* (2003). An ingested drug enters gut enterocytes, from where it may be metabolized, effluxed into the portal circulation, or effluxed back into the gut lumen. Similarly, a drug delivered to hepatocytes can be metabolized (possibly into a more reactive metabolite) and excreted into the bile, or returned to the systemic circulation, from where it can also be excreted, generally through biliary or renal routes. If the drug target is not located within the immediate reach of the vasculature, it may be impeded from its target by a number of further barriers, such as the general barrier between plasma and tissue and the often critical blood/brain barrier (see Graff and Pollack (2004) for a thorough review of this area). Finally, some drugs access intracellular molecular targets (e.g., nuclear targets), in which case uptake into, and efflux out of, the target cell are also likely to be key determinants of drug delivery and action. All of these further obstacles could limit drug access to certain cell populations, and in many cases are likely to be mediated by specific ADME genes. We review the main classes of ADME genes in the following sections.

Genes involved in drug metabolism

There are essentially two different classes of drug-metabolizing enzymes: phase I enzymes are involved in drug oxidation, reduction, hydrolysis and other transformations, and phase II enzymes conjugate drugs (e.g., sulphation, glucuronidation, and glutathione conjugation) to increase solubility, thereby aiding excretion. In man, these two classes are represented by at least 30 different protein families (Figure 19.6). Cytochrome (CYP) enzymes are the most ubiquitous phase I enzymes, CYP representing the largest family of CYP enzymes, and CYP2D6, an enzyme responsible for the metabolism of 20–30 per cent of prescription drugs, being the most studied among this family (Vermeulen, 2003). Several highly curated databases deconvolute the complexity of variation in CYP enzymes. Most highly recommended among these is the CYPAlleles database (Ingelman-Sundberg, 2002), which contains a detailed compilation of standardized CYP alleles, based on direct submission and surveys of the literature. A number of other resources exist in this area and are listed in Table 19.2.

Conjugation of drugs by phase II enzymes typically results in inactivation, detoxification, and enhanced likelihood of the excretion of a drug. The three most prevalent

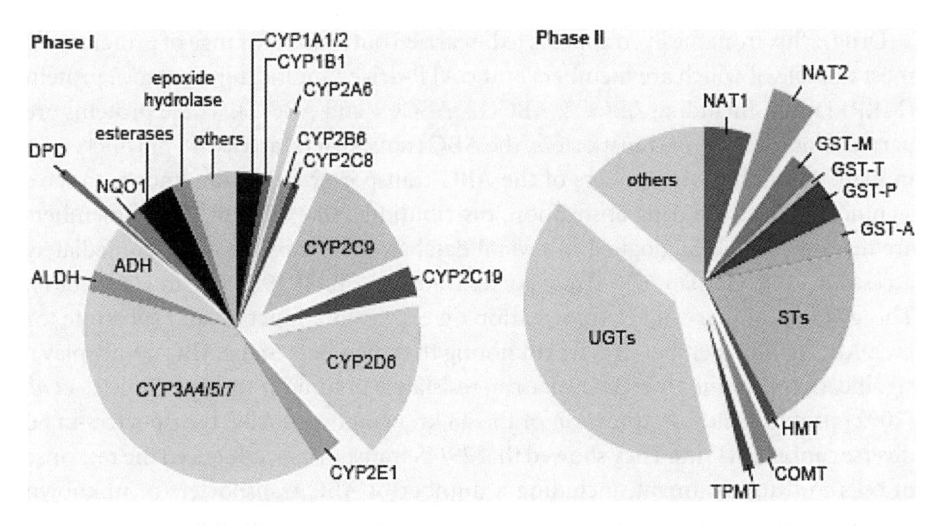

Figure 19.6 Phase I and phase II drug-metabolizing enzymes

classes of phase II metabolism are sulphation, mediated by members of the cytosolic sulfotransferase (SULT) superfamily; glucuronidation, mediated by the uridine diphosphate-glucuronosyl transferase (UGT) superfamily; and glutathionation, mediated by the glutathione-S-transferase (GST) superfamily. These conjugations occur directly on the parent compounds that contain appropriate structural motifs, or, more frequently, on functional groups added or exposed by phase I oxidation. The increase in molecular weight and water solubility that these conjugations cause tends to decrease membrane permeability dramatically, calling for active biliary or hepatic transport mechanisms to excrete these conjugates effectively (Zamek-Gliszczynski *et al.*, 2006). Unlike phase I enzymes, there are few specific databases for phase II enzymes; however, they are well represented in general PGx resources such as PharmGKB (Table 19.2). Alternatively, as always, a query on one member of a gene family with resources such as UniProt (http://www.uniprot.org/) can be used to determine other human family members, by utilizing UniProt links to other resources such as HoverGen and PFAM.

Genes involved in drug adsorption, distribution and excretion

Absorption, distribution and excretion of drugs and/or metabolites can sometimes be mediated by the same proteins; in the case of transport from blood to hepatocyte and vice versa, this can be mediated by bidirectional transporter families, such as the organic anion-transporting polypeptides (OATP) or organic anion transporters (OAT). In most cases, these proteins strongly favour inwardly directed transport into the liver and so are probably most comfortably placed within the absorption group in the ADME paradigm (Zamek-Gliszczynski *et al.*, 2006).

Drug efflux from the liver, appears to be carried out by a wide range of proteins, the most notable of which are members of the ATP-driven multidrug resistance protein (MRP) family, including ABCC2, ABCG2, ABCC3 and ABCC4. These proteins are part of a large family of transporters, the ABC transporters, at least 48 of which exist in man. A number of members of the ABC-transporter family are known to have significant effects on drug absorption, distribution and excretion. These members are reasonably well catalogued in several databases. Perhaps the most immediately accessible is the Human ABC-Transporter Database (Table 19.2) (Dean *et al.*, 2001). This gives a summary of key information on expression, function and substrate for each ABC family member. It is worth noting that members of the ABC family play a significant role in the emergence of drug resistance in tumour therapy. Szakács *et al.* (2002) profiled mRNA expression of the 48 known human ABC transporters in 60 diverse cancer cell lines. They showed that 29/48 transporters influenced the response of cells to drug treatment, including a number of ABC transporters of unknown function. Aside from the obvious relevance that this has to tumour therapy, it also illustrates the possible role that uncharacterized members of this family may play in ADME, and suggests that most of the 48 family members should be considered candidates for ADME phenotypes.

Several other transporter families are known to play a role in ADME. A good resource to identify these is the Human Membrane Transporter Database (HMTD) (Yan and Sadee, 2000), which indexes all known transporters, including pharmaceutically relevant members outside the ABC family, such as the serotonin transporter, SLC6A4. Indexing is performed in a number of ways, as by the substrate or drug transported. Alternatively, it is possible to reverse the query and ask which transporter is known to transport a given drug, using the drug as a query for the PharmGKB, or DrugBank databases.

19.3.4 The role of pharmacodynamics versus pharmacokinetics in PGx

One final word on the mechanisms that need to be investigated to study a PGx response. Variability in drug action may not be directly related to functional changes in ADME proteins (essentially a pharmacokinetic event); it may also arise through a pharmacodynamic mechanism. For example, the drug might interact with other (unintended) targets – so-called off-target effects. There may be variability in the function or expression level of the target, which might be caused by variation in the promoter or other regulatory regions. Finally, it is possible that other molecules (or drugs) might modulate the biological context within which the drug–target interaction takes place. Variation in any of the elements that control these types of processes can lead to variability in drug action, which might well confound the search for causative genes among the usual ADME and target-related candidates.

19.3.5 Using bioinformatics to gain understanding of adverse drug reaction (ADR)

One of the biggest concerns during the development of any medication is the possibility of unintended consequences in the patient. When harmful, these events are referred to as ADRs. While the nature of the intended benefit of a medication is usually known, ADRs can be both unprecedented and unpredictable. This problem mainly stems from the fact that phase I, II and III trials frequently do not have sufficient power to detect rare ADRs reliably, which may occur at rates of less than 1 in 10 000. A very large proportion of these ADRs are believed to have a genetic basis; however, by their intrinsic rarity, ADRs can be very challenging to study by genetic means. The problems of mounting such studies are obvious. Firstly, the difficulty of ascertainment of sufficient cases usually means a study will be seriously underpowered. Secondly, clinical trial populations are likely to show racial admixture, further reducing power and introducing the possibility of association signals due to population stratification (these might be resolved by tools such as STRAT; see Chapter 10). Finally, the mechanism of the ADR may be completely unknown, limiting the potential of a candidate gene approach. This combination of unknown mechanism, limited power and probable admixture makes effective bioinformatics more important than ever in genetic studies of ADRs.

The mechanisms of ADRs – immune mediation

As previously discussed, the mechanisms underlying ADRs vary considerably. Many may be due to the traditional target- and ADME-related mechanisms discussed earlier. However, there also appears to be a common, immune-mediated ADR mechanism (Bugelski, 2005). For a drug to elicit an immune-mediated response, it must be both immunogenic (i.e., able to sensitize the immune system) and antigenic (i.e., able to evoke a response from a sensitized immune system). Unlike protein therapeutics, small-molecule drugs are not usually immunogenic or antigenic. Immune-mediated ADRs are likely to be the result of complex interactions between drug-metabolizing enzymes, the metabolites of these enzymatic reactions (particularly reactive metabolites), and the ensuing immune sensitization and response that may result from these events. Teasing apart the aspects of this interplay calls for a full integration of data in this area and some quite demanding bioinformatics analysis.

Immune-mediated ADRs can affect many tissues and organs, including the skin, lungs, liver and kidneys. Most ADRs are mild (as in minor rashes); however, they can be severe, leading to organ failure or anaphylactic shock. Although poorly defined, it is clear that there is a major genetic component in these ADRs. Genetic polymorphisms have been identified in a range of immune-related genes and pathways, including immune receptors, heat-shock proteins, and components of the major histocompatibility complex (MHC). A good example of this is the association between the HLA-B*5701 haplotype and hypersensitivity to the anti-HIV drug abacavir in HIV patients (Hetherington *et al.*, 2002). Abacavir is a widely used nucleoside analogue with potent HIV-1 antiviral activity. Approximately 5 per cent of patients treated with abacavir develop a hypersensitivity reaction (HSR) characterized by multisystem involvement, which has proved fatal in rare cases. The symptoms, which usually appear within the first 6 weeks of treatment, include fever, rash and a range of less specific gastrointestinal symptoms; these symptoms improve within 72 h of discontinuation. Rechallenge with abacavir after an HSR episode can be fatal.

The abacavir HSR has been shown to have a very strong genetic and immune component, a number of strong associations being reported across the MHC region (Hetherington *et al.*, 2002; Mallal *et al.*, 2002). Most convincingly, the HLA-B*5701 ancestral haplotype strongly predicts abacavir hypersensitivity; 74 per cent of patients carrying this haplotype show a HSR when challenged with abacavir. It is difficult to determine the molecular basis of this association, as HLA-B*5701 extends across several hundred kilobases, encompassing a large number of immune-related genes. To add further complexity, an HLA-B*5701 haplotypic polymorphism within the tumour necrosis factor (TNF) promoter region (TNF-238A) has also been associated with the HSR, probably influencing the severity of the HSR by increasing TNF production (Hetherington *et al.*, 2002). Deconvoluting the molecular basis of an association to the MHC region is always a challenge, as the entire MHC region is highly duplicated and polymorphic, making it quite refractory to refinement of association by existing genetic analysis; however, some progress has been made in the case of abacavir hypersensitivity. Martin *et al.* (2004) used recombinant mapping techniques to narrow the HLA-B*5701-associated HSR susceptibility down to a 14-kb region containing the Hsp70, heat-shock protein cluster. A Met493Thr SNP in HSPA1L was found in combination with HLA-B*5701 in 94.4 per cent of hypersensitive cases and 0.4 per cent of controls. Martin *et al.* speculated that heat-shock proteins might be involved in hapten formation between reactive metabolites of abacavir and the HLA*B5701 peptide substrates. The reactive metabolite hapten hypothesis of ADRs is becoming well established. It is thought that CYP may bioactivate drugs to chemically reactive or toxic metabolites. These reactive metabolites may cause initial idiosyncratic hypersensitivity reactions, but further propagation of these reactions in serious ADRs seems to be mediated by the immune response of different individuals (Naisbitt *et al.*, 2001). By this hypothesis, haptenation of HSPA1L with reactive metabolites of abacavir and subsequent presentation of these haptens in the context of the general presentation of HLA-B*5701 antigens appears to induce vigorous T-cell response and HSR. This indicates that the abacavir HSR may follow the classic two-step process (immune sensitization, followed by response) mediated by at least two alleles: first, sensitization to a reactive metabolite, in this case, haptenation mediated by HSPA1L, and then the immune response, the severity of which appears to be mediated by the TNF-238A allele, which appears to increase the severity of the autoimmune HSR. This also illustrates an important point in the development of robust diagnostic tests for HSRs, especially in the case of drugs for life-threatening conditions. In the case of abacavir, not all HLA-B*5701 carriers show HSR. Martin *et al.* (2004) showed that the use of HLA-B*5701 alone, as a diagnostic marker, would inappropriately deny access of 1.6 per cent of their test population toabacavir. Testing for the combined presence of *HLA-B*5701* and the HSPA1LM493T variant reduced this percentage to 0.4 per cent, which, on a global scale, could account for significant additional clinical impact for a badly needed drug.

19.3.6 PGx and ethnicity

As a general rule, clinical trials in drug development are designed to capture the full range of variation in drug response within the patient population that is expected to be exposed to the drug. This usually involves a substantial proportion of Europeans, African-Americans and Asians – largely reflecting the demographics of the US population. This design may be relatively effective in assessing drug response in the expected patient populations, but may miss important variability in response in ethnic minority populations not included in the trial. Many well-documented interethnic differences are seen in response to drugs; perhaps one of the best known is that of response to cardiovascular drugs. In a number of studies, populations with European ancestry have been shown to respond significantly better to beta-blockers, ACE inhibitors and angiotensin-receptor antagonists than populations of African ancestry (Tate and Goldstein, 2004). Some have argued that this represents a difference in the pathogenesis of cardiovascular disease between these two populations. There is some evidence to support this; nitrous oxide bioactivity differs between Africans and Europeans, leading to the first FDA approval of a race-specific drug. BiDil is a nitric oxide enhancer for heart failure approved only for use in African populations (Senior, 2005). However, there is also a great deal of evidence that some of these observed differences are mediated by genetic variation between these populations, based on different polymorphism frequencies in ADME genes and drug targets (see Goldstein and Tate (2005) for a review of this area).

Considering these issues, and the likelihood that drug treatment could be tailored for greater effect where important genetic variants exist between ethnic groups, it may become important to incorporate this information into the drug development process. Most analysis of variation to date has focused on Caucasian populations; however, the HapMap project (see Chapter 3) has revolutionized the understanding of the differences between four ethnic groups, by generating data from Caucasian, Yoruban (Nigeria), Japanese and Chinese population samples. The number of data generated in these population samples may for the first time enable robust genome-wide studies of genetic variation related to drug response in these populations. Other population-specific resources are also being made available in this area; for example, JSNP (http://snp.ims.u-tokyo.ac.jp/) is a database of common gene variations in Japanese populations. The database contains 197 157 SNPs (1 July 2006), 84 612 with allele frequency; all SNPs are also deposited in dbSNP. Other databases are also emerging, including ThaiSNP (http://thaisnp.biotec.or.th:8080/thaisnp/db) and the Taiwan Han-Chinese SNP database (http://genepipe.ngc.sinica.edu.tw/thcsd/). Each of these databases could go a long way to answering some of the population-specific differences seen in drug response.

Now that ethnic variation data are available on a genome-wide scale, there are great opportunities for PGx. For a given drug, it should be possible to highlight common ethnic variation in all genes known to be affected by a drug or play a role in the ADME of a drug. In many cases, rare Caucasian variants will be seen at much higher frequencies in other ethnic groups and vice versa, highlighting the importance of considering all variants in PGx phenotypes, regardless of allele frequency. Genome-scale ethnic variation data also allow more esoteric (but powerful) analyses of the possible impact of natural selection on drug response. For example, individual variation in genes encoding CYP is already known, but their evolutionary origins in processing dietary toxins are just beginning to be appreciated (Jorge *et al.*, 1999). This suggests that different populations may be under different selective pressures, such as those of diet or environment, to give obvious examples.

It is possible to test for signatures of natural selection in specific ethnic groups. Carlson *et al.* (2005) and Voight *et al.* (2006) applied different analysis methods to SNP data generated in each of three ethnic groups to identify a number of regions showing evidence of strong, recent, selective sweeps. The term 'selective sweep' was coined because the alleles contained in the ancestral haplotype harbouring the selected allele are 'swept' along through generation, while the allele undergoing selection leaves a characteristic signature of reduced haplotype diversity in the region undergoing selection. One gene found by Carlson *et al.* to undergo strong positive selection in Caucasian populations (but not in African or Chinese populations) was the lactase gene. This gene has previously been shown to undergo positive selection, and has been proposed as an explanation of the observed predominance of lactose tolerance in Caucasian populations, in contrast to the predominant lactose intolerance in African and Asian populations. This strong positive selection is believed to have occurred within the past 5000–10 000 years in Caucasian populations, from the selective advantage that lactose tolerance afforded dairy farming populations (Bersaglieri *et al.*, 2004). Admittedly, this may not appear of immediate pharmaceutical relevance, but the principle of an enzyme/substrate interaction under selection is completely analogous to the drug/target paradigm. The data generated by Carlson *et al.* (2005) and Voight *et al.* (2006) are available on the Web. The former is available as a track (Tajima's D) in the UCSC genome browser. The latter is available in the stand-alone Haplotter tool to query by gene, genomic location or SNP (http://hg-wen.uchicago.edu/selection/haplotter.htm).

19.3.7 Key tools and databases for PGx

In the previous sections, some of the key mechanisms that form a common basis of many PGx phenotypes were reviewed. Gathering momentum in the field of PGx has led to the development of a number of detailed resources that capture much of this public knowledge of PGx. Table 19.2 lists all of the PGx tools and databases described in this chapter. It is not possible to review all of these in detail; however, two databases, PharmGKB and DrugBank, are worth some closer attention.

The pharmacogenomics knowledge base (PharmGKB)

One of the most extensive public domain resources focused on drug pharmacogentics and ADME is PharmGKB, the PharmacoGenomics Knowledge Base (Klein *et al.*, 2001). This database, which is driven by the NIH Pharmacogenomics Research Network, is a valuable, highly integrated resource that captures experimental and literature data on drug–gene interactions. The database is unique among public PGx databases, in that it provides genotype information related to human drug response, offering the potential for meta-analysis between related PGx phenotypes. PharmGKB includes extensive data on drug-metabolizing genes, drugs, diseases and drug pathways, all of which are linked to each other and to several primary data sets collected. The database contains partial information on 8860 human genes, of which 612 have drug phenotype or genotype data associated with them. Conversely, the database contains information on 3744 drugs, 82 of which have associated gene phenotype or genotype information. PharmGKB also contains information on 4076 diseases, 23 of which have associated drug phenotype/genotype information. As these figures illustrate, PharmGKB is quite data rich, although the high-value data can sometimes be buried under less directly relevant information curated from the literature. This is not a criticism of PharmGKB but simply an observation of some of the pitfalls in data mining – the more sensitive the approach, the higher the signal-to-noise ratio tends to be. However, we must credit the PharmGKB team with providing flexible query interfaces that allow us to exclude data from different evidence sources. PharmGKB also maintains an expanding range of fully interactive drug metabolism/action pathways (18 pathways; June 2006). These describe the molecular pathways for key drug classes, including ACE inhibitors, statins, antiarrhythmic agents and glucocorticoids.

DrugBank

While PharmGKB offers a highly integrated view of drug–gene and drug–pathway interactions, another quite complementary resource is DrugBank. This database is more drug-centric, focusing on 250 of the most frequently prescribed FDA-approved drugs. For each drug, DrugBank provides drug structure, generic and chemical names, 3-D structures, drug class, indication, and other aspects of pharmacology. Alongside the drug information, it also provides pharmacodynamic information, including the known protein target, 3-D structure, cellular localization and interacting partners. Finally, it provides drug pharmacokinetic information, including toxicity, metabolic fate and known metabolizing enzymes. The information in PharmGKB and DrugBank does not appear to overlap substantially; for example, DrugBank often records which enzyme is involved in the metabolism of a drug, where PharmGKB might only record other genes that are upregulated by drug treatment. The potential power of DrugBank probably lies in the wide range of query methods. We can use simple browsing, text queries and sequence queries (BLAST), but we can also do more complex queries based on chemical structure (using a structure-drawing applet). This flexibility allows users to scan DrugBank with a new chemical structure or a library of structures to identify the protein targets to which these compounds might bind or which phase I metabolizing enzymes might act on them. The power of a query to DrugBank is obviously limited by the quality and number of data in DrugBank, but as the database continues to develop, it is likely to become increasingly valuable.

19.3.8 Bioinformatics approaches to identify PGx candidate genes

We hope that this review has made the process of identification of candidate genes for PGx traits clearer to the reader. By a combination of the resources reviewed here, it should be possible to identify sets of genes in which polymorphisms might reasonably be expected to modulate PGx phenotypes. It should be possible to front-load a study with relevant PGx candidate genes or to filter a genome-wide approach by looking for genes with a PGx rationale.

Beyond the known PGx candidate approach, there is clearly a lot that is still unknown about PGx and ADME events; therefore, the candidate gene net should sometimes be spread wider. For example, in the case of a genetic association with a genomic region, it may be worthwhile considering all genes that are expressed in the tissue where the phenotype is observed. Usually, this is likely to include the liver, although in some ADRs the location may be more specific, as in the skin. More general bioinformatics tools and databases can be used to carry out these types of queries. For example, the GNF SymAtlas tool (http://symatlas.gnf.org/SymAtlas/) can be employed by non-profit users to identify all genes in a genomic region that are expressed in the liver. This is simply carried out by loading the list of gene symbols or accessions to SymAtlas; these should appear in the left-hand panel of the tool. Select the list of genes with the checkbox. Next, follow the Search Expression link and select the Human GeneAtlas GNF1H, MAS5. Then select Intersect with previous and select Liver > 2. After pressing the search button, the uploaded gene list will be modified to show only those genes expressed in liver (see Chapter 18 for more detailed coverage of this tool).

A number of public data resources are also being established to provide freely accessible microarray data on drug- and toxicity-related phenotypes. For example, the Chemical Effects in Biological Systems (CEBS) database (Mattes *et al.*, 2004) is a highly recommended resource that accommodates gene-expression profiles, and proteomics and metabolomics data and allows very complex queries across more than 100 experiments, mostly performed in rat liver. These experiments include data generated after exposure to members of key drug classes, including the antidiabetic, troglitazone (Rezulin); the antiepileptic, valproic acid; and the antidepressive, fluoxetine (Prozac) among other drugs (Mattes *et al.*, 2004). The CEBS interface allows the user to identify rat genes, which are differentially regulated by treatments with these drugs, and to overlay known pathways and gene ontologies. Another expanding database is the Edge^2, which, though currently limited to data on 29 experiments, may develop into a resource of value. The database contains mouse gene-expression profiles recorded in response to treatment with different toxic molecules, protein agents and drugs, including acetaminophen, TNF and carbamazepine (Hayes *et al.*, 2005).

19.4 Conclusions: toward 'personalized medicine'

One of the ultimate goals of genetics research in drug discovery is to develop the ability, based on *in vitro* and animal data, to predict *in vivo* drug efficacy and avoid adverse events in man. Good progress is already being made in this direction, with the cataloguing of genetic variants in targets and drug ADME genes. Evaluating the *in vitro* consequences of these variants and relating these consequences to clinical drug action is the next big challenge. As the fields of target genetics and PGx advance, the ideal of personalized medicine is becoming more tangible. How soon this ideal becomes reality will entirely depend on our understanding of the interplay between the almost unfathomable complexity of interactions between drug, patient and environment. Bioinformatics (alongside cheminformatics and mathematical modelling, to name a few of the other players) may be one of the integrative solutions to this problem.

References

Becker, K. G., Barnes, K. C., Bright, T. J. *et al.* (2004). The genetic association database. *Nat Genet* **36**, 431–432.

Bersaglieri, T., Sabeti, P. C., Patterson, N. *et al.* (2004). Genetic signatures of strong recent positive selection at the lactase gene. *Am J Hum Genet* **74**(6), 1111–1120.

Betz, U. A. (2005). How many genomics targets can a portfolio afford? *Drug Discov Today* **10**(15), 1057–1063.

Bugelski, P. J. (2005). Genetic aspects of immune-mediated adverse drug effects. *Nat Rev Drug Discov* **4**, 59–69.

Carlson, C. S., Thomas, D. J., Eberle, M. A. *et al.* (2005). Genomic regions exhibiting positive selection identified from dense genotype data. *Genome Res* **15**(11), 1553–1565.

Dean, M., Rzhetsky, A. and Allikmets, R. (2001). The human ATP-binding cassette (ABC) transporter superfamily. *Genome Res* **11**(7), 1156–1166.

Dishy, V., Landau, R., Sofowora, G. G. *et al.* (2004). Beta2-adrenoceptor Thr164Ile polymorphism is associated with markedly decreased vasodilator and increased vasoconstrictor sensitivity in vivo. *Pharmacogenetics* **14**(8), 517–522.

Garrod, A. E. (1902). The incidence of alkaptonuria: a study in chemical individuality. *Lancet* **II**, 1616–1620.

Goldstein, D. B., Tate, S. K. and Sisodiya, S. M. (2003). Pharmacogenetics goes genomic. *Nat Rev Genet* **4**(12), 937–947.

Gowen, M., Lazner, F., Dodds, R. *et al.* (1999). Cathepsin K knockout mice develop osteopetrosis due to a deficit in matrix degradation but not demineralization. *J Bone Miner Res* **14**(10), 1654–1663.

Graff, C. L. and Pollack, G. M. (2004). Drug transport at the blood–brain barrier and the choroid plexus. *Curr Drug Metab* **5**(1), 95–108.

Hayes, K. R., Vollrath, A. L., Zastrow, G. M. *et al.* (2005). EDGE: a centralized resource for the comparison, analysis, and distribution of toxicogenomic information. *Mol Pharmacol* **67**(4), 1360–1368.

Herbert, A., Gerry, N. P., McQueen, M. B. *et al.* (2006). A common genetic variant is associated with adult and childhood obesity. *Science* **312**, 279–283.

Hetherington, S., Hughes, A. R., Mosteller, M. *et al.* (2002). Genetic variations in HLA-B region and hypersensitivity reactions to abacavir. *Lancet* **359**(9312), 1121–1122.

Hopkins, A. L. and Groom, C. R. (2002). The druggable genome. *Nat Rev Drug Discov* **1**, 727–730.

Jorge, L. F, Eichelbaum, M., Griese, E. U. *et al.* (1999). Comparative evolutionary pharmacogenetics of CYP2D6 in Ngawbe and Embera Amerindians of Panama and Colombia: role of selection versus drift in world populations. *Pharmacogenetics* **9**(2), 217–228.

Kola, I. and Landis, J. (2004). Can the pharmaceutical industry reduce attrition rates? *Nat Rev Drug Discov* **3**(8), 711–715.

Kane, G. C. and Lipsky, J. J. (2000). Drug–grapefruit juice interactions. *Mayo Clin Proc* **75**(9), 933–942.

Klein, T. E., Chang, J. T., Cho, M. K. *et al.* (2001). Integrating genotype and phenotype information: an overview of the PharmGKB project. Pharmacogenetics Research Network and Knowledge Base. *Pharmacogenomics J* **1**(3), 167–170.

Lazarou, J., Pomeranz, B. H. and Corey, P. N. (1998). Incidence of adverse drug reactions in hospitalized patients: a meta-analysis of prospective studies. *JAMA* **279**(15), 1200–1205.

Lesko, L. J. and Woodcock, J. (2002). Pharmacogenomic-guided drug development: regulatory perspective. *Pharmacogenomics J* **2**(1), 20–24.

Lesko, L. J. and Woodcock, J. (2004). Translation of pharmacogenomics and pharmacogenetics: a regulatory perspective. *Nat Rev Drug Discov* **3**(9), 763–769.

Lindpaintner, K. (2003). Pharmacogenetics and the future of medical practice. *J Mol Med* **81**(3), 141–153.

Lipinski, C. A. (2004). Lead- and drug-like compounds: the rule-of-five revolution. *Drug Discov Technol* **1**, 337–341.

Mallal, S., Nolan, D., Witt, C. *et al.* (2002). Association between presence of HLA-B*5701, HLA-DR7, and HLA-DQ3 and hypersensitivity to HIV-1 reverse-transcriptase inhibitor abacavir. *Lancet* **359**(9308), 727–732.

Martin, A. M., Nolan, D., Gaudieri, S. *et al.* (2004). Predisposition to abacavir hypersensitivity conferred by HLA-B*5701 and a haplotypic Hsp70-Hom variant. *Proc Natl Acad Sci USA* **101**(12), 4180–4185.

Naisbitt, D. J., Williams, D. P., Pirmohamed, M. *et al.* (2001). Reactive metabolites and their role in drug reactions. *Curr Opin Allergy Clin Immunol* **1**(4), 317–325.

Orth, A. P., Batalov, S., Perrone, M. *et al.* (2004). The promise of genomics to identify novel therapeutic targets. *Expert Opin Ther Targets* **8**, 587–596.

Pettipher, R., Mangion, J., Hunter, M. G. *et al.* (2005). Identification of G-protein-coupled receptors involved in inflammatory disease by genetic association studies. *Curr Opin Pharmacol* **5**(4), 412–417.

Puente, X. S., Sanchez, L. M., Overall, C. M. *et al.* (2003). Human and mouse proteases: a comparative genomic approach. *Nat Rev Genet* **4**, 558.

Roden, D. M. and George, A. L., Jr. (2002). The genetic basis of variability in drug responses. *Nat Rev Drug Discov* **1**, 37–44.

Roses, A. D. (2002). Genome-based pharmacogenetics and the pharmaceutical industry. *Nat Rev Drug Discov* **1**(7), 541–549.

Roses, A. D., Burns, D. K., Chissoe, S. *et al.* (2005). Disease-specific target selection: a critical first step down the right road. *Drug Discov Today* **10**(3), 177–189.

Senior, K. (2005). Drugs tailored to race move a step closer to reality. *Drug Discov Today* **10**(16), 1076–1077.

Silverman, E., In, K. H., Yandava, C.*et al.* (1998). Pharmacogenetics of the 5-lipoxygenase pathway in asthma. *Clin Exp Allergy* **28** (Suppl. 5), 164–170.

Sorensen, J. M. (2002). Herb-drug, food-drug, nutrient-drug, and drug-drug interactions: mechanisms involved and their medical implications. *J Altern Complement Med* **8**(3), 293–308.

Southan, C. (2001). A genomic perspective on human proteases as drug targets. *Drug Discov Today* **6**, 681–688.

Southan, C. (2004). Has the yo-yo stopped? An assessment of human protein-coding gene number. *Proteomics* **4**, 1712–1726.

Sundram, S., Joyce, P. R. and Kennedy, M. A. (2003). Schizophrenia and bipolar affective disorder: perspectives for the development of therapeutics. *Curr Mol Med* **3**(5), 393–407.

Swaminath, G., Deupi, X., Lee, T. W. *et al.* (2005). Probing the beta2 adrenoceptor binding site with catechol reveals differences in binding and activation by agonists and partial agonists. *J Biol Chem* **280**(23), 22165–22171.

Szakacs, G., Annereau, J. P., Lababidi, S. *et al.* (2004). Predicting drug sensitivity and resistance: profiling ABC transporter genes in cancer cells. *Cancer Cell* **6**(2), 129–137.

Tate, S. K. and Goldstein, D. B. (2004). Will tomorrow's medicines work for everyone? *Nat Genet* **36**(Suppl. 11), S34–42.

Thorn, C. F., Klein, T. E. and Altman, R. B. (2005). PharmGKB: the pharmacogenetics and pharmacogenomics knowledge base. *Methods Mol Biol* **311**, 179–191.

Verkman, A. S. (2004). Drug discovery in academia. *Am J Physiol Cell Physiol* **286**, C465–474.

Voight, B. F., Kudaravalli, S., Wen, X. *et al.* (2006). A map of recent positive selection in the human genome. *PLoS Biol* **4**(3), e72.

Wishart, D. S., Knox, C., Guo, A. C. *et al.* (2006). DrugBank: a comprehensive resource for *in silico* drug discovery and exploration. *Nucleic Acids Res* **34**, 668–672.

Woolfe, A., Goodson, M., Goode, D. K. *et al.* (2005). Highly conserved non-coding sequences are associated with vertebrate development. *PLoS Biol* **3**, 0116–0130.

Yan, Q. and Sadee, W. (2000). Human Membrane Transporter Database: a Web-accessible relational database for drug transport studies and pharmacogenomics. *AAPS PharmSci* **2** (3).

Yu, H. and Adedoyin, A. (2003). ADME-Tox in drug discovery: integration of experimental and computational technologies. *Drug Discov Today* **8**, 852–861.

Zamek-Gliszczynski, M. J., Hoffmaster, K. A., Nezasa, K. I. *et al.* (2006). Integration of hepatic drug transporters and phase II metabolizing enzymes: mechanisms of hepatic excretion of sulfate, glucuronide, and glutathione metabolites. *Eur J Pharm Sci* **27**(5), 447–486.

Zheng, C., Han, L., Yap, C. W. *et al.* (2006). Progress and problems in the exploration of therapeutic targets. *Drug Discov Today* **11**, 420.

Appendix I

IUPAC nucleotide ambiguity codes

IUPAC code	Meaning	Complement
A	A	T
C	C	G
G	G	C
T/U	T	A
M	A or C	K
R	A or G	Y
W	A or T	W
S	C or G	S
Y	C or T	R
K	G or T	M
V	A or C or G	B
H	A or C or T	D
D	A or G or T	H
B	C or G or T	V
N	G or A or T or C	N

Bioinformatics for Geneticists, Second Edition. Edited by Michael R. Barnes
 ISBN 978-0-470-02619-9 (HB) ISBN 978-0-470-02620-5 (PB)

IUPAC amino-acid codes

IUPAC amino-acid code	Three-letter code	Amino acid
A	Ala	Alanine
C	Cys	Cysteine
D	Asp	Aspartate
E	Glu	Glutamate
F	Phe	Phenylalanine
G	Gly	Glycine
H	His	Histidine
I	Ile	Isoleucine
K	Lys	Lysine
L	Leu	Leucine
M	Met	Methionine
N	Asn	Asparagine
P	Pro	Proline
Q	Gln	Glutamine
R	Arg	Arginine
S	Ser	Serine
T	Thr	Threonine
V	Val	Valine
W	Trp	Tryptophan
Y	Tyr	Tyrosine

Human codon usage table

First codon	Second codon U	C	A	G	Last codon
U	Phe	Ser	Tyr	Cys	**U**
	Phe	Ser	Tyr	Cys	**C**
	Leu	Ser	**Stop**	**Stop**	**A**
	Leu	Ser	**Stop**	Trp	**G**
C	Leu	Pro	His	Arg	**U**
	Leu	Pro	His	Arg	**C**
	Leu	Pro	Gln	Arg	**A**
	Leu	Pro	Gln	Arg	**G**
A	Ile	Thr	Asn	Ser	**U**
	Ile	Thr	Asn	Ser	**C**
	Ile	Thr	Lys	Arg	**A**
	Met	Thr	Lys	Arg	**G**
G	Val	Ala	Asp	Gly	**U**
	Val	Ala	Asp	Gly	**C**
	Val	Ala	Glu	Gly	**A**
	Val	Ala	Glu	Gly	**G**

Appendix II Amino-Acid Substitution Matrices

More information on these matrices is available at the following site: http://www. russell.embl-heidelberg.de/aas.

All protein types

	ALA	ARG	ASN	ASP	CYS	GLN	GLU	GLY	HIS	ILE	LEU	LYS	MET	PHE	PRO	SER	THR	TRP	TYR	VAL
ALA		-2	0	0	-2	0	0	1	-1	-1	-2	-1	-1	-3	1	1	1	-6	-3	0
ARG	-2		0	-1	-4	1	-1	-3	2	-2	-3	3	0	-4	0	0	-1	2	-4	-2
ASN	0	0		2	-4	1	1	0	2	-2	-3	1	-2	-3	0	1	0	-4	-2	-2
ASP	0	-1	2		-5	2	3	1	1	-2	-4	0	-3	-6	-1	0	0	-7	-4	-2
CYS	-2	-4	-4	-5		-5	-5	-3	-3	-2	-6	-5	-5	-4	-3	0	-2	-8	0	-2
GLN	0	1	1	2	-5		2	-1	3	-2	-2	1	-1	-5	0	-1	-1	-5	-4	-2
GLU	0	-1	1	3	-5	2		0	1	-2	-3	0	-2	-5	-1	0	0	-7	-4	-2
GLY	1	-3	0	1	-3	-1	0		-2	-3	-4	-2	-3	-5	0	1	0	-7	-5	-1
HIS	-1	2	2	1	-3	3	1	-2		-2	-2	0	-2	-2	0	-1	-1	-3	0	-2
ILE	-1	-2	-2	-2	-2	-2	-2	-3	-2		2	-2	2	1	-2	-1	0	-5	-1	4
LEU	-2	-3	-3	-4	-6	-2	-3	-4	-2	2		-3	4	2	-3	-3	-2	-2	-1	2
LYS	-1	3	1	0	-5	1	0	-2	0	-2	-3		0	-5	-1	0	0	-3	-4	-2
MET	-1	0	-2	-3	-5	-1	-2	-3	-2	2	4	0		0	-2	-2	-1	-4	-2	2
PHE	-3	-4	-3	-6	-4	-5	-5	-5	-2	1	2	-5	0		-5	-3	-3	0	7	-1
PRO	1	0	0	-1	-3	0	-1	0	0	-2	-3	-1	-2	-5		1	0	-6	-5	-1
SER	1	0	1	0	0	-1	0	1	-1	-1	-3	0	-2	-3	1		1	-2	-3	-1
THR	1	-1	0	0	-2	-1	0	0	-1	0	-2	0	-1	-3	0	1		-5	-3	0
TRP	-6	2	-4	-7	-8	-5	-7	-7	-3	-5	-2	-3	-4	0	-6	-2	-5		0	-6
TYR	-3	-4	-2	-4	0	-4	-4	-5	0	-1	-1	-4	-2	7	-5	-3	-3	0		-2
VAL	0	-2	-2	-2	-2	-2	-2	-1	-2	4	2	-2	2	-1	-1	-1	0	-6	-2	

Bioinformatics for Geneticists, Second Edition. Edited by Michael R. Barnes
 ISBN 978-0-470-02619-9 (HB) ISBN 978-0-470-02620-5 (PB)

Extracellular proteins

	ALA	ARG	ASN	ASP	CYS	GLN	GLU	GLY	HIS	ILE	LEU	LYS	MET	PHE	PRO	SER	THR	TRP	TYR	VAL
ALA		0	0	-1	-4	0	0	0	0	0	0	0	0	-1	0	0	0	-2	-1	0
ARG	0		0	0	-5	0	0	0	0	0	-1	1	0	-1	0	0	0	-1	0	0
ASN	0	0		1	-6	0	0	0	0	-1	-2	0	-1	-2	0	0	0	-3	-1	-1
ASP	-1	0	1		-7	0	0	0	0	-2	-2	0	-2	-2	0	0	0	-3	-2	-1
CYS	-4	-5	-6	-7		-5	-6	-6	-5	-5	-5	-6	-5	-5	-6	-5	-5	-5	-4	-4
GLN	0	0	0	0	-5		0	0	0	-1	-1	0	0	-2	0	0	0	-1	-1	0
GLU	0	0	0	0	-6	0		-1	0	-1	-1	0	0	-2	0	0	0	-1	-1	0
GLY	0	0	0	0	-6	0	-1		0	-2	-2	-1	-2	-3	0	0	0	-2	-2	-2
HIS	0	0	0	0	-5	0	0	0		-1	-1	0	-1	-1	0	0	0	-1	0	-1
ILE	0	0	-1	-2	-5	-1	-1	-2	-1		1	0	0	0	-1	-1	0	-1	0	2
LEU	0	-1	-2	-2	-5	-1	-1	-2	-1	1		-1	1	0	0	-1	0	-2	-1	1
LYS	0	1	0	0	-6	0	0	-1	0	0	-1		-1	-2	0	0	0	-2	-1	0
MET	0	0	-1	-2	-5	0	0	-2	-1	0	1	-1		0	-1	-1	0	-1	-1	0
PHE	-1	-1	-2	-2	-5	-2	-2	-3	-1	0	0	-2	0		-2	-2	-1	1	2	0
PRO	0	0	0	0	-6	0	0	0	0	-1	0	0	-1	-2		0	0	-3	-1	0
SER	0	0	0	0	-5	0	0	0	0	-1	-1	0	-1	-2	0		1	-1	-1	-1
THR	0	0	0	0	-5	0	0	0	0	0	0	0	0	-1	0	1		-1	-1	0
TRP	-2	-1	-3	-3	-5	-1	-1	-2	-1	-1	-2	-2	-1	1	-3	-1	-1		1	-1
TYR	-1	0	-1	-2	-4	-1	-1	-2	0	0	-1	-1	-1	2	-1	-1	-1	1		0
VAL	0	0	-1	-1	-4	0	0	-2	-1	2	1	0	0	0	0	-1	0	-1	0	

Intracellular proteins

	ALA	ARG	ASN	ASP	CYS	GLN	GLU	GLY	HIS	ILE	LEU	LYS	MET	PHE	PRO	SER	THR	TRP	TYR	VAL
ALA		0	-1	-1	0	0	0	0	-1	0	0	0	0	-1	0	0	0	-2	-1	0
ARG	0		0	0	-1	0	0	0	0	-1	-1	1	0	-2	0	0	0	-1	-1	-1
ASN	-1	0		1	-1	0	0	0	0	-2	-2	0	-1	-2	-1	0	0	-2	-1	-2
ASP	-1	0	1		-2	0	1	0	0	-3	-3	0	-2	-3	0	0	0	-2	-2	-2
CYS	0	-1	-1	-2		-2	-2	-1	0	0	0	-1	0	0	-2	0	0	-1	0	0
GLN	0	0	0	0	-2		1	0	0	-2	-1	0	0	-2	0	0	0	-2	-1	-1
GLU	0	0	0	1	-2	1		-1	0	-2	-2	0	-1	-2	0	0	0	-2	-1	-1
GLY	0	0	0	0	-1	0	-1		-1	-3	-3	0	-2	-3	0	0	-1	-2	-2	-2
HIS	-1	0	0	0	0	0	0	-1		-2	-1	0	-1	-1	-1	0	0	0	1	-1
ILE	0	-1	-2	-3	0	-2	-2	-3	-2		2	-1	1	0	-2	-2	0	-1	0	2
LEU	0	-1	-2	-3	0	-1	-2	-3	-1	2		-1	2	1	-2	-2	-1	0	0	1
LYS	0	1	0	0	-1	0	0	0	0	-1	-1		0	-2	0	0	0	-1	-1	-1
MET	0	0	-1	-2	0	0	-1	-2	-1	1	2	0		1	-1	-1	0	0	0	0
PHE	-1	-2	-2	-3	0	-2	-2	-3	-1	0	1	-2	1		-2	-2	-1	1	2	0
PRO	0	0	-1	0	-2	0	0	0	-1	-2	-2	0	-1	-2		0	0	-2	-1	-1
SER	0	0	0	0	0	0	0	0	0	-2	-2	0	-1	-2	0		0	-2	-1	-1
THR	0	0	0	0	0	0	0	-1	0	0	-1	0	0	-1	0	0		-2	-1	0
TRP	-2	-1	-2	-2	-1	-2	-2	-2	0	-1	0	-1	0	1	-2	-2	-2		2	-1
TYR	-1	-1	-1	-2	0	-1	-1	-2	1	0	0	-1	0	2	-1	-1	-1	2		0
VAL	0	-1	-2	-2	0	-1	-1	-2	-1	2	1	-1	0	0	-1	-1	0	-1	0	

Transmembrane proteins

	ALA	ARG	ASN	ASP	CYS	GLN	GLU	GLY	HIS	ILE	LEU	LYS	MET	PHE	PRO	SER	THR	TRP	TYR	VAL
ALA		-1	-1	0	0	-2	0	1	-3	0	-2	-2	-1	-2	0	2	1	-4	-3	0
ARG	-1		2	1	-1	6	2	0	5	-3	-3	9	0	-4	-3	-1	-1	5	-1	-2
ASN	-1	2		6	-1	3	1	-2	3	-3	-4	5	-2	-4	-2	2	1	-3	-1	-3
ASP	0	1	6		-3	2	8	2	3	-3	-5	3	-3	-6	-2	0	0	-4	-2	-3
CYS	0	-1	-1	-3		-3	-3	-1	-1	-1	-1	-3	-1	1	-3	2	0	1	3	0
GLN	-2	5	3	2	-3		7	-2	7	-4	-2	6	-2	-4	0	-1	-2	0	0	-4
GLU	0	2	1	8	-3	7		3	2	-4	-5	1	-3	-6	-3	0	-1	-3	-5	-2
GLY	1	0	-2	3	-1	-1	3		-3	-2	-4	-1	-3	-5	-1	1	0	-2	-5	-1
HIS	-3	5	3	3	-1	7	2	-3		-4	-4	4	-3	-3	-4	-2	-2	-1	6	-4
ILE	0	-3	-3	-3	-1	-4	-4	-2	-4		1	-4	1	-1	-3	-1	0	-3	-4	2
LEU	-2	-3	-4	-5	-1	-2	-5	-4	-4	1		-4	1	1	-1	-2	-1	-2	-3	0
LYS	-2	9	5	3	-3	6	1	-1	4	-4	-4		-1	-5	-4	-1	-2	3	1	-4
MET	-1	0	-2	-3	-1	-2	-3	-3	-3	1	1	-1		0	-3	-2	0	-2	-3	1
PHE	-2	-4	-4	-6	1	-4	-6	-5	-3	-1	1	-5	0		-4	-1	-2	-3	2	-1
PRO	0	-3	-2	-2	-4	0	-3	-1	-4	-3	-1	-4	-3	-4		-1	-1	-6	-5	-3
SER	2	-1	2	0	1	-1	0	1	-2	-1	-2	-1	-2	-1	-1		2	-3	0	-1
THR	1	-1	1	0	0	-2	-1	0	-2	0	-1	-2	0	-2	-1	2		-4	-3	0
TRP	-3	5	-3	-4	1	0	-3	-2	-1	-3	-2	3	-2	-3	-6	-3	-4		-3	-2
TYR	-3	-1	-1	-2	3	0	-5	-5	6	-4	-2	1	-3	2	-5	0	-3	-3		-4
VAL	0	-2	-3	-3	0	-4	-2	-1	-4	2	0	-4	1	-1	-3	-1	0	-2	-4	

Index

Note: page numbers in *italics* refer to figures and tables

Bioinformatics for Geneticists, Second Edition. Edited by Michael R. Barnes

ISBN 978-0-470-02619-9 (HB) ISBN 978-0-470-02620-5 (PB)

Index compiled by Paul Nash